Microbiome-Gut-Brain Axis

R. Z. Sayyed • Mahejibin Khan
Editors

Microbiome-Gut-Brain Axis

Implications on Health

Springer

Editors
R. Z. Sayyed
Department of Microbiology
PSGVP Mandal's Arts
Science and Commerce College
Shahada, Maharashtra, India

Mahejibin Khan
Microbiology and Fermentation Technology
Central Food Technological Research Institute
Mysore, Karnataka, India

ISBN 978-981-16-1625-9 ISBN 978-981-16-1626-6 (eBook)
https://doi.org/10.1007/978-981-16-1626-6

This Springer imprint is published by the registered company Springer Nature Singapore Pte Ltd.
The registered company address is: 152 Beach Road, #21-01/04 Gateway East, Singapore 189721, Singapore

Preface

Gut-brain axis is a signaling mechanism between gut microbiota and the central nervous system. It is a bidirectional process wherein the brain influences the intestinal function and the gut microbiota affects brain health/functioning through the secretion of various compounds. Gut microbiota represents a vital supplementary organ of human beings. A stable gut microbiota is essential for normal gut physiology and contributes to appropriate signaling along the brain-gut axis and to the healthy status of the individual. Probiotic microorganisms are known to colonize the gastrointestinal tract and provide an array of health benefits to the host. Certain nondigestible oligosaccharides (also known as prebiotics) resist hydrolysis by human alimentary enzymes and could reach the colon intact. Colonic fermentation of dietary fiber by specialist microbes in the gut leads to the formation of a variety of health-promoting metabolites, e.g., short-chain fatty acids (SCFA), maintaining intestinal homeostasis, and strengthening the gut barrier function. Moreover, SCFA affect gut-brain signaling and selectively stimulate the growth and/or activity of one or a limited number of probiotic bacteria and also inhibit the growth of pathogenic bacteria.

Environmental factors, such as hygiene and the use of antibiotics, and various lifestyles together with the consumption of an imbalanced diet are linked with intestinal dysbiosis, which may adversely influence gut physiology leading to inappropriate brain-gut axis signaling. The perturbation of gut microbiota functions and play a regulatory role in several human abnormal health conditions, viz. IBD, neurological disorders, anxiety, mood, cognition, and pain. Based on the evidences, the therapeutic potential of prebiotics and probiotics and the importance of designing new functional foods containing prebiotics, probiotic, or synbiotics (prebiotics and probiotics together) are highlighted.

Recent advances in *gut-microbiome-brain axis* studies have shown a possible association between diet, composition of gut microbiota, and incidence of various diseases. Gut microbiome has recently become a target for live bacterial cell biotherapies for various chronic diseases including metabolic syndrome, diabetes, obesity, and neurodegenerative disease. Prebiotic and probiotic biotherapies are known to create a healthy gut environment by balancing bacterial populations and promoting their favorable metabolic action. Alteration of gut microbiota acts as a preventive as well as therapeutic measure in various disorders.

Jalgaon, Maharashtra, India R. Z. Sayyed
Mysore, Karnataka, India Mahejibin Khan

Contents

About the Editors

R. Z. Sayyed is the President of the India chapter of Asian PGPR Society, India, General Secretary of Asian PGPR Society (International), and a Professor and Head of the Department of Microbiology at PSGVP ASC College, Shahada (KBC North Maharashtra University, Jalgaon), India. He has 24 years of teaching and 21 years of research experience in Microbiology and Biotechnology in the field of Plant-Microbe Interaction and Microbial Biopolymers. He has authored and coauthored over 196 research papers and 39 book chapters in reputed national and international journals and books. He has authored and edited 74 books with international publishers like Springer-Nature and Cambridge Scholars Press, London. Prof. Sayyed has presented his research in several South East Asian and European Countries over the years. To his credit, he has received many prestigious awards from Government and NGO. He is an Associate Editor of *Environmental Sustainability* (Springer, Journal) and Guest Editor of a special issue of *Sustainability* (MDPI).

Mahejibin Khan is a Senior Scientist at the Central Food Technological Research Institute (CFTRI), Mysore, India, and Assistant Professor of India's Academy of Scientific and Innovative Research. She worked as an Assistant Professor in Microbiology at the Kumaun University, Nainital, India, and did her Postdoctoral Research at Japan's National Food Research Institute, Tsukuba Science City. Dr. Khan's research focuses on Metagenomics and Protein Modeling that identified specialty enzymes for food applications, which have appeared in peer-reviewed journal publications. Her publications in highly cited journals led to advances in the understanding of Human-Gut Microbiome, Nutraceuticals, and Functional foods and the role of intestinal commensals in the development of innate immunity. Dr. Khan's laboratory is working closely to develop resilient models for a platform to study neuro-entero-endocrine-based disease combating programs.

Part I

Gut-Brain Axis and Human Health

Gut Microbes: Influencers of Human Brain

1

Ankur Anavkar, Nimisha Patel, Ahmad Ali, Walhe Rajan, and Hina Alim

Abstract

Bidirectional interaction of gut microbiota and the brain has been shown to play an important role in the structural, behavioural development and also various functioning aspects of the brain. A number of factors affect and influence the colonization and alteration of gut microbiota through out life. There are several known mechanisms for the interaction of the brain and the gut microbiota (e.g. blood-brain barrier). Behavioural changes and neurological diseases can cause variations in the gut microbiota composition. Adaptive changes in diet and administration of various prebiotics and probiotics could possibly be a futuristic therapeutic treatment for depression, stress and various other neurological disorders. Also, the prebiotics and probiotics play a major role in balancing functional and behavioural aspects of both healthy and diseased individuals. The aim of this chapter is to discuss the role of prebiotics and probiotics as a determinant of composition of the gut microbiota and consequences of this on gut-brain axis.

Keywords

Blood-brain barrier · Gut microbiota · Neurological disorders · Prebiotics · Probiotics

A. Anavkar · N. Patel · A. Ali · H. Alim (✉)
Department of Life Sciences, University of Mumbai, Vidyanagari, Mumbai, Maharashtra, India
e-mail: hina.alim@lifesciences.mu.ac.in

W. Rajan
MES Abasaheb Garware College, Sadashiv Peth, Pune, Maharashtra, India

R. Z. Sayyed, M. Khan (eds.), *Microbiome-Gut-Brain Axis*,
https://doi.org/10.1007/978-981-16-1626-6_1

1.1 Introduction

The different interior parts of the human body serve as a niche for the microbiota forming a human microbiome (Kumari and Verma 2018). Human health can benefit and be affected by the gastrointestinal microorganisms called gut microbiota that not only consist of bacteria but also bacteriophages, viruses, fungi, protozoa and archaea (Mohajeri et al. 2018). Intestinal microbiota is indeed a diverse community that helps to sustain a dynamic ecological equilibrium (Wang and Wang 2016). The estimated number of bacteria present in the human body is 100 trillion, and out of them, 80% exist in the gut, which harbours more than 1000 species, responsible for performing various physiological functions (Kumari and Verma 2018). Density of microbiota has its highest levels in the ileum and colon where the anaerobic bacteria are dominating the environment in the presence of virus, archaea, protozoa and fungi (Mohajeri et al. 2018). The two main phylotypes present in the gastrointestinal tract (GIT) are anaerobic Firmicutes and Bacteroidetes that constitute 64% and 23%, respectively, while others such as Verrucomicrobia, Fusobacteria, Cyanobacteria, Proteobacteria, Actinobacteria, Spirochetes, and species of protozoa, fungi and viruses are relatively present in undersized amounts (Wang and Wang 2016). The microbiota present in each person is distinct and variable but has a conserved set of colonizers (Kumari and Verma 2018).

1.2 The Gut Microbiota

The most astonishing revelation of the twenty-first century was the total size of human gene pool that contained only 26,000 functioning genes as compared to the other simpler organisms. Scientists and researchers all over the world call this a "genomic complexity conundrum", and this led to a speculation regarding the role of microbial genes in relation to behavioural and physiological changes in the human body (Galland 2014). According to a statement given by the World Health Organization (WHO), probiotics have proven beneficial to human health by providing an improvement in the skin health, mood and behaviour; resistance to allergens; protection of molecules from oxidative damage; reduced pathogenic microorganisms and proper cognitive functioning; and immune system support (Mohajeri et al. 2018). It has also been speculated that adequate ingestion of probiotic bacteria can be useful for infectious, inflammatory, neoplastic and allergic disorders, but the ideal strain of probiotic has yet to be identified (Bravo et al. 2012).

1.2.1 Alterations and Factors Impacting the Gut Microbiota

Gut microbiota does not have a fixed composition. It changes dynamically with human development that is influenced by various factors, and the change is also not similar between two individuals, but the end effect is similar, i.e. macrobalance (Mohajeri et al. 2018). The change of the bacteria is not just influenced by the health of individuals but also by other factors such as illness, drug, infection and diet, which also change the composition of microbiota (Wang and Wang 2016). Also variability can be seen in gut microbiota of healthy individuals. The variability depends on factors such as genetic, physiological and psychological. Although there is an assumption that each person will have a unique microbiota, it is assumed that everyone might share a set of standard microbiota. Also the microbiota of the GI (gastrointestinal) tract may have similar colonization throughout life. Furthermore, in recent studies, it has been found that bacteria are also present in the placenta, amniotic fluid and meconium (Mohajeri et al. 2018).

Various environmental factors affect the inception of colonization and development of microbiota in the gut. For natural assembly of the gut microbiota in infants, mode of delivery plays an important role (Lu and Claud 2019). Gestational age (GA) and postmenstrual age (PMA) together play a pivotal role in development of microbiota in preterm infants. Currently, there are no likely established microbiotic factors that affect the delivery mode in infants born pre-term and infants born after complete term, and this is also one of many reasons that makes the assessments difficult (Dahl et al. 2018). Early studies of vaginally delivered infants who had completed their full term interpreted that pioneer colonizing bacteria are similar to that found in mother's vaginal tract. On the other hand, the newborns have similar microbiota to that of the skin of the mother, the room environment and the people around when they are delivered through C-section rather than the normal mode (Lu and Claud 2019). The mode of delivery, i.e. C-section or vaginal mode, does not significantly impute the gut microbiota; meanwhile another study had reported a connection between delivery mode and gut microbiota. The microbiota of the gut was shown evidently only when the vaginally delivered infants were compared with unlaboured C-sectioned infants (Chu et al. 2017). In addition, low levels of *Bacteroides* and *Bifidobacterium* were to be equally observed in neonates that were delivered by C-section as well as the vaginally delivered ones, thus making a disagreement with early studies. In addition, a research conducted on gut microbiota of pre-term infants agrees with previous studies. These studies have stated factors such as antibiotic exposure and delivery method of infants do not significantly affect the diversity of different bacteria (e.g. alpha) in pre-term born infants (Dahl et al. 2018). Furthermore, a study made a revelation that was agreeing with other studies regarding the following: the infants born after a complete term have a greater abundance of *Bacteroides* when vaginally delivered, but nevertheless, it made no difference on microbiota of the infants that were born pre-term. It is worth noting that the study performed was a little selective as there was a very limited number of babies born pre-term and naturally delivered (Lu and Claud 2019; Chu et al. 2017).

However, the major difference between pre-term infants and full-term infants can only be seen after a few months of delivery, and the confirmation was done by comparing age-dependent maturation of the pre-term and full-term infants microbiota. Another discovery made when the research was carried out further that there was a change in the microbiota of pre-term born infants, when the infants are around 6 weeks old, thus confirming that age depends on the maturation of the microbiota (Dahl et al. 2018). This proves that progression of bacteria in the gut of pre-term and full-term infants has no major difference. The anaerobic genera starts colonizing and increasing in number by the end of the first week of delivery of the infant (Hill et al. 2017). At PMA between 25 and 30 weeks, the microbiotas that have prominence are *Staphylococcus*; after that, there is dominance of Enterobacteriaceae, and it can be seen up to a PMA of 35 weeks. Another bacterium, Bifidobacterium, only begins to exhibit dominance in healthy babies that are born on time, but the dominance is only seen after a 30-week PMA (Lu and Claud 2019). Because of the extension of dominance even after a PMA of 35 weeks by Enterobacteriaceae, pre-term infants have a retarded progression of this bacterial population relative to term-born infants. Such evidences prove that while the bacterial progression at the gene level is the same, the bacterial dominance of pre-term and term-born infants differs considerably (Lu and Claud 2019; Hill et al. 2017).

Milk is another influential factor of gut microbiota; the composition of microbiota develops differently based on infants are naturally nursed or feeding on formulated milk. The recent studies do not give us any clue regarding the impact on microbiota congregation by factors including feeding on the milk of the mother or formulated milk because a very low number of pre-term infants are nursed naturally (Lu and Claud 2019). There is a noticeable increase in the number of bacteria belonging to the class Actinobacteria when the infants are nursed naturally, while the variety of bacterial composition can be seen in infants that have been feeding upon the formulated milk (Lu and Claud 2019; Hill et al. 2017). Although the diversity of intestinal microbiota is significantly lower in breastfed infants compared to formulated milk-fed infants, their microbial communities associate significantly more with host genes, and transcription activities are more related to immune defensive function and metabolism. For example, the infants nursed with breast milk develop anti-inflammatory genes and also the genes essential for the use of HMO, i.e. human milk oligosaccharides (Hill et al. 2017). The microbial genes are influenced by the epithelial cells of the host, in this case the infant. The usage of HMO (human milk oligosaccharides) can encourage growth of a certain strain of bacteria and other species that may lead to alteration of the microbiota (Dahl et al. 2018). Another study demonstrated that the gut microbiota group differently in 5-week-old infants born preterm when compared to infants born full-term and nursed naturally. But within 6 weeks of birth, the microbiota of pre-term infants congregate in the same fashion as those of the full-term infants nursed naturally (Lu and Claud 2019). Furthermore, a study also concluded that the microbiota of the infants born pre-term have low diversification of bacterial classes but on the other hand, it also had variation that could be seen between two individuals and also there is succession of different bacterial classes accordingly (Lu and Claud 2019; Dahl et al. 2018; Hill et al. 2017).

Intestinal microbiota is often impacted by influencing factors such as time and various forms of antibiotic exposure towards pre-term or full-term newborns. Intrapartum antibiotic prophylaxis causes a decrease in natural diversity of bacteria including change composition in of lactobacilli and bifidobacteria (Dahl et al. 2018). Some antibiotics are safe to use and do not cause any diversity change in the gut microbiota of infants born pre-term; these antibiotics include ampicillin, gentamicin and vancomycin, whereas antibiotics such as cefotaxime and meropenem belonging to the class of broad spectrum are not related to the diversity of the intestinal microbiota in pre-term infants. On the other hand, a broad-spectrum antibiotic decreases the abundance of bacterial species in the gut of infants born pre-term (Lu and Claud 2019). Pathogenic species gained more resistance to antibiotics after antibiotic treatment, thus escalating the risk for probable pathogenic bacterial supremacy (Dahl et al. 2018). The amount of Proteobacteria and other bacteria such as Actinobacteria and *Lactobacillus* shows a considerable increase and decrease, respectively, in term infants who were given ampicillin and gentamicin via non-oral route (Lu and Claud 2019). It is proved by the results verified for 4-week-old infants after the treatment in comparison with the controls. The levels of above mentioned bacteria are not affected, i.e. they remain the same; but Proteobacteria abundance is high when compared to the controls and the ones treated for 8 weeks (Chu et al. 2017). The gut microbiota development is also affected by factors such as economic status and also the environmental factor of the delivery room. The assembly of microbiota in the infant born pre-term is shaped by the environment of the care unit in which the infants are taken care of for days to months (Hill et al. 2017). The (ICU) intensive care unit is also one of the first environments where infants are exposed. The study demonstrated the relationship between the microbiota taxa of the neonatal intensive care unit and its two-way association with infants and the occupation of these taxa in the intestinal microbiota of newborns. The study also indicated a technique to modify the NICU (neonatal intensive care unit) microbiota, which could be the answer to microbiota modulation in pre-term infants (Lu and Claud 2019; Chu et al. 2017).

As per conducted research and analysed studies, it was concluded that the PMA and GA can influence the production of microbiota in infants born pre-term, mainly affecting the slowest rate of microbiota assembly for infants born between 25 and 30 weeks pre-term (Korpela et al. 2018). Further study reported that the GA is responsible for the assembly of microbiota and not at all dependent on other factors such as how the infant was delivered, i.e. C-section or vaginal delivery; how much time and how many times the infant was nursed; and if there was any exposure to antibiotics to the infants born pre-term (Lu and Claud 2019). For the establishment of the gut microbiota, the period important is the initial one. PMA, GA and environmental factors influence the grouping and functionality of gut microbiota in infants (Chu et al. 2017). Both the pre-term infants and full-term infants have differences and uniqueness in their gut microbiota. Premature birth has its flaws that can be minimized by the initial colonization and production of microbiota, which will contribute to maturation and thus counteract the negative effects of birth by pre-maturation (Lu and Claud 2019).

Due to diet changes when developing, the infant gut microbiota is continuously changing, and it only becomes consistent or stable at around 2 years of age when the infant starts to eat solid food. Compared to formulated milk-fed babies, naturally nursed infants have a diverse gut microbiota (Lu and Claud 2019). The gut microbiota of all premature and mature infants gets stabilized and starts to develop into the adult microbiota composition; this process is completed by 3 years of age. In contrary to infants, the microbiota of adults is mainly composed of few phyla but has a greater diversity (Hill et al. 2017). Depending on age, the main characteristic of human microbiota has different shifts; it becomes less diverse as there is a popular feature of high abundance of *Bacteroides* species in older people, although there would be low numbers of *Clostridium* classes (Chu et al. 2017). Adults have very much stable microbiota compared with infants and the elderly. The factors such as treatment by antibiotics, infection by pathogens, stress diet and, lastly, genetics of the individual influence the microbiota of the gut in a very short period of time (Lu and Claud 2019).

1.2.2 The Ways of Harnessing Human Microbiota

The examination of patient and control community microbiota is studied, and major changes in the composition and their functions are established. Patient and control group stool samples are usually utilized for microbiota analysis, but mostly colonic bacteria are not a reliable substitute for small intestine microbiota. There are three potential approaches to use human microbiota in preclinical and clinical trials to evaluate disease pathways and effects of likely therapies (Cryan et al. 2020).

1.2.2.1 Mechanism of Sample Study for Animal Experiments

Animals may be anthropomorphized through transplantation of a faecal microbiota; in this process, human faecal matter is used to replicate a rodent intestine with a constitution identical to that of a human donor (Cryan et al. 2020; Sampson et al. 2016). One can restructure a mouse phenotype similar to that of a human donor, which offers a form of clinical trial to examine the mechanisms that can or would otherwise be very difficult for human study. Functional analysis can be carried out effectively via a series of animal behavioural experiments, as well as in vitro tests with different methods, such as molecular and imaging techniques (Sampson et al. 2016).

1.2.2.2 Faecal Transplantation

If potential changes within the microbiota constitution are present, a faecal microbiota transplantation can be designed for individuals in the control group or by a particular bacterial group. Assured research trials have been published, including those identified with autism spectrum disorder (Kang et al. 2017).

1.2.2.3 Dietary Intervention

The type of diet, i.e. plant and animal based, plays an important role in the composition of gut microbiota. Ingestion of prebiotics, probiotics and fermented foods can help in maintaining the composition and also the abundance of the intestinal microbiota (Sampson et al. 2016). Multiple animal topic studies have used prebiotics and probiotics for the treatment of different diseases and have recently performed a small number of human trials that have produced some assuring outcomes, opening a new route for therapeutic interventions (Cryan et al. 2020).

1.2.3 Functional Characteristics of Gut Microbiota

Firstly, gut microbiota comprises the intestinal barrier, which promotes the continuous survival of gut microbiota and accelerates the intestinal epithelial cell regeneration, which produces mucus and provides nourishment to the membrane by formation of short-chain fatty acids, popularly known as SCFAs (short-chain fatty acids). SCFAs plays a significant role in energy production, epigenetics and have anti-inflammatory effects inside the body (Wang and Wang 2016). SCFA is an inhibitor and activator of two of the most important molecular signalling systems in our body, i.e. HDAC (histone deacetylase) and GPCR (G-protein-coupled receptors). One of the major causes of neurological diseases is the imbalance of HDAC; inhibitors of the SCFA may play a major role in the treatment and recovery of patients (Galland 2014). SCFA activates specifically the two ligands, GPR41 and GPR43, of GPCRs. The activation of GPR41 causes an outflow increase in the SNS (sympathetic nervous system) and BMR (basal metabolic rate) and helps in the control of obesity (Galland 2014). Intestinal flora is accountable for the growth and development of the immune system by triggering the innate immune response at the start of life. It will make a significant contribution to the growth and development of the intestinal lymphoid tissue by inducing a humoral and cellular immune reaction (Wang and Wang 2016). The major role in drug and poison elimination is played by the biosynthesis of various enzymes, hormones, vitamins, minerals, etc. The continuous activated state of immune system by intestinal microbiota, leads to a physiological inflammatory state, forming an effective and rapid defence mechanism against pathogens (Galland 2014). The clinical studies performed on humans are also comparatively limited in number, so as with advances in this field, more and more gut microbiota functions can unfold (Wang and Wang 2016).

1.3 The Functional Relation in Intestinal Microbiota and Brain Using Animal Models

The use of germ-free mice for various clinical studies has helped to establish the link of the BGM axis (brain-gut-microbiota axis). Results of various studies have been culminated, and an increase in myelination and neurogenesis in prefrontal and hippocampal regions, can be seen respectively. Also, concentration change of neurotransmitters could be observed clearly. Germ-free mice shows the role of intestinal microbiota in the functioning of the brain (Cryan et al. 2020).

The blood-brain barrier (BBB) and microglia cells do not function effectively in germ-free mice. The reason is decrease in expression of tight junction and immature phenotype of cells, respectively (Cryan et al. 2020). Increased permeability in BBB allows various translocations of metabolites and pathogens. In strain-dependent mouse, there are records of structural defects such as reduced myelination and hippocampus and also various neurological disorders. Studies using different external agents such as antibiotics, prebiotics and probiotics also plays crucial role in neuro development and behavioural shaping in germ-free mice (Cryan et al. 2020).

1.4 Brain Development and Microbiota

Recently, there is a rise in attention towards recognizing the part of the brain-gut-microbiota axis in processes of neuro development. Then again, the studies conducted are relatively less in newborns. Some of the studies contained a total number of 89 infants; the study concluded that the cognitive function of children aged 2 years was notably interconnected with intestinal microbiota composition 1 year prior (Gao et al. 2019). In a clinical study group of 39 infants intestinal microbiotas, alpha-diversity was linked with growth of functional connection between the supplementary motor area and the inferior parietal lobule. This was also related to cognitive functional consequences at age 2 (Lu and Claud 2019). The most positive confirmation of intestinal microbiota role in the development of the nervous system is an outcome of research conducted on germ-free mice. In these GF (germ-free) mice subjects, elementary neurological processes were critically dependent on the intestinal microbiota composition (Cryan et al. 2020; Gao et al. 2019).

The part which intestinal microbiota plays in initial development of the brain and behaviour could possibly be determined by germ-free mice. The experiments on GF mice reported activity increase in relative exploration and decreased levels of anxiety were also observed. The hypothalamus and pituitary system responded to stress on a very large scale (Claud et al. 2013). This study reported the actions of GF mice that can be normalized by certain pathogen-free microbiota, known as SPF mice (specific pathogen-free). This study, therefore, points out that changes in the gut microbiota population can affect the functioning of the brain (Lu and Claud 2019).

A method used to verify the effect of microbiota on the development of the brain is by colonizing an expecting GF mouse with a strain specific bacterium and observing the nervous system development of the infant (Lu and Claud 2019). Using this model, pregnant GF mice were colonized with an infant's microbiota that had poor growth, resulting in uneven development of the nervous system and its structures, which was shown by evidence that the expression of neuronal nuclei (NeuN) marker known for neuronal structure development was decreased (Frohlich et al. 2016). The myelination mechanism was also altered, and its proof was the decrease in the expression of myelination markers and basic myelin protein in infant milk feeding and food feeding period. Neuroinflammation along with local and systemic IGF-1 (insulin-like growth factor 1) mediate the effect of gut microbiota on brain development in pre-term infants. For testing the behavioural changes, offspring of GF mice were used for this study (Lu and Claud 2019). The offspring were first transferred with the faecal lysate of a pre-term infant who was suffering from a disease named NEC (necrotizing enterocolitis). This administration to the offspring results in a decline in the overall learning curve, a substantial decrease in memory and a loss of certain vital functions such as locomotion after the child ceases feeding on breast milk (Lu and Claud 2019; Frohlich et al. 2016).

The administration of antibiotics to premature infants is a basic practice in the intensive care unit because there are suspicions about infection from the intrauterine surface, the reason being the unplanned labour and untimely rupturing of membranes such as the chorion and amnion. However, there is a lack of studies on long-term consequences concerning the use of antibiotic treatment for infant growth during the initial period of life (Lu and Claud 2019). Animal studies have shown that there are short-term disturbances of gut microbiota due to the induction of antibiotics that cause harm to the animals cognitive output. Massive quantities of data on the impact of antibiotics, microbiota and their association with neuro development were all collected from mouse studies (Claud et al. 2013).

There are major variations in the metabolites that circulate in the body after mice have undergone antibiotic treatment, and there has also been distortion in their functioning. When the SPF mice were given antimicrobials via oral route, the composition of microbiota was altered, the exploratory behaviour and hippocampus-related expression of neurotrophic factor were enhanced in a short period of time (Lu and Claud 2019). A recent study has recorded a rise in the levels of *Proteobacteria* in faecal matter and liposaccharides due to imbalance in gut microbiota, although there is a substantial decrease in *Lactobacilli*. Another research recorded the brain showing an increase in cell migration to the brain's hippocampus, and there is also activated NF-κB (Nuclear Factor kappa-light-chain-enhancer of activated B cells) that indicates anxiety and neuronal inflammation due to the treatment of antibiotics (Jang et al. 2018). From the study of numerous animal models, it has been concluded that the continuously evolving microbiota plays a vital role in brain growth. It has been concluded from the study of various animal models that the constantly changing microbiota plays a vital role in the development of the brain (Claud et al. 2013). Another research was based on the effects of microbiota composition when pre-term and full-term babies with a specific brain processing deficiency were given antibiotic care. This will give us an insight into the optimization of a specific initial microbial community that can enable us to boost the neurological results compared to those that are the result of premature birth before (Lu and Claud 2019; Jang et al. 2018).

1.4.1 Types of Barriers

One of the barriers present is between the blood and brain, while the other one is found in the intestine that is useful for signalling between the brain, gut and microbiota axis. Due to the major three factors, amount of information exchanged with the brain is completely unpredictable, depending on the host state. The three major factors include gut microbiota, stress and inflammation (Martin et al. 2018).

1.4.2 Barriers in the Intestine

Intestinal barrier mainly consists of two different layers: the inner layer is made up of basal epithelium; it is single layered and attached to each-other through tight junction; the outer layer is made up of mucus whose viscosity and composition are not stable and keep changing over a period of time. Meanwhile, there are pattern recognition receptors in the GI (gastrointestinal) mucosa that only function when such microbes or their metabolites are encountered (Kelly et al. 2015). These receptors signal causing antimicrobial protection mechanism to strengthen, inflammation of the intestines being changed, while immunogenic tolerance ability is also enhanced. In the gut and also in the mucosa-associated lymphoid tissue, M cells are also known as microfold cells. When the conditions are homeostatic, variety of microorganisms and macro-sized molecules that can enter through these M cells, help the immune system to be active (Martin et al. 2018).

Another type of cells present in the intestine is known as the Paneth cells, which are responsible for stimulating various antimicrobial factors and also limiting penetration of bacteria into the host tissue with the help of a MyD88-dependent receptor (Myeloid differentiation primary response 88) with Toll-like activation that senses the bacteria. Various microbes and ligands extracted from the microbes help to maintain integrity-critical cell-to-cell junctions (Martin et al. 2018). By administering probiotics, the normalization of intestinal barrier defects caused by stress or some unrelated mechanisms can be brought back to normal. The second factor that looks beyond the function of the intestinal barrier is the layer of intestinal mucosa (Kelly et al. 2015). The mucus found in the colon is divided into two separate layers: first, the dense but loose outer layer is exposed, and an inner layer is closely bound to the epithelium. Commensal microbes live in the outer layer, an essential environment for the production of biofilms and also a reliable source of energy rich in glycoproteins that the microbiota depends on when it is deprived of dietary fibre, successively causing an increase in pathogen susceptibility (Martin et al. 2018). The mucosal layer consists of antimicrobial peptides and IgA which makes the layer bacteria-free and protects the epithelial cells from microbial contact physically (Martin et al. 2018; Kelly et al. 2015).

1.4.3 Blood-Brain Barrier (BBB)

The vascular blood-brain barrier is made up of specific endothelial cells of the brain that causes stoppage of unobstructed plasma protein leak into the nervous system, as the BBB regulates the interface, while coordinating various functions such as nutrition, homeostasis and communication (Logsdon et al. 2018). Various proteins that make up the tight junction are characteristic of brain endothelial cells and help by restricting the metabolite diffusion into the blood and brain. Harmful substances in the circulating fluid can enter into the CNS (central nervous system) due to disruption of tight junctions in BBB. Various neurological disorders and diseases have been linked to disruption of structural and functional aspects of the BBB (Logsdon et al. 2018).

Dysfunctionality and disruption in the BBB are mainly caused by pathogenic constituents. Studies of sepsis, meningitis and inflammatory response support the current available knowledge. When the infection of sepsis originates from perforations in the intestine, the microbiota present in the gut plays an important role (Opp et al. 2015). In model organisms such as rodents, sepsis is instigated by puncturing and ligating the caecum which causes changes in the BBB. This includes increase in molecular abundance of cell adhesion metabolites, increase of concentration of immune cells and permeability (Logsdon et al. 2018).

Pathogenic bacteria can cross the BBB easily, inducing transcytosis by pili or components of the bacterial cell wall. Different types of CNS-tropic bacteria have the ability to cross the BBB even when there is no disruption, while some bacteria require disruption of the BBB (Logsdon et al. 2018). The endothelial cells of the brain have a high expression of TLRs (Toll-like receptors), which has responses based on the cell wall components of bacteria; for example, cell wall components such as LPS (lipopolysaccharide) and LTA (lipoteichoic acid) are present on gram-negative and gram-positive bacteria, respectively (Tang et al. 2017). Both the cell wall components are responsible for various mechanisms, and that mechanism can alter the functionality of the BBB. LPS directly or indirectly affects various functioning processes of BBB. It causes disruption in expression of tight junction protein which provides accessibility to various pathogenic and non-pathogenic metabolites (Tang et al. 2017).

A study was conducted by comparing the BBB of GF (germ-free) mice and pathogen-free mice and validated with evidence that there was decrease in expression of tight junction proteins and had structural deficits in the BBB. These changes are only possible when the brain endothelium shows absence in changing vascular density and pericyte coverage (Logsdon et al. 2018). Furthermore, BBB functions in GF mice were restored by colonizing the GF mice with flora from pathogen-free mice. During one of the trials, it was found that administration of penicillin (low dose) during an early period of life can cause changes in the intestinal microbiota by upregulating tight junction proteins and having long-lasting effects (Tang et al. 2017). Another study has put forth the theory that live bacterium could successively influence BBB and relatable functions. The study says that live bacterium may have potent mechanisms to influence the nervous system functioning through the BBB. Metabolite production is a potential mechanism by which live bacteria influence BBB by altering the CNS functions (Braniste et al. 2014). The functional change in gut microbiota due to various stimuli affects cohesiveness in BBB and functioning of the CNS. The GF mice models have given positive results for interaction between the gut microbiota, brain and also the role of BBB in it. But the interpreted results are not fully reliable (Braniste et al. 2014).

1.5 Brain-Gut-Microbiota Axis

1.5.1 Diagnostic Evidences

Various experiments have been conducted to have a clear idea about the role of intestinal microbiota on the brain-gut-microbiota axis. This includes not just antibiotic treatment but also the administration of various types of strains of probiotics, colonization of either the synthetic or the human microbes or the gut organ microbial system and also the transplantation of faecal microbiota into GF (germ-free) animal models (Bravo et al. 2012). A great progress has been made in spite of various restrictions such as the standard gut microbiota being absent in early life which cause stress compassion majorly in adults. But on the other hand, it could be partly undone by colonizing the gut with standard intestinal microbiota. There are various relatable neurochemical changes such as changes in cortical and hippocampal brain-derived neurotrophic factor levels and reduction of hippocampal serotonin (5-HT) receptor 1A expression (Martin et al. 2018). On the other hand, increase could be seen in monoamine levels of the striatal region and lessening gene expression of synaptic plasticity, which shows how diversifying and impactful the microbiota can be on the phenotype of the nervous system (Wang and Wang 2016). Therefore, the gut microbiota not just affects the responsiveness of stress but also behaviour during stress and depression, response from the pain receptors, preferences in taste and consequences during metabolic activities, feeding patterns and related physiologies (Martin et al. 2018). Various limitations were recognized in the GF models, and these were achieved by recolonizing with the SPF (specific pathogen-free) bacteria, human-derived bacteria, synthetic bacteria causing reversal of phenotype, which lead to some conclusion. This explanatory role of microbiota tells about its significance in the nervous system development as well as in the neurogenesis (Wang and Wang 2016).

Another way to approach the GF model is through the use of broad-spectrum antibiotics to generate impermanent changes in the composition and variability of microbiota in mice faeces, and antibiotic influence on the microbial community that is associated with mucosa is not completely understood till date (Martin et al. 2018). Soon the antimicrobials might also start working together immediately with the host physiology mechanisms. This leads to a number of different documented neurotoxic effects independently without the microbiota. Nevertheless, broad-spectrum antibiotic therapy endures to be a strong way of identifying the influence of the intestinal microbiota on the CNS (Mohle et al. 2016). Mice having an SPF microbiota were orally given an antibiotic; results showed an increase in exploring behaviour, and also increased expression by the hippocampus of the developed brain-derived neurotrophic factor can be seen. The outcome from antibiotic induction in GF mice could be replicated; thus, this implied that alterations in the CNS are not due to the improper communication of antibiotics, but the alteration in model GF animal makes the finding totally inconclusive (Martin et al. 2018; Mohle et al. 2016). Neurogenesis of the hippocampal region was affected by an antibiotic treatment, and this has been used for a long time. The basic task-performing abilities such as object recognition were affected because the hippocampus region was also affected. Voluntary exercise, probiotic treatment and accepted transfer of Ly-6C^{hi} monocytes can help in decreasing these phenotypes (Martin et al. 2018).

During an experiment, another approach considered was depleting of the gut microbiota completely or partially just to recognize the influence of gut microbiota on the nervous system. Thus, when probiotics were introduced to the standard model, it causes normal development and risk such as the off-target effects were very less (Galland 2014). There are also probable chances of having an immune response by the host for any short-term exposure of probiotics not having any resident communities. Oral administration of probiotics causes reduction in the basal or induced anxiety-like behaviour, improvement in inflammation-associated sickness behaviour, weakening of the induced obsessive-compulsive-like behaviour and also normalization of the developmental routes of emotion relating to behaviour after early-life stress (Beilharz et al. 2018). Diet can also play a vital role in the structural and functional composition of intestinal microbiota in both mice and human beings. Diet plays an influential role in changing intestinal microbiota and may have an effect on brain functioning efficiency (Galland 2014). Summarizing, diagnostic studies have been able to identify unambiguously the evidences that influence intestinal microbiota has on the nervous system (mainly the CNS). But then again, problems such as reproducibility need continuous improvement in the approach of experiments for better results (Martin et al. 2018).

1.6 Effect of Prebiotics and Probiotics on the BGM (Brain-Gut-Microbiota)

Various methodologies related to clinical studies have come forwarded to establish a connection between intestinal microbiota and brain functioning on a small scale and also observe the effects of probiotics and prebiotics. In the past few years, various medications have come forward to bring the composition of intestinal microbiota back to normal and have a healthy impact on the BGM axis (Galland 2014). But very few medications have possibly come forward that provide the safest route and promising results. Some of them include administration of prebiotics and probiotics or of antibiotics (Mohajeri et al. 2018).

The administration of probiotics has shown progressive results in clinical trials of various diseases, indicating a possible future prospect for its inclusion in treatment or recovery models of diseases (Kumari and Verma 2018). Clinical studies on patients having IBS (irritable bowel syndrome) were done by administering the strain *Bifidobacterium infantis* 35624, and the results were very much promising (Table 1.1), which included discomfort and pain reduction in the abdominal region, easiness in defecation and, lastly, reduction of cardinal symptoms. Another study on normosensitive and hypersensitive rats was done with administration of *Bifidobacterium infantis* 25624, and this study resulted in lowering of visceral pain, which was the major symptom of IBS (Table 1.1) (Bravo et al. 2012). In a study with patients having stress related to IBS, when they were given probiotics

Table 1.1 Effects of microbial probiotic strain on unhealthy individual's brain and physiological functions

Strain of probiotics	Effects
Bifidobacterium infantis 35624	Pain reduction in the abdominal region, easiness in defecation
Bifidobacterium infantis 25624	
Lactobacillus acidophilus Rosell-52	
Bifidobacterium longum Rosell-175	
B. longum 1714	Decrease in stress response, boost in cognitive function
B. infantis 35624	Normalization of behaviour immune response, noradrenaline concentration
Lactobacillus rhamnosus JB-1	Increase in T cells, inhibition of cardiovascular autonomic response, bloating of the abdominal area, increase in stimulation of neurons
Lactobacillus helveticus R0052	Improved cognitive and psychological aspects
Bifidobacterium longum R0175	
L. acidophilus, *B. bifidum*, *L. casei*	Reduced depressive behaviour
L. helveticus IDCC3801	Increase of cognitive functions in the elderly
L. acidophilus, *L. casei*, *B. bifidum*, *L. fermentum*	Improvement in cognitive function of patients suffering from Alzheimer's disease
Bacteroides fragilis; *Lactobacillus reuteri*	Improved behavioural and gastrointestinal conditions in patients suffering from ASD

such as *Lactobacillus acidophilus* Rosell-52 and *Bifidobacterium longum* Rosell-175, promising results have been observed such as abdominal pain and nausea reduction (Table 1.1). No effects have been observed related to psychological symptoms and sleep deprivation (Galland 2014). One of the clinical studies used *Trichuris muris* (non-invasive parasite) to infect the model mice, and the characteristic symptoms such as inflammation, anxiety and expression of BDNF (brain-derived neurotrophic factor) caused by this parasite were reduced (Bravo et al. 2012). When the mice were given *Bifidobacterium longum* NCC3001, the heightened cytokine levels did not reduce. This study was also conducted by using another bacterium strain, *Lactobacillus rhamnosus* NCC4007, which did not show any positive results (Kumari and Verma 2018).

A study was conducted on rats having stress in their early stages of life, and after administration of *B. infantis* 35624, it gave fantastic results such as normalization of behaviour and immune response and normalization of noradrenaline concentration (Table 1.1) (Bravo et al. 2012). Another study conducted on the same model organisms showed positive data on normalization of dysfunctionality of the colon and corticosterone release. On the other hand, consumption of prebiotics like galacto-oligosaccharide (commercially known as Bimuno) reduced the cortisol response and attention response after 3 weeks (Mohajeri et al. 2018). Meanwhile, the administration of another prebiotic, fructo-oligosaccharides did not show any specific effect indicating the effects could be strain specific. Preclinical studies using

B. longum 1714 were procured, and results showed decrease in stress response, while a boost in cognitive function in accordance to modulated activity in EEG (electroencephalogram) was observed (Table 1.1) (Mohajeri et al. 2018; Galland 2014).

Administration of Lactobacillus rhamnosus JB-1 causes various changes in the body such as increase in T cells and inhibition of cardiovascular autonomic response and bloating of the abdominal area (Mohajeri et al. 2018). *L. rhamnosus* JB-1 had the diet which increases the stimulation of neurons belonging to the dorsal root ganglion (Table 1.1). But when the probiotic was given for more than 9 days, prevention of excitability could be seen, and this excitability was mainly caused by colorectal distension (Kumari and Verma 2018). Thus, results suggested about the excitability state of neurons by the administered probiotic, and also when another strain of bacteria was used, the results could not be reproduced, indicating a specific strain result. One of the studies used BALB/c (Bagg Albino Inbred Research Mouse Strain) mice, and during this study, these healthy adult mice fed on *L. rhamnosus* JB-1 strain (Bravo et al. 2012).

As a result, these mice had reduction in anxiety, depressive behaviour, and also the changes in abilities such as cognitive and emotional in accordance with aversive stimulus could be seen clearly. Also, combinations of probiotics have been known to improve cognitive and psychological aspects in healthy subjects, e.g. *Lactobacillus helveticus* R0052 and *Bifidobacterium longum* R0175 (Galland 2014). When a similar combination of *L. helveticus* R0052 and *B. longum* R0175 was given to rats, induced with experimental MI (Myocardial infraction), increased permeability in intestines and reduction of apoptosis in the cerebral region could be observed (Mohajeri et al. 2018). Consumption of probiotics has been known to increase concentration of tryptophan in blood and induce change in the levels of serotonin and dopamine metabolites. In short, it elevates moods in humans (Bravo et al. 2012). In another study, the role of the vagus nerve as a pathway of communication between the intestinal bacteria and the brain has been investigated. This study was supported by *B. longum* NC3001 and concluded that effects in accordance with the induction of colitis were dependent on the vagus nerve. However, this theory is not being completely proved because when animals (having their vagus nerve removed) were given antibiotics, behavioural changes could be seen in them (Kumari and Verma 2018).

A 6-week study was conducted on petrochemical workers that consumed probiotic yogurt or a capsule containing multispecies probiotic, and as a result, it caused improvement in the general health and also in the areas such as anxiety, depression and stress (Fig. 1.1) (Galland 2014). Another 8-week probiotic study conducted on patients with depressive behaviour showed improvement after taking the probiotics consisting of *L. acidophilus*, *B. bifidum* and *L. casei*. A study included the use of *L. helveticus* IDCC3801 on elderly individuals who were healthy; the results of the study showed a recuperation of cognitive functions in the subjects (Bravo et al. 2012). One clinical study used probiotic milk product, which was fermented having a combination of bacterial strains, and a healthy woman consumed it for a period of 4 weeks, resulting in modulation of brain activity. This was tested by functional MRI

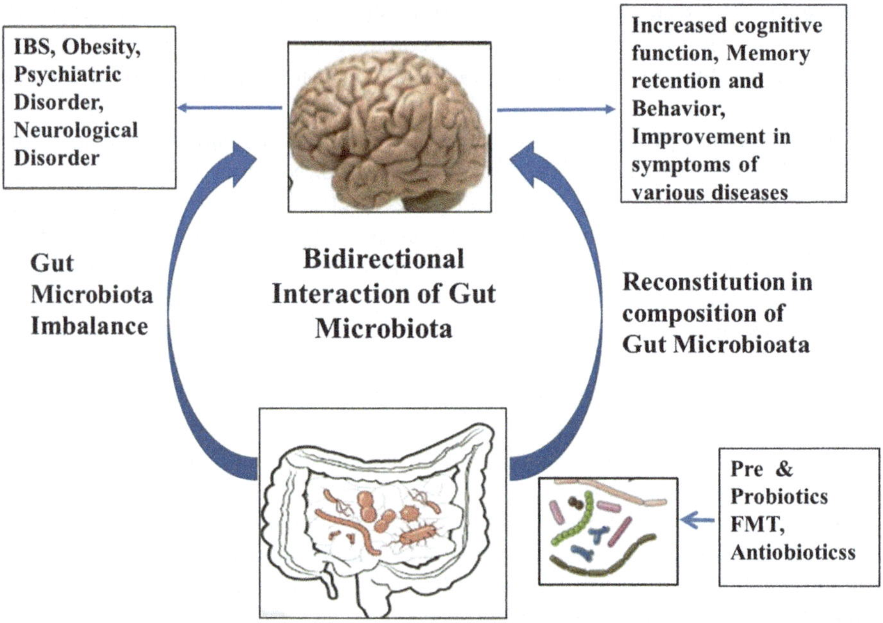

Fig. 1.1 Bidirectional interaction of gut microbiota and reconstitution of gut microbiota by adaptive changes

(magnetic resonance imaging) scans on emotional response basis, and it also showed greater connectivity in mid-brain (Mohajeri et al. 2018).

One study was conducted on patients suffering from Alzheimer's disease, and probiotic milk was given to the patients for a period of 12 weeks. Positive results such as improvement of cognitive functions were observed (Lu and Claud 2019). This probiotic milk consisted of *L. acidophilus*, *L. casei*, *B. bifidum* and *L. fermentum*. Another study was conducted on babies (up to 6 months old) with a sample size of 27 individuals (Mohajeri et al. 2018). One of the groups was given *L. rhamnosus*, while the other group was given placebo. At the age of 13, less amount of *Bifidobacterium* and higher diagnosis rate in ADHD (attention deficit hyperactivity disorder) could be seen (Galland 2014). However, more studies are needed to discover the link between ADHD and the role of the intestinal microbiota. A study was conducted on humans and animal models, and at least one of the bacteria, *Bacteroides fragilis* and *Lactobacillus reuteri*, was administered. This resulted in positive reports such as reversal of behavioural and gastrointestinal conditions in both models having ASD (autism spectrum disorder) (Fig. 1.1) (Cryan et al. 2020). Also, another clinical study was conducted by transferring a specific bacterial consortium into ASD patients (children) and reported promising results regarding reduction of various symptoms in ASD such as constipation, diarrhoea, pain in the abdominal region and non-digestion while there is improvement in behaviour. All the changes lasted for around 2 years from the time of transfer (Lu and Claud 2019; Cryan et al. 2020).

Sometimes, probiotics also have a negative impact on our body system. This information was revealed by a study conducted on healthy elderly men and women who were given a beverage containing probiotics for a period of 3 weeks. Afterwards, results showed improvement in their mood, but their memory performance was affected (Bravo et al. 2012). Thus, administration of probiotics may have unexplained composition and functional change of the intestinal microbiota. The strain present in the probiotics does not colonize and thus remains nontraceable after few weeks (Galland 2014). On the other hand, diet quickly brings changes in the composition of intestinal microbiota. Also, unhealthy diet is antagonistic in nature to probiotics and could lead to depression (Lu and Claud 2019). After antibiotic therapy, diet inclusive of probiotics also took longer time (about 5 months) for the microbiota to come back to normal composition. Thus, many questions still need to be answered before probiotics can be routinely used for treatment purposes. Nonetheless, the efficiency of probiotics does not depend only on strain and disease specificity but many other factors also (Galland 2014; Lu and Claud 2019).

1.7 Gut Microbiota Interaction and the Immune System Response

The distal part of the intestine in mammals accommodates an augmented extremely varied ecosystem of bacteria which encompasses most amount of intestinal microbiota. A major portion of the defence system of the body is also present in the intestine (Logsdon et al. 2018.) Therefore, for the maintenance of host immunity, communications are vital between gut microbiota and the host system. The greater part of intestinal microbiota inhabits in the lumen of the intestine and forms an interface for host-microbe interactions (Wang and Wang 2016). During the formation, barrier lining is done by epithelial cells. Interference in the epithelial barrier present in the gut may authorize an unfettered entry in the lamina propria by the intestinal microbiota, where the cells of the defence system is located (Logsdon et al. 2018). Cells of the immune system reside in systematized arrangements in the intestine, jointly known as gut-associated lymphoid tissues (GALT). GALT is exceedingly flexible and is colonized by bacteria. Immune system is highly functional within the intestine. Large amounts of macrophages and lymphocytes are spread all over the lamina propria and present upto basal epithelium (Wang and Wang 2016). Macrophages that inhabit the intestine are mostly insensitive to bacteria and their constituents, as there is absence of lipopolysaccharide (LPS) co-receptor. Their proinflammatory reactions are thereby suppressed with anti-inflammatory cytokines formed by GALT (Logsdon et al. 2018).

The innate immune system gets induced or interacts directly with the T cells present in the GALT (Wekerle 2017). Dendritic cells come across translocated microbial antigens, T cells and B cells then come in contact with stimulative antigens that causes differentiation and maturation. The immune system has become extremely specific as it is exposed to various antigens on daily basis through various routes. The differentiation of two important cells (Th17 and helper T) and their entry into the brain are facilitated by the gut microbiota (Wekerle 2017; Logsdon et al. 2018).

The intestinal microbiota not just influences the T cells but also the other immune cells. A study uncovered that antibiotic-induced imbalance of gut microbiota caused intensified levels of Ly-6C^{hi} monocytes, and this caused increase neurogenesis, which directly resulted in memory retention (Logsdon et al. 2018). Pathogenic invasion that causes intestinal microbiota imbalance resulted in change in the functioning of the immune system. *Clostridium difficile* infection is a primary example of infection causing imbalance in the intestinal microbiota, and due to the extensive use of antibiotics nowadays, the risk and probability of such infection have been increased (Logsdon et al. 2018). The usage of antibiotics weakens the immune system and affects the healthy microbes present in the gut, and this can be observed currently in elderly people. After infection with HIV (human immunodeficiency virus), the initial phase consists of reduction in CD4+ T cells, which continues even after the administration of retroviral drugs (Dillon et al. 2016). Correlation of reduction in T-cell abundance is observed with imbalance of the intestinal microbiota, interruption in epithelial barrier and immunogenic metabolite leak into the circulatory system. Defence to the central nervous system is mainly provided by the blood-brain barrier against systemic circulation (Dillon et al. 2016).

1.8 Role of BGM in Diseases

1.8.1 Gastrointestinal Disorders

A study was conducted in which there were a total of 827 subjects. Out of these subjects, only 22 reported that there was a microbial shift in composition of faecal microbiota community amongst controls, and these were healthy individuals and IBS patients based on various disease subtypes (Wang and Wang 2016). These disease subtypes include constipation-related IBS, diarrhoea-related IBS and varied types of IBS based on the phase of life, i.e. paediatric against adult, and also based on the compartment (mucosa against stool) (Martin et al. 2018). Recent finding suggested that there were at least two patient subgroups in IBS which fall into the Rome criteria on the structural basis of intestinal microbial community, regardless of similar gastrointestinal symptoms. One subgroup matches to the healthy control subjects. Another study demonstrated the imbalance in microbiota of IBS subgroup and balanced group when compared variation could be seen in the brain volumes (Martin et al. 2018).

An alternative study could not find any variation even though the IBS symptoms were related to the imbalance of the gut microbiota. There is no clear agreement regarding the relation of gut microbiota alterations in IBS, healthy subjects and the changes in microbiota that may be linked to the disease (Cryan et al. 2020).

1.8.2 Food Addiction

In this current scenario, obesity becomes an epidemic, and there is a need to stop food addiction, which has become vitally important nowadays. Eating behaviours and metabolites production is controlled by the gut microbiota (Martin et al. 2018). Transplanting faecal microbiota of obese mice to healthy GF mice caused food addiction and mass gain in the recipient. Furthermore, the microstructural changes in the encephalon during obesity have been considerably related to gut microbiota; also the microbial brain signatures of obese and lean subjects are distinctly variable (Cryan et al. 2020). The gut microbiota generates a number of compounds that are neurologically active. These activated compounds include a number of metabolites containing 5-HT and indole. Probiotic administration causes modifiction in brain functionality and metabolites such as GABA (gamma aminobutyric acid) and glutamate (Cryan et al. 2020). Also, bariatric surgery studies showed a noteworthy change in the makeup of intestinal microbiota. Interestingly when transplantation of faecal bacteria from a patient recovering from bariatric surgery took place into a GF non-operated animal, the recipient of the microbiota could be seeing weight loss and less intake of food (Martin et al. 2018).

1.8.3 Psychiatric Diseases Associated with Brain

1.8.3.1 Depression and Anxiety

The two often comorbid conditions in patients with IBS are depression and anxiety. Some of the diagnostic studies have demonstrated the ability of intestinal microbiota to control emotional behaviours and manipulate parameters that are important in depression and the severity of pathogenesis (Martin et al. 2018). Various studies have reported different association of gut microbiota in depressed individuals when compared to healthy people. The three diversifying studies suggested the cause and effect of microbiota; for example, when there is transplantation of depressed human faecal microbiota in the rodents, then there was a depressive behaviour observed in rodents (Zhang et al. 2019). Similar studies reported decrease in anxiety, anhedonia and tryptophan level increase in the subjects (Kelly et al. 2016). The healthy subject had an improvement in their mood due to administration of prebiotics and probiotics (Cryan et al. 2020; Martin et al. 2018).

Depression is described basically as losing interest in everyday life and being unhappy; stress plays a major role here; other factors such as genetics, environment and diet also affecst the psychological health of a person (Simpson et al. 2020; Xu et al. 2021). A case study of older woman suffering from depression reportedly showed improvement in health after an FMT transmission. Improvement in sleep, diet, BMI (body mass index), and overall behaviour of the patient was also observed (Xu et al. 2021; Cai et al. 2019). A study reported depression causing a disruption of gut microbiota composition, thus forming a feedback loop between them. Anxiety is described as a feeling of uneasiness without any particular reason and dysfunctionality in the ANS (autonomic nervous system) (Xu et al. 2021). Some of the studies have showed increase in anxiety levels due to administration of antibiotics, and it is also the reason for depression. Various studies conducted on nearly 1500 subjects which includs targeted microbiota therapy resulted in lowering the severity of patients suffering from anxiety. Another study using mice models concluded that diseases such as IBS are prominent reasons for anxiety and depression also affect the congregation pattern of the gut microbiota (Xu et al. 2021; Yang et al. 2019).

1.8.3.2 Autism Spectrum Disorder

Furthermore, along with the core symptoms of autism spectrum disorder (ASD), i.e. problem in being socially active and having underdeveloped communicative behaviour or a repetitive behaviour, also there are various gastrointestinal symptoms that are very general and contribute appreciably to sickness (Cryan et al. 2020). One of the clinical trials transplanted the gut microbiota of human patients suffering from ASD to GF mice which induced autistic behaviours in the animal models (Vuong and Hsiao 2017).

Some recent studies have shown possible composition change in gut microbiota and its metabolites of the patients when compared to healthy subjects (De Angelis et al. 2015). Another study on maternal diet has established a relation between the diet, gut microbiota and social behaviour. Female rats born were fed with diet inclusive of high fat content and *Lactobacillus reuteri*, which caused an improvement in gut microbial composition and social behaviour (Buffington et al. 2016; Xu et al. 2021). Various clinical trials conducted using FMT have resulted in improving various symptoms regarding digestive problems, behaviour and decreasing the brain oxidative stress response (Kang et al. 2019; Xu et al. 2021).

1.8.4 Neurological Disorders

1.8.4.1 Parkinson's Disease

Even though the hallmark symptoms of Parkinson's disease are motor deficits, some other symptoms such as gastrointestinal symptoms and non-motor deficiency add more problems to the quality of life of the patient. Symptoms of non-motor deficiency consist of dysfunctionality in autonomic and enteric nervous systems (Cryan et al. 2020). It has been observed that irregular bowel movements and constipation raise the risk of developing a disease like Parkinson's disease. Symptoms such as constipation could be observed much before (15 years) than other symptoms like motor dysfunction (Martin et al. 2018). Various studies are still undergoing to find relation between gut microbiota and Parkinson's disease. A prominent study suggested that α-synuclein, one of the protein metabolites that is present in Parkinson's disease patient, can be transferred to the brain through one of the cranial nerves (vagus nerve). Also, patients whose vagus nerve was removed were found to have protection against Parkinson's disease. A recently validated proof demonstrated that when the microbiota of a Parkinson's disease patient was transplanted into the rodent model, it caused damages to the brain and caused causality, while there was no causality seen in the healthy subjects (Martin et al. 2018). Thus, the gut microbiota can be used as a tool for diagnosis of the disease. Another study revealed that intestinal microbiota may have a function by interacting with drug molecules administered during treatment (Cryan et al. 2020).

1.8.4.2 Brain Stroke and Injury

Various risk factors are found to affect the brain injury and stroke caused by microbiota. Risk factors of peripheral origin are found to worsen the reaction caused due to brain injury (Cussotto et al. 2019). Trimethylamine N-oxide, one of the metabolic products produced by a gut microbe, is found to be associated with both gestational diabetes and Alzheimer's disease (Vogt et al. 2018). This indicates that it is possible to develop a drug by modulation of the type of gut microbiota. Altered microbiota has been observed in cerebral ischemia models where it was found that alteration in the composition of microbiota could worsen the situation (Singh et al. 2016). The use of antibiotics for the treatment process is generally responsible for dysregulation of the overall microflora that ultimately results in reduction of IL-17, $\gamma\delta$ T cells and IL-17-associated chemokine expression (Cowan et al. 2018). From the above-mentioned factors, it could be understood that gut microbiota has a tremendous influence on the neuroinflammation after stroke and alters the T-cell trafficking to the brain (Cryan et al. 2020).

1.8.4.3 Alzheimer's Disease

Alzheimer's disease is caused due to the formation of neurofibrillary tangles and accumulation of misfolded plaque of amyloid β protein and leads to the decline in cognitive abilities due to loss of neurons and synapses (Andrews et al. 2020). It is reported that patients with cognitive impairment and cerebral amyloidosis have more inflammatory cytokines in circulation that leads to higher levels of proinflammatory bacteria in their excretory wastes (Xu et al. 2021). Patients with cerebral amyloidosis (Amy+) show less abundance of *Eubacterium rectale* and more abundance of *Escherichia* and *Shigella* than normal healthy individuals and patients with (Amy-). Thus there is a direct relationship between proinflammatory factors and abundance of *Escherichia* and *Shigella* (Maqsood and Stone 2016). Studies have shown that mouse model with AD results in depletion of gut microbiota along with inflammation in neurons (Dodiya et al. 2019). A study conducted on an 82-year-old AD male patient using FMT from his wife reported a negative stool test for *Clostridium difficile* infection. Also the FMT improved his memory retention capacity and behaviour (Hazan 2020).

1.8.4.4 Epilepsy

Epilepsy is a chronic disease categorized by the abrupt abnormal discharge from cerebral neurons, leading to transient brain dysfunction. Genetic and environmental factors are associated with the person's susceptibility to the disorder (Xie et al. 2017). Composition and distribution of gut microbiota in patients with intractable epilepsy are different from the healthy controls (Lum et al. 2020). From the studies, it was observed that intestinal Firmicutes/*Bacteroides* ratio and α-diversity were considerably high in drug-resistant patients compared to drug-sensitive patients. The α-diversity was high in healthy individuals due to increase in rare bacterial genera (Peng et al. 2018). Reduction in seizures was observed by intake of ketogenic diet, increase in Lactobacillus and Bifidobacterium spp. was also observed (Dahlin and Prast-Nielsen 2019). Transplantation of ketogenic microbiota reduced the number of seizures in mice at a higher threshold. A probiotic strain shows positive results in patients suffering from epilepsy. Studies conducted on a 17-year-old patient suffering from epilepsy using FMT observed that epileptic recurrence was reduced with treatment (He et al. 2017).

The increase number of evidences shows the gut microbiota involvement in the developmental and functional aspects of the nervous system including various acute and chronic diseases influenced by the gut microbiota. From various clinical models, it can be concluded that intestinal microbiota also plays an important role in not just assisting the pathogens but also in increasing the pathogenicity of a disease (Sampson et al. 2016).

In the recent years there has been exemplifying progress in connecting the gut microbiota and nervous system. But still there are considerate amounts of questions remaining. The role of gut microbiota in treatment of various physiological disorders is still unclear (Martin et al. 2018). Various clinical experiences suggest patients suffering from diseases does not affect their brain functionality. Removal of any portion of the colon due to ulcerative colitis does not affect the brain functioning is a prime example. (Galland 2014). Various techniques have been used to study the microbial as well as the host data and will help to understand the interactions in a much better way. A number of attempts have been made continuously to determine the structure, functions of microbial community and identification of each taxa. But due to technological difficulties, other communities such as viromes and mycobiomes have been ignored for a long time (Martin et al. 2018).

As the cost of novel multiomic analysis decreased, it will help to identify patients having a unique pattern in the imbalance of gut microbiota, and the response of these patients towards various treatments involving diet, probiotics and also prebiotics could be recorded (Wang and Wang 2016). Still there is requirement of large-scale studies that shows the reasons of imbalance in gut microbiota and also explain the changes individually. In the previous years, various interactons between organs and microorganisms have been discovered; opening up various possibilities for treatment of a range of diseases (Martin et al. 2018).

References

Andrews SJ, Fulton-Howard B, Goate A (2020) Interpretation of risk loci from genome-wide association studies of Alzheimer's disease. Lancet Neurol 19(4):326–335

Beilharz JE, Kaakoush NO, Maniam J, Morris MJ (2018) Cafeteria diet and probiotic therapy: cross talk among memory, neuroplasticity, serotonin receptors and gut microbiota in the rat. Mol Psychiatry 23(2):351–361

Braniste V, Al-Asmakh M, Kowal C, Anuar F, Abbaspour A, Tóth M, Korecka A, Bakocevic N, Kundu P, Gulyás B, Halldin C, Hultenby K, Nilsson H, Hebert H, Volpe BT, Diamond B,

Pettersson S (2014) The gut microbiota influences blood-brain barrier permeability in mice. Sci Transl Med 6(263):263

Bravo JA, Julio-Pieper M, Forsythe P, Kunze W, Dinan TG, Bienenstock J, Cryan JF (2012) Communication between gastrointestinal bacteria and the nervous system. Curr Opin Pharmacol 12(6):667–672

Buffington SA, Di Prisco GV, Auchtung TA, Ajami NJ, Petrosino JF, Costa-Mattioli M (2016) Microbial reconstitution reverses maternal diet-induced social and synaptic deficits in offspring. Cell 165(7):1762–1775

Cai T, Shi X, Yuan LZ, Tang D, Wang F (2019) Fecal microbiota transplantation in an elderly patient with mental depression. Int Psychogeriatr 31(10):1525–1526

Chu DM, Ma J, Prince AL, Antony KM, Seferovic MD, Aagaard KM (2017) Maturation of the infant microbiome community structure and function across multiple body sites and in relation to mode of delivery. Nat Med 23(3):314–326

Claud EC, Keegan KP, Brulc JM, Lu L, Bartels D, Glass E, Chang EB, Meyer F, Antonopoulos DA (2013) Bacterial community structure and functional contributions to emergence of health or necrotizing enterocolitis in preterm infants. Microbiome 1(1):20

Cowan C, Hoban AE, Ventura-Silva AP, Dinan TG, Clarke G, Cryan JF (2018) Gutsy moves: the amygdala as a critical node in microbiota to brain signaling. BioEssays 40(1):10

Cryan JF, O'Riordan KJ, Sandhu K, Peterson V, Dinan TG (2020) The gut microbiota in neurological disorders. Lancet Neurol 19(2):179–194

Cussotto S, Strain CR, Fouhy F, Strain RG, Peterson VL, Clarke G, Stanton C, Dinan TG, Cryan JF (2019) Differential effects of psychotropic drugs on microbiome composition and gastrointestinal function. Psychopharmacology 236(5):1671–1685

Dahl C, Stigum H, Valeur J, Iszatt N, Lenters V, Peddada S, Bjørnholt JV, Midtvedt T, Mandal S, Eggesbø M (2018) Preterm infants have distinct microbiomes not explained by mode of delivery, breastfeeding duration or antibiotic exposure. Int J Epidemiol 47(5):1658–1669

Dahlin M, Prast-Nielsen S (2019) The gut microbiome and epilepsy. E Bio Med 44:741–746

De Angelis M, Francavilla R, Piccolo M, De Giacomo A, Gobbetti M (2015) Autism spectrum disorders and intestinal microbiota. Gut Microbes 6(3):207–213

Dillon SM, Frank DN, Wilson CC (2016) The gut microbiome and HIV-1 pathogenesis: a two-way street. AIDS 30(18):2737–2751

Dodiya HB, Kuntz T, Shaik SM, Baufeld C, Leibowitz J, Zhang X, Gottel N, Zhang X, Butovsky O, Gilbert JA, Sisodia SS (2019) Sex-specific effects of microbiome perturbations on cerebral Aβ amyloidosis and microglia phenotypes. J Exp Med 216(7):1542–1560

Frohlich EE, Farzi A, Mayerhofer R, Reichmann F, Jačan A, Wagner B, Zinser E, Bordag N, Magnes C, Fröhlich E, Kashofer K, Gorkiewicz G, Holzer P (2016) Cognitive impairment by antibiotic-induced gut dysbiosis: analysis of gut microbiota-brain communication. Brain Behav Immun 56:140–155

Galland L (2014) The gut microbiota and the brain. J Med Food 17(12):1261–1272

Gao W, Salzwedel AP, Carlson AL, Xia K, Azcarate-Peril MA, Styner MA, Thompson AL, Geng X, Goldman BD, Gilmore JH, Knickmeyer RC (2019) Gut microbiome and brain functional connectivity in infants-a preliminary study focusing on the amygdala. Psychopharmacology 236(5):1641–1651

Hazan S (2020) Rapid improvement in Alzheimer's disease symptoms following fecal microbiota transplantation: a case report. J Int Med Res 48(6):300060520925930

He Z, Cui BT, Zhang T, Li P, Long CY, Ji GZ, Zhang FM (2017) Fecal microbiota transplantation cured epilepsy in a case with Crohn's disease: the first report. World J Gastroenterol 23 (19):3565–3568

Hill CJ, Lynch DB, Murphy K, Ulaszewska M, Jeffery IB, O'Shea CA, Watkins C, Dempsey E, Mattivi F, Tuohy K, Ross RP, Ryan CA, O'Toole PW, Stanton C (2017) Evolution of gut microbiota composition from birth to 24 weeks in the INFANTMET Cohort. Microbiome 5(1):4

Jang HM, Lee HJ, Jang SE, Han MJ, Kim DH (2018) Evidence for interplay among antibacterial-induced gut microbiota disturbance, neuro-inflammation, and anxiety in mice. Mucosal Immunol 11(5):1386–1397

Kang DW, Adams JB, Gregory AC, Borody T, Chittick L, Fasano A, Khoruts A, Geis E, Maldonado J, McDonough-Means S, Pollard EL, Roux S, Sadowsky MJ, Lipson KS, Sullivan MB, Caporaso JG, Krajmalnik-Brown R (2017) Microbiota Transfer Therapy alters gut ecosystem and improves gastrointestinal and autism symptoms: an open-label study. Microbiome 5 (1):10

Kang DW, Adams JB, Coleman DM, Pollard EL, Maldonado J, McDonough-Means S, Caporaso JG, Krajmalnik-Brown R (2019) Long-term benefit of Microbiota Transfer Therapy on autism symptoms and gut microbiota. Sci Rep 9(1):5821

Kelly JR, Kennedy PJ, Cryan JF, Dinan TG, Clarke G, Hyland NP (2015) Breaking down the barriers: the gut microbiome, intestinal permeability and stress-related psychiatric disorders. Front Cell Neurosci 9:392

Kelly JR, Borre Y, O'Brien C, Patterson E, El Aidy S, Deane J, Kennedy PJ, Beers S, Scott K, Moloney G, Hoban AE, Scott L, Fitzgerald P, Ross P, Stanton C, Clarke G, Cryan JF, Dinan TG (2016) Transferring the blues: depression-associated gut microbiota induces neurobehavioural changes in the rat. J Psychiatr Res 82:109–118

Korpela K, Blakstad EW, Moltu SJ, Strømmen K, Nakstad B, Rønnestad AE, Brække K, Iversen PO, Drevon CA, De Vos W (2018) Intestinal microbiota development and gestational age in preterm neonates. Sci Rep 8(1):2453

Kumari R, Verma N (2018) Implication of gut microbiota in brain chemistry and behavior. Ann Biol Sci 6(1):18–22

Logsdon AF, Erickson MA, Rhea EM, Salameh TS, Banks WA (2018) Gut reactions: how the blood-brain barrier connects the microbiota and the brain. Exp Biol Med 243(2):159–165

Lu J, Claud EC (2019) Connection between gut microbiota and brain development in preterm infants. Dev Psychobiol 61(5):739–751

Lum GR, Olson CA, Hsiao EY (2020) Emerging roles for the intestinal microbiome in epilepsy. Neurobiol Dis 135:104576

Maqsood R, Stone TW (2016) The gut-brain axis, BDNF, NMDA and CNS disorders. Neurochem Res 41(11):2819–2835

Martin CR, Osadchiy V, Kalani A, Mayer EA (2018) The brain-gut-microbiota axis. Cell Mol Gastroenterol Hepatol 6(2):133–148

Mohajeri MH, La Fata G, Steinert RE, Weber P (2018) Relationship between the gut microbiota and brain function. Nutr Rev 76(7):481–496

Mohle L, Mattei D, Heimesaat MM, Bereswill S, Fischer A, Alutis M, French T, Hambardzumyan D, Matzinger P, Dunay IR, Wolf SA (2016) Ly6C(hi) Monocytes provide a link between antibiotic-induced changes in gut microbiota and adult hippocampal. neurogenesis. Cell Rep 15(9):1945–1956

Opp MR, George A, Ringgold KM, Hansen KM, Bullock KM, Banks WA (2015) Sleep fragmentation and sepsis differentially impact blood-brain barrier integrity and transport of tumor necrosis factor-α in aging. Brain Behav Immun 50:259–265

Peng A, Qiu X, Lai W, Li W, Zhang L, Zhu X, He S, Duan J, Chen L (2018) Altered composition of the gut microbiome in patients with drug-resistant epilepsy. Epilepsy Res 147:102–107

Sampson TR, Debelius JW, Thron T, Janssen S, Shastri GG, Ilhan ZE, Challis C, Schretter CE, Rocha S, Gradinaru V, Chesselet MF, Keshavarzian A, Shannon KM, Krajmalnik-Brown R, Wittung-Stafshede P, Knight R, Mazmanian SK (2016) Gut microbiota regulate motor deficits and neuroinflammation in a model of Parkinson's disease. Cell 167(6):1469–1480

Simpson CA, Schwartz OS, Simmons JG (2020) The human gut microbiota and depression: widely reviewed, yet poorly understood. J Affect Disord 274:73–75

Singh V, Roth S, Llovera G, Sadler R, Garzetti D, Stecher B, Dichgans M, Liesz A (2016) Microbiota dysbiosis controls the neuroinflammatory response after stroke. J Neurosci 36 (28):7428–7440

Tang AT, Choi JP, Kotzin JJ, Yang Y, Hong CC, Hobson N, Girard R, Zeineddine HA, Lightle R, Moore T, Cao Y, Shenkar R, Chen M, Mericko P, Yang J, Li L, Tanes C, Kobuley D, Võsa U, Whitehead KJ, Kahn ML (2017) Endothelial TLR4 and the microbiome drive cerebral cavernous malformations. Nature 545(7654):305–310

Vogt NM, Romano KA, Darst BF, Engelman CD, Johnson SC, Carlsson CM, Asthana S, Blennow K, Zetterberg H, Bendlin BB, Rey FE (2018) The gut microbiota-derived metabolite trimethylamine N-oxide is elevated in Alzheimer's disease. Alzheimer Res Ther 10(1):124

Vuong HE, Hsiao EY (2017) Emerging roles for the gut microbiome in autism spectrum disorder. Biol Psychiatry 81(5):411–423

Wang HX, Wang YP (2016) Gut microbiota-brain axis. Chin Med J 129(19):2373–2380

Wekerle H (2017) Brain autoimmunity and intestinal microbiota: 100 trillion game changers. Trends Immunol 38(7):483–497

Xie G, Zhou Q, Qiu CZ, Dai WK, Wang HP, Li YH, Liao JX, Lu XG, Lin SF, Ye JH, Ma ZY, Wang WJ (2017) Ketogenic diet poses a significant effect on imbalanced gut microbiota in infants with refractory epilepsy. World J Gastroenterol 23(33):6164–6171

Xu HM, Huang HL, Zhou YL, Zhao HL, Xu J, Shou DW, Liu YD, Zhou YJ, Nie YQ (2021) Fecal microbiota transplantation: a new therapeutic attempt from the gut to the brain. Gastroenterol Res Pract 2021:6699268

Yang B, Wei J, Ju P, Chen J (2019) Effects of regulating intestinal microbiota on anxiety symptoms: a systematic review. Gen Psychiat 32(2):e100056

Zhang Y, Huang R, Cheng M, Wang L, Chao J, Li J, Zheng P, Xie P, Zhang Z, Yao H (2019) Gut microbiota from NLRP3-deficient mice ameliorates depressive-like behaviors by regulating astrocyte dysfunction via circHIPK2. Microbiome 7(1):116

Current Insights on the Modulation of Gut Microbiome and Its Effect on Human Health

2

Devaraja Gayathri, Chakra Siddappa Prashantkumar, and Mahanthesh Vasudha

Abstract

The gut of human beings is inhabited by a diverse group of microorganisms, around trillions, which makes it a new essential endocrine organ and shows a symbiotic connection with a host, and they metabolize the food ingested and produce diverse bioactive and dietary compounds. This may include organic acids, bacteriocins, and short-chain fatty acids, which provide potential to impact on physiological and pathological conditions of the host and maintain homeostasis. In recent times, due to rapid advancement in technology, our understanding about microbiome has also expanded. The modulation of the microbiome leads to disturbance in homeostasis, which causes imbalance and leads to dysbiosis, and the gut barrier integrity gets disturbed and immunological reaction leads to inflammation. This chapter reviews the current insights on various diseases and gastroenterological disorders associated with the modulation of the gut microbiome and how probiotics help in maintaining the healthy gut with intact gut barrier by regulating the expression of tight junction proteins, perhaps leading to good human health.

Keywords

Colon cancer · Dysbiosis · Gut microbiome · Irritable bowel disease/syndrome · Obesity · Probiotics

D. Gayathri (✉) · C. S. Prashantkumar · M. Vasudha
Department of Studies in Microbiology, Davanagere University, Shivagangothri, Davanagere, Karnataka, India

R. Z. Sayyed, M. Khan (eds.), *Microbiome-Gut-Brain Axis*,
https://doi.org/10.1007/978-981-16-1626-6_2

2.1 Introduction

Mammalian gut is inhabited by a wide range of microbial communities, which are described as the gut microbiome. Due to advancement in technology and with omics study, our understanding toward the organisms colonizing the gut, their functionality, and their roles in healthy humans and gut-related diseases has significantly advanced (Schmidt et al. 2018). Like conventional culture-dependent techniques, culture-independent approaches like the use of next-generation sequencing technology and 16S rRNA sequencing revealed the abundance of the microbial community residing in the gut (Eckburg et al. 2005). In the gut, the microbial cell concentration exceeds 10^{11} cells/g contents, which makes up 1–2 kg of our body mass. It accounts for more than five million different genes, and 1000 to 1500 different species like Bacteroidetes, Firmicutes, Actinobacteria, Proteobacteria, Fusobacteria, Cyanobacteria, and Verrucomicrobia are well represented (D'Argenio and Salvatore 2015). So the complex interplay between the gut microbiome and the host physiology has become a hot topic of research (Ghaisas et al. 2016).

Genetic inheritance and environmental factors affect the gut microbiome, and each individual has a substantially diverse microbiome, which mainly helps in maintaining the homeostasis of the host, as there is a symbiotic relationship; the host is benefitted by colonic fermentation; The barrier effect can be observed against opportunistic pathogen colonization, as well as the growth of the gut immune system and also they synthesize beneficial metabolites and nutrients for the host. In turn, the host provides shelter and nutrients for the microbial complex in the gut environment.

Due to the diverse range of environmental, host, and immunological factors, there would be an imbalance or maladaptation in gut microbiota; such imbalance could be termed dysbiosis. In developed countries, improved hygiene, the decrease in the count of vaginal deliveries, the low rate of breastfeeding, and the pervasive use of antibiotics affect the indigenous gut microbiota. As these children grow, there will be a shift in healthy symbiotic microorganisms to enteric pathogens leading to immune and inflammatory disorders (Jain and Walker 2015). Further, we shall discuss various diseases/disorders caused because of gut dysbiosis and current research issues to overcome these clinical conditions with probiotics intervention.

2.2 Probiotics in the Treatment of Gut Diseases

The phrase "probiotic" is derived from the Greek language; it means "for life" since ancient times mankind knew about the use of fermented food for maintaining good health; even in the Old Testament there is a description of consumption of sour milk by Abraham for longevity. The first reference or idea of the probiotic concept was given by Nobel Laureate Elie Metchnikoff and proposed that to displace putrefactive and pathogenic intestinal organism, one must consume fermented milk containing lactobacilli, which maintain good health and increase longevity. In the year 1954, Ferdinand Vergin published an article on "Anti-und Probiotika" in which he used the term "probiotic" where a list of several useful microorganisms and effects of the antibacterial agents and antibiotics on intestinal microbiota was published. Later, after few years, according to Lilly and Stillwell (1965), microorganisms that are beneficial as well as those that produce growth-promoting factors for the organisms were described as probiotics, and the term has been customized over a period of time. Now the widely accepted definition is the following: "probiotics are live strains of microorganisms which when administered in adequate amounts confer a health benefit on the host" by the Food and Agriculture Organization (FAO) and the World Health Organization (WHO), followed by the International Scientific Association for Probiotics and Prebiotics (ISAPP). Many probiotic strains that are extensively studied for human use and many probiotic products in the market developed using different species of bacteria are listed in Table 2.1 (Azad et al. 2018).

The mechanism of action of probiotics has been postulated in many ways; however, the exact mechanism of action was sparsely deciphered; some postulates are as follows: when taken in the adequate amount, probiotics fight for epithelial cell receptors and form colonizing barriers, and the other pathogenic strains cannot adhere to gut cells and compete for survival with one other. Besides, these probiotic organisms produce a class of antimicrobial compounds called bacteriocins, along with certain other metabolic products like lactic acid, diacetyl, short-chain fatty acids, and hydrogen peroxide, which kills pathogenic bacteria. These organisms have a symbiotic relationship with the gut and, in turn, stimulate immune response and show some immunomodulatory effects by increasing the secretion of immunoglobulin A and enhance the activity of natural killer cells, macrophages, and other immune cells (Khalighi et al. 2016). Due to these properties, many researchers propose probiotics as one of the alternative treatment aids for gastrointestinal disorders. Beneficial aspects of probiotics are depicted in Fig. 2.1.

2.2.1 Obesity

Obesity is a rising epidemic worldwide and the fifth leading cause for death. An estimate says that at least 2.8 million adults die due to obesity (Ahmed et al. 2014). It is defined as atypical or unbalanced fat buildup that may impair health. Obesity is the result of an energy imbalance between calories consumed and utilized. Consumption of high-fat/high-sucrose diets is one of the primary causes of obesity, and it is measured in terms of body mass index (BMI). If an individual is having a BMI less than 25 kg/m^2, the person is considered as normal; if the BMI is above 25 kg/m^2, then the individual is considered as overweight, and if the BMI crossed 30 kg/m^2, then the individual is considered as obese. Further, obesity patients are divided into three categories based on the BMI: first-degree obesity, if a person has a BMI between 30 and 35 kg/m^2; second-degree obesity, if a person has a BMI between 35 and 40 kg/m^2; and third-degree obesity, if BMI is between 40 and 45 kg/m^2. Even waist circumference is used for classifying obesity wherein visceral/central or subcutaneous obesity has been taken into account. If women have a waist circumference of over 88 cm and if men have a waist circumference of over 102 cm, then they are said to be having visceral/central or subcutaneous obesity. Waist-to-hip

Table 2.1 List of commonly used organisms as probiotics

Lactobacillus spp.	*Bifidobacterium* spp.	*Bacillus* spp.	*Streptococcus* spp.	*Enterococcus* spp.	*Saccharomyces* spp.
L. acidophilus	*B. breve*	*B. coagulans*	*S. thermophilus*	*E. faecium*	*S. cerevisiae*
L. plantarum	*B. infantis*				
L. rhamnosus	*B. longum*				
L. paracasei	*B. bifidum*				
L. fermentum	*B. thermophilum*				
L. reuteri	*B. adolescentis*				
L. johnsonii	*B. animalis*				
L. brevis	*B. lactis*				
L. casei					
L. lactis					
L. delbrueckii					
L. gasseri					

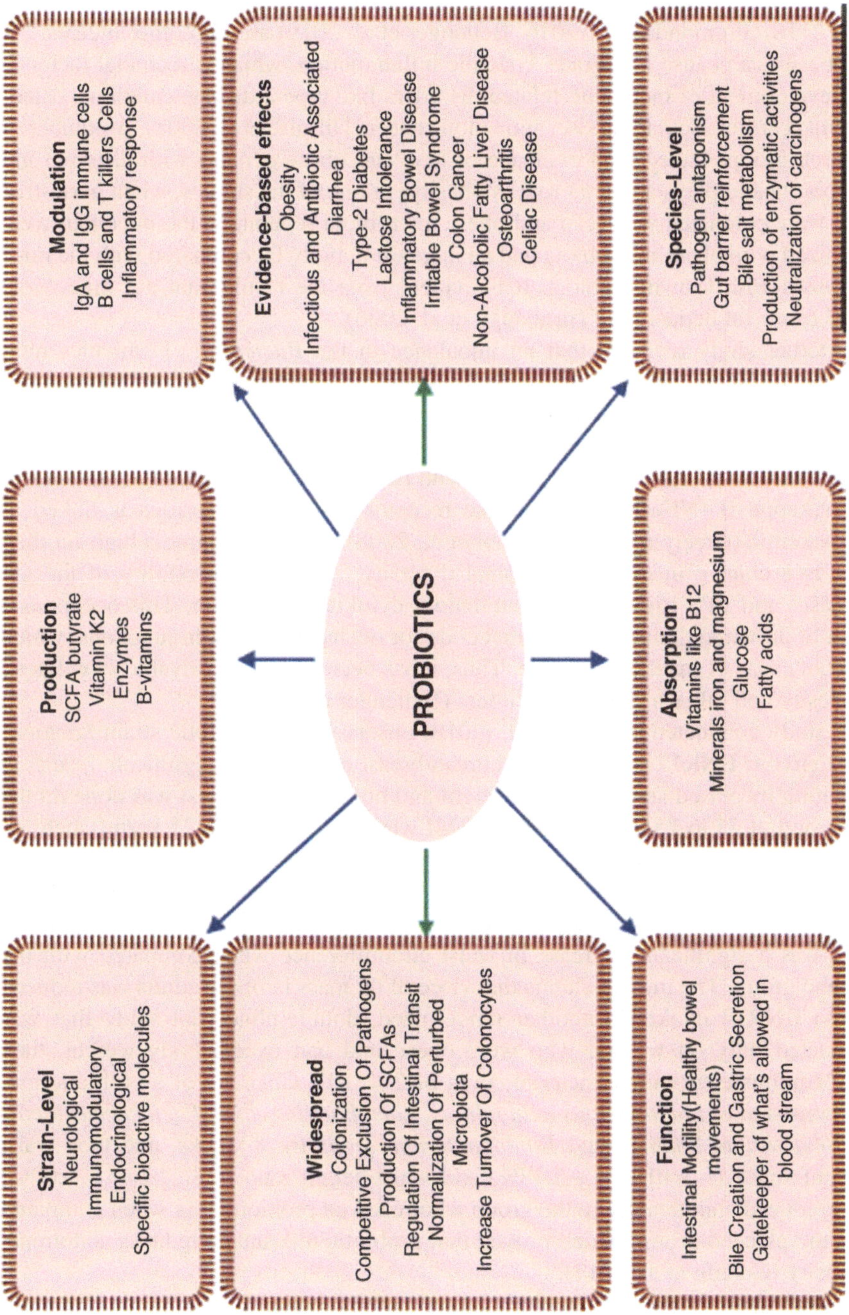

Fig. 2.1 Different beneficial aspects of probiotics, created with BioRender.com

ratio, body adiposity index, waist-to-height ratio, Rohrer's ponderal index, neck circumference, Benn's index, skinfold thickness, and fat mass index (Rouxinol-Dias et al. 2016; Simmonds et al. 2016; Bellenger et al. 2019) are also other indices.

Obesity can cause low-grade systemic inflammation, which is a crucial factor in the development of metabolic-related disorders like type 2 diabetes mellitus, osteoarthritis, cardiovascular disease, and colon cancer (DiBaise et al. 2008). A number of research groups have given evidence of lean and obese persons with different gut microbiota and plays a crucial role in obesity. In one of the studies, when germ-free mice were colonized with lean and obese microbiota separately, the mice that were colonized with obese microbiota had an increase in body fat compared with the mice colonized with lean microbiota. It is evident from the results that gut microbiota could cause fat deposition (Turnbaugh et al. 2006).

Another study revealed that an imbalance in the *Bacteroides*/Firmicutes ratio leads to dysbiosis and increase in the Gram-negative bacteria count and its component. In this, mainly lipopolysaccharides found on the outer membrane are the triggering factor of metabolic endotoxemia, which leads to obesity and clinical diabetes. High-energy and high-fat diet with rising concentration along with nuclear factor kappa B (NF-κB) and Toll-like receptor (TLR4) expressions is linked to endotoxemia (Creely et al. 2007; Cani et al. 2008). As we consume a high-fat diet, there is a change in the gut microbial diversity, and the permeability of the gut increases and LPS triggers an inflammatory cytokine expression. This occurs as a result of downregulation of genes that code for occluding and tight junctions, which is a root cause of Insulin resistance. Thus, it has been proved that dysbiosis will lead to obesity and other metabolic disorders (Bellenger et al. 2019).

A study conducted by Kim et al. (2018) observed that probiotic strain *Lactobacillus gasseri* BNR17 isolated from human breast milk showed significant results in inhibiting increased adipose tissue weight and body weight. A test was done on the age group of 25 to 75 years, and their BMI was between 25 and 35 kg/m^2 such 90 volunteers were randomized, double-blind, placebo-controlled trial for 12 weeks. The patients were given high-dose BNR-H 10^{10} CFU/day and low-dose BNR-L 10^9 CFU/day; both groups showed decreased visceral adipose tissue (VAT), and also there was a significant decrease in waist circumference when compared with the placebo group. The finding claims that visceral fat mass in obese adults was reduced due to BNR17 intake. In another randomized, double-blind trial study that was conducted only on women who were obese and had excess body weight, they were administered with a bacterial count of 2×10^{10} CFU/day of a probiotic mix that consisted of *Lactobacillus acidophilus*, *Lactobacillus casei*, *Lactococcus lactis*, *Bifidobacterium bifidum*, and *Bifidobacterium lactis* for 8 weeks, and then waist circumference (−3.40% to −5.48%) and waist-height ratio (−3.27% to −5.00%) were reduced significantly in the group who received probiotic mix when compared with the placebo group, which proved that probiotic mix intake reduces abdominal adiposity (Corado et al. 2017).

2.2.2 Infectious Diarrhea

Infectious diarrhea is also known as gastroenteritis and stomach flu. It is the inflammation of the stomach and small intestine due to infection along with symptoms like diarrhea, nausea, vomiting, abdominal pain, fever, and dehydration, and if this diarrhea due to infection lasts for 14 or less than 14 days, it can be called acute diarrhea. If it continues for more than 4 weeks, then it is said to be chronic diarrhea. Around 2.5 million deaths worldwide occur due to diarrhea, and there were around 1.3 million deaths and more than two billion cases of gastroenteritis in the year 2015. Children, especially those under the age of 5, were among the most affected groups, mainly due to unhygienic conditions. Diarrhea is generally caused by rotavirus, while in adults, it is caused by norovirus. However, bacteria, fungi, and parasites can also cause gastroenteritis. (Barr and Smith 2014; Fernández-Bañares et al. 2016).

Normal intestinal physiology is disturbed during infectious diarrhea by fluid and electrolyte secretion; for example, *Vibrio cholerae*, a Gram-negative comma-shaped bacillus produces cholera toxin (CT), an oligomeric protein made up of two subunits, A and B. Subunit B helps subunit A to enter into gut epithelial cells, and then subunit A regulates adenylate cyclase by ADP ribosylation and increases cAMP production. This in turn activates PKA, which phosphorylates CFTR channel and increases Cl^- secretion. The activity of two sodium transporters, NHE2 and NHE3, is decreased by increased cAMP production and affects Na^+ absorption; therefore, NaCl levels increase in the lumen of the intestine (Cheng et al. 1991).

In addition, *Clostridium difficile* is a causal agent of diarrhea, acute colitis, and inflammation and produces exotoxins, namely, toxin A and toxin B (TcdA and TcdB) along with binary toxin (CDT). TcdA adheres to the host cell via glycoprotein (gp) 96, whereas TcdB gains entry via dissolved tight junctions between two cells and binds to an unknown receptor. These are cytotoxic enzymes that disrupt cytoskeletal integrity by glycosylating *Rho* protein and also affect ion transport and cause accumulation of sodium, chloride, and potassium ions. Apart from this, these toxins initiate an inflammatory response by activating IL-6, IL-1, and TNF-α in the infective tissue (Hodges and Gill 2010). The presence of microorganisms, especially pathogens, may lead to disruption in gut homeostasis and inflammation and continues as a chronic condition leading to inflammatory bowel disease (IBD), irritable bowel syndrome (IBS), and other metabolic disorders. If diarrhea persists for a longer time and is not treated properly, especially in children, it may lead to dehydration and finally death.

Probiotic strains such as *Lactobacillus reuteri*, *Bifidobacterium lactis* Bb12, *L. casei* Shirota, and *Lactobacillus rhamnosus* GG can be used in acute rotavirus diarrhea. These strains shorten the duration of diarrhea (Isolauri 2003). Basu et al. (2007) reported that *Lactobacillus rhamnosus*(LGG) can decrease the frequency as well as the duration of diarrhea-related symptoms and reduce the stay time in the hospital for the patients suffering from persistent diarrhea (PD). They took 235 patients from North Bengal, India, and divided them into two groups. The test group was given oral rehydration solution and LGG, whereas the control group was given oral rehydration solution only. It was found that the mean duration of diarrhea was reduced in the test group compared to the control group.

2.2.3 Type 2 Diabetes

In the twenty-first century, type 2 diabetes mellitus (T2DM) is among the very common disorders and has become the third-largest disease after cancer. The frequency of T2DM has increased worldwide in recent years. According to the International Diabetes Federation, it has been estimated that by the year 2040, approximately 642 million people would suffer from T2DM. T2DM is depicted by hyperinsulinemia in target organs and insulin deficiency caused by the destruction of β-cells in the pancreas. It has been observed that this destruction is partly contributed due to obesity. Overweight induces low-grade inflammation in obese individuals that leads to insulin resistance, and the increased level of inflammatory cytokine causes oxidative stress and destroys the β-cells in the pancreas. Genetics, lifestyle, stress, and environmental factors play a key role in the epidemiology of T2DM (Guariguata et al. 2014; Chatterjee et al. 2017).

A study carried out by Larsen et al. (2010) revealed gut dysbiosis in T2DM patient group compared to the healthy glucose tolerance group. In their experiment, a total of 36 male adult subjects were taken and they were divided into two groups among them, 18 were healthy controls, and another 18 were T2DM patients, and they were of different age group and BMI. It was found that the gut microbiota of the normal healthy person was different from the T2DM patient's microbiota. Another recent study made use of 16S rRNA-based high-throughput sequencing of three different groups of human subjects with T2DM diabetes, prediabetes, and healthy glucose tolerance. The results clearly stated that there is a structural modulation of the gut microbiome in the T2DM patient group which may lead to dysbiosis compared with the tolerance group (Egshatyan et al. 2016).

T2DM and obesity are inflammatory diseases. The patients show elevated circulating levels of haptoglobin, serum amyloid A, C-reactive protein (CRP), sialic acid, cytokines, chemokines, and interleukins like interleukin-6 (IL-6), IL-1β, IL-1, and TNF, and its receptor antagonist IL-1RA concentration was elevated in the patients with prediabetes, T2DM, and obesity. CRP is the biomarker for T2DM-associated cardiovascular disease. The mechanism involved is induced expression of pro-angiogenic and inflammatory genes in macrophages due to hypoxia. The main metabolic pathway involved in the process is *NF-κB and JNK pathway*. The expression of NF-κβ target genes such as cytokines including TNFα, IL-6, and IL-1β promotes insulin resistance that has been initially produced in the adipose tissue and liver and then migrates through circulation to other parts of the body like the kidneys and circulating leukocytes, and vessel walls, skeletal and cardiac muscle induce insulin resistance in T2DM patients (Donath and Shoelson 2011).

Another most crucial pathway plays a significant role in cellular processes, and cellular physiology such as cell survival, proliferation, protein synthesis, lipid metabolism, and maintaining glucose homeostasis is done by phosphoinositide

3-kinase/protein kinase B (PI3K/AKT) pathway. In the host, almost 90% of glucose utilization has been observed in skeletal muscles, which are insulin stimulated. Knockdown experiments of AKT protein showed decrease in insulin-induced glucose uptake, whereas overexpression led to increased uptake of glucose. AKT functions by phosphorylating AS160 and by activating GagAKT. The glucose transporter GLUT4 is activated by AS160, which transports glucose from stored vesicles into skeletal cells, whereas GagAKT promotes glycogen synthesis. Likewise in the pancreas, activation of this pathway increases the synthesis of insulin from β-cells in obese and T2DM patients. PI3K/AKT pathway is blocked or dysfunctional, which affects β-cells, and insulin production is reduced. This further affects other tissues by insulin resistance (Huang et al. 2018).

Further, in a study, 50 volunteers were selected for a double-blind, randomized, placebo-controlled trial for 6 weeks, in which the test group was given 120 g/d of fermented milk containing probiotic strains *L. acidophilus* LA-5 and *B. animalis* subsp. *lactis* BB-12. After 6 weeks, test patients showed improved glycemic control, the levels of fructosamine and hemoglobin A_1c lowered significantly, and also inflammatory cytokines were decreased (Tonucci et al. 2017).

Another study conducted by Firouzi et al. (2017) reported a decrease in fasting insulin level in patients who were enrolled in a randomized, double-blind, parallel-group, controlled clinical trial, where 136 patients aged between 30 and 70 years suffering from T2DM were selected, and among them, the test group received a probiotic strain mixture of *Lactobacillus and Bifidobacterium* for 12 weeks. A significant reduction in glycated hemoglobin and fasting insulin was observed in the test group compared to the placebo group.

Shah and Swami (2017) carried out a meta-analysis and found that fasting blood glucose, HbA1c, and HOMA-IR showed a significant reduction in T2DM patients; on the other hand, there was no reduction in serum insulin concentration. These results were obtained by 12 randomized controlled trials involving 770 T2DM patients, and they were treated with probiotics; perhaps probiotics are an alternative way for controlling T2DM.

2.2.4 Lactose Intolerance

A large section of the world population is intolerant to many food items, in that 70% of adults suffer from lactose intolerance with clinical symptoms like abdominal pain and distention, flatus, and diarrhea after consumption of lactose-containing food. Lactose intolerance is an inability to digest lactose (LI) due to deficiency of lactase or β-galactosidase enzyme in the small intestine (Harrington et al. 2008). In lactose intolerance, undigested lactose in the colon could be fermented by some gut bacteria, producing acid and gas, leading to the development of lactose intolerance symptoms (Horner et al. 2011; Savaiano et al. 2011). Probiotic bacteria provide health benefits to the host gut, like protection from pathogen colonization, restoration of the gut microbiome composition, and prevention of gastrointestinal disorders (Matthews et al. 2005; Heyman 2006; Gayathri and Vasudha 2018). Many probiotic bacteria have been used in the treatment of lactose intolerance, mainly genera *Bifidobacterium* and *Lactobacillus*.

A number of research teams have been working on probiotics for the lactose intolerance treatment. In a study, Almeida et al. (2012) examined LI patients supplemented with yogurt containing *L. casei* Shirota and *B. breve* for 4 weeks, and it showed proof of reduced symptom scores. In another study, Li et al. (2012) conducted an experiment on post-weaning Balb/c mice with LI symptoms, which were orally administered with 1×10^8 CFU of *L. lactis* for 4 weeks and compared with control mice for diarrhea test and showed suppressed intestinal motility after lactose challenge.

Additionally, a comparative study was conducted by Ojetti et al. (2009). They treated LI patients with *L. reuteri*, and they showed a reduction in gastrointestinal symptoms after consumption of lactose. He et al. (2008) examined LI patients treated with *B. longum* supplementation and showed the β-galactosidase enzyme activity during and after supplementation. Later, Luyer et al. (2010) studied the effect of *B. animalis and B. longum* supplementation for the modification of gut microbial composition and β-galactosidase activity. Overall studies revealed that the consumption of probiotic bacteria improves the lactose digestion and alleviates the lactose intolerance symptoms and other gastrointestinal disorders.

2.2.5 Inflammatory Bowel Disease (IBD)

In genetically susceptible host, the immunological response of commensal microflora mediated a complex disease known as ulcerative colitis (UC) and Crohn's disease (CD). Together these disorders are called inflammatory bowel disease (IBD). The incidence of IBD is high, affecting 1.5 million Americans and 2.2 million Europeans and several hundred thousand people in Asia and other developed countries. IBD is a consequence of a complex interaction between microbial, genetic, immunologic, and environmental factors, which arises due to fast-track lifestyle, fast food, behavior, smoking, lack of physical exercise, sleep, stress, and genetic susceptibility, and gut dysbiosis caused by excessive use of antibiotics also plays a significant role in disease pathogenesis. Patients experience diarrhea, rectal bleeding, abdominal pain, inflammation, and weight loss in both CD and UC (Ananthakrishnan 2015).

Gut dysbiosis is one of the significant causes of IBD as studies found that patients with IBD have decreased Firmicutes proportions and increased *Bacteroides* and Enterobacteriaceae. CD patients have abundant *Enterococcus* spp., *Clostridium difficile*, *Escherichia coli*, *Shigella flexneri*, and *Listeria* spp. compared to the healthy individual. These alterations in the homeostasis of the gut lead to an inflammatory environment in the gut. Therefore, innate and adaptive immune cells produce pro-inflammatory cytokines such as IFN-γ, IL-1β, TNF-α, and IL-17, which enhance the inflammation leading to gut epithelial damage (Venegas et al. 2019).

There is a continuous interaction among the gut microbiota and the immune cells. Some of these interactions are beneficial. The interaction with capsulated

Bacteroides fragilis, coated with polysaccharide A, when they evade host defense and come in contact with immune cells like regulatory T cells and dendritic cells, produces anti-inflammatory cytokines like IL-10 and TGF-β, which help in maintaining tissue homeostasis in IBD patients. Further, the anti-inflammatory response genes like *ATG16L1* and *NOD2* are dysregulated, which impairs the sensitivity of receiving protective signals. Enterobacteriaceae induces TH17-dependent inflammation by producing IL-17 cytokine in CD and UC (Shamoon et al. 2019), and the study shows the association of gut microbiota and their role in IBD.

Sood et al. (2009) reported that a commercial probiotic mixture named VSL#3 contains eight different species, namely, *Lactobacillus casei, Lactobacillus delbrueckii, Bifidobacterium infantis, Bifidobacterium longum, Streptococcus salivarius, Lactobacillus plantarum, Lactobacillus acidophilus*, and *Bifidobacterium breve*; when used as a treatment strategy on patients with mild-to-moderate UC, VSL#3 showed an effective response in achieving clinical response and remissions in the patients, and they found that there was 50% decrease in Ulcerative Colitis Disease Activity Index (UCDAI) within 6 weeks.

Another study conducted by Steed et al. (2010) used a synbiotic approach, wherein 35 patients who were suffering from Crohn's disease were subjected to a double-blind, placebo-controlled trial and the test group was treated with a synbiotic mixture of *Bifidobacterium longum* and Synergy 1. A significant decrease in TNF-α and the clinical symptoms of Crohn's disease was observed.

2.2.6 Irritable Bowel Syndrome

Irritable bowel syndrome (IBS) is one of the most general chronic functional gastrointestinal disorders; it is defined as a functional chronic disorder distinguished by abdominal pain or tribulation coupled with altered bowel habits. The prevalence of IBS is around 14%. Generally, women are more affected than men; in the UK alone, the prevalence rate of IBS is 10 to 20%, and roughly one in five people suffers from IBS. If a person recovered from acute bacterial gastroenteritis, the chance of that patient getting IBS was found to be 30%. The criteria that are used in the diagnosis of IBS are according to Rome IV, and it has divided IBS into four subtypes: IBS-C, IBS with predominant constipation; IBS-D, IBS with predominant diarrhea; IBS-M, IBS with mixed bowel habits; and IBS-U, unclassified IBS. Symptoms of IBS persist for more than 6 months, and the patient may not have any structural gut abnormalities. Symptoms may include pain in the abdomen or discomfort, bloating, and distorted bowel habit. The pain in the abdomen is reduced when stool and mucus change or pass from the rectum, or when the patient defecates completely, but the pain is intensified when the patient takes food. Also the rate of anxiety, stress, and depression is higher in the patient suffering from IBS (Sutcliffe 2019).

IBS is a complex disease, and the fundamental cause is not understood completely. It is suggested in the literature that in addition nonspecific pathogenic factors like food intolerance, genetic influence, gut microbiota, and intestinal dysbiosis perhaps induce IBS (Chong et al. 2019). Food tolerance is one of the major causes of IBS in 89% of patients. Certain food like lactose-containing food, vegetables, fat-rich foods, and artificial sweeteners triggers the symptoms of IBS; food containing FODMAPs (fermentable oligosaccharides, disaccharides, monosaccharides, and polyols) worsens IBS symptoms due to the fermentation as well as osmotic effects. In allergic patients, gluten/lactose-containing diet can also worsen the patient condition. Several studies have shown that a patient suffering from IBS has SCN5A, a sodium channel gene that got mutated, which has been associated with abdominal pain and prolonged QT interval. However, many genes are linked to IBS pathogenesis, genes coding for immune regulation, epithelial barrier function, serotonin signalling, and bile acid synthesis; cannabinoid receptors, glutamate receptor, ionotropic, delta 2 interacting protein (GRID2IP), and KDEL endoplasmic reticulum protein retention receptor 2 (KDELR2) are associated with the risk of IBS development. One of the studies involving 110 IBS patients revealed that patients suffering from IBS have a decreased level of beneficial bacteria such as bifidobacteria, *Bacteroides*, Methanobacteriales, and *Prevotella* species and increased pathogenic strains like *Streptococcus* spp., *Enterococcus faecalis*, *Clostridium difficile*, and *Giardia duodenalis* and other gas-producing organisms which lead to the activation of immune cells, infiltration of immune cells, and the release of inflammatory cytokines such as IL-6, TNF-a, and IL-1 are all caused by abdominal distension and gut permeability. Stress, which affects the gut-brain axis and increases the release of pro-inflammatory cytokines like IL-8 and IL-6, is another major component that leads to dysbiosis and causes IBD and IBS. Secretion of these pro-inflammatory cytokines activates hypothalamic-pituitary-adrenal (HPA) and hypothalamic-autonomic nervous system axes and triggers the release of corticotropin-releasing factor (CRF), adrenocorticotropic hormone, and cortisol (Strege et al. 2018; Lazaridis and Germanidis 2018; Chong et al. 2019; Hadjivasilis et al. 2019).

Agrawal et al. (2009) prepared fermented milk using the strain *Bifidobacterium lactis*, and 64 female patients suffering from IBS-C were selected for study. They divided them into two groups: the test group and control group. The test group was administered with fermented milk for 4 weeks, and the results were compared to those of the control group (without *Bifidobacterium lactis*); the test group showed improvement in individual symptoms of IBS like abdominal pain, bloating, urgency, incomplete evacuation, straining, and gas. O'Mahony et al. (2005) reported that *Bifidobacterium infantis* 35,624 was a better candidate than *Lactobacillus salivarius* UCC4331 for treating IBS and the test group showed improvement in the symptoms of IBS.

2.2.7 Colon Cancer

Colon cancer, also known as colorectal cancer (CRC), is cancer that develops in the colon, a section of the large intestine. It encompasses colon and rectal cancer of the digestive tract's lower end and is the third biggest cause of cancer-related fatalities. Due to CRC, there were 147,950 new cases, and 53,200 estimated deaths were reported in the United States (Pothuraju et al. 2020). The cases of CRC are increasing to pandemic scale by subsequent morbidity and mortality; the annual rate of CRC in India is about 35,000 out of 3.5 million cancer cases (Velayutham and Velayutham 2019). Risk factors for colon cancer include personal history of CRC or IBD, family history, lifestyle (especially in dietary habits), lack of physical activity, alcohol consumption, smoking, etc. (Cassiem and de Kock 2019).

CRC can be catalogued into familial, inherited, and sporadic based on the origin of the mutation. Worldwide familial CRC accounts for 35% of cases; Inheritance, genetic factors, and environmental factors all play a role in CRC in these patients. There are about 5% CRC cases of inherited cancer, and they are classified into two groups: non-polyposis and polyposis cancer. Non-polyposis cancer is called hereditary non-polyposis colorectal cancer (HNPCC); it occurs due to mutation in the DNA repair mechanisms. Lynch syndrome is the chief cause of HNPCC; if there is a mutation in an allele of protein-coding genes for DNA repair like MLH1, MSH2, PMS1, PMS2, and MSH2, it can lead to Lynch syndrome in HNPCC group of patients. Inherited CRC is caused by the growth of numerous malignant polyps in the colon, a condition known as familial adenomatous polyposis (FAP). Sporadic malignancies, which account for 70% of CRC cases, are caused by point mutations. It starts with polyps or non-malignant adenomas due to mutation in a tumor suppressor gene, adenomatous polyposis coli (APC), followed by a mutation in TP53, KRAS, and DCC. These mutations transform the polyps into a carcinoma state (Armelao and de Pretis 2014; Mármol et al. 2017).

There is no clear evidence on how dysbiosis induces CRC, but patients suffering from IBD and chronic inflammation are at increased risk of getting CRC, and even the secondary metabolites produced from altered gut microbiota can damage the DNA and would induce malignancy. In a study conducted on European patients, three CRC human subjects were compared with healthy patients, and it was found that *F. nucleatum* count was high in CRC patients and they had more adenomas than the healthy patients. Similar results were also reported in CRC patients in the USA and China (Kostic et al. 2013; Li et al. 2016).

Two groups of *B. fragilis* colonize the gut; one has a symbiotic relationship with the host, and the other can produce a toxin called BFT, which induces inflammation by stimulating the production of IL-18 cytokine, and it even disturbs epithelial homeostasis resulting in CRC (Sears et al. 2014). Some strains of *E.coli* produce the bacteriocin colibactin, which has pro-tumor properties and causes DNA double-stranded breaks and chromosome instability, and it is found at a high level in CRC patients (Allen-Vercoe and Jobin 2014).

Asha and Gayathri (2012) conducted in vivo study on mice wherein the combination of probiotic strains *L. fermentum* and *L. plantarum* along with vincristine was used in the feed given to mice, and results showed that there was a marked decrease in ammonia concentration and β-glucuronidase enzyme activity and also a significant reduction in aberrant crypt foci (ACF) when compared to the control.

Liu et al. (2011) conducted a study on 100 patients with a control group of 50 members and a test group of 50 members, and the probiotic group was given encapsulated bacteria containing *Lactobacillus plantarum*, *Lactobacillus acidophilus*, and *Bifidobacterium longum* orally for 6 days before the operation and continued to receive the encapsulated bacteria for 10 days after the operation. They found improvement in postoperative complications such as infection-related complications, the incidence of diarrhea, decrease in enteropathogenic bacteria, and enhancement in the expression of proteins of the mucosal tight junction.

A study was conducted on selected patients (52) who were diagnosed with colorectal cancer and underwent surgery 4 weeks before the trial. Twenty-five patients received placebo, and 27 patients received a mixture of six viable strains: *Bifidobacterium bifidum* BCMC® 02290, *Lactobacillus lactis* BCMC® 12,451, *Lactobacillus casei* subsp. BCMC® 12,313, *Bifidobacterium longum* BCMC® 02120, *Lactobacillus acidophilus* BCMC® 12,130, and *Bifidobacterium infantis* BCMC® 02129 two times daily for about 6 months. Results showed inhibition of surgical infections along with a significant reduction in pro-inflammatory cytokines in treated groups (Zaharuddin et al. 2019).

2.2.8 Non-Alcoholic Fatty Liver Disease

Non-alcoholic fatty liver disease (NAFLD) is a disruption of systematic functioning in which extra fat is stored in the liver; it is one of the growing concerns worldwide, leading to chronic liver diseases from fibrosis, cirrhosis, steatosis to non-alcoholic steatohepatitis (NASH) and ultimately hepatocellular carcinoma. The pervasiveness of NAFLD is high in the Middle East with 32%, followed by South America with 31%, and Asia with 27% (Safari and Gérard 2019).

There is no clear knowledge of the pathophysiology of NAFLD, although elements that are thought to play a role in NAFLD pathogenesis include nutrition, interaction with the environment, lifestyle, lipid and glucose metabolism, and biochemical and immunological abnormalities, as well as the significant role of gut bacteria. Since there is a link between the liver and gut through the portal vein that supplies blood, nutrient metabolites produced from gut microbiota and even bacteria move to the liver. In dysbiosis condition, lipopolysaccharides (LPS), also termed endotoxins, are also passed to the liver as there is a dysfunction in the tight junction of the gut cells; when these metabolites enter into the liver, they generate a response from Kupffer cells. Toll-like receptor interacts with the foreign particles of bacteria and bacteria themselves which leads to inflammation response and NAFLD in the liver (Quesada-Vázquez et al. 2020).

Some of the bacteria are directly related to the progression of NAFLD. Recent studies on *Bilophila wadsworthia*, a Gram-negative proteobacterium, revealed that it raises key cytokines like serum amyloid A and IL-6 by producing endotoxin such as LPS, which elicits inflammation in the liver. It also disrupts the gut barrier tight junction proteins and their expression, which in turn affects bile production and causes dysbiosis (Everard et al. 2013; Feng et al. 2017). When the gut has increased the number of *Klebsiella pneumonia*, a Gram-negative proteobacterium, the production of alcohol exceeds the detoxification capacity of the liver which produces reactive oxygen species leading to inflammation and steatohepatitis; several genes overexpress and fat storage increases, and synthesis of unsaturated fatty acid and other metabolites leads to NAFLD similar to alcoholic fatty liver disease (Yuan et al. 2019). *Helicobacter pylori* infection elicits several inflammatory cytokines like TNF-α, IL-8, IL-6, and IL-1β that elicit inflammation, and insulin resistance is caused by the release of leptin from infectious tissues, which causes fat deposition and leads to NAFLD (Ning et al. 2019).

A study carried out by Ahn et al. (2019) involved 68 obese and nonalcoholic fatty liver disease (NAFLD) patients. They were divided into control and test groups; a probiotic mixture of six bacterial strains (*Pediococcus pentosaceus*, *Lactobacillus rhamnosus*, *Lactobacillus paracasei*, *Bifidobacterium lactis*, *Lactobacillus acidophilus*, and *Bifidobacterium breve*) was given to the test group for 12 weeks, and change in intrahepatic fat (IHF) and visceral fat area (VFA) fraction was measured. After 12 weeks, there was a mean difference of -2.61%, indicating IHF, and bodyweight of the test group was considerably lower in NAFLD patients as compared to a control group.

A batch of 64 obese children with NAFLD was chosen for a 12-week randomized triple-blind trial, and they were given a probiotic supplement. The test group received a capsule containing four different probiotic strains, namely, *Lactobacillus rhamnosus* DSMZ 21690, *Bifidobacterium lactis* DSMZ 32269, *Bifidobacterium bifidum* ATCC SD6576, and *Lactobacillus acidophilus* ATCC B3208. On the other hand, the control group received a similar capsule without probiotics. After a period of 12 weeks, a decrease in enzymes aspartate aminotransferase and alanine aminotransferase and a significant decrease in low-density lipoprotein-C, cholesterol, and triglycerides were observed in the test group. It was also observed that although there was a decrease in the waist circumference, there was no change in body mass index z-score. Sonography of the liver after the trial was reported in 17 (53.1%) and 5 (16.5%) of patients in the intervention and placebo groups, respectively (Famouri et al. 2017).

2.2.9 Osteoarthritis

Osteoarthritis (OA) is a degenerative disorder portrayed by the dynamic weakening of the articular ligament, bringing about agony and all-out joint incapacity at propelled stages. In the USA, 31 million people suffer from OA; global estimate prevalence exceeds 250 million. Illness movement can be subject to hereditary and epigenetic factors, sex, ethnicity, and age. The major factor is obesity, and a strong relationship can be seen between body mass index and knee (Szychlinska et al. 2019). A few dietary components, alongside quality and amount of supplement consumption, have been found to be engaged with the pathogenesis of OA. Among these, nutrients, unsaturated fats, and magnesium appear to assume a key job. It has been demonstrated that low admission of vitamin D and vitamin C is a potential hazard factor for knee OA, while certain nourishing food, for example, milk and dairy items, meat, and poultry, are beneficial for knee OA (Musumeci et al. 2015). Also, wrong propensities (smoking, stationary life, liquor misuse) and unfortunate dietary propensities (quick and greasy nourishment) may incline individuals to stoutness and accordingly to numerous different inconveniences which may prompt the advancement of extreme metabolic dysfunctions. It is additionally notable that inactive conduct is related to an expanded danger of building up a few incessant infections. It is an independent hazard factor for both dreariness and mortality that has been proven (Warren et al. 2010).

One of the breakthrough facts revealed by the research study is that some of the gut microbiomes were found in the cartilage of the knee and hip samples from patients, and 16 s RNA gene deep sequencing analysis stated that OA patients had Betaproteobacteria compared to normal control who had Actinobacteria and Clostridia; increasing Betaproteobacteria is the marker for patients suffering from the metabolic disorder due to dysbiosis condition (Schott et al. 2018). Besides adipose tissue surrounding the joint cells produce adipokines such as leptin which indirectly increases inflammation. Patients with OA have a higher level of leptins in there synovial fluid compared to normal patients; this leads to elevated levels of IL-6 through various signalling pathways like PI3K/AKT, p38 MAPK, and JAK2/STAT3 pathways and the release of certain other cytokines and other factors like MMP9, MMP13, TNFα, and IL-1 which induce inflammation and damage to the cartilage in the joints (Zeddou 2019).

Lei et al. (2017) conducted a double-blind, placebo-controlled trial on 537 osteoarthritis patients; the test group was given skimmed milk containing *Lactobacillus casei* Shirota (LcS) for 26 weeks, and the control group was given placebo. The result showed that probiotics promote bone metabolism which reduced inflammatory response and pain, and there was a change in serum levels of high sensitivity C-reactive protein. Similarly, Lyu et al. (2020) carried out an experiment using a probiotic strain TCI633 (*Streptococcus thermophilus*) on 80 patients with osteoarthritis for 12 weeks, and improvement in serum collagen type II C-telopeptide (sCTX-II) and serum C-reactive protein (sCRP) was observed when compared with the placebo group. Therefore, the selection of ideal probiotic isolate may provide an alternate approach for osteoarthritis even though more thorough investigations need to be conducted for concrete conclusions.

2.2.10 Celiac Disease

Celiac disease (CD) is a type of persistent enteropathy with a multifactorial disorder that mainly causes small intestinal injuries and malabsorption of minerals and nutrition. The wheat protein gluten and related cereal proteins that escape human digestive enzyme activity are the primary cause of the disease. Further, the complete exclusion of gluten from the diet is the only remedy available for CD patients (Gayathri and Rashmi 2014). Several studies suggested the use of microorganisms for the preparation of gluten-free/reduced foods (De Angelis et al. 2006a, b; Gass et al. 2005). Currently, CD is a common condition that may be diagnosed at any age, but formerly CD was considered as a rare malabsorption syndrome of infancy. According to World Gastroenterology Organisation (WGO) data, out of 100, one person is diagnosed with CD in the Western population, whereas out of 300, one will suffer from CD in other parts of the world, and it is predominant in females than males with a ratio of 2:1. CD has become a focal and universally distributed, according to studies, and serological diagnosis in India, Africa, and the Middle East revealed the same prevalence rate as in Western countries. The presence of human leukocyte antigens HLA-DQ2 or HLA DQ-8 molecules gets triggered from gluten protein of wheat and other cereals, and that generates circulating autoantibodies to tissue transglutaminase (tTG) which leads to the disease (Visser et al. 2009; Lorand and Graham 2003). Not all people develop CD; only 1 to 4% of people develop CD if they have aberrated HLA-DQ2 or HLA-DQ8 genes (Sollid 2000). Although the strong association between CD and HLA DQ2/DQ8 has been well documented, CD may not be present at the time of birth or before the introduction of gluten in the diet (Green and Cellier 2007; Dube et al. 2005) and usually does not manifest before the age of 2 years even in the individuals expressing HLA DQ2/DQ8 (Hill 2006; Ludvigsson et al. 2001). The analysis was based on the morphological assessment of the small intestinal mucosa obtained at three distinct conditions:

(a) Initial flat mucosa when the patient has ingested gluten.
(b) Upon withdrawal of gluten from the diet, there must be an improvement in the small intestinal mucosa.
(c) Deterioration of the mucosa is seen due to gluten.

Further, antibodies such as endomysial antibodies (EMA), tissue transglutaminase antibodies (tTGA), and antibodies against gliadin (AGA) of the IgA class are also significant diagnostic tools for CD. Among them, EMA and tTGA are widely used. The role of HLA DQ2/DQ8 in the development of CD has opened genetic tests involving HLA typing. However, in a majority of the cases, HLA DQ2/DQ8 carriers do not develop CD, and therefore, genetic tests for CD diagnosis have limited application (Liu et al. 2005).

Treatment of CD can be done using several non-dietary strategies like the use of larazotide acetate, which is a tight junction regulator, and it decreases the intestinal tight junction permeability; furthermore, the use of corticosteroids like budesonide for inhibition of tTG activity and sequestering polymers help in changing the structure of gliadin, which in turn reduces the tissue damage and symptoms of CD (Ciacci et al. 2009; Liang et al. 2009; Paterson et al. 2007). Strategies for the long-term treatment of CD raise a concern regarding issues like safety and efficacy as the response to these strategies is not the same for each individual CD patient and is found to be unsatisfactory(Gayathri and Rashmi 2014). As there is a concern over the non-dietary alternative strategies, by creating genetically modified crops of wheat, rye, and barley or by breeding the less immunogenic crop varieties, the gluten content in the diet can be reduced. Otherwise, the patient must adopt altered gluten polypeptides or exclude gluten from their diet for the rest of their lives, or they can use probiotics for fermentation, where gluten is digested, as an alternative. (De Angelis et al. 2006a, b; Gass et al. 2005).

De Angelis et al. (2006a, b) further reported the effectiveness of VSL#3, a combination of eight strains: *Bifidobacterium breve*, *B. longum*, *B. infantis*, *Lactobacillus plantarum*, *L. acidophilus*, *L. casei*, *L. delbrueckii* subsp. *bulgaricus*, and *Streptococcus thermophilus*. The enzymes produced by VSL#3 during the fermentation process of the dough digest gliadin polypeptides, epitopes of gliadin, and digest proline-rich peptides.

Lindfors et al. (2008) conducted an in vitro study on Caco-2 cells derived from the human colon, where the effect of the probiotic strain *Lactobacillus fermentum* or *Bifidobacterium lactis*, separately; they found that *B. lactis* inhibited increased epithelial gliadin-induced permeability and also increased the expression of Zonula occludens-1 protein. In another in vitro study (Laparra and Sanz 2010), there was reduced expression of pro-inflammatory biomarkers which proved that the gliadin was altered by *Bifidobacterium* spp.

2.3 Conclusion

Gut microbiome modulation and its effect on human health and diseases are interrelated. Further research is certainly required to disentangle these complexities, and advancement in pre- and probiotic research and management of the gut microbiome along with this correlation and careful analysis would certainly impact the gut microbiome and improve human health.

References

Agrawal A, Houghton LA, Morris J, Reilly B, Guyonnet D, Goupil Feuillerat N, Schlumberger A, Jakob S, Whorwell PJ (2009) Clinical trial: the effects of a fermented milk product containing Bifidobacterium lactis DN-173 010 on abdominal distension and gastrointestinal transit in irritable bowel syndrome with constipation. Aliment Pharm Ther 29(1):104–114

Ahmed HG, Ginawi IA, Elasbali AM, Ashankyty IM, Al-Hazimi AM (2014) Prevalence of obesity in hail region, KSA: in a comprehensive survey. J Obesity 2014:961861

Ahn SB, Jun DW, Kang BK, Lim JH, Lim S, Chung MJ (2019) Randomized, double-blind, placebo-controlled study of a multispecies probiotic mixture in nonalcoholic fatty liver disease. Sci Rep 9(1):1–9

Allen-Vercoe E, Jobin C (2014) Fusobacterium and Enterobacteriaceae: important players for CRC? Immunol Lett 162(2):54–61

Almeida CC, Lorena SLS, Pavan CR, Akasaka HMI, Mesquita MA (2012) Beneficial effects of long-term consumption of a probiotic combination of Lactobacillus casei Shirota and Bifidobacterium breve Yakult may persist after suspension of therapy in lactose-intolerant patients. Nutr Clin Pract 27(2):247–251

Ananthakrishnan AN (2015) Epidemiology and risk factors for IBD. Nat Rev Gastroenterol Hepatol 12(4):205

Armelao F, de Pretis G (2014) Familial colorectal cancer: a review. World J Gastroenterol 20 (28):9292

Asha DG, Gayathri D (2012) Synergistic impact of Lactobacillus fermentum, Lactobacillus plantarum and vincristine on 1, 2-dimethylhydrazine-induced colorectal carcinogenesis in mice. Exp Ther Med 3(6):1049–1054

Azad M, Kalam A, Sarker M, Li T, Yin J (2018) Probiotic species in the modulation of gut microbiota: an overview. Biomed Res Int 2018:9478630

Barr W, Smith A (2014) Acute diarrhea in adults. Am Fam Physician 89(3):180–189

Basu S, Chatterjee M, Ganguly S, Chandra PK (2007) Effect of Lactobacillus rhamnosus GG in persistent diarrhea in Indian children: a randomized controlled trial. J Clin Gastroenterol 41 (8):756–760

Bellenger J, Bellenger S, Escoula Q, Bidu C, Narce M (2019) N-3 polyunsaturated fatty acids: an innovative strategy against obesity and related metabolic disorders, intestinal alteration and gut microbiota dysbiosis. Biochimie 159:66–71

Cani PD, Bibiloni R, Knauf C, Waget A, Neyrinck AM, Delzenne NM, Burcelin R (2008) Changes in gut microbiota control metabolic endotoxemia-induced inflammation in high-fat diet–induced obesity and diabetes in mice. Diabetes 57(6):1470–1481

Cassiem W, de Kock M (2019) The anti-proliferative effect of apricot and peach kernel extracts on human colon cancer cells in vitro. BMC Complement Altern Med 19(1):32

Chatterjee S, Khunti K, Davies MJ (2017) Type 2 diabetes. Lancet 389(10085):2239–2251

Cheng SH, Rich DP, Marshall J, Gregory RJ, Welsh MJ, Smith AE (1991) Phosphorylation of the R domain by cAMP-dependent protein kinase regulates the CFTR chloride channel. Cell 66 (5):1027–1036

Chong PP, Chin VK, Looi CY, Wong WF, Madhavan P, Yong VC (2019) The microbiome and irritable bowel syndrome–a review on the pathophysiology, current research and future therapy. Front Microbiol 10:1136

Ciacci C, Maiuri L, Russo I, Tortora R, Bucci C, Cappello C, Santonicola A, Luciani A, Passananti V, Iovino P (2009) Efficacy of budesonide therapy in the early phase of treatment of adult coeliac disease patients with malabsorption: an in vivo/in vitro pilot study. Clin Exp Pharmacol Physiol 36:1170–1176

Corado A, de Sousa M, Graziany R, Borges P, Gomes N, Leticia T, Oliveira P, Felipe J (2017) The additional effects of a probiotic mix on abdominal adiposity and antioxidant status: a double-blind, randomized trial. Obesity 25(1):30–38

Creely SJ, McTernan PG, Kusminski CM, Fisher FM, Da Silva NF, Khanolkar M et al (2007) Lipopolysaccharide activates an innate immune system response in human adipose tissue in obesity and type 2 diabetes. Am J Physiol Endocrinol Metab 292(3):E740–E747

D'Argenio V, Salvatore F (2015) The role of the gut microbiome in the healthy adult status. Clin Chim Acta 451:97–102

De Angelis M, Coda R, Silano M, Minervini F, Rizzello CG, Di Cagno R, Vicentini O, De Vincenzi M, Gobbetti M (2006a) Fermentation by selected sourdough lactic acid bacteria to decrease the intolerance to rye and barley flours. J Cereal Sci 43:301–314

De Angelis M, Rizzello CG, Fasano A, Clemente MG, De Simone C, Silano M, De Vincenzi M, Losito I, Gobbetti M (2006b) VSL#3 probiotic preparation has the capacity to hydrolyze gliadin polypeptides responsible for celiac sprue probiotics and gluten intolerance. Biochim Biophys Acta 1762(1):80–93

DiBaise JK, Zhang H, Crowell MD, Krajmalnik-Brown R, Decker GA, Rittmann BE (2008) Gut microbiota and its possible relationship with obesity. In: Mayo clinic proceedings, vol 83. Elsevier, Amsterdam, pp 460–469

Donath MY, Shoelson SE (2011) Type 2 diabetes as an inflammatory disease. Nat Rev Immunol 11 (2):98–107

Dube C, Rostom A, Sy R, Cranney A, Saloojee N, Garritty C, Sampson M, Zhang L, Yazdi F, Mamaladze V, Pan I, Macneil J, Mack D, Patel D, Moher D (2005) The prevalence of celiac disease in average-risk and at-risk Western European populations: a systematic review. Gastroenterology 128:S57–S67

Eckburg PB, Bik EM, Bernstein CN, Purdom E, Dethlefsen L, Sargent M et al (2005) Diversity of the human intestinal microbial flora. Science 308:1635–1638

Egshatyan L, Kashtanova D, Popenko A, Tkacheva O, Tyakht A, Alexeev D et al (2016) Gut microbiota and diet in patients with different glucose tolerance. Endocr Connect 5(1):1–9

Everard A, Belzer C, Geurts L, Ouwerkerk JP, Druart C, Bindels LB et al (2013) Cross-talk between Akkermansiamuciniphila and intestinal epithelium controls diet-induced obesity. Proc Natl Acad Sci 110(22):9066–9071

Famouri F, Shariat Z, Hashemipour M, Keikha M, Kelishadi R (2017) Effects of probiotics on nonalcoholic fatty liver disease in obese children and adolescents. J Pediatr Gastroenterol Nutr 64(3):413–417

Feng Z, Long W, Hao B, Ding D, Ma X, Zhao L, Pang X (2017) A human stool-derived Bilophilawadsworthia strain caused systemic inflammation in specific-pathogen-free mice. Gut Pathog 9(1):59

Fernández-Bañares F, Accarino A, Balboa A, Domènech E, Esteve M, Garcia-Planella E et al (2016) Chronic diarrhoea: definition, classification and diagnosis. Gastroenterol Hepatol 39 (8):535–559

Firouzi S, Majid HA, Ismail A, Kamaruddin NA, Barakatun-Nisak MY (2017) Effect of multi-strain probiotics (multi-strain microbial cell preparation) on glycemic control and other diabetes-related outcomes in people with type 2 diabetes: a randomized controlled trial. Eur J Nutr 56 (4):1535–1550

Gass J, Ehren J, Strohmeier G, Isaacs I, Khosla C (2005) Fermentation, purification, formulation, and pharmacological evaluation of a prolyl endopeptidase from Myxococcusxanthus: implications for celiac sprue therapy. Biotechnol Bioeng 92(6):674–684

Gayathri D, Rashmi BS (2014) Development of celiac disease; pathogenesis and strategies to control: a molecular approach. J Nutr Food Sci 4:310. https://doi.org/10.4172/2155-9600.1000310

Gayathri D, Vasudha M (2018) Lactose intolerance with special emphasis on probiotics for management. EC Nutr 13(5):325–332

Ghaisas S, Maher J, Kanthasamy A (2016) Gut microbiome in health and disease: linking the microbiome–gut–brain axis and environmental factors in the pathogenesis of systemic and neurodegenerative diseases. Pharmacol Ther 158:52–62

Green PH, Cellier C (2007) Celiac disease. New england J Med 357(17):1731–1743

Guariguata L, Whiting DR, Hambleton I, Beagley J, Linnenkamp U, Shaw JE (2014) Global estimates of diabetes prevalence for 2013 and projections for 2035. Diabetes Res Clin Pract 103(2):137–149

Hadjivasilis A, Tsioutis C, Michalinos A, Ntourakis D, Christodoulou DK, Agouridis AP (2019) New insights into irritable bowel syndrome: from pathophysiology to treatment. Ann Gastroenterol 32(6):554

Harrington LK et al (2008) A re-appraisal of lactose intolerance. Int J Clin Pract 62(10):1541–1546

He T, Priebe MG, Zhong Y, Huang C, Harmsen HJM, Raangs GC et al (2008) Effects of yogurt and bifidobacteria supplementation on the colonic microbiota in lactose-intolerant subjects. J Appl Microbiol 104(2):595–604

Heyman MB (2006) Lactose intolerance in infants, children, and adolescents. Pediatrics 118 (3):1279–1286

Hill ID (2006) Management of celiac disease in childhood and adolescence: unique challenges and strategies. Curr Treat Options Gastroenterol 9:399–408

Hodges K, Gill R (2010) Infectious diarrhea: cellular and molecular mechanisms. Gut Microbes 1 (1):4–21

Horner TW et al (2011) β-Galactosidase activity of commercial lactase samples in raw and pasteurized milk at refrigerated temperatures. J Dairy Sci 94(7):3242–3249

Huang X, Liu G, Guo J, Su Z (2018) The PI3K/AKT pathway in obesity and type 2 diabetes. Int J Biol Sci 14(11):1483

Isolauri E (2003) Probiotics for infectious diarrhoea. Gut 52(3):436–437

Jain N, Walker WA (2015) Diet and host–microbial crosstalk in postnatal intestinal immune homeostasis. Nat Rev Gastroenterol Hepatol 12(1):14
Khalighi A, Behdani R, Kouhestani S (2016) Probiotics: a comprehensive review of their classification, mode of action and role in human nutrition. InTech, London, pp 19–39
Kim J, Yun JM, Kim MK, Kwon O, Cho B (2018) Lactobacillus gasseri BNR17 supplementation reduces the visceral fat accumulation and waist circumference in obese adults: a randomized, double-blind, placebo-controlled trial. J Med Food 21(5):454–461
Kostic AD, Chun E, Robertson L, Glickman JN, Gallini CA, Michaud M, El-Omar EM (2013) Fusobacterium nucleatum potentiates intestinal tumorigenesis and modulates the tumor-immune microenvironment. Cell Host Microbe 14(2):207–215
Laparra JM, Sanz Y (2010) Bifidobacteria inhibit the inflammatory response induced by gliadins in intestinal epithelial cells via modifications of toxic peptide generation during digestion. J Cell Biochem 109(4):801–807
Larsen N, Vogensen FK, Van Den Berg FW, Nielsen DS, Andreasen AS, Pedersen BK et al (2010) Gut microbiota in human adults with type 2 diabetes differs from non-diabetic adults. PLoS One 5(2):e9085
Lazaridis N, Germanidis G (2018) Current insights into the innate immune system dysfunction in irritable bowel syndrome. Ann Gastroenterol 31(2):171
Lei M, Guo C, Wang D, Zhang C, Hua L (2017) The effect of probiotic Lactobacillus casei Shirota on knee osteoarthritis: a randomised double-blind, placebo-controlled clinical trial. Benef Microbes 8(5):697–703
Li J, Zhang W, Wang C, Yu Q, Dai R, Pei X (2012) Lactococcuslactis expressing food-grade β-galactosidase alleviates lactose intolerance symptoms in post-weaning Balb/c mice. Appl Microbiol Biotechnol 96(6):1499–1506
Li YY, Ge QX, Cao J, Zhou YJ, Du YL, Shen B et al (2016) Association of Fusobacterium nucleatum infection with colorectal cancer in Chinese patients. World J Gastroenterol 22 (11):3227
Liang L, Pinier M, Leroux JC, Subirade M (2009) Interaction of alpha-gliadin with poly (HEMA-co-SS): structural characterization and biological implication. Biopolymers 91:169–178
Lilly DM, Stillwell RH (1965) Probiotics: growth-promoting factors produced by microorganisms. Science 147(3659):747–748
Lindfors K, Blomqvist T, Juuti-Uusitalo K, Stenman S, Venäläinen J, Mäki M, Kaukinen K (2008) Live probiotic Bifidobacterium lactis bacteria inhibit the toxic effects induced by wheat gliadin in epithelial cell culture. Clin Exp Immunol 152(3):552–558
Liu Z, Li N, Neu J (2005) Tight junctions, leaky intestines, and pediatric diseases. Acta Paediatr 94 (4):386–393
Liu Z, Qin H, Yang Z, Xia Y, Liu W, Yang J, Jiang Y, Zhang H, Yang Z, Wang Y, Zheng Q (2011) Randomised clinical trial: the effects of perioperative probiotic treatment on barrier function and post-operative infectious complications in colorectal cancer surgery–a double-blind study. Aliment Pharm Ther 33(1):50–63
Lorand L, Graham RM (2003) Transglutaminases: crosslinking enzymes with pleiotropic functions. Nat Rev Mol Cell Biol 4(2):140–156
Ludvigsson JF, Falth-Magnusson K, Ludvigsson J (2001) Tissue transglutaminase auto-antibodies in cord blood from children to become celiacs. Scand J Gastroenterol 36:1279–1283
Luyer BL, Makhoul G, Duhamel JF (2010) A multicentric study of a lactose free formula supplemented with Saccharomyces boulardii in children with acute diarrhea. Arch Pédiatr 17 (5):459–465
Lyu JL, Wang TM, Chen YH, Chang ST, Wu MS, Lin YH, Lin YH, Kuan CM (2020) Oral intake of Streptococcus thermophilus improves knee osteoarthritis degeneration: a randomized, double-blind, placebo-controlled clinical study. Heliyon 6(4):e03757
Mármol I, Sánchez-de-Diego C, PradillaDieste A, Cerrada E, Rodriguez Yoldi MJ (2017) Colorectal carcinoma: a general overview and future perspectives in colorectal cancer. Int J Mol Sci 18 (1):197

Matthews SB et al (2005) Systemic lactose intolerance: a new perspective on an old problem. Postgrad Med J 81(953):167–173

Musumeci G, Aiello FC, Szychlinska MA, Di Rosa M, Castrogiovanni P, Mobasheri A (2015) Osteoarthritis in the XXIst century: risk factors and behaviours that influence disease onset and progression. Int J Mol Sci 16(3):6093–6112

Ning L, Liu R, Lou X, Du H, Chen W, Zhang F et al (2019) Association between helicobacter pylori infection and nonalcoholic fatty liver disease: a systemic review and meta-analysis. Eur J Gastroenterol Hepatol 31(7):735–742

O'Mahony L, McCarthy J, Kelly P, Hurley G, Luo F, Chen K, O'Sullivan GC, Kiely B, Collins JK, Shanahan F, Quigley EM (2005) Lactobacillus and bifidobacterium in irritable bowel syndrome: symptom responses and relationship to cytokine profiles. Gastroenterology 128(3):541–551

Ojetti V, Gigante G, Ainora ME, Gabrielli M, Migneco A, Gasbarrini G, Silveri NG, Gasbarrini A (2009) S1213 the effect of oral supplementation with Lactobacillus reuteri or tilactase in lactose-intolerant patients: a placebo controlled study. Gastroenterology 136(5):214

Paterson BM, Lammers KM, Arrieta MC, Fasano A, Meddings JB (2007) The safety, tolerance, pharmacokinetic and pharmacodynamic effects of single doses of AT-1001 in coeliac disease subjects: a proof of concept study. Aliment Pharmacol Ther 26(5):757–766

Pothuraju R, Krishn SR, Gautam SK, Pai P, Ganguly K, Chaudhary S et al (2020) Mechanistic and functional shades of mucins and associated glycans in colon cancer. Cancer 12(3):649

Quesada-Vázquez S, Aragonès G, Del Bas JM, Escoté X (2020) Review diet, gut microbiota and non-alcoholic fatty liver disease: three parts of the same axis. Cell 9(1):2073–4409

Rouxinol-Dias AL, Pinto AR, Janeiro C, Rodrigues D, Moreira M, Dias J, Pereira P (2016) Probiotics for the control of obesity–its effect on weight change. Porto Biomed J 1(1):12–24

Safari Z, Gérard P (2019) The links between the gut microbiome and non-alcoholic fatty liver disease (NAFLD). Cell Mol Life Sci 76(8):1541–1558

Savaiano D et al (2011) Lactose intolerance: an unnecessary risk for low bone density. In: Nestle nutrition workshop series pediatric program, vol 67, pp 161–171

Schmidt TS, Raes J, Bork P (2018) The human gut microbiome: from association to modulation. Cell 172(6):1198–1215

Schott EM, Farnsworth CW, Grier A, Lillis JA, Soniwala S, Dadourian GH et al (2018) Targeting the gut microbiome to treat the osteoarthritis of obesity. JCI Insight 3(8):e95997

Sears CL, Geis AL, Housseau F (2014) Bacteroidesfragilis subverts mucosal biology: from symbiont to colon carcinogenesis. J Clin Invest 124(10):4166–4172

Shah NJ, Swami OC (2017) Role of probiotics in diabetes: a review of their rationale and efficacy. Diabetes 5:104–110

Shamoon M, Martin NM, O'Brien CL (2019) Recent advances in gut Microbiota mediated therapeutic targets in inflammatory bowel diseases: Emerging modalities for future pharmacological implications. Pharmacol Res 148:104344

Simmonds M, Llewellyn A, Owen CG, Woolacott N (2016) Predicting adult obesity from childhood obesity: a systematic review and meta-analysis. Obes Rev 17(2):95–107

Sollid LM (2000) Molecular basis of celiac disease. Annu Rev Immunol 18:53–81

Sood A, Midha V, Makharia GK, Ahuja V, Singal D, Goswami P, Tandon RK (2009) The probiotic preparation, VSL#3 induces remission in patients with mild-to-moderately active ulcerative colitis. Clin Gastroenterol Hepatol 7(11):1202–1209. https://doi.org/10.1016/j.cgh.2009.07.016

Steed H, Macfarlane GT, Blackett KL, Bahrami B, Reynolds N, Walsh SV, Cummings JH, Macfarlane S (2010) Clinical trial: the microbiological and immunological effects of synbiotic consumption—a randomized double-blind placebo-controlled study in active Crohn's disease. Aliment Pharm Ther 32(7):872–883

Strege PR, Mazzone A, Bernard CE, Neshatian L, Gibbons SJ, Saito YA et al (2018) Irritable bowel syndrome patients have SCN5A channelopathies that lead to decreased NaV1. 5 current and mechanosensitivity. Am J Physiol Gastrointest Liver Physiol 314(4):G494–G503

Sutcliffe M (2019) Irritable bowel syndrome. InnovAiT 12(9):497–500. https://doi.org/10.1177/1755738019855099

Szychlinska MA, Di Rosa M, Castorina A, Mobasheri A, Musumeci G (2019) A correlation between intestinal microbiota dysbiosis and osteoarthritis. Heliyon 5(1):e01134

Tonucci LB, Dos Santos KMO, de Oliveira LL, Ribeiro SMR, Martino HSD (2017) Clinical application of probiotics in type 2 diabetes mellitus: a randomized, double-blind, placebo-controlled study. Clin Nutr 36(1):85–92

Turnbaugh PJ, Ley RE, Mahowald MA, Magrini V, Mardis ER, Gordon JI (2006) An obesity-associated gut microbiome with increased capacity for energy harvest. Nature 444(7122):1027

Velayutham P, Velayutham S (2019) A comprehensive study on colorectal malignancies. Int Surg J 6(5):1500–1504

Venegas DP, Marjorie K, Landskron G, González MJ, Quera R, Dijkstra G et al (2019) Short chain fatty acids (SCFAs)-mediated gut epithelial and immune regulation and its relevance for inflammatory bowel diseases. Front Immunol 10:277

Visser J, Rozing J, Sapone A, Lammers K, Fasano A (2009) Tight junctions, intestinal permeability, and autoimmunity: celiac disease and type 1 diabetes paradigms. Ann N Y Acad Sci 1165:195–205

Warren TY, Barry V, Hooker SP, Sui X, Church TS, Blair SN (2010) Sedentary behaviors increase risk of cardiovascular disease mortality in men. Med Sci Sports Exerc 42(5):879

Yuan J, Chen C, Cui J, Lu J, Yan C, Wei X et al (2019) Fatty liver disease caused by high-alcohol-producing Klebsiella pneumoniae. Cell Metab 30(4):675–688

Zaharuddin L, Mokhtar NM, Nawawi KNM, Ali RAR (2019) A randomized double-blind placebo-controlled trial of probiotics in post-surgical colorectal cancer. BMC Gastroenterol 19(1):131

Zeddou M (2019) Osteoarthritis is a low-grade inflammatory disease: obesity's involvement and herbal treatment. Evid Based Complement Alternat Med 2019:2037484

Trust Your Gut: The Human Gut Microbiome in Health and Disease

3

Hebatallah H. AboNahas, Amira M. G. Darwish, Hanaa F. Abd EL-kareem, Yousef H. AboNahas, Safaa A. Mansour, Yasmin H. Korra, R. Z. Sayyed, Ahmed M. Abdel-Azeem, and Essa M. Saied

Abstract

Since Joshua Lederberg defined the gut microbiome and its collective genetic material present in the gastrointestinal tract (GIT), the gut microbiome attracted the attention of researchers worldwide. The human gut microbiota is divided into

H. H. AboNahas (✉)
Zoology Department, Faculty of Science, Suez Canal University, Ismailia, Egypt

A. M. G. Darwish
Food Technology Department, Arid Lands Cultivation Research Institute (ALCRI), City of Scientific Research and Technological Applications (SRTA-City), Alexandria, Egypt

H. F. Abd EL-kareem
Zoology Department, Faculty of Science, Ain Shams University, Cairo, Egypt

Y. H. AboNahas
Microbiology Department, Faculty of Medicine, Helwan University, Cairo, Egypt

S. A. Mansour
Botany and Microbiology Department, Faculty of Science, Suez Canal University, Ismailia, Egypt

Y. H. Korra
Faculty of Pharmacy, Pharos University, Alexandria, Egypt

R. Z. Sayyed
Department of Microbiology, PSGVP Mandal's Arts, Science and Commerce College, Shahada, Maharashtra, India

A. M. Abdel-Azeem
Botany and Microbiology Department, Faculty of Science, University of Suez Canal, Ismailia, Egypt

E. M. Saied (✉)
Chemistry Department, Faculty of Science, Suez Canal University, Ismailia, Egypt

Institute for Chemistry, Humboldt Universität zu Berlin, Berlin, Germany
e-mail: saiedess@hu-berlin.de

R. Z. Sayyed, M. Khan (eds.), *Microbiome-Gut-Brain Axis*,
https://doi.org/10.1007/978-981-16-1626-6_3

many phyla in which the gut microbiota is comprised primarily of four main phyla that include Firmicutes, Bacteroidetes, Actinobacteria, and Proteobacteria. Recent studies focused on the microbe-host interactions which included their effects on the metabolism and immunity. In addition, the gut microbiome plays an important role in the absorption of nutrients and minerals; the biosynthesis of enzymes, vitamins, and amino acids; and the production of short-chain fatty acids (SCFAs). In this chapter, we shed the light on different groups of gut microbiomes and their effects on human health and diseases.

Keywords

Gut microbiome · Microbiota development · Microbiota immunosenescence · Human health · Immunity · Obesity

3.1 Introduction

The "holobiont" as the collective commitment of the prokaryotic and eukaryotic partners to the multicellular organism presents a complex definition of individuality enabling a new encyclopedic view of human evolution. The human intestinal tract accommodates a diverse and complex microbial combination that plays a master role in human health. It has been estimated that our gut contains about 1000 bacterial species and 100-fold more genes than are found in the human genome (Qin et al. 2010). This bacterial community is commonly indicated to as our hidden metabolic "organ" due to their massive impact on human health, including host metabolism, nutrition, immune function, and physiology. It is presently clear that our gut microbiome coevolves with us (Ley et al. 2008) and that changes to this populace can have major ramification, both useful and harmful, for human health. Actually, it has been proposed that disturbance of the gut microbiota (or dysbiosis) can be significant with regard to pathological intestinal conditions like obesity (Zhang et al. 2005) and malnutrition (Kau et al. 2011). Systematic diseases such as diabetes (Qin et al. 2012) and chronic diseases, e.g., encompassing ulcerative colitis (UC) and Crohn's disease (CD) (inflammatory bowel diseases (IBD) were recorded (Frank et al. 2007).

The major function of the gut microbiome in human health and disease is getting clearer much appreciated to high-throughput sequencing (HTS) technologies as well as parallel recent developments in non-genomic techniques (Guinane and Cotter 2013). Although worthwhile efforts have centered on cataloguing the adult human gut microbiome and its relationship to complex diseases (Karlstrom and Lindgren 2013), investigations on the infant gut microbiota (bacteria) have been limited on culture-based enumeration, 16S-based profiling, and small sample sizes (Subramanian et al. 2014). Therefore, factors that figure the gut microbiota in early life have not been adequately examined. From an ecological perspective, colonization of the newborn's gut represents the de novo gathering of a microbial community (Costello et al. 2012) and is impacted by dietary and medical therapeutic component (La Rosa et al. 2014). However, it is not obvious how dietary and therapeutic component contributes to the overall composition and function of the newborn's gut microbiome and how various microbes contribute or compete with one another as the gut environment changes (Bäckhed et al. 2015). Here, we provide snapshots of the advancing microbe-host associations and relations during distinct milestones across the lifespan of a human being.

3.2 The Nature and Nurture of Gut Microbiome

Numerous symbiotic microorganisms that play an important role in maintaining health are colonized by the human body. This so-called microbiota has the capacity to liberate the host's indigestible energy and nutrients from fibers. This fermenting capacity makes them "keystone species" that play a vital role in the formation and functioning of the intestinal microbial community to serve the host a beneficial and protective role. To release energy from the complex carbohydrate structures, the other microbiota members depend on these "keystone organisms." The host creates structures rich in carbon and nitrogen, such as mucus, dietary fibers, and oligosaccharides, e.g., glycans that exert biological roles, i.e., prebiotic, anti-adhesive, or anti-inflammatory effects. Metabolically, diverse members of this microorganisms' ecosystem depend on their glycan fermentation. Colonization of these "keystone species" in early life is essential for immune and metabolic imprinting. Permanent symbiotic colonization of the mucosal layer in later life contributes to continuous regulation of immune and metabolic processes and host health maintenance (de Vos et al. 2017).

3.3 Origin, Composition, Metabolism, and Development of Gut Microbiome

The human gut is populated by trillions of microorganisms that are considered potential key health factors (Schroeder and Bäckhed 2016; Lynch and Pedersen 2016). The composition of the gut microbiota can be affected by factors such as diet, lifestyle, and genetics and is commonly shared by individuals of comparable ethnic origin (Lynch and Pedersen 2016; Goodrich et al. 2014). Most research investigating how ethnicity relates to intestinal microbiota so far has compared small groups residing in distinct geographical locations (Goodrich et al. 2014; Rothschild et al. 2018).

The understanding of the sources of the infant microbiota is another very significant feature of human gut microbial ecology. In this sense, Palmer et al. (2007) explored potential relations between the composition of the vaginal microbiota, the stool microbiota, mother's breast-milk microbiota, and the formation of the infant gastrointestinal tract (GIT) microbiota. In this analysis, it was shown that a single infant's GIT microbiota clustered with the mother's vaginal microbiota, but only during the first day after birth. Palmer and his colleagues identified for the first time breast-milk bacteria, including mixtures of enteric bacterial species belonging to

Bacteroides, *Haemophilus*, *Pseudomonas*, *Streptococcus*, and *Veillonella*. These bacteria seemed to be entirely different from those found in infants' GIT microbiota fed with the same samples of breast milk, indicating that there is no relation between mother and child microbiota. This result opened up new thoughts on the real contribution of the GIT microbiota of the mothers to that of the child. The presence of bacteria in breast milk seems to be confirmed by other studies (Martín et al. 2003; Sinkiewicz and Nordström 2005; Gueimonde et al. 2006). This may be caused by improper sampling methods that contribute to the breast-milk sample being contaminated by bacteria such as *Lactobacillus* and *Staphylococcus* that are present on the skin of the mother.

Microbial colonization of the infant gastrointestinal (GI) tract is an important process in our life cycle, as encounters with microbiota host have a significant effect on human health and disease. The theory that fetuses are sterile in utero and that microbial colonization of the newborn occurs during and after birth has been generally accepted since the studies of Tissier (Rodríguez et al. 2015) concerning the acquisition of the infant gut microbiota. More than a century later, the theory that the placenta barrier keeps embryos sterile during a healthy pregnancy remains a general dogma, and as a consequence, the presence of any bacteria in the uterus is commonly regarded as a potential fetal threat. This view comes from the fact that microbiological studies of pregnancy-related biological samples (chorioamnion, amniotic fluid, and meconium) have been performed for decades only in cases where there is evidence or suggestion of an intrauterine infection. Numerous studies have shown that premature delivery is highly associated with intrauterine infections (Goldenberg et al. 2008; DiGiulio et al. 2008), the leading cause of infant mortality worldwide (Blencowe et al. 2013). Relatively few studies have investigated the uterine microbiota associated with healthy term pregnancies, partially because of the enduring impact of the sterile womb paradigm and also because of the technical and ethical problems of obtaining samples before birth from stable pregnancies. However, bacteria in placenta tissue (Aagaard et al. 2014), umbilical cord blood (Jiménez et al. 2005), amniotic fluid (Shan et al. 2013), and fetal membranes (Rautava et al. 2012a, b; Steel et al. 2005) have been identified by research into the potential for bacterial transmission across the placental barrier from healthy newborns without any sign of infection or inflammation.

In general, several species of microorganisms, including bacteria, yeast, and viruses, are composed of gut microbiota. Bacteria are categorized according to phyla, classes, orders, families, genera, and species, taxonomically. There are only a few phyla represented, including more than 160 species (Laterza et al. 2016). Firmicutes, Bacteroidetes, Actinobacteria, Proteobacteria, Fusobacteria, and Verrucomicrobia are the dominant gut microbial phyla, with both phyla Firmicutes and Bacteroidetes (Arumugam et al. 2011) accounting for 90% of gut microbiota. More than 200 separate genera, such as *Lactobacillus*, *Bacillus*, *Clostridium*, *Enterococcus*, and *Ruminococcus*, form the phylum Firmicutes. Ninety-five percent of the phylum Firmicutes are *Clostridium* genera. Bacteroidetes consist of genera such as *Bacteroides* and *Prevotella* that are predominant. The phylum Actinobacteria is proportionally less common and mostly represented by the genus

Bifidobacterium (Arumugam et al. 2011). Microbial contact during prenatal life may influence the microbiota and immune system of the offspring in preparation for the much larger inoculum transferred during vaginal delivery and breastfeeding. When introduced to the postnatal environment, a remarkably wide variety of bacteria will colonize the infant, indicating a high inter-individual diversity in the intestinal microbiota composition of newborns (Palmer et al. 2007; Avershina et al. 2013). Advances in metagenomic technologies have revealed the composition of the human gut microbiota from early infancy (Palmer et al. 2007) along with old age (Claesson et al. 2010). After birth, a number of microbes quickly colonize the human intestine, and factors known to affect colonization include gestational age, mode of delivery (vaginal birth vs. assisted delivery), diet (breast milk vs. formula), hygiene, and treatment with antibiotics (Adlerberth and Wold 2009; Marques et al. 2010). First colonizers, facultative anaerobes, establish a new environment that enhances colonization of strict anaerobes such as *Clostridium*, *Bacteroides*, and *Bifidobacterium* sp. Low diversity and relative dominance of the phyla Proteobacteria and Actinobacteria characterize the intestinal microbiota of neonates, with the microbiota becoming more complex as time after birth increases with the appearance and dominance of Firmicutes and Bacteroidetes (Rodríguez et al. 2015). Infants have an individually distinct microbial profile by the end of the first year of life, converging toward the characteristic microbiota of an adult, so that the microbiota completely resembles that of an adult in terms of composition and diversity by 25 years of age (Palmer et al. 2007; Koenig et al. 2010; Yatsunenko et al. 2012). The first 3 years of life, therefore, represent the most important time frame for dietary interventions to enhance the growth and development of children. This is the time in which the intestinal microbiota, a critical asset for health and neurodevelopment, is developed (Rodríguez et al. 2015), and its modification has the potential to have a profound effect on host health and development throughout this period. The development of gut microbiome has been studied even more than in other body habitats, and microbial colonization processes for oral, skin, and respiratory infants are still being uncovered. The development of gut microbiota is influenced by various factors, such as delivery mode, diet, genetics, health status, and gestational age (Rodríguez et al. 2015). After birth, Enterobacteriaceae and *Staphylococcus* transiently dominate the gut microbiota of a newborn (Matsuki et al. 2016). Subsequently, *Bifidobacterium* and certain lactic acid bacteria dominate the gut microbiota of an infant (Mitsuoka 2014). *Bifidobacterium*-dominated microbiota, called "Bifidus biota," is established until solid food is introduced (Vallès et al. 2014; Palmer et al. 2007). After weaning, Bifidus biota is outcompeted by microorganisms of the adult type, characterized mainly by bacteria of the genera *Bacteroides*, *Veillonella*, *Prevotella*, *Ruminococcus*, and *Clostridium*, which colonize the intestines of an infant (Vallès et al. 2014). Eventually, a normal adult-like gut microbiota is formed at around 3 years of age (Yatsunenko et al. 2012). The roles of the gut microbiota also change significantly before and after the introduction of weaning foods. During the first year of life, the functional repertoire of the microbiota of an infant changes as the early microbiota before weaning is enriched in bacteria with genes that encourage lactate consumption, whereas solid foods promote the growth of bacteria enriched in genes coding after weaning to enable the use of a wider variety of carbohydrates, vitamin synthesis, and xenobiotic degradation (Tanaka and Nakayama 2020).

The microbiota of the human gut is a complex ecosystem of microorganisms that occupy and critically maintain gastrointestinal (GI) tract homeostasis (Thursby and Juge 2017). Many of the contributions to the physiology of the human superorganism made by the gut microbiota are linked to microbial metabolism (Li et al. 2014), with bacteria being the biggest contributors to the functioning of the ecosystem in terms of relative genetic content (Li et al. 2014). In general, the direct advantage is the microbial metabolism of both exogenous and endogenous substrates to nutrient that can be used by the host, but metabolites can also function to modulate the immune system by influencing host cell physiology and gene expression (Turnbaugh et al. 2007; Belkaid and Hand 2014; Spiljar et al. 2017). The key site of this fermentation is the colon, since its relatively high transit time and pH, combined with low cell turnover and redox potential, present more favorable conditions for bacterial proliferation (Hillman et al. 2017).

However, this does not preclude the value of microbiota at other sites, as it has been shown, for example, that small intestinal microbiota controls the absorption and metabolism of nutrients conducted by the host (Martinez-Guryn et al. 2018). In addition, the presence of various metabolic activities will enable the microbiota to maximally occupy the available ecological niches and to competitively inhibit pathogen colonization at all sites (Sommer et al. 2017; Stecher and Hardt 2011). High concentrations of mainly acidic fermentation by-products often minimize pH locally to generate a more inhospitable atmosphere for these invaders (Stecher and Hardt 2011). Specific fermentation pathways performed by intestinal microbes can, however, lead to the development of toxic compounds that can damage the host epithelium and cause inflammation (Fan et al. 2015; Yao et al. 2016). Due to the inherent structural complexity of particular biomolecules, the three macronutrients consumed in the human diet, carbohydrates, proteins, and fat, can enter the colon after either escaping primary digestion until the amount consumed exceeds the rate of digestion or resisting primary digestion altogether (Yao et al. 2016; Morales et al. 2016). Digestive efficiency can be influenced by several factors, which in turn modulate the substrates available for consumption by intestinal microbiota, including the shape and size of food particles (affected by cooking and processing), meal composition (affected by relative macronutrient ratios and the presence of anti-nutrients such as alpha-amylase inhibitors), and transit time (Wong and Jenkins 2007). Transit time has shown to increase the richness and change the composition of fecal microbial communities (Roager et al. 2016), resulting from many factors, including diet, physical activity, genetics, medications (e.g., caffeine and alcohol), and psychological status (Degen and Phillips 1996). Intestinal microbial metabolic processes can also affect the bioavailability of micronutrients to the host. Colonic bacteria can endogenously synthesize important cofactors, such as B vitamins (Biesalski 2016), for host energy metabolism and gene expression control. Another example is the biotransformation by the intestinal microbiota of exogenous plant-derived polyphenols that have antioxidant, anticancer, and/or anti-inflammatory properties, enhancing their host uptake (Ozdal et al. 2016). Primary degradation of dietary polysaccharides can be interlinked in complex ways through a diverse array of bonds between monosaccharide units, reflecting the sheer number of carbohydrate-activating enzymes recorded in the human gut microbiome (Bhattacharya et al. 2015). For instance, in its genome alone, *Bacteroides thetaiotaomicron* has 260 glycoside hydrolases (Xu 2003), which illustrates the evolutionary need for adaptation in order to optimize the use of resistant starch and the variety of fibers available as part of the human diet. In comparison, very few of these enzymes are produced by human cells (although they produce amylase to extract alpha-linked sugar units from starch and can use sugars in the small intestine such as glucose, fructose, sucrose, and lactose) and so rely on gut microbes to derive energy from the remaining complex carbohydrates (Wong and Jenkins 2007; Singh et al. 2017).

Human beings may be viewed as superorganisms whose metabolism is a combination of microbial and human characteristics. The human microbiome makes it possible to greatly increase the metabolism of glycans, amino acids, and xenobiotics; methanogenesis; and vitamin and isoprenoid 2-C-methyl-D-erythritol-4-phosphate pathway-mediated biosynthesis. Fermentation of dietary fiber or host-derived glycans, such as host mucus, needs the support of bacterial groups linked to the trophic chain. In this respect, glycans are degraded by primary fermenters to short-chain fatty acids (SCFAs) (e.g., acetate, butyrate, and propionate) plus other gases, such as hydrogen and carbon dioxide. The SCFAs are adsorbed by the host providing energy (e.g., around 10% of calories extracted by the daily diet). Enrichment of key genes responsible for producing acetate, butyrate, lactate, and succinate was shown by COG studies of the human intestinal microbiome (Gill et al. 2006). In addition, the distal gut microbiome is enriched by methanogenesis for different COGs that represent central genes in the methanogenic pathway that enhance the removal of hydrogen from the distal gut ecosystem. In addition, the gut microbiome is enriched with a wide variety of COGs that are involved in the anabolism of amino acids and vitamins that are considered essential for human health, including thiamine and pyridoxal, i.e., vitamins B1 and B6, respectively (Rodriguez-Concepcion 2002). Finally, β-glucuronidase activity that could be induced by glucuronide conjugates of xenobiotics and bile salts (Mallett et al. 1983) is enriched by the human distal microbiome.

3.4 Features and High-Throughput Diversity of the Gastrointestinal Tract Microbiota

For more than a decade, it has been known that a major part of the human GI tract microbiota has not been characterized by cultivation. In studying the human GI tract microbiota diversity, this has led to the implementation of sequence analysis of SSU rRNA and its corresponding gene, and this has provided a considerable expansion of our knowledge about the ecology of the GI tract. These approaches, however, have shown that the microbiota of the GI tract is individual and location specific and that its diversity is massive, with thousands of new microbial species to be discovered (Rajilić-Stojanović et al. 2007). This argues for the introduction of novel high-throughput and comprehensive technologies for the study of microbiota in the human GI tract. The implementation of high-throughput phylogenetic microarrays enables thousands of microbes to be studied simultaneously in a single experiment and is therefore very desirable for studying the population dynamics of the GI microbiota tract in health and disease. This resulted in the discovery that the development of the microbiota in infants is initially chaotic but stabilizes after 1 year into an adult-like community (Palmer et al. 2007). In addition, phylogenetic microarray analysis showed that the human microbiota fluctuates around an individual center of stable colonizers. Last but not least, important relations have been identified between the existences or abundance of specific groups of microbes and disorders of the GI tract such as irritable bowel syndrome (IBS) and IBD (M. Rajilic-Stojanovic 2007). Therefore, it is already clear that in the near future, the implementation of phylogenetic microarrays in GI tract research will increase our knowledge. However, the up-to-date status of phylogenetic microarrays will always depend on the discovery of new inhabitants of the GI tract, and therefore, ongoing sequence SSU rRNA gene libraries and cultivation of the novel GI tract inhabitants are indispensable. It is clear that phylogenetic microarrays will allow us to establish links between host characteristics and microbial groups. Nevertheless, as the majority of GI tract microbes are identified only as a partial SSU rRNA gene, this will not lead to extrapolation of microbial functions. Therefore, to gain insight into the genetic potential and behavior of GI tract microbiota, metagenomics and other meta-"omics" approaches are required. The field of these meta-"omics" is hardly 5 years old, and even younger is their application in the study of the human GI tract. Therefore, the analysis and interpretation of data obtained from methods of meta-"omics" are still in infancy.

In the near future, we expect many novel technological procedures to be developed and improved, not only with regard to wet lab technologies, like pyrosequencing and functional screening methods, but also in the field of bioinformatics, in order to analyze the enormous mass of data obtained with such approaches. Additionally, the first implementations of meta-"omics" approaches have already demonstrated their power, and we expect this area to explosively expand in the near future. In the coming years, only an integration of both reductionist and meta-"omics" methods can provide adequate understanding of the GI tract microbiota, as these methods complement each other by presenting various pieces of the puzzle of the GI tract. The study of different cell cultures, for instance, will contribute to the discovery of novel genes and functions of individual organisms and will thus serve as landmarks for meta-"omics" approaches. This incorporation is important to understand the collected data from meta-"omics" analysis, as the limitations in our predictive capacity were seen in the first metaproteomics study of the microbiota of the human GI tract (Klaassens et al. 2007). Overall, researchers of the GI tract will have a challenging future. With the implementation of the novel high-throughput technologies, it will be possible, for the first time, to obtain statistically meaningful relations between microbial phylotypes and activities and human health. This will eventually contribute to the discovery of biomarkers that will enable us to recognize and predict the microbial existence in our intestine, which is a microbial ecologist's dream of the GI tract (Zoetendal et al. 2008).

3.5 The Early Years of Life Influenced by Our Intestinal Microbiome

Lim et al. (2016) mentioned in their study that billions of microbes colonize mucosal surfaces of the human body and play significant roles in different pathophysiological and physiological processes that influence host health throughout the life cycle. Communication between gut microbes and host cells is a vital step during the developmental route. Moreover, blatant evidence has appeared that microbiome colonizing the human intestinal tract may apply essential impacts on infant development and the maturation of their immune system (Gensollen et al. 2016).

These discoveries and investigations propose that the risk of diseases may be programmed during the fetal development and early life, making it mandatory to explore the role of the human microbiome in early life (Donaldson et al. 2018). For the intestinal microbial organisms, during the first 2.5 years of life, the plenitude of Bacteroidetes continues to increase. Meanwhile, the number of genes related to carbohydrate utilization, vitamin biosynthesis, and xenobiotic degradation increases, together with an elevated quantities of fecal short-chain fatty acids. Eventually, a steadier gut microbial community structure is gradually formed (Koenig et al. 2011). Assuming that the establishment of intestinal microbial community for an adult-like during early life and that the early microbiota plays a dominant role in future health as mentioned by (Kundu et al. 2017).

3.5.1 Gut Microbiota Before Birth

The placenta is believed to act as a barrier that selectively prevents maternal toxic compounds, antibodies, and microorganisms from translocating into the fetal blood, thus providing the "sterility" of the growing offspring. Recent findings, however, challenge this sterility belief by suggesting that diverse microbial population may exist in human semen and in the cervix (Verstraelen et al. 2016). Moreover, other studies showed that parental microbiome may colonize various niches, even in normal embryonic development, involving the placenta (Zheng et al. 2015), the amniotic fluid (Rautava et al. 2012a), and the umbilical cord (Jiménez et al. 2005), demonstrating a maternal-to-offspring in uterine primary colonization of selected microbes in the fetus development.

The presence of transitory microbial community in the meconium (the first stool of an infant) supported a possible uterine route of colonization (Gosalbes et al. 2013). Comparing several microbiome samples from mother-progeny pairs revealed shared microbial signatures among placenta, amniotic fluid, and meconium, which prove that early gut colonization is initiated prenatally by a distinct route of maternal microbiome (Collado et al. 2016). Recently, a shotgun metagenomic study of placental specimen's genome collected from 320 specimen under sterile conditions shows a unique placental microbial biota that involves nonpathogenic symbiotic microbes from the phyla Firmicutes, Bacteroidetes, Tenericutes, Proteobacteria, and Fusobacteria (Aagaard et al. 2014).

Nevertheless, evidence indicates that the fetus might inherit mother's microbiome before birth; there have been arguments about the microbial colonization before birth. The "in utero colonization" hypothesis may need to be reevaluated since these studies were mostly conducted utilizing molecular methods, which are not appropriate for the study of low-abundance microbial communities, due to the lack of appropriate controls to evaluate contamination (Perez-Muñoz et al. 2017).

3.5.2 Development of Microbiome During Labor

Dominguez-Bello et al. (2016) carried a study on 18 maternal/newborn twosomes; the maternal and neonate microbiomes in three groups were compared: vaginally delivered newborns, cesarean-delivered newborns with standard postoperative treatment, and newborns born by means of cesarean who were exposed to maternal vaginal fluids instantly following birth. Within about 2 min of birth, the final group with newborn had their mouth, face, and body swabbed with a dressing pad that had been incubated for an hour in mothers' vagina. This microbiome inhabiting intestine, oral, and skin of these newborns were more similar to vaginal delivered newborns than to other newborns who experienced the standard cesarean delivery. This similitude continued through 1 month of life. These findings are harmonious with population-based studies showing that children born by means of elective cesarean birth (no labor) are at higher risk for health problems like asthma in comparison with children who had some exposure to their mother's vaginal microbiome during delivery, even if the delivery ended (Kristensen and Henriksen 2016). Cesarean birth may impact the neonatal microbiome seeding process by postponed breastfeeding initiation. Compared to vaginally delivered newborns, newborns born by means of cesarean are nearly half as likely to start breastfeeding before hospital discharge and are more likely to have breastfeeding troubles (Karlstrom and Lindgren 2013). Hence, microbiota transmission by breastfeeding is postponed or eliminated in many cesarean-born babies. Antibiotics given to almost all women having cesarean birth decrease the counts in breast milk of *Bifidobacterium* species, which are known to prevent infection and give anticarcinogenic capabilities (Dunn et al. 2017).

3.5.3 Development of the Microbiota After Birth

The neonatal and one's early day periods are vital stages in the intestinal microbial community setting up (Rautava et al. 2012b). At the time of birth, microbes colonize the newborn. Vaginally delivered neonates are passed through the maternal vagina, and the fecal microbiota of these is dominated by *Prevotella* spp. and *Lactobacillus*. Neonates born by means of cesarean delivery are not directly expose to the maternal vaginal microbial population and are subsequently more likely to have a microbiome dominated by microbes, like *Staphylococcus*, *Corynebacterium*, and *Propionibacterium* spp., that are acquired from the maternal skin and the hospital staff (Bäckhed et al. 2015). During the first week of neonatal life, a dominance of Actinobacteria mainly (*Bifidobacterium*) has been observed for vaginally delivered neonates, whereas Firmicutes has been noted as the most common microbial population for cesarean-delivered newborns. Besides, the frequency of bifidobacteria regulary increased in both vaginally delivered and cesarean-delivered newborns over time. In terms of food admissions, the intestinal microbiota of newborns is essentially affected by the feeding mode, and differences in intestinal microbes between solely breast-fed and formula-fed neonates have been well recorded (Zhuang et al. 2019). As recorded by numerous studies, the stools of breast-fed newborns contain more lactobacilli and bifidobacteria and fewer potential pathogens than the stools of formula-fed newborns, which contain a more diverse intestinal microbiome dominated by *Bacteroides*, enterococci, enterobacteria, staphylococci, clostridia, and *Atopobium* (Martin et al. 2016). Oligosaccharides (like human milk oligosaccharides (HMOs)), of which human milk is a wealthy source, are considered to be natural prebiotics and can effectively enhance the growth of specific microbial species, like bifidobacteria, in the infant intestinal microbiome (LoCascio et al. 2009). With the withdrawal of breast milk and the presentation of solid foods, the diversity of the intestinal microbiota elevates, with Proteobacteria and Actinobacteria becoming the dominant components of the newborn microbiome (koenig et al. 2011). Moreover, delivery and feeding style and other factors, including the gestational age at neonate birth, family lifestyle, host genetics, geographical location, and the use of antibiotics, are responsible for newborns' gut microbiota colonization (Fig. 3.1).

Preterm newborns routinely present with immature gastrointestinal, neurological, and respiratory and immune systems. So preterm infants are often exposed to medicine treatments, particularly the broad use of antibiotics. These infants usually need long-term hospitalization and get parenteral nutrition and mechanical ventilation, which could influence the natural process of the colonization of the microbiome and may result in a deviation in the foundation of the gut microbiome or an unusual composition of the gut microbiome (Milani et al. 2017). In preterm infants, the gut colonization of commensal anaerobic microbes is postponed. Consequently, the feces of preterm neonates contains significantly higher levels of *Enterococcus*, Enterobacteriaceae, and opportunistic pathogens than the feces of term neonates (Cong et al. 2016).

3.5.4 Functional Maturation of the Gut Microbiome

To govern how the functional magnitude of the neonatal gut microbiota developed at the first year of life, Bäckhed and his coworker analyzed vaginally delivered infants'

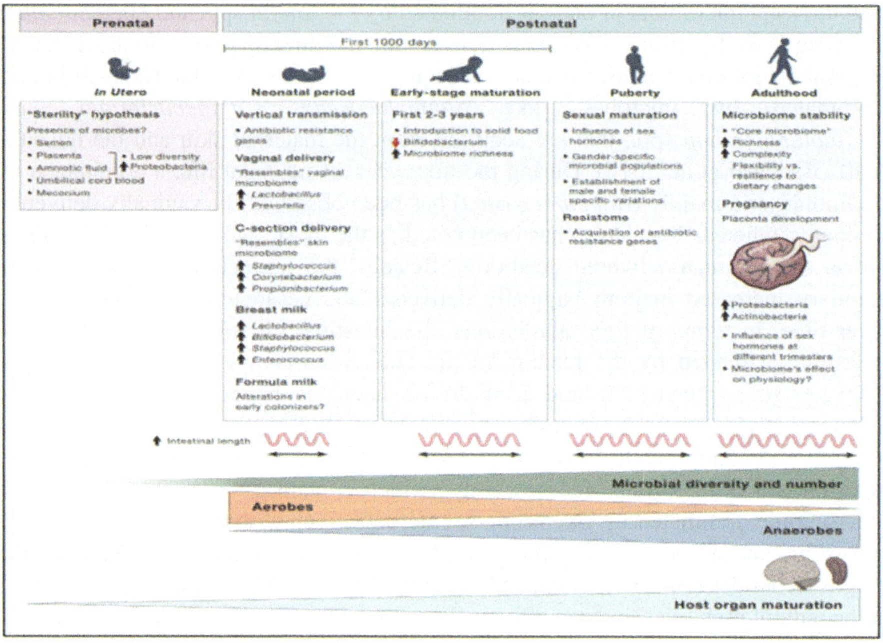

Fig. 3.1 Gut microbiome alteration from infancy to adulthood (Kundu et al. 2017)

gut microbiome using the KEGG orthology (KOs) groups. In the first year of life, the infants' comparatively simple gut microbiome transformed into a more complicated configuration in adult (Yatsunenko et al. 2012). In this study, they reported an increase in functional similarity with the mother's gut metagenome and diminished interindividual differences for the 1-year samples. The resistome is human gut microbiota reservoir for antibiotic resistance genes (Forslund et al. 2013). Here, they recorded the presence of antibiotic resistance genes already in the newborn microbiome, possibly a ramification of the relatively high plenitude of Proteobacteria DNA, whose genomes contain relatively high levels of antibiotic resistance genes. The newborn microbiome had over 90% prevalence of genes included in resistance against tetracycline, bacitracin, and macrolides; the resistance against this antibiotic was also prevalent in the adult gut (Forslund et al. 2013). Prevalence for resistance against antibiotics such as kanamycin elevated with age, with the highest prominence in the mother microbiome. Five of the newborn children got ten antibiotic treatments within 4 months after delivery, which produced a minor shift in the microbiota composition at 4 months but did not influence the pool of antibiotic resistance genes in the 12-month microbiome. The microbiome of newborns delivered by cesarean, however, tended to contain a greater involvement of antibiotic resistance genes compared to that of vaginally delivered neonates. The metagenomic investigations also revealed special energy source utilization in the neonate's gut at the sampled time points. In particular, phosphotransferase system

(PTS) genes for carbohydrate utilization were enriched in the neonate's microbiome, while the lactose-specific transporter was most prominent in 4-month-old infants, proportionate with a diet dominated by milk. The microbiomes of neonates and 4-month-old infants were enriched with breast-milk sugar degradation gene. On other hand, β-glucoside-specific transporters were most prominent in 4-month-old and 12-month-old infants. Accordingly, the 1-year microbiome was enriched in genes included in the degradation of complex sugars and starch and associated with the presence of *B. thetaiotaomicron*, known to have a broad repertoire of glycan-degrading enzymes (Sonnenburg et al. 2005), and of modules included in carbohydrate metabolism. Pectinesterase and *B. thetaiotaomicron*, the essential enzymes in pectin degradation, were most prominent in 1-year-old children, possibly due to the expanded intake at this age of pectin-enriched food. As a result of the succession of bacterial metabolic functions in the maturing infant gut, they reported that *Methanobrevibacter smithii* and *Desulfovibrio* sp. were present in mothers and absent in infants, except in two 12-month-old infants that were colonized by *M. smithii*. This finding is in agreement with the microbiome's elevated capacity for methane production in the mothers, which is related with increased fermentative capacity in the adult microbiome that requires disposal of hydrogen as methane (Charalampopoulos and Rastall 2009). The microbiome is exposed to a larger difference of dietary substrates as the infant grows older, which is connected to enrichment of genes in the central carbon metabolism. Like KO modules for pyruvate metabolism, the pyruvate-ferredoxin oxidoreductase catalyzing pyruvate to acetyl-CoA conversation was enriched in 4-month-old and 12-month-old infants versus neonates. In contrast, the relatively oxidized gut environment of neonates enables gut microbes to abuse TCA cycle for metabolism and energy production, as shown by the enrichment of KO modules for TCA cycle in neonates compared with 4-month-old and 12-month-old infants and the mothers. Taken together, the results revealed that the microbiome adapts to the accessibility of energy substrates as the infant grows older. The gut microbiome is a significant producer of vitamins. All newborns in Sweden get prophylactic vitamin K injections to avoid classic hemorrhagic disease. We observed enriched levels of genes for vitamin K_2 (menaquinone) synthesis in neonates, which correlated with the high prominence of *Bacteroides* and *Escherichia/Shigella*, known vitamin K_2 producers (Wang et al. 2013). Vitamin K_2 is significant for heart and bone health, and the microbiome was recently described to modulate homeostasis of the bone (Sjögren et al. 2012). Metabolism of retinol was also most enriched in the newborns, with numerous implications in essential developmental processes such as teeth, bone, and vision. Vitamins B complex is required for the body to convert nutrients into glucose and produce energy. Folate (vitamin B_9) is one of the fundamental B vitamins responsible for DNA synthesis and repair. Folate biosynthetic genes were significantly enriched in newborns. Genes for pyridoxal (vitamin B_6) and biotin (vitamin B_7) synthesis were too essentially enriched in newborns. In contrast, cobalamin, pantothenate, and thiamine (vitamins B_{12}, B_5, and B_1, respectively) biosynthetic genes elevated with age (Yatsunenko et al. 2012). However, modules for vitamin B_{12} transport system were greatly elevated in the neonate's metagenome but diminished with age. Essentially, transporters for heme, hemin, and iron, which are connected to vitamin B_{12} synthesis and significant for iron metabolism, were too expanded in the microbiome of neonates. The representation of KOs analysis for amino acid metabolism is diverse with age. The transport systems for all essential amino acids were elevated in the neonate's microbiome and remained high until the age of 4 months. Protein necessitates, calculated per kilogram of total body weight, continuously diminished with age until weaning (Boudry et al. 2010), which in parallel could influence the requirement for amino acid transport systems in agreement with KOs data analysis. The pathway for lysine biosynthesis, leucine and tryptophan metabolism, and methionine degradation elevated with time in comparable to those found in the mothers by 12 months of age. Eventually, the genes for synthesis and metabolism of the amino acid neurotransmitter GABA (gamma-aminobutyrate) and the hormone melatonin appeared differential enrichment in neonates, 4-month-old, and 12-month-old. In humans, melatonin plays a function in entraining the circadian system (Reiter 1991). The fact that newborns do not have an established circadian melatonin pattern, which appears later, at 3–4 months of age, and matures in childhood (Bäckhed et al. 2015).

3.6 The Human Gut Bifidobacteria

The establishment of a stable gut microbial consortium is critical to the development of the infant's gastrointestinal tract (GIT) (Stappenbeck et al. 2002; Rakoff-Nahoum et al. 2004). Breast-fed infants develop colonic microbiota that is frequently marked by a high concentration of bifidobacteria (Favier et al. 2003; Kleessen et al. 1995). This numerical advantage confers a substantial health benefit to the neonate by hindering pathogen colonization through competitive exclusion. Candidate milk-borne growth factors include oligosaccharides; these human milk oligosaccharide (HMO) structures possess a vast combinatorial potential, as they incorporate carbohydrate monomers via possible glycosidic linkages to assemble molecules containing three to 32 units. Therefore, the characterization of only approximately 200 soluble structures within the human milk glycome strongly suggests a functional role for these sugars (Ninonuevo et al. 2006). Of the 200 known compositions, five low-molecular-weight oligosaccharides abundant in breast milk are the preferred substrates of *Bifidobacterium longum* subsp. *infantis* ATCC15697, the archetypal HMO-utilizing bacterium (Locascio et al. 2007). At present, only two phylotypes (i.e., *B. longum* subsp. *longum* and *B. adolescentis*) have been fully sequenced and deposited into public databases.

In conclusion, our gut microbiota evolves with us and plays a pivotal role in human health and disease. We now know that the resident microbiota influence host metabolism, physiology, and immune system development, while perturbation of the microbial community can result in chronic GI disease. While the revolution in molecular technologies has provided us with the tools necessary to more accurately study the gut microbiota, we now need to more accurately elucidate the relationships between the gut microbiota and several intestinal pathologies. Understanding the part that microbial populations play in GI disease is fundamental to the ultimate development of appropriate therapeutic approaches. The concept of altering our gut community by microbial intervention in an effort to improve GI health is currently a topic that is receiving considerable interest. The targeting of specific components of the gut microbiome will potentially allow the removal of the harmful organisms and enrich the beneficial microbes that contribute to our health.

3.6.1 Probiotics, Prebiotics, and Synbiotics

Probiotic, prebiotic, and synbiotic are words of the modern era. Probiotics are live microorganisms, which, when administered in adequate amounts, confer a health benefit on the host. Probiotics have become a very important element to everyday health food products, and their global market is increasing. Consumers are very concerned of chemical preservatives and processed foods, even though they provide a grade of safety and food diversity never seen before. However, consumers accept easily lactic acid bacteria (LAB) as a natural way to preserve food and promote their health. The target of probiotic food products is to have up to 107 CFU/g at the end of their shelf life; but probiotic LAB must endure some stresses to ensure they reach the adequate numbers in the target location to elicit their effect. The human gastrointestinal tract contains up to 1013–1014 cells. It is a complex ecosystem combining the gastrointestinal epithelium, immune cells, and resident microbiota. The three major sections of the human gastrointestinal tract are the stomach, the small intestine, and the large intestine. Every section has its own distinct microbiota. The stomach is primarily inhabited by aerobic gram-positive microorganisms (<103 CFU/g). The small intestine is inhabited by the genera *Lactobacillus*, *Bifidobacterium*, *Bacteroides*, and *Streptococcus* (103–104 CFU/g), while the large intestine is populated by the genera *Bacteroides*, *Fusobacterium*, *Lactobacillus*, *Bifidobacterium*, and *Eubacterium* in large numbers (1011–1012 CFU/g) (Quinto et al. 2014).

The GIT is also rich in many molecules that can be used as nutrients by microbes. Hence, the GIT has the potential to be heavily colonized by various bacteria, both harmful and beneficial. The mucosa of the gastrointestinal tract is continuously exposed to an environment that is rich in foreign substances, such as food particles and antigens of microbial origin. Particular changes in the intestinal ecosystem might contribute to the development of certain illness. There is therefore a need for an exhaustive review on the functions of the gut microbiota, occurrence of gut dysbiosis (alteration or imbalance of the micro biota), how these intestinal bacteria trigger development of disease once the normal biota of a healthy individual is imbalanced, exploiting this intricate and interwoven ecosystem for understanding human health, development of bio-therapeutics, and future perspectives (Vyas and Ranganathan 2012).

The human intestinal tract has been colonized by thousands of species of bacteria during the coevolution of man and microbes. Gut-borne microbes outnumber the total number of body tissue cells by a factor often. Recent metagenomic analysis of the human gut microbiota has revealed the presence of some 3.3 million genes, as compared to the mere 23,000 genes present in the cells of the tissues in the entire human body. Evidence for various beneficial roles of the intestinal microbiota in human health and disease is expanding rapidly. Perturbation of the intestinal microbiota may lead to chronic diseases such as autoimmune diseases, colon cancers, gastric ulcers, cardiovascular diseases, functional bowel diseases, and obesity. Restoration of the gut microbiota may be difficult to accomplish, but the use of probiotics has led to promising results in a large number of well-designed (clinical) studies. Microbiomics has spurred a dramatic increase in scientific, industrial, and public interest in probiotics and prebiotics as possible agents for gut microbiota management and control. Genomics and bioinformatics tools may allow us to establish mechanistic relationships among gut microbiota, health status, and the effects of drugs in the individual. This will hopefully provide perspectives for personalized gut microbiota management (Vyas and Ranganathan 2012).

Gut microbes and microbial products influence host functions both locally in the gastrointestinal (GI) tract and systemically. The gut microbiome contains a vast amount of genetic material that dwarfs the human genome, and its contents can be modified far more easily than human genes. Thus, it is not surprising that the gut microbiota has emerged as a promising therapeutic target. Among the most recognized of microbiota-modifying strategies is the use of fecal microbiota transplantation (FMT). However, FMT lacks precision; microbiome-targeting therapies that are well-defined are preferred, given their potential for more favorable safety profiles, large-scale production, and patient acceptance (Edwards et al. 2020).

Human nutrition has already been using probiotics in fermented products, especially in dairy products. Lactic acid bacteria (LAB) are a group of gram-positive, non-sporulating, anaerobic, or facultative aerobic cocci or rods, which produce lactic acid as one of the main fermentation products of the metabolism of carbohydrates. The present classification of lactic acid bacteria (LAB) is according to the following criteria: cellular morphology, mode of glucose fermentation, range of growth temperature, and sugar utilization patterns. Four genera were recognized as LAB: *Lactobacillus*, *Leuconostoc*, *Pediococcus*, and *Streptococcus*. Molecular biological methods have increased the number of genera included in this group. The current taxonomic classification includes the LAB group in the phylum Firmicutes, class Bacilli, and order Lactobacillales. The different families and genera can be searched in the NCBI Taxonomy browser or in the UniProt Taxonomy browser (Quinto et al. 2014).

The lactic acid bacteria (LAB) are natural biota of fermented foods, Generally Recognized as Safe (GRAS), and contribute to the bouquet and particular flavor of different kinds of fermented foods and suggested for use as a starter in the production of different types of functional food products. Some species involved are *Lactobacillus acidophilus*, *L. casei*, *L. johnsonii*, *L. fermentum*, *L. rhamnosus*, *L. plantarum*, *L. reuteri*, *L. salivarius*, *L. paracasei*, *L. delbrueckii* subsp. *bulgaricus*, *Saccharomyces boulardii*, *Streptococcus thermophilus*, *Bifidobacterium lactis*, *B. longum*, and *B. breve*. Probiotics presumably exert a dual effect, preventing or decreasing the intestinal colonization with pathogen microorganisms, or interacting with the gut-associated lymphoid tissue (GALT) to prevent inflammatory responses and promote a state of tolerance to themselves and possibly to foods. The beneficial effects of probiotics are often disparate and strain specific. Some species conferred beneficial effects, such as the treatment of acute diarrhea associated with rotavirus, ulcerative colitis, *Clostridium difficile*-associated diarrhea, and *Helicobacter pylori* infection. Furthermore, they have preventive effects, such as the prevention of antibiotic-associated diarrhea in children and improvement in lactose digestion. Other effects are still under investigation (Quinto et al. 2014).

Lactococcus sp., for example, is one of the best known and characterized species of lactic acid bacteria (LAB). These bacteria are present in the natural environment, including products of spontaneous milk or plant fermentation. They are relevant to the production of various dairy products and several fresh cheeses such as Brazilian Minas frescal cheese and cottage cheese (Felicio et al. 2016; Jesus et al. 2016). *Lactococci* play an important role in flavor formation through their proteolytic and amino acid conversion pathways (Zycka-Krzesinska et al. 2015). Recent efforts have focused on screening and evaluating new LAB strains isolated from wild ecological niches and good quality raw milk and traditional cheeses. These strains could be used as starters to increase biodiversity, diversify flavor, and restore the unique characteristics of traditional cheese varieties (De La Plaza et al. 2006; Allam et al. 2017; Darwish et al. 2018).

Some of LAB are recognized as nosocomial pathogens such as *Enterococcus* spp.; from this consideration, it is important to determine safety before using enterococci for probiotic preparations. On the basis, the proper selection of enterococci is possible, and they can be used as ideal probiotics (Saavedra et al. 2003; Bhardwaj et al. 2008). Animal nutrition is another area for successful application of probiotics, and the *Enterococcus* spp. seems to be the most utilized as additive microorganisms. Several probiotics have received temporary approval in the European Union, but their modes of action, which lead to beneficial effects, are only partly known (Aukováa et al. 2008; Allam et al. 2016). Positive effects were reported earlier for *Enterococcus* spp. such as ability to inhibit the growth of foodborne pathogens, stimulating weight gain, hypocholesterolemic and reduction of LDL-cholesterol, enhancing gut microbiota, immune modulatory by increasing lymphocytes, and bacteriocin-producing (Surono et al. 2011; Pajarillo et al. 2015; Rieger et al. 2015; El-ghaish et al. 2016; Hyrslova et al. 2016).

Prebiotics are substrates selectively utilized by host microorganisms, conferring a health benefit. Synbiotics are selected combinations of probiotics and prebiotics that synergize to confer a health benefit. Although the menu of potential therapeutic combinations is vast, the current therapeutic strategy for these agents is empirical; formulations are chosen more often for historical reasons than for mechanistic potential. As we gain new insights into how the microbiome impacts disease pathogenesis, we hope to develop rationally selected microbiota-targeting therapies for individual patients or well-defined clinical scenarios.

Efforts were exhorted for production of new types of healthy fermented products using beneficial probiotics and natural source additives as prebiotics in order to serve synbiotic functional foods with high nutritional value, safe to all age groups, and provide many benefits. Darwish et al. (2020) produced yogurt and frozen yogurt-like products using common yogurt starter culture (*Streptococcus thermophilus* and *Lactobacillus delbrueckii* subsp. *bulgaricus*) either as single or mixed with *Bifidobacterium bifidum* as source of probiotics. Talbina, a gentle soup made from barley bran flour that has considerable therapeutic value prescribed as a cure for many diseases in Sunnah, was applied as a natural prebiotic.

Burlington and Classification (2006) produced synbiotic dairy beverage, soy-based beverage, oats-based beverage, frozen food products, and yogurt-like products. The beverage is made by the integration of both probiotic (live microbial food cultures) and prebiotic (non-digestible carbohydrates) supplements that beneficially affect the host by improving the survival and implantation of live microbial dietary supplements in the gastrointestinal tract.

Information on prebiotics and synbiotics is quite a few. Hopefully in the coming years, there will be further studies on combinations of probiotics and prebiotics and further development of synbiotics. However, the health claims made needs to be substantiated and firmly established by properly designed large-scale clinical trials on the human body (Maftei 2012). Further investigations to evaluate the best dose-response effect and the length of probiotic and synbiotic supplementation are also needed to evaluate if the persistence of their potential beneficial effects is maintained after interruption or if continuous supplementation should be used for an efficient treatment or disease prevention (Ferrarese et al. 2018). Therefore, current focus is on evaluating new strains of probiotics, new prebiotics, and new synbiotic products and their applicability in biomedical/clinical research, paving a new direction for exploration and exploitation of probiotics, prebiotics, and synbiotics aimed at improving human health.

3.7 Function of Microbiota in Human Health

3.7.1 Function of Microbiota in Human Breast Milk

Breast milk contains a rich microbiota composed of wealthy skin and non-skin bacteria. The extent of the breast-milk microbiota diversity has been revealed through modern culture-independent studies utilizing microbial DNA signatures. However, the extent to which the breast-milk microbiota is transported from mother to newborn and the role of these breast-milk microbiota for the infant are not completely elucidated. Regarding the formation of breast-milk microbiota, including retrograde infant-to-mother transport and enteromammary trafficking, through a complex, highly evolved process in the early stages of discovery, mothers transfer the breast-milk microbiota to their infants to impact infant growth and development (Latuga et al. 2014). Breast milk supplies optimal nutrition for newborns and decreases their risk of having irresistible diseases. Additionally, breast milk could be a vehicle for transmission of bacteria and viruses from mother to infant. However, the variable dictating the composition of the breast-milk microbiota and the function of the breast-milk microbiota for infant is still not elucidated. Breast milk typically contains both skin microbiota and what are naturally considered enteric organisms.

Proposed speculations for the microbiome composition of breast milk involve retrograde flow from the infant's oral cavity, transfer of organisms from maternal skin, and transport of microbiota from the maternal enteric tract to the mammary gland. Numerous studies have pointed out a higher relative prominence of bifidobacteria and *Lactobacillus* in breastfed newborns' microbiome (Knol et al. 2005) compared with formula-fed newborns. It is intriguing to observe these differences in the microbiota composition still even after breast-feeding is discontinued (Rautava et al. 2012b). Culture-dependent and culture-independent investigations of breast-milk samples revealed the existence of *Staphylococcus* and *Streptococcus*, which correspond to early colonizers of the gut (Collado et al. 2009). *Lactobacillus* and bifidobacteria are also detected, suggesting a significant role of breast milk as a delivering system for beneficial microbiome (Fernández et al. 2013). Other differences in formula-fed infants' microbiota have been illustrated by Le Huërou-Luron et al. (2010) and revealed that facultative anaerobes, clostridia and *Bacteroides*, appear at higher level and frequency than in breastfed newborns; the main species in clostridia is *C. perfringens* instead of *C. difficile* that is the most abundant *Clostridium* in breastfed newborns. Moreover, breastfed, vaginally delivered term newborns recorded low level in the microbiota rate of *C. difficile* and *E. coli* but a high frequency in beneficial bacteria, such as *Bifidobacterium* spp. In reality, the microbial composition of breast milk is also dependent on the type of delivery. A different and less diverse community of microorganisms has been recorded in milk samples from mothers who had cesarean section compared to mothers who had vaginal delivery (Rautava et al. 2012b).

Feeding with human milk improves immune system development in full-term newborns compared with formula-fed newborns; breastfed infants show elevated T helper type 1 (Th1) activity, higher proliferative T-cell response to tetanus toxoid (Stephens et al. 1986), and declined counts of CD4 using flow cytometry (Carver et al. 1991). Recent investigation revealed that preterm newborns, newborns nourished by breast milk also had low B-cell counts compared with formula-fed newborns (Tarcan et al. 2004). Because of the immaturity of the neonatal humoral immunity, newborns confronting infection concurring to maternal antibodies and a powerful cytotoxic T helper type 1 (Th1) response. Enhanced cytotoxic function in newborns fed with breast milk may be supported by bacterial ligands in breast milk (Donnet-Hughes 2008). To demonstrate this hypothesis, in vitro stimulation of dendritic cells with lipopolysaccharide boosted differentiation of T cell (Spörri and Sousa 2005). Animals raised in a germ-free (GF) environment had lasting impedances in their immunologic function. Therefore, microbiota in breast milk stimulates cytotoxic Th cell maturation and enhances their capacities to prevent infection. In the intestinal tract, the microbiota participates in nutrient synthesis and metabolism. "Enterotype" clarifies the collective functional digestive and nutritive potentiality of the enteric microbiota. Enterotypes may be linked with body habitus, diet, or geography (Arumugam et al. 2011). While enterotypes have not yet been completely characterized in human infants, there is evidence of a breastfed newborn children enterotype. In the feces of eight breastfed newborn children compared with ten formula-fed infants, metagenomic examination recommended an elevation in amino acid and nitrogen metabolism, carbohydrate metabolism, and cobalamin synthesis (Yatsunenko et al. 2012). Additionally, the breast-milk metagenome is enriched for oxidative stress response, membrane transport, and nitrogen metabolism. In term and premature infants fed with breast milk and formula, metagenomic examinations of fecal samples have revealed an improved harmfulness potential with the presence of bacteriophage and genes encoding for type III and IV secretion systems. These information are authenticated with an animal model in which there is expanded oxidative stress and a diminished in protein production utilized in cell adhesion with formula feedings in comparison with breast-milk feedings (LaTuga et al. 2014). The breast-milk microbiota may also help in improving intestinal barrier protection. Animal studies have revealed that enteric colonization is essential for stimulation of antimicrobial peptide defenses, upregulation of epithelial junctional complexes, and expression of key enzymes for detoxification enzymes like alkaline phosphatase to mitigate over-stimulation by bacterial lipopolysaccharide ligands (Bates et al. 2007). In an animal model, heat shock protein 70 (HSP70) in breast milk diminished bacterial translocation from the intestinal lumen (Liedel et al. 2011). It is conceivable that microbiota in breast milk may elevate HSP70 levels in the intestinal lumen and contribute to epithelial barrier function in neonates (Arvans et al. 2005). Breast-milk oligosaccharides have a vital relationship with microbiota in breast milk and the intestinal tract. Basically, these are complex glycans found in human breast milk. Traditionally, oligosaccharides were thought to act as a substrate for the growth of intestinal bacteria in the distal enteric tract (Marcobal et al. 2011). Recent information reveals more complex relationship, through which oligosaccharides in breast milk are not consumed by microbiota but still alter the growth of microbiota (Hunt et al. 2012). In a rat model, oligosaccharide levels were decreased in the small intestine and differentially emitted into urine, proposing selective absorption of oligosaccharides conceivably in concert with differing microbiota throughout the intestinal tract (Jantscher-Krenn et al. 2013). Additionally, oligosaccharides have independent immune function in neonates (Eiwegger et al. 2010). Eventually, oligosaccharides may work synergistically with breast milk and enteric microbiota to reinforce barrier function. In transiting or colonizing the infant enteric tract, the breast-milk microbiota may have broader developmental ramification for the newborn. Microbiota in breast milk may, moreover, build up a typical gut-brain axis. Animals raised in GF environments have diminished intestinal peristalsis that can be reestablished with the introduction of enteric microbiome from animals with conventional microbial exposure (Husebye et al. 1994). In a comparison of GF, specific pathogen-free (SPF), and gnotobiotic animals, GF animals showed an exaggerated stress response compared with SPF mice. This response could be switched with early exposure to *Bifidobacterium infantis*, which is identified in breast milk. Oral antibiotic administration to animals raised in an SPF environment alters enteric microbiota, upregulates brain-derived neurotrophic factor, and elevates exploratory behavior (LaTuga et al. 2014).

3.7.2 Microbial Diversity for Human Health

Gut microbiota plays a major role in the alimentation of homeostasis of host physiology, metabolism, development, and immunity. Profiling the gastrointestinal microbiome composition with authentic strategy is of considerable interest to yield novel insights into the pathogenesis of numerous diseases as well as to define new therapeutic and prophylactic interventions. Modern developments in metagenomics have provided researchers with the tools required to open the "black box" of microbiome science. These novel technologies have empowered the foundation of correlations between dysbiosis microbial communities and several diseases. Expanded approaches and thoroughgoing data interpretation will be significant for resolution of these discrepancies. In this context, diagnostic tools for research purposes are required to support both preclinical developments and clinical studies in humans using laboratory rodents. The growing requires to survey the tremendous microbial diversity in a culture-independent manner, which leads to the advancement of molecular methods through sequence profiling of part of conserved genes such as 16S rDNA in different scientific fields. Next-generation sequencing technologies providing unprecedented throughput of data are now routinely used to assess bacterial community composition in complex samples. Depending if rough/basic bacterial signature or extensive resolution of taxonomic assignment of organisms is required, the time and costs for 16S rRNA profiling versus full genome analysis or bacterial RNA sequencing may vary from 1 to 50. Knowing the composition of the microbial community alone does not essentially lead to an understanding of its role, and functional metagenomic and metabolomic might be required. Nevertheless, such analyses are accommodated to clarify discrepancies between individuals considering the microbial determinant of biochemical singularity. Without playing the role of Cassandra, pointing out the tremendous technical bias and both inter-/intra-individual varieties as well as time-related changes of gut microbial composition, the separate profiling of bacterial communities is of main interest for scientists and clinicians. It permits advance stratification of distinct responders both in demonstrating immune and infectious diseases and for personalized therapeutic interventions. Collectively, this process is useful for the diagnostic of dysbiosis states and the follow-up of diet and treatments in clinical studies. Additionally, it may clearly serve as cornerstone for research purposes in microbiota-presumed diseases modeling in rodents, the latter being more practical and thus fitting the 3Rs ethical rules (Richmond 2000). Although the microbiome science needs a healthy dose of skepticism (Hanage 2014), it also requires reliable and consistent tools for gold standard metagenomic analysis.

3.7.3 Function of Microbiome in Immune System Development, Host Protection, and Metabolism Homeostasis

As the gut microbiota encode a substantively larger number of genes than their human host, it follows that they are able to undertake a variety of metabolic functions that humans are unable to do or are only able to do in a limited capacity. The gut bacteria are able to produce a variety of vitamins, synthesize all essential and nonessential amino acids, and carry out biotransformation of bile (Vyas and Ranganathan 2012). In addition, the microbiome provides the vital biochemical pathways for the metabolism of nondigestible carbohydrates; large polysaccharides, such as starches, cellulose, pectins, and gums; some oligosaccharides that escape digestion; unabsorbed sugars and alcohols from the diet (Cummings et al. 1987); and host-derived mucins (Koropatkin et al. 2012). This functionality results in the recovery of energy and absorbable substrates for the host and a supply of energy and nutrients for bacterial growth and proliferation (Guarner and Malagelada 2003). Metabolism of carbohydrates is a major source of energy in the colon.

Many intestinal bacteria produce antimicrobial compounds and compete for nutrients and sites of attachment in the gut lining, thereby preventing colonization by pathogens. This action is known as the barrier or competitive-exclusion effect. Host cells in the gut wall have attachment sites that can be used by pathogenic bacteria to enter the epithelial cells. Further, because bacteria compete for nutrients in their immediate surroundings and maintain their collective habitat by administering and consuming all resources, the enteric microbiota can outcompete pathogenic bacteria for resources by sheer force of numbers. In addition, bacteria can inhibit the growth of their competitors by producing antimicrobial substances known as bacteriocins, and the ability to synthesize these bacteriocins is widely distributed among gastrointestinal bacteria (Guarner and Malagelada 2003).

In metabolism homeostasis, the gut's microbial population provides host physiology with major benefits. An obvious relationship has now been identified between gut microbiota and host metabolism, in which secretion of microbial-mediated gut hormone plays a significant role. Bacteria generate a series of metabolites within the gut lumen and provide structural components that act as signalling molecules to a number of types of cells within the mucosa. Enteroendocrine cells inside the gut mucosal lining synthesize and secrete a variety of hormones, such as CCK, PYY, GLP-1, GIP, and 5-HT, that have regulatory functions in key metabolic activities such as insulin sensitivity, glucose tolerance, fat accumulation, and appetite. The release of these hormones can be affected by the involvement of bacteria and their metabolites inside the gut, and therefore, the release of microbiota-mediated gut hormones is an essential component of host metabolism microbial control. Consequently, dietary or pharmacological treatments that change the gut microbiome present possible therapies for the treatment of human metabolic disorders (Martin et al. 2019).

The absence of gut-derived 5-HT, as a result of pharmacological inhibition or genetic ablation of the rate-limiting enzyme for 5-HT synthesis in the gut, tryptophan hydroxylase 1 (TPH1), demonstrates protection from diet-induced obesity in mice (Crane et al. 2015). In addition, circulating 5-HT in obese humans is elevated and is positively associated with body mass index (Young et al. 2018) and impaired glycemic regulation (Takahashi et al. 2002).

Hence, the capacity of gut microbiota to affect PYY secretion has important implications for obesity and metabolic disease growth. In two EE model cell lines and in primary human colonic cell cultures, microbial SCFAs, especially butyrate, trigger a dose- and time-dependent boost in PYY gene expression (Larraufie et al. 2018). Peptide tyrosine-tyrosine (PYY), in addition to GLP-1, is synthesized and secreted by L cells and is mainly expressed in the lower small intestine and colon. PYY controls the intake and satiety of foods by activating central G protein-coupled Y2 receptors in the hypothalamic arcuate nucleus on neuropeptide Y (NPY) and AgRP neurons (Dumont et al. 1995).

By modulating gut microbial populations utilizing a range of nutrients such as prebiotics and probiotics, synbiotics, FMT (fecal microbiota transplantation), and postbiotics, the host-microbiome field moves toward enhancing metabolism and weight maintenance. Prebiotics are foods or nutritional supplements that promote the growth of saccharolytic bacteria, which metabolize nondigestible carbohydrates such as inulin and oligofructose (Martinez et al. 2016).

Recent research has shown that prebiotics improve complications associated with metabolic disorders, such as obesity and insulin resistance (Wang et al. 2015). Different mechanisms have been identified to explain these beneficial effects, including the development of SCFA, stimulation of intestinal gluconeogenesis, epithelial integrity, secretion of PYY and GLP-1 hormones to enhance satiety and insulin sensitivity, increased antimicrobial peptide expression, and altered gut microbial community structure (Wang et al. 2015).

Another widely used strategy and intensively researched supplement is the use of probiotics, which are live microorganisms supplied individually or in hybrids, such as VSL#3, that produce better health outcomes in the host (Mennigen et al. 2009). Consideration of the composition of probiotic formulations is critical since each strain can have a different effect, for example, on microbial structure/function or on the host immune system response. Wang and his coworkers (Wang et al. 2015) demonstrated that in mice three bacterial strains (*Lactobacillus paracasei* CNCM I-4270, *Lactobacillus rhamnosus* CNCM I-3690, and *Bifidobacterium animalis* subsp. *lactis* CNCM I-2494) independently reduced body weight and increased glucose tolerance but through various mechanisms. Regular gavage of probiotic yeast *Saccharomyces boulardii* (Biocodex) has induced improvements in intestinal microbiota, indicating a less obesogenic condition and improving the metabolic activity of genetically obese and diabetic db/db mice (Everard et al. 2014).

In immune system development, under normal circumstances, the fetal gastrointestinal tract is assumed to be sterile, with the immune system's first exposure to commensals that occur during the passage in the birth canal. These early encounters are deemed to set long-term tone of the mucosal and systemic immune system. The process by which neonate tissues respond to the formidable challenge of microbial colonization is incompletely understood, but some of these early responses to commensals are assumed to be determined by factors found in maternal milk. Colostrum and breast milk also contain live bacteria, metabolites, IgA, immune cells, and cytokines as well. To form the breastfed infant microbiota and the host's response to these microbes, these factors synergize. For example, maternal IgA limits immune reaction and microbial attachment through binding nutritional and microbial antigens, and the presence of metabolites like oligosaccharides in mother's milk enhances the expansion of established microbiota constituents such as *Bifidobacterium* (Marcobal et al. 2010; Marcobal and Sonnenburg 2012). Throughout pregnancy and lactation, bacterial translocation from the mouse gut increases, and bacterially charged dendritic cells in the milk have been suggested to lead to neonatal immune imprinting by affecting the form of the immune response to commensal antigens (Perez et al. 2007).

The ability to tolerate the microbiota can also be demonstrated by the relative immaturity at birth of the neonate immune system and by the tolerogenic environment defining early mammalian existence. Admittedly, blunted inflammatory cytokine production and distorted development of T and B cells in favor of regulatory responses are characteristic of the developing immune system (PrabhuDas et al. 2011; Siegrist 2001). This regulatory environment guarantees microbiota is formed without overt inflammation. Recent reports reveal that a given population of neonatally enriched erythroid cells contributes to the maintenance of this immunoregulatory environment and limits mucosal inflammation after microbiota colonization (Elahi et al. 2013). The host's early exposure to commensals can also suppress cells involved in the activation of inflammatory responses such as invariant natural killer T (iNKT) cells, an impact that has long-term implications for the host's ability to develop inflammatory diseases (Olszak et al. 2012).

Early findings in GF (germ-free) mice showed that for the maturation of the immune system, the host microbiota is crucial (Clavel et al. 2017; Pickard et al. 2017). GF mice have multiple immunological deficiencies in the absence of a microbiota, including decreased lymphoid cell numbers and function (Fiebiger et al. 2016). GF mice, for instance, have fewer T helper type 1 (Th1) cells than their conventionalized counterparts (Wu and Wu 2012). In order to target intracellular pathogens, Th1 cells facilitate cell-mediated immune responses and phagocyte-dependent inflammation (Damsker et al. 2010). Th1 responses in GF mice can be restored via host colonization with a wide range of microbes, including the well-studied pathogen *Listeria monocytogenes*, which enhances Th1 development through macrophage production of the T cell-stimulating factor, interleukin 12 (IL-12) (Hsieh et al. 1993). Intracellular bacteria like *L. monocytogenes* explicitly stimulate Th1 responses in the gut (Atarashi et al. 2015). In addition, GF mice have a decreased number of T helper type 17 (Th17) cells. In general, Th17 cells are pro-inflammatory, but they stimulate IL-17 development and mediate resistance against extracellular pathogens and autoimmune disease (Wu and Wu 2012; Damsker et al. 2010). Adherent bacteria, like clostridia-related segmented filamentous bacteria (SFB), induce the production of Th17 cells in the small intestine by causing serum amyloid A release from the intestinal epithelial cells (IECs). Serum amyloid A release leads to the development of cytokines of the group 3 innate lymphoid cells (ILC3) that upregulate the Th17 response (Ivanov et al. 2009). To achieve immune tolerance toward the host microbiota, fine-tuning of Th1 and Th17 responses is important, as seen in the case of IBD, where disordered populations of Th1 and Th17 cells result in increased pathology (Gálvez 2014). Underdevelopment of these responses can implicate the progress of other chronic inflammation-related diseases, such as cancer (Bailey et al. 2014; Vinay et al. 2015).

The identification of conserved microbe-associated molecular patterns (MAMPs) mediates one of the key modes of communication between the host and the microbiota. These signals are incorporated into the neonate innate immune system in a special way to facilitate stable microbial colonization. For example, while neonate innate cells express ligands such as Toll-like receptors (TLR), their response to microbial ligands is different from that of adult cells with significant impairment in the release of inflammatory mediators such as oxygen radicals and increased production of regulatory cytokines such as IL-10 (Kollmann et al. 2012). Part of this occurrence comes as the result of the activity of the microbiota itself. In fact, early reactions to microbial ligands such as LPS, the endotoxin present in the outer membrane of gram-negative bacterial walls, cause gut epithelial cells to become hypo-responsive to corresponding TLR stimulation (Lotz et al. 2006; Chassin et al. 2010).

Commensals also help in postnatal immune system development, which in turn contributes to its containment. Studies conducted in animals raised in the absence of live microbes known as germ-free (GF) showed that microbiota plays an important role in the production of secondary and lymphoid structures. This effect is especially evident in the smaller Peyer patch-size gastrointestinal tract and a reduced number of plasma-producing CD4+ T cells and IgA (Bauer et al. 1963; Talham et al. 1999; Hamada et al. 2002; Macpherson et al. 2001; Mazmanian et al. 2005; Smith et al. 2007). In the intestines, due to commensal contact, tertiary lymphoid structures such as isolated lymphoid follicles or crytopatches are caused after birth (Bouskra et al. 2008; Ohnmacht et al. 2011). Commensals may also contribute to strengthening of the intestinal barrier by various mechanisms including enhancing epithelial cell maturation and angiogenesis (Hooper et al. 2001; Stappenbeck et al. 2002).

The high regulatory tone of the neonate immune system and the activity of commensals in the development and training of this system results in a lasting, homeostatic host/commensal relationship being established when operating properly. These key interactions between the host immune system and the microbiota have far-reaching and long-term human health effects. Admittedly, epidemiological studies have revealed that microbiota alterations in moms or neonates can predispose to diseases associated with dysregulated barrier responses like asthma (Ege et al. 2011).

Commensals are an important and essential inducer of regulatory responses. Following stabilization, the core human microbiota primarily includes the following phyla: Bacteroidetes and Firmicutes, with a lower abundance of Actinobacteria, Proteobacteria, and Verrucomicrobia (Arumugam et al. 2011). Noticeably, in the absence of signals originating from gut biota, the development of tolerance—the aggressive suppression of inflammatory responses to food and other orally consumed antigens—could not be induced (Wannemuehler et al. 1982; Kiyono et al. 1982; Sudo et al. 1997; Weiner et al. 2011). Though immunological tolerance is likely to be achieved by multiple and recurrent mechanisms (Weiner et al. 2011), Foxp3+ regulatory T (Treg) cells have taken center stage in our comprehension of this process in recent years. All throughout the host's lifetime, these cells sustain both peripheral and mucosal homeostasis, and impairment in the homeostasis of these cells leads to a loss of oral tolerance and production of abnormal effector reactions in the gut (Josefowicz et al. 2012; Worbs et al. 2006; Mucida et al. 2005).

Defense of the host from exogenous pathogens by commensal bacteria was identified more than five decades ago as a phenomenon called colonization resistance (Van der Waaij et al. 1971; Buffie and Pamer 2013). The need for all species of organisms to fight for the same ecological niche is one of the main ways of interaction between the microbiota and the invading microbes. Subsequently, in a process called colonization resistance, commensals have been shown to restrict pathogen colonization through competition for established metabolites (Kamada et al. 2013). Altering the availability of nutrients by the host microbiota can also have significant effects on the expression of virulence genes and the growth rate of pathogens such as enterohemorrhagic *E. coli* or *C. difficile* (Pacheco et al. 2012; Ng et al. 2013).

3.7.3.1 Immunosenescence and Microbiota

Although changes in the innate immune system have been noted with aging, there are more prominent long-term effects of microbiota on adaptive responses (Linton and Dorshkind 2004; Castelo-Branco and Soveral 2014). For long-term defense from environmental insults and invading pathogens, the adaptive immune system is used. As such, this subsystem's long-term education may have additive effects on immunosenescence. A predominant cell type of the adaptive immune system is B cells. The population of antigen-experienced B cells is split into plasma cells and memory B cells. Plasma cells generate pathogen-specific antibodies, whereas memory B cells offer long-term antigen recognition through their ability to reactivate rapidly after subsequent antigen exposures (Eibel et al. 2014). Peripheral blood extracted from the elderly (aged 86–94 years) showed a decline in B-cell population diversity due to a reduction in B cells in memory (Gibson et al. 2009). This reduction in the diversity of B cells was associated with increased frailty and can be used as a general health status indicator (Gibson et al. 2009). A decrease in memory B-cell numbers can trigger an inappropriate immune response to the microbiota because B cells are necessary to differentiate between pathogenic and commensal bacteria (Eibel et al. 2014). The decrease in age-accompanying memory B-cell numbers can encourage inappropriate immune responses to the microbiota, promote microbial dysbiosis, and increase the risk of disease.

T cells are the adaptive immune system's second main cell type and are known as either conventional or unconventional T lymphocytes (Roberts and Girardi 2008). This classification is based upon specific surface markers of T cells, functional capacity and localization of body sites (Roberts and Girardi 2008). T cells are typically activated by the binding of an antigen that is seen on the surface of antigen-presenting cells (APC) (Jin et al. 2012). Conventional T cells can carry out a wide range of functions once triggered, from stimulating long-term immunity to killing infected cells (Jin et al. 2012). Throughout immunosenescence, the decrease of CD28+ T cells is one of the most notable changes to occur in conventional T-cell populations (Tu and Rao 2016). CD28 is a co-stimulatory protein displayed on naive T cells and is essential for T-cell activation, regulation, and sustainability. A drop in CD28+ T cells can also reduce the activation of T cells, leading to increased vulnerability to pathogens (Bour-Jordan and Bluestone 2002). In addition, the loss of CD28 can decrease self-antigen and microbiota tolerance as it is also a negative immune response regulator (Perez et al. 1997; Bour-Jordan and Bluestone 2002).

Although a broad range of functions are performed by conventional T cells, their position during immunosenescence is complex and not fully explained. The microbiota and immunosenescence, on the other hand, are highly affected by a class of unconventional T cells known as natural killer T (NKT) cells. In separate studies, populations of T cells separated from the peripheral blood of elderly patients showed a decrease in the proportion of NKT cells versus cells isolated from young patients (DelaRosa et al. 2002; Jing et al. 2007). In addition, liver isolated NKT cells of aged mice (aged >20 months old) showed a decrease in cytotoxic effector activity and decreased release of cytokine versus young mice isolated NKT cells (aged 2 months old) (Mocchegiani et al. 2004). This decrease in the number of NKT cells and immunological functioning will intensify the development of disease by deteriorating the response of the host to pathogens and decreasing immunotolerance toward the microbiota.

In early life, the host microbiota starts the development of the immune system. Even so, the immune response must be fine-tuned and properly trained over the lifetime to maintain pace with a lifelong antigenic burden. It remains a mystery, despite these findings, how age-related immunological changes affect cellular crosstalk and general immunocompetence. Besides that, it remains to be clarified how the microbiota influences the immune system during immunosenescence. As demonstrated by illnesses such as IBD, changes to the immune system are likely to lead to an abnormal response to commensal microbes (Sun et al. 2015; Sartor and Wu 2017). Improper reactions to the native microbiota and decreased ability to regulate invasive pathogens can lead to chronic inflammation and the initiation of age-related diseases like cancer (Tilg et al. 2018).

3.7.4 Obesity Alters Gut Microbial Ecology

Obesity is described as irregular or extreme fat accumulation which will ruin health. It can be determined by calculating the body mass index (BMI) (Smith 2015). BMI is associated index of the weight relative to the height of the person that is typically a familiar reason of overweight and obesity in adults. Its formula is individual's weight in kilograms divided by the square of his height in meters (kg/m^2). As shown on Table 3.1, we can classify the adults according to their BMI into underweight, normal weight, overweight, class 1 obesity, class 2 obesity, and class 3 obesity (severe obesity). The main explanation of obesity or overweight is an energy distinction between calories consumed and calories used (Sanmiguel et al. 2015).

The increase in BMI is a major risk factor for some diseases such as heart diseases, diabetes, and some cancers.

Table 3.1 The categories of the BMI for adults who are at least 20 years old (Chaudhary et al. 2019)

BMI	Class
18.5 or under	Underweight
18.5 to <25.0	"Normal" weight
25.0 to <30.0	Overweight
30.0 to <35.0	Class 1 obesity
35.0 to <40.0	Class 2 obesity
40.0 or over	Class 3 obesity (severe obesity)

3.7.4.1 Gut Microbiome and Obesity

The bacteria in the gut belong mainly to five phyla that inhabit in the large intestine. Approximately 90% of bacterial species belong to the phyla Firmicutes and Bacteroidetes; also there are other important phyla, which are Actinobacteria, Proteobacteria, and Verrucomicrobia (Malele et al. 2018).

In the human intestine, there are many microorganisms that help us in removing the calories from the undigested polysaccharides in the diet, and this is because the constituents of the microbiota can organize a large range of glycoside hydrolases and polysaccharide lyases that do not encode in the human genome. According to certain studies, eating a polysaccharide-rich meal allows for the removal of energy to be stored in adipocytes via microorganisms of the intestinal epithelial that regulate the expression of the fasting-induced adipocyte protein which inhibits lipoprotein lipase (Ley et al. 2005).

Besten et al. (2013) mentioned that the fermentation of the polysaccharides in the diet to monosaccharides and short-chain fatty acid in the distal gut by the intestinal microbiota provokes the synthesis of the triglycerides in the liver. In the intestinal epithelium, when the microbial suppression of fasting-induced adipocyte protein occurs, this leads to decrease the level of the circulation lipoprotein lipase inhibitor and increase both the lipoprotein lipase activity and the storage of liver-derived triacylglycerols in adipocytes. Eating a high-fat and high-sugar diets changes the gut microbiome, correlated with adiposity and changes in the metabolic syndrome, as hepatic and cardiovascular diseases. Many studies found that high caloric diet, especially high-fat diet, prompted dysbiosis causing obesity-related metabolic disorders and may alter richness of the microbiota (Daniel et al. 2014).

As we mention before, the main cause of obesity is that the calorie intake is more than the consumed, but also there is a hypothesis that there is another reason that may cause the obesity which is the gut microbial ecology that affects the energy homeostasis (Riedl et al. 2017). The microbiome of the gut in the obese person may produce more storage of energy from a given diet than the thin person (Davis 2016).

Obesity has been linked to a variation in the presence of Firmicutes and Bacteroidetes in the gut microbiota, since their relative abundance has been used as a marker in the development of obesity (Castaner et al. 2018). The ratio of Firmicutes to Bacteroidetes in some animals when increased may cause adiposity and may limit the energy uptake storage by decreasing the ability to ferment polysaccharides. Some studies mentioned that there are differences between the obese and lean individuals in there gut microbiomes, as that *Roseburia* and *Mogibacterium* are significantly increased in obese individuals whereas *Anaerovorax*, *Oscillibacter*, *Pseudoflavonifractor*, and *Clostridium cluster IV* are decreased. Moreover, obesity is accompanied by high levels of short-chain fatty acids (Delzenne et al. 2020).

The time of eating, eating patterns, and variations in the diet affect gut bacterial structure, bacterial richness, and abundance in the gastrointestinal tract (Conlon and Bird 2015).

The diets that offered to lose weight in humans and are based on a high intake of protein and low intake of fermentable carbohydrate may vary microbial activity and bacterial populations in the large intestine and so influence on gut health (Diether and Willing 2019). One study mentioned that *Roseburia* spp. and *Eubacterium rectale* decreased as carbohydrate intake decreased in healthy obese individuals (Duncan et al. 2007).

Nearly all kinds of dietary fibers are fermentable, entirely or to some degree. Some fibers are quickly fermented by the colonic microbiota, whereas others are fermented more slowly. Dietary fibers may adjust microbiota richness, diversity, and metabolism (Myhrstad et al. 2020).

There is a significant role of dietary lipids on the gut microbiota composition and function, as the high-fat diet can exert adverse effects on the gut microbiota and is related to metabolic disorders (Schoeler and Caesar 2019).

Short-chain fatty acid is considered as a microbiota-related marker of obesity in humans because their colonic manufacture is associated well with the BMIs and the levels of short-chain fatty acids are suitable for the alteration of gut bacteria (Chambers et al. 2018). Despite that, short-chain fatty acids are usually known as markers of carbohydrate fermentation in the colon and are therefore normally considered as beneficial to health. Han and Hang Xiao (2020) stated that the modifications of gut microbiota in obese rodents and humans were induced by dietary intervention with numerous whole fruits and vegetables.

3.7.5 What Is Gut-Brain Axis and Microbiome-Gut-Brain Axis?

The human body is occupied with microbes (which are one of the most abundant ecosystems in life), which inhabit many parts of the human body, including the gastrointestinal tract, where a symbiotic relationship between the microbes and the human occurs, as microbes play a vital role in health and disease (Kumar and Chordia 2017). The intestinal microbiota is essential in keeping intestinal health, helps in digestion process, regulates the metabolic homeostasis by regulating the breakdown of the different nutrients, helps in energy absorption and appetite, and improves the intestinal immune system and keeps the individual from any infection. Moreover, the microbiota has an important task in controlling the manufacture and function of the central nervous system (Zheng et al. 2020), as they organize the size of dendrites, the mass of spines of the dendrites, and the thickness of the myelin sheath (Sharon et al. 2016). Gut microbiota is important to release the stress hormones in a suitable manner and restore the intestinal ecosystem to reverse irregular stress response (Foster et al. 2017).

Chakaroun et al. (2020) stated that the intestinal microbiota was considered as the intestinal barrier and regulated the passage of antigens by the para-cellular way where the gut-associated lymphoid tissue (GALT) is sited. The structure of intestinal microbiota is affected by diet and environmental stressors such as drugs and also affects the permeability of the intestinal barrier and allows the molecules to move through the para-cellular way to blood vessels, as the GALT protects against the pathogenic microorganism. Furthermore, this microbiota is able to generate hormones, immune modulators, and neurotransmitters to affect many cellular functions. Microbes are able to mediate the metabolism of the amino acid tryptophan and can synthesize dopamine, noradrenaline, and acetylcholine.

The gut microbiota can produce pro- and anti-inflammatory cytokines that can signal to the brain by circulation (Wang et al. 2020). There is a link between the brain and the gut that contain the microbiota resulting in the microbiota-gut-brain axis by the vagus nerve.

Changes in the environment of the GI tract such as disturbances to the microbiota of the gut may change its conformation or its variety and so affect the health; this change is called dysbiosis (Gagliardi et al. 2018). Dysbiosis is known as a change in structure and function of the gut microbiota. Definitely, the progression of some diseases such as autoimmune diseases, metabolic syndromes and neurological conditions related to the dysbiosis, and so this may be a great risk factor of stroke (Kumar and Wong 2020).

Liu and Dai (2020) mentioned that there are factors that induce systemic inflammation and increase the risk of cardiovascular diseases, stroke, or obesity such as the reduction of *Akkermansia* bacteria and increase levels of trimethylamine N-oxide (derived metabolite of pro-thrombotic bacteria). Changes in the gut microbiota have been accompanied with neurodegenerative diseases in addition to mood disruption and depression, so in some psychological health changes, this microbiota are noticed in many GI diseases. As shown in Table 3.2, there are changes in the microbiota and its metabolites according to different psychiatric conditions (Skonieczna-Zydecka et al. 2018). When there is activation of the gut-associated lymphoid tissue, dysbiosis takes place, causing a disturbance to the gut barrier integrity leading to increase in the intestinal permeability (Brandl and Schnabl 2015).

The communication between the cells of the intestinal barrier offers the ability to absorb and secrete certain materials that prevent the translocation of microorganisms and the diffusion of toxic substance and other destructive antigens; consequently, any acute or chronic stress may cause damage to the integrity of the intestinal barrier and changes in gut microbiota structure. Temper and psychiatric distortions besides some gastrointestinal tract ailments are associated with chronic inflammation; these inflammations with nervousness and depression affect the gut microbial composition. Any changes in the structure of the gut microbiota in turn affect the peripheral

Table 3.2 Microbiota and its metabolite alterations in various psychiatric conditions (Skonieczna-Zydecka et al. 2018)

Disease	Disease microbiota-related fingerprint
Depression	**Increase in** *Bacteroidetes*, *Proteobacteria*, *Actinobacteria*, *Enterobacteriaceae*, *Alistipes*, propionic, isobutyric, and isovaleric acids
	Decrease in *Faecalibacterium*, *Bifidobacterium*, *Lactobacillus*, serotonin, noradrenaline, short-chain fatty acid, kynurenic acid, kynurenine
Schizophrenia	**Increase in** *Coriobacteriaceae*, *Prevotella*, *Succinivibrio*, *Collinsella*, *Megasphaera*, *Klebsiella*, *Methanobrevibacter*, and *Clostridium*
	Decrease in *Blautia*, *Coprococcus*, and *Roseburia*
Bipolar disorder	**Increase in** *Bacteroides*, *Actinobacteria*, and *Coriobacteria*
	Decrease in *Faecalibacterium*, *Roseburia*, and *Alistipes*
Parkinson's disease	**Increase in** *Bacteroides* and *Roseburia*
	Decrease in *Blautia*, *Coprococcus*, *Dorea*, *Oscillospira*, and *Akkermansia*
Autism spectrum disorder	**Increase in** *Streptococcus*, Clostridiales, Comamonadaceae, *Akkermansia*, *Rhodococcus*, *Oscillospira*, *Desulfovibrio*, *Burkholderia*, *Collinsella*, *Corynebacterium*, *Dorea*, *Lactobacillus*, acetic and propionic acid, *p*-cresol, and glutamate
	Decrease in *Firmicutes*, *Faecalibacterium*, *Ruminococcus*, *Proteobacteria*, Fusobacteria, Verrucomicrobia, *Bifidobacterium*, *Neisseria*, *Alistipes*, *Bilophila*, *Dialister*, *Parabacteroides*, *Veillonella*, butyric acid, tryptophan, and kynurenic acid
Alzheimer's disease	**Increase in** *Blautia*, *Phascolarctobacterium*, *Gemella*, *E. coli*, and *Shigella*
	Decrease in Ruminococcaceae, Turicibacteraceae, Peptostreptococcaceae, Clostridiaceae, Mogibacteriaceae, and the genus *SMB53* (family, Clostridiaceae) *Dialister*, *Clostridium*, *Turicibacter*, and cc115 (family Erysipelotrichaceae)
Anorexia nervosa	**Increase in** *Methanobrevibacter smithii*

tolerance and consider risk factors of stroke and also may change the brain's reaction to the stroke (Tan et al. 2020).

3.7.5.1 Stroke Development and Microbiota Gut-Brain Axis

The main regulator of the immune system after acute stroke is the microbiota gut-brain axis (Arya and Hu 2018). Stroke (cerebral infarction) is responsible for nearly 10% of deaths all over the world, and it is most predominant in individuals over the age of 55 years (Ojaghihaghighi et al. 2017). Stroke occurs if there is an unexpected block of blood supply to a part of the brain, depriving it of oxygen and nutrients, or if there is a sudden rupture of blood vessels discharging blood into the brain, causing acute stroke and hemorrhagic stroke, respectively (Ojaghihaghighi et al. 2017). Any type of stroke causes damage to the surrounding neurons and may cause stroke. Arya and Hu (2018) found that changes in the microbial structure in the gut took place after a stroke or some neurological disorders. Transformation in the immune function and decreasing metabolism during aging allow any microbial changes in the gut of older individuals to make them have a high risk of stroke.

Moreover, having an unhealthy diet can lead to atherosclerosis, type 2 diabetes, obesity, and cardiovascular disease; all of them are risk factors of stroke.

Moreover, Yamashiro et al. (2017) found that stroke caused high incidence of *Lactobacillus ruminis* that is associated with serum levels of interleukin-6 and increased concentrations of valeric acid in the fecal metabolites and related to the level of high-sensitivity C-reactive protein and white blood cell count. From these findings, gut dysbiosis in patients with stroke is accompanied with change in the metabolism of the host. Tissue injury induced by stroke is traditionally localized and limited to the brain, but the brain works to reduce the harm caused by inflammation. Giacinto et al. (2005) revealed that antibiotics stimulate gut dysbiosis that is able to change immune homeostasis in the small intestine, causing an increase in regulatory T cells, which sequentially cause a decrease in the effector T cells from the gut to the brain after stroke and then decrease stroke infarct size. The metabolites of the mucosal microbiota affect the movement of the gut, and it can either attack the host or allow epithelial healing responses to restore gut barrier and host immunity after stroke.

References

Aagaard K, Ma J, Antony KM, Ganu R, Petrosino J, Versalovic J (2014) The placenta harbors a unique microbiome. Sci Transl Med 6(237):237ra65. https://doi.org/10.1126/scitranslmed.3008599

Adlerberth I, Wold A (2009) Establishment of the gut microbiota in Western infants. Acta Paediatr 98(2):229–238. https://doi.org/10.1111/j.1651-2227.2008.01060.x

Allam MGM, Darwish AMG, Ayad EHE et al (2016) In vivo evaluation of safety and probiotic traits of isolated Enterococcus faecium strain KT712. Res J Microbiol 11:169–177. https://doi.org/10.3923/jm.2016.169.177

Allam MGM, Darwish AMG, Ayad EHE et al (2017) Lactococcus species for conventional Karish cheese conservation. LWT Food Sci Technol 79:625–631. https://doi.org/10.1016/j.lwt.2016.11.032

Arumugam M, Raes J, Pelletier E, Le Paslier D, Yamada T, Mende DR, Bertalan M (2011) Enterotypes of the human gut microbiome. Nature 473(7346):174–180. https://doi.org/10.1038/nature09944

Arvans DL, Vavricka SR, Ren H, Musch MW, Kang L, Rocha FG et al (2005) Luminal bacterial flora determines physiological expression of intestinal epithelial cytoprotective heat shock proteins 25 and 72. Am J Physiol Gastrointest Liv Physiol 288(4):G696–G704

Arya AK, Hu B (2018) Brain–gut axis after stroke. Brain Circ 4(4):165–173

Atarashi K, Tanoue T, Ando M, Kamada N, Nagano Y, Narushima S et al (2015) Th17 cell induction by adhesion of microbes to intestinal epithelial cells. Cell 163(2):367–380

Aukováa AL, Ákováa MMAŇ, Trompfováa VS, Uwehandb ACO (2008) Probiotic potential of enterococci isolated from canine feed. Folia Microbiol 53:84–88

Avershina E, Storrø O, Øien T, Johnsen R, Pope P, Rudi K (2013) Major faecal microbiota shifts in composition and diversity with age in a geographically restricted cohort of mothers and their children. FEMS Microbiol Ecol 87(1):280–290. https://doi.org/10.1111/1574-6941.12223

Bäckhed F, Roswall J, Peng Y, Feng Q, Jia H, Kovatcheva-Datchary P, Li Y, Xia Y, Xie H, Zhong H, Khan MT (2015) Dynamics and stabilization of the human gut microbiome during the first year of life. Cell Host Microbe 17(5):690–703

Bailey SR, Nelson MH, Himes RA, Li Z, Mehrotra S, Paulos CM (2014) Th17 cells in cancer: the ultimate identity crisis. Front Immunol 5:276

Bates JM, Akerlund J, Mittge E, Guillemin K (2007) Intestinal alkaline phosphatase detoxifies lipopolysaccharide and prevents inflammation in zebrafish in response to the gut microbiota. Cell Host Microbe 2(6):371–382
Bauer H, Horowitz RE, Levenson SM, Popper H (1963) The response of the lymphatic tissue to the microbial flora. Studies on germfree mice. Am J Pathol 42(4):471
Belkaid Y, Hand TW (2014) Role of the microbiota in immunity and inflammation. Cell 157 (1):121–141. https://doi.org/10.1016/j.cell.2014.03.011
Besten G, Eunen K, Groen AK, Koen Venema K, Dirk-Jan Reijngoud D, Bakker BM (2013) The role of short-chain fatty acids in the interplay between diet, gut microbiota, and host energy metabolism. J Lipid Res 54(9):2325–2340
Bhardwaj A, Malik RK, Chauhan P (2008) Functional and safety aspects of enterococci in dairy foods. Indian J Microbiol 48:317–325. https://doi.org/10.1007/s12088-008-0041-2
Bhattacharya T, Ghosh T, Mande S (2015) Global profiling of carbohydrate active enzymes in human gut microbiome. PLoS One 10(11):e0142038. https://doi.org/10.1371/journal.pone.0142038
Biesalski H (2016) Nutrition meets the microbiome: micronutrients and the microbiota. Ann N Y Acad Sci 1372(1):53–64. https://doi.org/10.1111/nyas.13145
Blencowe H, Cousens S, Chou D, Oestergaard M, Say L, Moller A et al (2013) Born too soon: the global epidemiology of 15 million preterm births. Reprod Health 10(Suppl 1):S2. https://doi.org/10.1186/1742-4755-10-s1-s2
Boudry G, David ES, Douard V, Monteiro IM, Le Huërou-Luron I, Ferraris RP (2010) Role of intestinal transporters in neonatal nutrition: carbohydrates, proteins, lipids, minerals, and vitamins. J Pediatr Gastroenterol Nutr 51(4):380–401
Bour-Jordan H, Bluestone JA (2002) CD28 function: a balance of costimulatory and regulatory signals. J Clin Immunol 22(1):1–7
Bouskra D, Brézillon C, Bérard M, Werts C, Varona R, Boneca IG, Eberl G (2008) Lymphoid tissue genesis induced by commensals through NOD1 regulates intestinal homeostasis. Nature 456(7221):507–510
Brandl K, Schnabl B (2015) Is intestinal inflammation linking dysbiosis to gut barrier dysfunction during liver disease? Expert Rev Gastroenterol Hepatol 9(8):1069–1076
Buffie CG, Pamer EG (2013) Microbiota-mediated colonization resistance against intestinal pathogens. Nat Rev Immunol 13(11):790–801
Burlington S, Classification P (2006) Symbiotic food products comprising oats and methods for manufacturing the same. Patent No. US 2006/0141097 A1. 1–16
Carver JD, Pimentel B, Barness LA, Wiener DA, Lowell NE (1991) Infant feeding effects on flow cytometric analysis of blood. J Clin Lab Anal 5(1):54–56
Castaner O, Goday A, Park Y, Lee S, Magkos F, Shiow STE, Schröder H (2018) The gut microbiome profile in obesity: a systematic review. Int J Endocrinol 2018:4095789. https://doi.org/10.1155/2018/4095789
Castelo-Branco C, Soveral I (2014) The immune system and aging: a review. Gynecol Endocrinol 30(1):16–22
Chakaroun RM, Massier L, Kovacs P (2020) Gut microbiome, intestinal permeability, and tissue bacteria in metabolic disease: perpetrators or bystanders? Nutrients 12(4):1082. https://doi.org/10.3390/nu12041082
Chambers ES, Preston T, Frost G, Morrison DJ (2018) Role of gut microbiota-generated short-chain fatty acids in metabolic and cardiovascular health. Curr Nutr Rep 7:198–206
Charalampopoulos D, Rastall RA (eds) (2009) Prebiotics and probiotics science and technology, vol 1. Springer Science and Business Media, New York, NY
Chassin C, Kocur M, Pott J, Duerr CU, Gütle D, Lotz M, Hornef MW (2010) miR-146a mediates protective innate immune tolerance in the neonate intestine. Cell Host Microbe 8(4):358–368
Chaudhary S, Contag S, Yao R (2019) The impact of maternal body mass index on external cephalic version success. J Matern Fetal Neonatal Med 32(13):2159–2165

Claesson M, Cusack S, O'Sullivan O, Greene-Diniz R, de Weerd H, Flannery E et al (2010) Composition, variability, and temporal stability of the intestinal microbiota of the elderly. Proc Natl Acad Sci 108(Suppl 1):4586–4591. https://doi.org/10.1073/pnas.1000097107
Clavel T, Gomes-Neto JC, Lagkouvardos I, Ramer-Tait AE (2017) Deciphering interactions between the gut microbiota and the immune system via microbial cultivation and minimal microbiomes. Immunol Rev 279(1):8–22
Collado MC, Delgado S, Maldonado A, Rodríguez JM (2009) Assessment of the bacterial diversity of breast milk of healthy women by quantitative real-time PCR. Lett Appl Microbiol 48 (5):523–528
Collado MC, Rautava S, Aakko J, Isolauri E, Salminen S (2016) Human gut colonisation may be initiated in utero by distinct microbial communities in the placenta and amniotic fluid. Sci Rep 6:23129
Cong X, Xu W, Janton S, Henderson WA, Matson A, McGrath JM, Maas K, Graf J (2016) Gut microbiome developmental patterns in early life of preterm infants: impacts of feeding and gender. PLoS One 11(4):e0152751
Conlon MA, Bird AR (2015) The impact of diet and lifestyle on gut microbiota and human health. Nutrients 7:17–44
Costello EK, Stagaman K, Dethlefsen L, Bohannan BJ, Relman DA (2012) The application of ecological theory toward an understanding of the human microbiome. Science 336 (6086):1255–1262
Crane JD, Palanivel R, Mottillo EP, Bujak AL, Wang H, Ford RJ, Collins A, Blümer RM, Fullerton MD, Yabut JM, Kim JJ, Ghia JE, Hamza SM, Morrison KM, Schertzer JD, Dyck JR, Khan WI, Steinberg GR (2015) Inhibiting peripheral serotonin synthesis reduces obesity and metabolic dysfunction by promoting brown adipose tissue thermogenesis. Nat Med 21(2):166–172. https://doi.org/10.1038/nm.3766
Cummings JH, Pomare EW, Branch WJ, Naylor CP, Macfarlane GT (1987) Short chain fatty acids in human large intestine, portal, hepatic and venous blood. Gut 28(10):1221–1227
Damsker JM, Hansen AM, Caspi RR (2010) Th1 and Th17 cells: adversaries and collaborators. Ann N Y Acad Sci 1183:211
Daniel H, Gholami AM, Berry D, Desmarchelier C, Hahne H, Loh G, Mondot S, Lepage P, Rothballer M, Walker A, Bohm C, Wenning M, Wagner M, Blaut M, Schmitt-Kopplin P, Kuster B, Haller D, Clavel T (2014) High-fat diet alters gut microbiota physiology in mice. ISME J 8:295–308
Darwish AMG, Allam MG, Ayad EH (2018) Physicochemical profile and Lactic Acid Bacteria genera inhabit Egyptian raw camel, sheep, goat, buffalo and cow milks. Microb Biosyst J 3:12–24. https://doi.org/10.21608/mb.2018.12355
Darwish AMG, Ayad EHE, El-Souda MA, Darwish SM (2020) Symbiotic dairy products contain Talbina and methods of manufacturing. Patent No. EG 2020/29742. 1–45
Davis CD (2016) The gut microbiome and its role in obesity. Nutr Today 51(4):167–174
De La Plaza M, Rodríguez A, Fernández De Palencia P et al (2006) Discrepancies between the phenotypic and genotypic characterization of Lactococcus lactis cheese isolates. Lett Appl Microbiol 43:637–644. https://doi.org/10.1111/j.1472-765X.2006.02012.x
Degen L, Phillips S (1996) Variability of gastrointestinal transit in healthy women and men. Gut 39 (2):299–305. https://doi.org/10.1136/gut.39.2.299
DelaRosa O, Tarazona R, Casado JG, Alonso C, Ostos B, Peña J, Solana R (2002) Vα24+ NKT cells are decreased in elderly humans. Exp Gerontol 37(2–3):213–217
Delzenne NM, Rodriguez J, Olivares M, Neyrinck AM (2020) Microbiome response to diet: focus on obesity and related diseases. Rev Endocr Metab Disord 21:369–380
Diether NE, Willing BP (2019) Microbial fermentation of dietary protein: an important factor in diet–microbe–host interaction. Microorganisms 7:19. https://doi.org/10.3390/microorganisms7010019
DiGiulio D, Romero R, Amogan H, Kusanovic J, Bik E, Gotsch F et al (2008) Microbial prevalence, diversity and abundance in amniotic fluid during preterm labor: a molecular and

culture-based investigation. PLoS One 3(8):e3056. https://doi.org/10.1371/journal.pone.0003056

Dominguez-Bello MG, De Jesus-Laboy KM, Shen N, Cox LM, Amir A, Gonzalez A, Bokulich NA, Song SJ, Hoashi M, Rivera-Vinas JI, Mendez K (2016) Partial restoration of the microbiota of cesarean-born infants via vaginal microbial transfer. Nat Med 22(3):250

Donaldson GP, Ladinsky MS, Yu KB, Sanders JG, Yoo BB, Chou WC, Conner ME, Earl AM, Knight R, Bjorkman PJ, Mazmanian SK (2018) Gut microbiota utilize immunoglobulin A for mucosal colonization. Science 360(6390):795–800

Donnet-Hughes A (2008) Protective properties of human milk. In: Duggan D (ed) Nutrition in pediatrics, 4th edn. Canada Decker Publishing, Toronto, ON, pp 355–362

Dumont Y, Fournier A, St-Pierre S, Quirion R (1995) Characterization of neuropeptide Y binding sites in rat brain membrane preparations using [125I][Leu31,Pro34]peptide YY and [125I] peptide YY3-36 as selective Y1 and Y2 radioligands. J Pharmacol Exp Ther 272(2):673–680

Dunn AB, Jordan S, Baker BJ, Carlson NS (2017) The maternal infant microbiome: considerations for labor and birth. MCN. Am J Matern Child Nurs 42(6):318

Duncan SH, Belenguer A, Holtrop G, Alexandra M, Johnstone AM, Harry J, Flint HJ, Lobley GE (2007) Reduced dietary intake of carbohydrates by obese subjects results in decreased concentrations of butyrate and butyrate-producing bacteria in feces. Appl Environ Microbiol 73(4):1073–1078

Edwards PT, Kashyap PC, Preidis GA (2020) Microbiota on biotics: probiotics, prebiotics, and synbiotics to optimize growth and metabolism. Am J Physiol Liv Physiol 19:382–390. https://doi.org/10.1152/ajpgi.00028.2020

Ege MJ, Mayer M, Normand AC, Genuneit J, Cookson WO, Braun-Fahrländer C et al (2011) Exposure to environmental microorganisms and childhood asthma. N Engl J Med 364 (8):701–709

Eibel H, Kraus H, Sic H, Kienzler AK, Rizzi M (2014) B cell biology: an overview. Curr Allergy Asthma Rep 14(5):434

Eiwegger T, Stahl B, Haidl P, Schmitt J, Boehm G, Dehlink E et al (2010) Prebiotic oligosaccharides: in vitro evidence for gastrointestinal epithelial transfer and immunomodulatory properties. Pediatr Allergy Immunol 21(8):1179–1188

Elahi S, Ertelt JM, Kinder JM, Jiang TT, Zhang X, Xin L et al (2013) Immunosuppressive CD71+ erythroid cells compromise neonatal host defence against infection. Nature 504(7478):158–162

El-ghaish S, Khalifa M, Elmahdy A (2016) Antimicrobial impact for Lactococcus Lactis subsp. lactis A15 and Enterococcus Faecium A15 isolated from some traditional Egyptian dairy products on some pathogenic bacteria. J Food Biochem 41:e12279. https://doi.org/10.1111/jfbc.12279

Everard A, Matamoros S, Geurts L, Delzenne NM, Cani PD (2014) Saccharomyces boulardii administration changes gut microbiota and reduces hepatic steatosis, low-grade inflammation, and fat mass in obese and type 2 diabetic db/db mice. MBio 5(3):e01011

Fan P, Li L, Rezaei A, Eslamfam S, Che D, Ma X (2015) Metabolites of dietary protein and peptides by intestinal microbes and their impacts on gut. Curr Protein Pept Sci 16(7):646–654. https://doi.org/10.2174/1389203716666150630133657

Favier C, de Vos W, Akkermans A (2003) Development of bacterial and bifidobacterial communities in feces of newborn babies. Anaerobe 9:219–229

Felicio TL, Esmerino EA, Vidal VAS, Cappato LP, Garcia RKA, Cavalcanti RN, Freitas MQ, Conte Junior CA, Padilha MC, Silvaa MC, Raices RSL, Arellano DB, Bollini HMA, Pollonio MAR, Cruz AG (2016) Physico-chemical changes during storage and sensory acceptance of low sodium probiotic Minas cheese added with arginine. Food Chem 196:628–637. https://doi.org/10.1016/j.foodchem.2015.09.102

Fernández L, Langa S, Martín V, Maldonado A, Jiménez E, Martín R, Rodríguez JM (2013) The human milk microbiota: origin and potential roles in health and disease. Pharmacol Res 69 (1):1–10

Ferrarese R, Ceresola ER, Preti A, Canducci F (2018) Probiotics, prebiotics and synbiotics for weight loss and metabolic syndrome in the microbiome era. Eur Rev Med Pharmacol Sci 23:7–9. https://doi.org/10.26355/eurrev

Fiebiger U, Bereswill S, Heimesaat MM (2016) Dissecting the interplay between intestinal microbiota and host immunity in health and disease: lessons learned from germfree and gnotobiotic animal models. Eur J Microbiol Immunol 6(4):253–271

Forslund K, Sunagawa S, Kultima JR, Mende DR, Arumugam M, Typas A, Bork P (2013) Country-specific antibiotic use practices impact the human gut resistome. Genome Res 23(7):1163–1169

Foster JA, Rinaman L, Cryanc JF (2017) Stress and the gut-brain axis: regulation by the microbiome. Neurobiol Stress 7:124–136

Frank DN, Amand ALS, Feldman RA, Boedeker EC, Harpaz N, Pace NR (2007) Molecular-phylogenetic characterization of microbial community imbalances in human inflammatory bowel diseases. Proc Natl Acad Sci 104(34):13780–13785

Gagliardi A, Totino V, Cacciotti F, Iebba V, Neroni B, Bonfiglio G, Trancassini M, Passariello C, Pantanella F, Schippa S (2018) Rebuilding the gut microbiota ecosystem. Int J Environ Res Public Health 15(8):1679. https://doi.org/10.3390/ijerph15081679

Gálvez J (2014) Role of Th17 cells in the pathogenesis of human IBD. ISRN Inflamm 2014:928461. https://doi.org/10.1155/2014/928461

Gensollen T, Iyer SS, Kasper DL, Blumberg RS (2016) How colonization by microbiota in early life shapes the immune system. Science 352(6285):539–544

Giacinto CD, Marinaro M, Sanchez M, Strober W, Boirivant M (2005) Probiotics ameliorate recurrent th1-mediated murine colitis by inducing il-10 and il-10-dependent tgf-β-bearing regulatory cells. J Immunol 174(6):3237–3246

Gibson KL, Wu YC, Barnett Y, Duggan O, Vaughan R, Kondeatis E et al (2009) B-cell diversity decreases in old age and is correlated with poor health status. Aging Cell 8(1):18–25

Gill SR, Pop M, DeBoy RT, Eckburg PB, Turnbaugh PJ, Samuel BS et al (2006) Metagenomic analysis of the human distal gut microbiome. Science 312(5778):1355–1359

Goldenberg R, Culhane J, Iams J, Romero R (2008) Epidemiology and causes of preterm birth. Lancet 371(9606):75–84. https://doi.org/10.1016/s0140-6736(08)60074-4

Goodrich J, Waters J, Poole A, Sutter J, Koren O, Blekhman R et al (2014) Human genetics shape the gut microbiome. Cell 159(4):789–799. https://doi.org/10.1016/j.cell.2014.09.053

Gosalbes MJ, Llop S, Valles Y, Moya A, Ballester F, Francino MP (2013) Meconium microbiota types dominated by lactic acid or enteric bacteria are differentially associated with maternal eczema and respiratory problems in infants. Clin Exp Allergy 43(2):198–211

Guarner F, Malagelada JR (2003) Gut flora in health and disease. Lancet 361(9356):512–519

Gueimonde M, Sakata S, Kalliomäki M, Isolauri E, Benno Y, Salminen S (2006) Effect of maternal consumption of lactobacillus GG on transfer and establishment of fecal bifidobacterial microbiota in neonates. J Pediatr Gastroenterol Nutr 42(2):166–170. https://doi.org/10.1097/01.mpg.0000189346.25172.fd

Guinane CM, Cotter PD (2013) Role of the gut microbiota in health and chronic gastrointestinal disease: understanding a hidden metabolic organ. Ther Adv Gastroenterol 6(4):295–308

Hamada H, Hiroi T, Nishiyama Y, Takahashi H, Masunaga Y, Hachimura S et al (2002) Identification of multiple isolated lymphoid follicles on the antimesenteric wall of the mouse small intestine. J Immunol 168(1):57–64

Han Y, Hang Xiao H (2020) Whole food–based approaches to modulating gut microbiota and associated diseases. Annu Rev Food Sci Technol 11:14.1–14.25

Hanage WP (2014) Microbiology: microbiome science needs a healthy dose of scepticism. Nat News 512(7514):247

Hillman E, Lu H, Yao T, Nakatsu C (2017) Microbial ecology along the gastrointestinal tract. Microbes Environ 32(4):300–313. https://doi.org/10.1264/jsme2.me17017

Hooper LV, Wong MH, Thelin A, Hansson L, Falk PG, Gordon JI (2001) Molecular analysis of commensal host-microbial relationships in the intestine. Science 291(5505):881–884

Hsieh CS, Macatonia SE, Tripp CS, Wolf SF, O'Garra A, Murphy KM (1993) Development of TH1 CD4+ T cells through IL-12 produced by Listeria-induced macrophages. Science 260 (5107):547–549

Hunt KM, Preuss J, Nissan C, Davlin CA, Williams JE, Shafii B, McGuire MA (2012) Human milk oligosaccharides promote the growth of staphylococci. Appl Environ Microbiol 78 (14):4763–4770

Husebye E, Hellström PM, Midtvedt T (1994) Intestinal microflora stimulates myoelectric activity of rat small intestine by promoting cyclic initiation and aboral propagation of migrating myoelectric complex. Dig Dis Sci 39(5):946–956

Hyrslova I, Krausova G, Bartova J, Kolesar L (2016) Characterization of Enterococcus faecium CCDM 922 in respect of its technological and probiotic properties. Int J Curr Microbiol App Sci 5:474–482

Ivanov II, Atarashi K, Manel N, Brodie EL, Shima T, Karaoz U et al (2009) Induction of intestinal Th17 cells by segmented filamentous bacteria. Cell 139(3):485–498

Jantscher-Krenn E, Marx C, Bode L (2013) Human milk oligosaccharides are differentially metabolized in neonatal rats. Br J Nutr 110(4):640

Jesus ALT, Fernandes MS, Kamimura BA, Prado-Silva L, Silva R, Esmerino EA, Cruz AG, Sant' Ana AS (2016) Growth potential of Listeria monocytogenes in probiotic cottage cheese formulations with reduced sodium content formulations with reduced sodium content. Food Res Int J 81:180–187. https://doi.org/10.1016/j.foodres.2015.12.030

Jiménez E, Fernández L, Marín M, Martín R, Odriozola J, Nueno-Palop C et al (2005) Isolation of commensal bacteria from umbilical cord blood of healthy neonates born by cesarean section. Curr Microbiol 51(4):270–274. https://doi.org/10.1007/s00284-005-0020-3

Jin B, Sun T, Yu XH, Yang YX, Yeo AE (2012) The effects of TLR activation on T-cell development and differentiation. Clin Dev Immunol 2012:836485

Jing Y, Gravenstein S, Chaganty NR, Chen N, Lyerly KH, Joyce S, Deng Y (2007) Aging is associated with a rapid decline in frequency, alterations in subset composition, and enhanced Th2 response in CD1d-restricted NKT cells from human peripheral blood. Exp Gerontol 42 (8):719–732

Josefowicz SZ, Niec RE, Kim HY, Treuting P, Chinen T, Zheng Y et al (2012) Extrathymically generated regulatory T cells control mucosal TH 2 inflammation. Nature 482(7385):395–399

Kamada N, Chen GY, Inohara N, Núñez G (2013) Control of pathogens and pathobionts by the gut microbiota. Nat Immunol 14(7):685–690

Karlstrom A, Lindgren H (2013) Maternal and infant outcome after caesarean section without recorded medical indication: findings from a Swedish case. Int J Obstet Gynaecol 120:479

Kau AL, Ahern PP, Griffin NW, Goodman AL, Gordon JI (2011) Human nutrition, the gut microbiome and the immune system. Nature 474(7351):327–336

Kiyono HIROSHI, McGHEE JR, Wannemuehler MJ, Michalek SM (1982) Lack of oral tolerance in C3H/HeJ mice. J Exp Med 155(2):605–610

Klaassens ES, De Vos WM, Vaughan EE (2007) Metaproteomics approach to study the functionality of the microbiota in the human infant gastrointestinal tract. Appl Environ Microbiol 73 (4):1388–1392

Kleessen B, Bunke H, Tovar K, Noack J, Sawatzki G (1995) Influence of two infant formulas and human milk on the development of the faecal flora in newborn infants. Acta Paediatr 84:1347–1356

Knol J, Scholtens P, Kafka C, Steenbakkers J, Gro S, Helm K, Klarczyk M, Schöpfer H, Böckler HM, Wells J (2005) Colon microflora in infants fed formula with galacto-and fructo-oligosaccharides: more like breast-fed infants. J Pediatr Gastroenterol Nutr 40(1):36–42

Koenig J, Spor A, Scalfone N, Fricker A, Stombaugh J, Knight R et al (2010) Succession of microbial consortia in the developing infant gut microbiome. Proc Natl Acad Sci 108(Suppl 1):4578–4585. https://doi.org/10.1073/pnas.1000081107

Koenig JE, Spor A, Scalfone N, Fricker AD, Stombaugh J, Knight R, Angenent LT, Ley RE (2011) Succession of microbial consortia in the developing infant gut microbiome. Proc Natl Acad Sci 108(Suppl 1):4578–4585

Kollmann TR, Levy O, Montgomery RR, Goriely S (2012) Innate immune function by Toll-like receptors: distinct responses in newborns and the elderly. Immunity 37(5):771–783

Koropatkin NM, Cameron EA, Martens EC (2012) How glycan metabolism shapes the human gut microbiota. Nat Rev Microbiol 10(5):323–335

Kristensen K, Henriksen L (2016) Cesarean section and disease associated with immune function. J Allergy Clin Immunol 137(2):587–590

Kumar A, Chordia N (2017) Role of microbes in human health. Appl Microbiol Open Access 3:131. https://doi.org/10.4172/2471-9315.1000131

Kumar KP, Wong CHY (2020) Imbalance in the force: the dark side of the microbiota on stroke risk and progression. Curr Opin Neurobiol 62:10–16

Kundu P, Blacher E, Elinav E, Pettersson S (2017) Our gut microbiome: the evolving inner self. Cell 171(7):1481–1493

La Rosa PS, Warner BB, Zhou Y, Weinstock GM, Sodergren E, Hall-Moore CM, Stevens HJ, Bennett WE, Shaikh N, Linneman LA, Hoffmann JA (2014) Patterned progression of bacterial populations in the premature infant gut. Proc Natl Acad Sci 111(34):12522–12527

Larraufie P, Martin-Gallausiaux C, Lapaque N, Dore J, Gribble FM, Reimann F, Blottiere HM (2018) SCFAs strongly stimulate PYY production in human enteroendocrine cells. Sci Rep 8 (1):74. https://doi.org/10.1038/s41598-017-18259-0

Laterza L, Rizzatti G, Gaetani E, Chiusolo P, Gasbarrini A (2016) The gut microbiota and immune system relationship in human graft-versus-host disease. Medit J Hematol Infect Dis 8:2016025. https://doi.org/10.4084/mjhid.2016.025

Latuga MS, Stuebe A, Seed PC (2014) A review of the source and function of microbiota in breast milk. Semin Reprod Med 32(1):68–73. https://doi.org/10.1055/s-0033-1361824

Le Huërou-Luron I, Blat S, Boudry G (2010) Breast-v. formula-feeding: impacts on the digestive tract and immediate and long-term health effects. Nutr Res Rev 23(1):23–36

Ley RE, Backhed F, Turnbaugh P, Lozupone CA, Knight RD, Gordon JI (2005) Obesity alters gut microbial ecology. PNAS 102(31):11070–11075

Ley RE, Hamady M, Lozupone C, Turnbaugh PJ, Ramey RR, Bircher JS, Schlegel ML, Tucker TA, Schrenzel MD, Knight R, Gordon JI (2008) Evolution of mammals and their gut microbes. Science 320(5883):1647–1651

Li J, Jia H, Cai X, Zhong H, Feng Q, Sunagawa S et al (2014) An integrated catalog of reference genes in the human gut microbiome. Nat Biotechnol 32(8):834–841. https://doi.org/10.1038/nbt.2942

Liedel JL, Guo Y, Yu Y, Shiou SR, Chen S, Petrof EO et al (2011) Mother's milk-induced Hsp70 expression preserves intestinal epithelial barrier function in an immature rat pup model. Pediatr Res 69(7):395–400

Lim ES, Wang D, Holtz LR (2016) The bacterial microbiome and virome milestones of infant development. Trends Microbiol 24(10):801–810

Linton PJ, Dorshkind K (2004) Age-related changes in lymphocyte development and function. Nat Immunol 5(2):133–139

Liu Y, Dai M (2020) Trimethylamine N-oxide generated by the gut microbiota is associated with vascular inflammation: new insights into atherosclerosis. Mediat Inflamm 2020:4634172. https://doi.org/10.1155/2020/4634172

Locascio R et al (2007) Glycoprofiling of bifidobacterial consumption of human milk oligosaccharides demonstrate strain specific, preferential consumption of small chain glycans secreted in early human lactation. J Agric Food Chem 55:8914–8915

LoCascio RG, Niñonuevo MR, Kronewitter SR, Freeman SL, German JB, Lebrilla CB, Mills DA (2009) A versatile and scalable strategy for glycoprofiling bifidobacterial consumption of human milk oligosaccharides. Microb Biotechnol 2(3):333–342

Lotz M, Gütle D, Walther S, Ménard S, Bogdan C, Hornef MW (2006) Postnatal acquisition of endotoxin tolerance in intestinal epithelial cells. J Exp Med 203(4):973–984

Lynch S, Pedersen O (2016) The human intestinal microbiome in health and disease. N Engl J Med 375(24):2369–2379. https://doi.org/10.1056/nejmra1600266

Macpherson AJ, Hunziker L, McCoy K, Lamarre A (2001) IgA responses in the intestinal mucosa against pathogenic and non-pathogenic microorganisms. Microbes Infect 3(12):1021–1035

Maftei N-M (2012) Probiotic, prebiotic and synbiotic products in human health. In: Frontiers and new trends in the science of fermented food and beverages amounts. Intech Open, London, pp 1–19

Malele I, Nyingilili H, Lyaruu E, Tauzin M, Bernard Ollivier B, Cayol JL, Fardeau ML, Geiger A (2018) Bacterial diversity obtained by culturable approaches in the gut of Glossina pallidipes population from a non-sleeping sickness focus in Tanzania: preliminary results. BMC Microbiol 18(1):164

Mallett A, Bearne C, Rowland I (1983) Metabolic activity and enzyme induction in rat fecal microflora maintained in continuous culture. Appl Environ Microbiol 46(3):591–595. https://doi.org/10.1128/aem.46.3.591-595.1983

Marcobal A, Sonnenburg JL (2012) Human milk oligosaccharide consumption by intestinal microbiota. Clin Microbiol Infect 18:12–15

Marcobal A, Barboza M, Froehlich JW, Block DE, German JB, Lebrilla CB, Mills DA (2010) Consumption of human milk oligosaccharides by gut-related microbes. J Agric Food Chem 58 (9):5334–5340

Marcobal A, Barboza M, Sonnenburg ED, Pudlo N, Martens EC, Desai P, Sonnenburg JL (2011) Bacteroides in the infant gut consume milk oligosaccharides via mucus-utilization pathways. Cell Host Microbe 10(5):507–514

Marques T, Wall R, Ross R, Fitzgerald G, Ryan C, Stanton C (2010) Programming infant gut microbiota: influence of dietary and environmental factors. Curr Opin Biotechnol 21 (2):149–156. https://doi.org/10.1016/j.copbio.2010.03.020

Martín R, Langa S, Reviriego C, Jimínez E, Marín M, Xaus J et al (2003) Human milk is a source of lactic acid bacteria for the infant gut. J Pediatr 143(6):754–758. https://doi.org/10.1016/j.jpeds.2003.09.028

Martin R, Makino H, Cetinyurek Yavuz A, Ben-Amor K, Roelofs M, Ishikawa E, Kubota H, Swinkels S, Sakai T, Oishi K, Kushiro A (2016) Early-life events, including mode of delivery and type of feeding, siblings and gender, shape the developing gut microbiota. PLoS One 11(6): e0158498

Martin AM, Sun EW, Rogers GB, Keating DJ (2019) The influence of the gut microbiome on host metabolism through the regulation of gut hormone release. Front Physiol 10:428. https://doi.org/10.3389/fphys.2019.00428

Martinez KB, Pierre JF, Chang EB (2016) The gut microbiota: the gateway to improved metabolism. Gastroenterol Clin 45(4):601–614

Martinez-Guryn K, Hubert N, Frazier K, Urlass S, Musch M, Ojeda P et al (2018) Small intestine microbiota regulate host digestive and absorptive adaptive responses to dietary lipids. Cell Host Microbe 23(4):458–469.e5. https://doi.org/10.1016/j.chom.2018.03.011

Matsuki T, Yahagi K, Mori H, Matsumoto H, Hara T, Tajima S et al (2016) A key genetic factor for fucosyllactose utilization affects infant gut microbiota development. Nat Commun 7(1):11939. https://doi.org/10.1038/ncomms11939

Mazmanian SK, Liu CH, Tzianabos AO, Kasper DL (2005) An immunomodulatory molecule of symbiotic bacteria directs maturation of the host immune system. Cell 122(1):107–118

Mennigen R, Nolte K, Rijcken E, Utech M, Loeffler B, Senninger N, Bruewer M (2009) Probiotic mixture VSL# 3 protects the epithelial barrier by maintaining tight junction protein expression and preventing apoptosis in a murine model of colitis. Am J Physiol Gastrointest Liv Physiol 296:G1140

Milani C, Duranti S, Bottacini F, Casey E, Turroni F, Mahony J, Belzer C, Palacio SD, Montes SA, Mancabelli L, Lugli GA (2017) The first microbial colonizers of the human gut: composition, activities, and health implications of the infant gut microbiota. Microbiol Mol Biol Rev 81(4): e00036

Mitsuoka T (2014) Development of functional foods. Biosci Microb Food Health 33(3):117–128. https://doi.org/10.12938/bmfh.33.117

Mocchegiani E, Giacconi R, Cipriano C, Gasparini N, Bernardini G, Malavolta M et al (2004) The variations during the circadian cycle of liver CD1d-unrestricted NK1. 1+ TCRγ/δ+ cells lead to successful ageing. Role of metallothionein/IL-6/gp130/PARP-1 interplay in very old mice. Exp Gerontol 39(5):775–788

Morales P, Fujio S, Navarrete P, Ugalde J, Magne F, Carrasco-Pozo C et al (2016) Impact of dietary lipids on colonic function and microbiota: an experimental approach involving orlistat-induced fat malabsorption in human volunteers. Clin Transl Gastroenterol 7(4):e161. https://doi.org/10.1038/ctg.2016.20

Mucida D, Kutchukhidze N, Erazo A, Russo M, Lafaille JJ, de Lafaille MAC (2005) Oral tolerance in the absence of naturally occurring Tregs. J Clin Invest 115(7):1923–1933

Myhrstad MCW, Tunsjo H, Charnock C, Telle-Hansen VH (2020) Dietary fiber, gut microbiota, and metabolic regulation—current status in human randomized trials. Nutrients 12(3):859

Ng KM, Ferreyra JA, Higginbottom SK, Lynch JB, Kashyap PC, Gopinath S, Naidu N, Choudhury B, Weimer BC, Monack DM, Sonnenburg JL (2013) Microbiota-liberated host sugars facilitate post-antibiotic expansion of enteric pathogens. Nature 502(7469):96–99. https://doi.org/10.1038/nature12503

Ninonuevo M et al (2006) A strategy for annotating the human milk glycome. J Agric Food Chem 54:7471–7480

Ohnmacht C, Marques R, Presley L, Sawa S, Lochner M, Eberl G (2011) Intestinal microbiota, evolution of the immune system and the bad reputation of pro-inflammatory immunity. Cell Microbiol 13(5):653–659

Ojaghihaghighi S, Vahdati SS, Mikaeilpour A, Ramouz A (2017) Comparison of neurological clinical manifestation in patients with hemorrhagic and ischemic stroke. World J Emerg Med 8 (1):34–38

Olszak T, An D, Zeissig S, Vera MP, Richter J, Franke A et al (2012) Microbial exposure during early life has persistent effects on natural killer T cell function. Science 336(6080):489–493

Ozdal T, Sela D, Xiao J, Boyacioglu D, Chen F, Capanoglu E (2016) The reciprocal interactions between polyphenols and gut microbiota and effects on bioaccessibility. Nutrients 8(2):78. https://doi.org/10.3390/nu8020078

Pacheco AR, Curtis MM, Ritchie JM, Munera D, Waldor MK, Moreira CG, Sperandio V (2012) Fucose sensing regulates bacterial intestinal colonization. Nature 492(7427):113–117. https://doi.org/10.1038/nature11623

Pajarillo EAB, Chae JP, Balolong MP et al (2015) Effects of probiotic Enterococcus faecium NCIMB 11181 administration on swine fecal microbiota diversity and composition using barcoded pyrosequencing. Anim Feed Sci Technol 201:80–88. https://doi.org/10.1016/j.anifeedsci.2015.01.011

Palmer C, Bik E, DiGiulio D, Relman D, Brown P (2007) Development of the human infant intestinal microbiota. PLoS Biol 5(7):e177. https://doi.org/10.1371/journal.pbio.0050177

Perez VL, Van Parijs L, Biuckians A, Zheng XX, Strom TB, Abbas AK (1997) Induction of peripheral T cell tolerance in vivo requires CTLA-4 engagement. Immunity 6(4):411–417

Perez PF, Doré J, Leclerc M, Levenez F, Benyacoub J, Serrant P et al (2007) Bacterial imprinting of the neonatal immune system: lessons from maternal cells? Pediatrics 119(3):e724–e732

Perez-Muñoz ME, Arrieta MC, Ramer-Tait AE, Walter J (2017) A critical assessment of the "sterile womb" and "in utero colonization" hypotheses: implications for research on the pioneer infant microbiome. Microbiome 5(1):48

Pickard JM, Zeng MY, Caruso R, Núñez G (2017) Gut microbiota: role in pathogen colonization, immune responses, and inflammatory disease. Immunol Rev 279(1):70–89

PrabhuDas M, Adkins B, Gans H, King C, Levy O, Ramilo O, Siegrist CA (2011) Challenges in infant immunity: implications for responses to infection and vaccines. Nat Immunol 12 (3):189–194

Qin J, Li R, Raes J, Arumugam M, Burgdorf KS, Manichanh C, Nielsen T, Pons N, Levenez F, Yamada T, Mende DR (2010) A human gut microbial gene catalogue established by metagenomic sequencing. Nature 464(7285):59–65

Qin J, Li Y, Cai Z, Li S, Zhu J, Zhang F, Liang S, Zhang W, Guan Y, Shen D, Peng Y (2012) A metagenome-wide association study of gut microbiota in type 2 diabetes. Nature 490 (7418):55–60

Quinto EJ, Jiménez P, Caro I et al (2014) Probiotic lactic acid bacteria: a review. Food Nutr Sci 05:1765–1775. https://doi.org/10.4236/fns.2014.518190

Rajilic-Stojanovic M (2007) Diversity of the human gastrointestinal microbiota: novel perspectives from high throughput analyses

Rajilić-Stojanović M, Smidt H, de Vos W (2007) Diversity of the human gastrointestinal tract microbiota revisited. Environ Microbiol 9(9):2125–2136. https://doi.org/10.1111/j.1462-2920.2007.01369.x

Rakoff-Nahoum S, Paglino J, Eslami-Varzaneh F, Edberg S, Medzhitov R (2004) Recognition of commensal microflora by toll-like receptors is required for intestinal homeostasis. Cell 118:229–241

Rautava S, Collado M, Salminen S, Isolauri E (2012a) Probiotics modulate host-microbe interaction in the placenta and fetal gut: a randomized, double-blind, placebo-controlled trial. Neonatology 102(3):178–184. https://doi.org/10.1159/000339182

Rautava S, Luoto R, Salminen S, Isolauri E (2012b) Microbial contact during pregnancy, intestinal colonization and human disease. Nat Rev Gastroenterol Hepatol 9(10):565

Reiter RJ (1991) Pineal melatonin: cell biology of its synthesis and of its physiological interactions. Endocr Rev 12(2):151–180

Richmond J (2000) Scand J Lab Anim Sci 27:84–92

Riedl RA, Atkinson SN, Burnett CML, Grobe JL, Kirby JR (2017) The gut microbiome, energy homeostasis, and implications for hypertension. Curr Hypertens Rep 19(4):27

Rieger J, Janczyk P, Hünigen H et al (2015) Intraepithelial lymphocyte numbers and histomorphological parameters in the porcine gut after Enterococcus faecium NCIMB 10415 feeding in a Salmonella Typhimurium challenge. Vet Immunol Immunopathol 164:40–50. https://doi.org/10.1016/j.vetimm.2014.12.013

Roager H, Hansen L, Bahl M, Frandsen H, Carvalho V, Gøbel R et al (2016) Colonic transit time is related to bacterial metabolism and mucosal turnover in the gut. Nat Microbiol 1(9):16093. https://doi.org/10.1038/nmicrobiol.2016.93

Roberts S, Girardi M (2008) Conventional and unconventional T cells. In: Clinical and basic immunodermatology. Springer, London, pp 85–104

Rodríguez J, Murphy K, Stanton C, Ross R, Kober O, Juge N et al (2015) The composition of the gut microbiota throughout life, with an emphasis on early life. Microb Ecol Health Dis 26:26050. https://doi.org/10.3402/mehd.v26.26050

Rodriguez-Concepcion M (2002) Elucidation of the methylerythritol phosphate pathway for isoprenoid biosynthesis in bacteria and plastids. A metabolic milestone achieved through genomics. Plant Physiol 130(3):1079–1089. https://doi.org/10.1104/pp.007138

Rothschild D, Weissbrod O, Barkan E, Kurilshikov A, Korem T, Zeevi D et al (2018) Environment dominates over host genetics in shaping human gut microbiota. Nature 555(7695):210–215. https://doi.org/10.1038/nature25973

Saavedra L, Taranto MP, Sesma F, De Valdez GF (2003) Homemade traditional cheeses for the isolation of probiotic Enterococcus faecium strains. Int J Food Microbiol 88:241–245. https://doi.org/10.1016/S0168-1605(03)00186-7

Sanmiguel C, Gupt A, Mayer EA (2015) Gut microbiome and obesity: a plausible explanation for obesity. Curr Obes Rep 4(2):250–261

Sartor RB, Wu GD (2017) Roles for intestinal bacteria, viruses, and fungi in pathogenesis of inflammatory bowel diseases and therapeutic approaches. Gastroenterology 152(2):327–339

Schoeler M, Caesar R (2019) Dietary lipids, gut microbiota and lipid metabolism. Rev Endocr Metab Disord 20:461–472

Schroeder B, Bäckhed F (2016) Signals from the gut microbiota to distant organs in physiology and disease. Nat Med 22(10):1079–1089. https://doi.org/10.1038/nm.4185

Shan M, Gentile M, Yeiser J, Walland A, Bornstein V, Chen K et al (2013) Mucus enhances gut homeostasis and oral tolerance by delivering immunoregulatory signals. Science 342 (6157):447–453. https://doi.org/10.1126/science.1237910

Sharon G, Sampson TR, Geschwind DH, Mazmanian SK (2016) The central nervous system and the gut microbiome. Cell 167(4):915–932

Siegrist C-A (2001) Neonatal and early life vaccinology. Vaccine 19(25–26):3331–3346

Singh R, Chang H, Yan D, Lee K, Ucmak D, Wong K et al (2017) Influence of diet on the gut microbiome and implications for human health. J Transl Med 15(1):73. https://doi.org/10.1186/s12967-017-1175-y

Sinkiewicz G, Nordström E (2005) 353 occurrence of Lactobacillus Reuteri, Lactobacilli and Bifidobacteria in human breast milk. Pediatr Res 58(2):415–415. https://doi.org/10.1203/00006450-200508000-00382

Sjögren K, Engdahl C, Henning P, Lerner UH, Tremaroli V, Lagerquist MK, Bäckhed F, Ohlsson C (2012) The gut microbiota regulates bone mass in mice. J Bone Miner Res 27(6):1357–1367

Skonieczna-Zydecka K, Marlicz W, Misera A, Koulaouzidis A, Łoniewski I (2018) Microbiome-the missing link in the gut-brainaxis: focus on its role in gastrointestinal and mental health. J Clin Med 7(12):521. https://doi.org/10.3390/jcm7120521

Smith UIF (2015) Abdominal obesity: a marker of ectopic fat accumulation. J Clin Invest 125 (5):1790–1792

Smith, K., McCoy, K. D., and Macpherson, A. J. (2007). Use of axenic animals in studying the adaptation of mammals to their commensal intestinal microbiota. In Seminars in immunology (19, 2, pp. 59-69). Academic Press New York, NY.

Sommer F, Anderson J, Bharti R, Raes J, Rosenstiel P (2017) The resilience of the intestinal microbiota influences health and disease. Nat Rev Microbiol 15(10):630–638. https://doi.org/10.1038/nrmicro.2017.58

Sonnenburg JL, Xu J, Leip DD, Chen CH, Westover BP, Weatherford J, Buhler JD, Gordon JI (2005) Glycan foraging in vivo by an intestine-adapted bacterial symbiont. Science 307 (5717):1955–1195

Spiljar M, Merkler D, Trajkovski M (2017) The immune system bridges the gut microbiota with systemic energy homeostasis: focus on TLRs, mucosal barrier, and SCFAs. Front Immunol 8:1353. https://doi.org/10.3389/fimmu.2017.01353

Spörri R, Sousa CR (2005) Inflammatory mediators are insufficient for full dendritic cell activation and promote expansion of CD4+ T cell populations lacking helper function. Nat Immunol 6 (2):163–170. 9

Stappenbeck TS, Hooper LV, Gordon JI (2002) Developmental regulation of intestinal angiogenesis by indigenous microbes via Paneth cells. Proc Natl Acad Sci 99(24):15451–15455

Stecher B, Hardt W (2011) Mechanisms controlling pathogen colonization of the gut. Curr Opin Microbiol 14(1):82–91. https://doi.org/10.1016/j.mib.2010.10.003

Steel J, Malatos S, Kennea N, Edwards A, Miles L, Duggan P et al (2005) Bacteria and inflammatory cells in fetal membranes do not always cause preterm labor. Pediatr Res 57(3):404–411. https://doi.org/10.1203/01.pdr.0000153869.96337.90

Stephens S, Brenner MK, Duffy SW, Lakhani PK, Kennedy CR, Farrant J (1986) The effect of breast-feeding on proliferation by infant lymphocytes in vitro. Pediatr Res 20(3):227–231

Subramanian S, Huq S, Yatsunenko T, Haque R, Mahfuz M, Alam MA et al (2014) Persistent gut microbiota immaturity in malnourished Bangladeshi children. Nature 510(7505):417–421

Sudo N, Sawamura SA, Tanaka K, Aiba Y, Kubo C, Koga Y (1997) The requirement of intestinal bacterial flora for the development of an IgE production system fully susceptible to oral tolerance induction. J Immunol 159(4):1739–1745
Sun M, He C, Cong Y, Liu Z (2015) Regulatory immune cells in regulation of intestinal inflammatory response to microbiota. Mucosal Immunol 8(5):969–978
Surono IS, Koestomo FP, Novitasari N et al (2011) Novel probiotic Enterococcus faecium IS-27526 supplementation increased total salivary sIgA level and bodyweight of pre-school children: a pilot study. Anaerobe 17:496–500. https://doi.org/10.1016/j.anaerobe.2011.06.003
Takahashi T, Yano M, Minami J, Haraguchi T, Koga N, Higashi K, Kobori S (2002) Sarpogrelate hydrochloride, a serotonin2A receptor antagonist, reduces albuminuria in diabetic patients with early-stage diabetic nephropathy. Diabetes Res Clin Pract 58(2):123–129. https://doi.org/10.1016/s0168-8227(02)00105-5
Talham GL, Jiang HQ, Bos NA, Cebra JJ (1999) Segmented filamentous bacteria are potent stimuli of a physiologically normal state of the murine gut mucosal immune system. Infect Immun 67 (4):1992–2000
Tan BYQ, Paliwal PR, Sharma VK (2020) Gut microbiota and stroke. Ann Indian Acad Neurol 23 (2):155–158
Tanaka M, Nakayama J (2020) Development of the gut microbiota in infancy and its impact on health in later life. Allergol Int 66:515
Tarcan A, Gürakan B, Tiker F, Özbek N (2004) Influence of feeding formula and breast milk fortifier on lymphocyte subsets in very low birth weight premature newborns. Neonatology 86 (1):22–28
Thursby E, Juge N (2017) Introduction to the human gut microbiota. Biochem J 474 (11):1823–1836. https://doi.org/10.1042/bcj20160510
Tilg H, Adolph TE, Gerner RR, Moschen AR (2018) The intestinal microbiota in colorectal cancer. Cancer Cell 33(6):954–964
Tu W, Rao S (2016) Mechanisms underlying T cell immunosenescence: aging and cytomegalovirus infection. Front Microbiol 7:2111
Turnbaugh P, Ley R, Hamady M, Fraser-Liggett C, Knight R, Gordon J (2007) The human microbiome project. Nature 449(7164):804–810. https://doi.org/10.1038/nature06244
Vallès Y, Artacho A, Pascual-García A, Ferrús M, Gosalbes M, Abellán J, Francino M (2014) Microbial succession in the gut: directional trends of taxonomic and functional change in a birth cohort of Spanish infants. PLoS Genet 10(6):e1004406. https://doi.org/10.1371/journal.pgen.1004406
Van der Waaij D, Berghuis-de Vries JM, Lekkerkerk-Van der Wees JEC (1971) Colonization resistance of the digestive tract in conventional and antibiotic-treated mice. Epidemiol Infect 69 (3):405–411
Verstraelen H, Vilchez-Vargas R, Desimpel F, Jauregui R, Vankeirsbilck N, Weyers S, Verhelst R, De Sutter P, Pieper DH, Van De Wiele T (2016) Characterisation of the human uterine microbiome in non-pregnant women through deep sequencing of the V1-2 region of the 16S rRNA gene. PeerJ 4:e1602
Vinay DS, Ryan EP, Pawelec G, Talib WH, Stagg J, Elkord E et al (2015) Immune evasion in cancer: mechanistic basis and therapeutic strategies. In: Seminars in cancer biology, vol 35. Academic Press, New York, NY, pp S185–S198
de Vos W, Smidt H, van der Oost J, Stams F, Kengen S (2017) Microbiology Centennial Symposium 2017: 'Exploring Microbes for the Quality of Life': 18–20 October 2017, Wageningen: Book of Abstracts. https://doi.org/10.18174/424506
Vyas U, Ranganathan N (2012) Probiotics, prebiotics, and synbiotics: gut and beyond. Gastroenterol Res Pract 2012:872716. https://doi.org/10.1155/2012/872716
Wang M, Song F, Wu R, Allen KN, Mariano PS, Dunaway-Mariano D (2013) Co-evolution of HAD phosphatase and hotdog-fold thioesterase domain function in the menaquinone-pathway fusion proteins BF1314 and PG1653. FEBS Lett 587(17):2851–2859

Wang J, Tang H, Zhang C, Zhao Y, Derrien M, Rocher E et al (2015) Modulationof gut microbiota during probiotic-mediated attenuation of metabolic syndrome in high fat diet-fed mice. ISME J 9(1):1–15

Wang J, Chen W, Wang Y (2020) The relationship between gut microbiota and inflammatory diseases: the role of macrophages. Front Microbiol 11:1065. https://doi.org/10.3389/fmicb.2020.01065

Wannemuehler MJ, Kiyono H, Babb JL, Michalek SM, McGhee JR (1982) Lipopolysaccharide (LPS) regulation of the immune response: LPS converts germfree mice to sensitivity to oral tolerance induction. J Immunol 129(3):959–965

Weiner HL, da Cunha AP, Quintana F, Wu H (2011) Oral tolerance. Immunol Rev 241(1):241–259

Wong J, Jenkins D (2007) Carbohydrate digestibility and metabolic effects. J Nutr 137(11):2539S–2546S. https://doi.org/10.1093/jn/137.11.2539s

Worbs T, Bode U, Yan S, Hoffmann MW, Hintzen G, Bernhardt G et al (2006) Oral tolerance originates in the intestinal immune system and relies on antigen carriage by dendritic cells. J Exp Med 203(3):519–527

Wu HJ, Wu E (2012) The role of gut microbiota in immune homeostasis and autoimmunity. Gut Microbes 3(1):4–14

Xu J (2003) A genomic view of the human-bacteroides thetaiotaomicron symbiosis. Science 299 (5615):2074–2076. https://doi.org/10.1126/science.1080029

Yamashiro K, Tanaka R, Urabe T, Ueno Y, Yamashiro Y, Nomoto K, Takahashi T, Tsuji H, Asahara T, Hattor N (2017) Gut dysbiosis is associated with metabolism and systemic inflammation in patients with ischemic stroke. PLoS One 12(2):e0171521. https://doi.org/10.1371/journal.pone.0171521

Yao CK, Muir JG, Gibson PR (2016) Insights into colonic protein fermentation, its modulation and potential health implications. Aliment Pharmacol Ther 43(2):181–196

Yatsunenko T, Rey FE, Manary MJ, Trehan I, Dominguez-Bello MG, Contreras M, Magris M, Hidalgo G, Baldassano RN, Anokhin AP, Heath AC (2012) Human gut microbiome viewed across age and geography. Nature 486(7402):222–227. https://doi.org/10.1038/nature11053

Young RL, Lumsden AL, Martin AM, Schober G, Pezos N, Thazhath SS, Isaacs NJ, Cvijanovic N, Sun E, Wu T, Rayner CK, Nguyen NQ, Fontgalland D, Rabbitt P, Hollington P, Sposato L, Due SL, Wattchow DA, Liou AP, Jackson VM et al (2018) Augmented capacity for peripheral serotonin release in human obesity. Int J Obes (2005) 42(11):1880–1889. https://doi.org/10.1038/s41366-018-0047-8

Zhang B, Ohtsuka Y, Fujii T, Baba H, Okada K, Shoji H, Nagata S, Shimizu T, Yamashiro Y (2005) Immunological development of preterm infants in early infancy. Clin Exp Immunol 140 (1):92–96

Zheng J, Xiao X, Zhang Q, Mao L, Yu M, Xu J (2015) The placental microbiome varies in association with low birth weight in full-term neonates. Nutrients 7(8):6924–6937. https://doi.org/10.3390/nu7085315

Zheng D, Liwinski T, Elinav E (2020) Interaction between microbiota and immunity in health and disease. Cell Res 30:492–506

Zhuang L, Chen H, Zhang S, Zhuang J, Li Q, Feng Z (2019) Intestinal microbiota in earlylife and its implications on childhood health. Genom Proteom Bioinform *17*(1):13–25

Zoetendal E, Rajilic-Stojanovic M, de Vos W (2008) High-throughput diversity and functionality analysis of the gastrointestinal tract microbiota. Gut 57(11):1605–1615. https://doi.org/10.1136/gut.2007.133603

Zycka-Krzesinska J, Boguslawska J, Aleksandrzak-Piekarczyk T et al (2015) Identification and characterization of tetracycline resistance in *Lactococcus lactis* isolated from Polish raw milk and fermented artisanal products. Int J Food Microbiol 211:134–141. https://doi.org/10.1016/j.ijfoodmicro.2015.07.009

Virome and Microbiome Interaction and Their Implication in the Human Gut Health

4

Hebatallah H. AboNahas, Yousra A. El-Maradny, Sara A. Gezaf, Fatma A. Abo Nouh, Yousef H. AboNahas, Daniela Landa-Acuña, Alejandra Gil-Polo, Essa M. Saied, and Ahmed M. Abdel-Azeem

H. H. AboNahas (✉)
Zoology Department, Faculty of Science, University of Suez Canal, Ismailia, Egypt

Y. A. El-Maradny
Microbiology Department, High Institute of Public Health, Alexandria University, Alexandria, Egypt

S. A. Gezaf
Botany Department, Faculty of Science, AL-Arish University, North Sinai, Egypt

F. A. Abo Nouh
Botany and Microbiology Department, Faculty of Science, Suez Canal University, Ismailia, Egypt

Y. H. AboNahas
Microbiology Department, Faculty of Medicine, Helwan University, Cairo, Egypt

D. Landa-Acuña
Laboratory of Microbial Ecology and Biotechnology, Department of Biology, Faculty of Sciences, National Agrarian University La Molina (UNALM), Lima, Peru

Professional Career in Environmental Engineering, Faculty of Engineering, Private University of the North Lima—Los Olivos Campus, Lima, Peru

A. Gil-Polo
Laboratory of Microbial Ecology and Biotechnology, Department of Biology, Faculty of Sciences, National Agrarian University La Molina (UNALM), Lima, Peru

E. M. Saied (✉)
Chemistry Department, Faculty of Science, Suez Canal University, Ismailia, Egypt

Institute for Chemistry, Humboldt Universität zu Berlin, Berlin, Germany
e-mail: saiedess@hu-berlin.de

A. M. Abdel-Azeem
Botany and Microbiology Department, Faculty of Science, Suez Canal University, Ismailia, Egypt

R. Z. Sayyed, M. Khan (eds.), *Microbiome-Gut-Brain Axis*,
https://doi.org/10.1007/978-981-16-1626-6_4

Abstract

As a complex ecosystem, human gut microbiota have several functions integrated in the host organism through various activities including metabolism, immunity, absorption, etc. One of the human microbiota is viruses, whose composition has not been completely described up till now. Vaious studies proved that the human gut harbors different types of viruses like plant-derived viruses, giant viruses, and bacteriophages. Recently metagenomic methods have allowed to reconstitute entire viral genomes from the genetic material spread in the human gut, opening new avenue on the understanding of the gut virome composition, the importance of gut microbiome, and potential clinical applications. This chapter aims to shed light on the latest evidence on human gut "virome" composition, interaction, its function, and possible future therapeutic applications in human health.

Keywords

Bacteriophages · Diabetes · Homeostasis · Gut microbiome · Obesity

4.1 Introduction

The human microbiome contains communities of commensal, symbiotic, and pathogenic bacteria, archaea, viruses, and small eukaryotes that actively interact with one other and/or with the host to maintain homeostasis (Nicholson et al. 2012). The microbiome acquired at birth is shaped as a result of multiple factors, including the mode of delivery, breastfeeding, solid food, and other environmental exposures (Dominguez-Bello et al. 2010). The human gut microbiome harbors genes that are involved in nutrient synthesis, amino acid metabolism (Lin et al. 2017), carbohydrate metabolism (Rowland et al. 2018), and lipid metabolism (Wang et al. 2016). Other factors such as host genetics (Blekhman et al. 2015), dietary habits (David et al. 2014), lifestyle (Clemente et al. 2015), comorbidities (Nagpal et al. 2018), chemotherapy (Montassier et al. 2015), and antibiotics (Francino 2016) can also disrupt the microbiome balance and may promote disease. The research on microbiome is actively directed to understand the microbial association, function, structure, and trans-kingdom interactions, as well as their cause and effect in the context of health and disease. The techniques used for the isolation and characterization of bacterial communities are relatively well-developed and standardized compared to those of the archaeal, eukaryal, and viral communities, and therefore, most of microbiome studies have been focused on the bacterial component of the human microbiome (Pollock et al. 2018). Nevertheless, characterization of viral communities can be more challenging than bacteria, archaea, and eukaryotes due to the absence of phylogenetically conserved genes. The shotgun metagenomic sequencing is the potential approach to characterize viral communities, a concept known as virome (Santiago-Rodriguez and Hollister 2019). The human virome is defined as a collection of all viruses that are found in human. The human virome comprises eukaryotic and prokaryotic viruses; viruses that cause acute, persistent, or latent infection; and viruses that can integrate themselves into the human genome (Fields et al. 2007). The human intestinal microbiota represents one of the most complex microbial ecosystems that is comprised of bacteria, viruses, fungi, multicellular parasites, and archaea (Moeller et al. 2016), which provide a heterogeneous surface area (>200 m^2) for microbial life (Hooper and Macpherson 2010). It is estimated that the human gut contains between 30 and 400 trillion microorganisms from a variety and abundant bacterial hosts that can support a correspondingly rich and varied phage population (Ahmed et al. 2007; Zhang et al. 2014). The DNA and RNA viruses that collectively make up the intestinal virome are at least equivalent in number to the bacterial cells (Reyes et al. 2012), although on gut mucosal surfaces and within the mucus layers they may outnumber bacterial cells by 20:1 (Barr et al. 2013). This viral community encompasses an abundant and diverse collection of viruses that not only infect every domain of life (Eukaryota, Archaea, and Bacteria) but also dominated by viruses that infect and replicate within bacterial cells (bacteriophage or phage) (Reyes et al. 2010). The gut microbial community is now considered to be intimately involved in our health, providing a wide range of beneficial functions such as extraction of additional energy from our diet, shaping the immune systems, and protection from invading pathogens. Additionally, it has emerging roles in modulating mood, behavior, neurocognitive development, and even the aging process (Zoetendal et al. 2006).

4.1.1 Emerging View of the Human Virome

Until quite lately, human viruses were regarded as pathogens that are able to cause human pandemics and a large variety of illnesses that can cause high mortality rate. Recently, new human-associated viruses, known as human virome, have arisen with the advent of new sequencing technologies that enabled the study of the global viral population (DNA and RNA) in humans (Greninger et al. 2009; Allander 2008). Most of these high-throughput sequencing techniques, however, were conducted using filters with pore sizes varying from 0.2 to 0.45μm, which filter larger viruses, resulting in the human virome's technical bias. The viral resources and biodiversity in the human body were commonly underestimated under nonpathological conditions. The microbial biodiversity of the human gut is regulated by human-associated viruses (Stern et al. 2012; Marinelli et al. 2012). The very basis of our being, our genome, is influenced by viruses. Reminiscences of human-viral ancestral cohabitation are imprinted in about 100,000 documents for endogenous viral fragments in our genome, comprising about 8% of our genome (Belshaw et al. 2004). Eventually, major physiological roles, such as mammal placental morphogenesis, have been associated with endogenous viral proteins (Mi et al. 2000; Blaise et al. 2003).

Phage-bacteria-human interactome deciphering has only been lately started to appear. The viral metagenomics study of the oral cavity of healthy individuals, reported by Willner et al., showed that phages constitute an essential reservoir for genes of bacterial virulence. These results, therefore, indicated that phages play a dual role in regulating the bacterial population while also contributing through horizontal gene transfer to bacterial pathogenicity and resistance (Willner et al. 2011). Several studies identified the persistent viral scattering from the gastrointestinal tract of dsDNA viruses of the *Polyomaviridae* family. In healthy children and adults, the polymerase chain reaction (PCR)-based identification of the BK, JC, and SV40 viruses has been reported (Vanchiere et al. 2009). Compared with adults, viral identification was more common in stool samples from infants. These results highlight the hypothesis that the gastrointestinal tract could be a site of persistence of polyomavirus with a potential fecal-oral viral transmission pathway.

In the normal gut viral flora, several RNA viruses, commonly regarded as human pathogens, have also been identified. The existence of many eukaryotic viral families, such as *Astroviridae* (Gabbay et al. 2005; Méndez-Toss et al. 2004), *Caliciviridae* (Barreira et al. 2010; Ayukekbong et al. 2011), *Picornaviridae*, *Reoviridae*, and *Picobirnaviridae*, as well as plant viral families, such as *Virgaviridae*, has been discovered by PCR-based or by metagenomic studies on "healthy" human feces. *Reoviridae* and *Picobirnaviridae* are two gastroenteritis-responsible families of dsRNA viruses, but both may occur in healthy humans. For instance, in developing countries, rotaviruses (*Reoviridae*, genus *Rotavirus*) are the major cause of death among children below the age of 5. Some genotypes, such as G10P strains, have commonly been correlated in India with asymptomatic neonatal infections (Gómara et al. 2004). The existence of plant viruses in the human gut indicates that, as shown for bacteria, the virome would differ between individuals based on diet (Turnbaugh et al. 2009). Gut virome can also rely on environmental factors, including geography, eating patterns, or ethnic differences, resulting in heterogeneity between individuals. A persistent need for blood transfusions and medical care is reflected by human blood and related products. Blood, however, is also a major viral reservoir, and certain viruses can be pathogenic. Therefore, it has direct implications for public health to identify the viral flora in the blood. A growing body of evidence suggests (Nishizawa et al. 1997) that the blood is not sterile in healthy individuals and can carry several viral organisms. Most of "normal" blood viral flora consists of the most commonly detected ssDNA viruses of the *Anelloviridae* family of Torque teno viruses (TTVs). TTVs, originally found in a post-transfusion hepatitis patient in Japan, are small viruses of icosahedral symmetry that are non-enveloped and have a high level of genetic diversity. The first genus of *Anelloviridae*, *Alphatorquevirus*, currently comprises 29 species of TTV. Globally distributed TTVs are now known to be commensal (Virgin et al. 2009; Breitbart and Rohwer 2005; Biagini et al. 1998). Although replicative forms of TTV DNA have been discovered in peripheral mononuclear blood cells (Okamura et al. 1999), the bone marrow, lung, spleen, and liver have been described with viral loads higher than those in the blood (Okamoto et al. 2001). There have also been studies recording TTV mother-to-child transmission (Bagaglio et al. 2002).

The *Parvoviridae* family is another ssDNA virus family. The human parvovirus (PARV) was initially detected in the plasma of a person at risk for HIV infection (Jones et al. 2005). However, regular plasma identification of PARV4 and PARV5 has been documented in healthy blood donors as well as symptomatic individuals (Fryer et al. 2007). In blood donors, eukaryotic dsDNA viruses have also been identified. By analyzing blood from 400 donors, Egli et al. confirmed the prevalence of BK and JC polyomaviruses (Egli et al. 2009). Intriguingly, with respect to virus-host interaction and epidemiology, they reported crucial variations between the BK and JC viruses. In addition, lymphotropic polyomavirus and human bocavirus (HBoV) have also been frequently detected in immunocompromised and seemingly healthy individuals in the peripheral blood (Delbue et al. 2010; Bonvicini et al. 2011). Nevertheless, several studies reported the presence of some viral species, as the viral flora, in many parts of the human body such as the respiratory tract, teguments, the nervous system, and the genitourinary tract (Popgeorgiev et al. 2013).

4.1.2 Human Gut Microbiota Composition

The human intestine is colonized by several microbial strains after birth that fluctuate and change during our life span according to anatomical, dietary, and nutritional status changes (e.g., obese, anorexic, lean nutritional status) and environmental (e.g., climate, familial composition, lifestyle, working place, etc.), pathological (e.g., gastrointestinal and systemic infections), and pharmacological factors (e.g., use of antibiotics, prokinetics, laxatives, and probiotics). The main components of gut microbiota are bacteria, fungi, yeasts, archaea, viruses, and other Eukarya (such as Blastocystis and Amoebozoa) (Mai and Draganov 2009). Bacteria reach more than 1 kg of weight and account for more than 1100 species. Bacteroidetes and Firmicutes are the predominant phyla in adults, followed by Actinobacteria and Proteobacteria (Lozupone et al. 2012). Approximately 10^{13} bacterial cells and an average of $\sim$160 distinct species may reside in the adult human alimentary tract (predominantly in the colon), with over 1000 different bacterial species in total associated with the human gut microbiome (Qin et al. 2010). Little information is still known about commensal fungi, archaea, and protozoa. However, some emerging microbiological data on yeast composition and functions have clarified their subsequent clinical use in the modulation of gut microflora. In fact, *Saccharomyces boulardii* is currently used with significant efficacy over placebo in the treatment of post-infectious and post-antibiotic diarrhea (Dinleyici et al. 2012).

The last three decades of microbiological/clinical research have helped to understand how food, pre−/probiotics, and antibiotics can modulate the intestinal bacteria qualitative/quantitative pattern resulting in different microbial-host functions (Devaraj et al. 2013). The observations of an obese/lean gut microbiota associated with overweight or lean status clarified how microbiota manipulation by diet was possible and how microbiota could be responsible not only for overweight but also for the chronic inflammatory state typical of the metabolic syndrome (Met S) (Geurts et al. 2014). The diet and gut microbiota's role in obesity pathogenesis is not simply causative as was initially expected. A recent report by Ridaura et al. (2013) showed how co-housing mice with an obese twin's microbiota and with mice containing the lean co-twin's microbiota prevented the development of increased body mass and obesity-associated metabolic phenotypes (greater polysaccharide metabolism and protein degradation) in obese cage mates. In a study by Qin et al., the role of diet in gut microbiota modulation showed to be strengthened by the recent metagenome-wide association in type 2 diabetic patients. Indeed, in a diet-associated insulin resistance status, the authors showed that these patients have a peculiar decrease in some butyrate-producing bacteria, an increase in various opportunistic pathogens, and an enrichment of other microbial functions conferring sulfate reduction and oxidative stress resistance (Qin et al. 2012). Based on these results, the possible functions of gut microbiota were quickly related to other organs. The previous association between spontaneous bacterial peritonitis and small bowel bacterial overgrowth in liver cirrhosis (Guarner and Soriano 2005) has led to the understanding of the microbial molecular patterns triggering inflammation and fibrosis in liver diseases such as non-alcoholic fatty liver disease (NAFLD) and its complications, i.e., non-alcoholic steatohepatitis (NASH), liver cirrhosis, and hepatocellular carcinoma (HCC) (Friedman 2013). Actually, gut bacteria seem to interact with the central nervous system (CNS) via the enteric nervous system through the endocannabinoid system. Thus, gut microbiota can affect the neuro-psychiatric state of the host, and also the CNS is able to affect its composition through food intake regulation (Cani et al. 2014).

4.2 Human Gut Virome Composition: Main Players

The concept of the existence of a "gut virome" has recently discovered (Mai and Draganov 2009), although the presence of viruses as pathogenic organisms in human intestine has been known and documented for more than a century (Lozupone et al. 2012). Recent studies described the temporal dynamics of the human gut virome. It appears that the symbiotic relationships between host and virome develop at a young age, with specific variations occurring during the first 2 years of birth, coinciding with environmental and dietary changes. As a result, individuals on the same diet showed similar gut virome composition (Minot et al. 2013). Norwalk virus, *Rotavirus*, and enterovirus are the well-known agents of gastroenteritis in man (Lagier et al. 2012). The reason we consider linking these pathogens with the gut virome is that the infection of the gut is responsible for enterocyte and bacterial microflora changes. These pathogens can affect the host in the acute phase of the infection with gastrointestinal complaints such as nausea, vomiting, diarrhea, and weight loss and also in the long-term persistence of symptoms and the possible eliciting of functional gastrointestinal disorders such as functional dyspepsia and post-infectious irritable bowel syndrome (Beatty et al. 2014). Although the viruses are the most numerous (about 10^{31} viral particles on earth and approximately 10^8 to 10^9 per gram of feces) and diverse microbial entities, there are relatively few studies that have focused on the association and function of viruses as part of the human microbiome (Robinson and Pfeiffer 2014). This is due to the challenges encountered in viral isolation, nucleic acid extraction, sequencing, and analysis pipelines (Mukhopadhya et al. 2019). The advent of high-throughput sequencing technologies permitted further insights into the human-microbe complex relationship that revealed the significant associations between microbial ecosystem shifts and disease (termed dysbiosis) and highlighted the diverse and abundant retinue of viruses intimately associated with the human gut microbiome (Breitbart et al. 2003). This human gut virome may be defined as the total population of viruses [or virus-like particles (VLPs)] associated with the underlying gut microbial community. In keeping with the dominance of bacteria in the gut microbiome, the gut virome appears to be predominated by prokaryotic viruses (bacteriophage) (Reyes et al. 2010). Both eukaryotic and prokaryotic viruses share lytic or latent life cycles, which allow different virome/host interactions and promote virus survival and evolution (Virgin 2014). As a result, human eukaryotic viruses can affect host physiology, mainly when chronically infecting particular sites, and virus-derived genetic elements can modify host gene and protein expression once integrated into host chromosomes (Foxman and Iwasaki 2011).

4.2.1 Eukaryotic Viruses

There are far fewer eukaryotic viruses than bacteriophages in the gut (Lim et al. 2015). According to metagenomic methods, novel enteric eukaryotic viruses were found to be responsible for acute diarrhea in children's small bowel enteropathy in developing areas of Australia (Scarpellini et al. 2015). These new data were confirmed by quantitative real-time polymerase chain reaction (qRT-PCR), which showed that diarrhea in children contains a higher abundance of viruses, many of them not previously known to be pathogenic (Holtz et al. 2014). Sequencing of eukaryotic viral communities in fecal samples from children has identified *Picobirnaviridae*, *Adenoviridae*, *Anelloviridae*, and *Astroviridae* family members and species such as *bocaviruses*, *enteroviruses*, *rotaviruses*, and *sapoviruses* (Minot et al. 2013). Despite being fewer in number, these viruses also have significant effects on human health, both in healthy and immunocompromised subjects, causing acute gastroenteritis, acute enteritis, or colitis (Eckardt and Baumgart 2011). Picobirnaviruses have been found in stool samples from individuals with diarrhea of unknown etiology (Banyai et al. 2003), as well as in healthy subjects (Kapusinszky et al. 2012), leaving their pathogenic capability up for discussion. Among the RNA viruses found in the gut, a prevalence of plant viruses has been introduced in the diet (Minot et al. 2011). In addition, disease-associated viruses such as herpesviruses, polyomaviruses, anelloviruses, adenoviruses, papillomaviruses, polyomaviruses, hepatitis B virus, hepatitis C virus, and human immunodeficiency virus (HIV) are also present in the intestinal viromes of some individuals, indicating that the gastrointestinal (GI) tract contains viruses capable of infecting host cells. As the majority of humans remain asymptomatic, it has been proposed that these pathogenic viruses (pathobionts) have become part of the metagenome of normal individuals, with the majority rarely causing disease and remaining dormant within the host (Hunter 2013). According to experiments in germ-free and antibiotic-treated mice, bacterial microbiome can promote the replication and, in some cases, persistence of enteric viruses (Pfeiffer and Virgin 2016), with the efficient transmission of mouse mammary tumor virus requiring intestinal bacteria (Kane et al. 2011). Thus, the interactions between viruses and bacteria, and other constituents of the intestinal microbiome, are important in influencing the course and outcome of virus infections (Almand et al. 2017).

4.2.2 Bacteriophages

In recent years, studies of the human gut virome have mainly focused on the analysis of virus-like particles (VLPs) purified from fecal samples and the application of high-throughput metagenomic approaches to characterize these viruses (Reyes et al. 2012). Several studies have provided insights into the diversity and structure of the gut virome, which is likely to reflect the underlying diversity of the bacterial microbiome (Qin et al. 2010). Intestinal bacteriophages (prokaryotic viruses) were recently considered as the main component of the gut virome, accounting for about 90% of its composition (Lozupone et al. 2012). Bacteriophages can be quite literally defined as "viruses of bacteria." Bacteriophages are commonly known as "bacterial parasites" that inject their genome in their host, integrating with its genetic material (prophage state) and inducing synthesis of other phage particle with bacterial cell lysis (lytic state) (Mokili et al. 2012). Double-stranded DNA phages from the order *Caudovirales* (*Podoviridae, Siphoviridae*, and *Myoviridae*) as well as single-stranded DNA phages from the family *Microviridae* constituted the bacteriophage component of the gut virome (or phageome) (Minot et al. 2011). The *Microviridae* family was initially considered as secondary players in the environmental viral community because of their modest genome size. *Microviridae* are small icosahedral viruses with circular single-stranded DNA genomes, and their members are divided into microviruses (genus *Microvirus*) and gokushoviruses (subfamily *Gokushovirinae*) (Wegley et al. 2007). These viruses have been retrieved in bacteria belonging to two genera of the phylum Bacteroidetes: *Prevotella* and *Bacteroides* (Roux et al. 2012). The virotypes mostly infect bacteria belonging to the most prevalent phyla within the gut, comprising members of Firmicutes, Bacteroidetes, Proteobacteria, and Actinobacteria (Manrique et al. 2017). The adult gut virome may be dominated by just one or a few different virotypes and is characterized by a high degree of stability in terms of its structure over time, with temporal tracking of gut virotypes, revealing the retention of between 80 and 95% of virotypes over a period of 1–2.5 years (Reyes et al. 2012).

Microbial viruses modulate their bacterial hosts directly through affecting their mortality and through horizontal gene transfer and indirectly by reprogramming host metabolism (Dalmasso et al. 2014). The human GI tract contains an estimated 10^{15} bacteriophages (phages; the phageome) that may represent the richest concentration of biological entities on earth (Dalmasso et al. 2014). Phages can be functionally categorized based on their life cycle after infecting host cells (Weinbauer 2004). Lytic (virulent) phages lyse the infected cells by hijacking the host cell's replication mechanism to package and produce more phages and lytic enzymes that cause cell lysis to release the newly formed phages (Mukhopadhya et al. 2019). After that, temperate phages incorporate their genetic material into the host cell chromosome as prophages and replicate alongside the host cell (Girons et al. 2000). Recently, analysis of the viral fraction of existing metagenomic studies identified a DNA phage called crAssphage that is highly abundant in the gut microbiome. It has been predicted, based on host co-occurrence profiling, that crAssphage infects *Bacteroides* species (Guerin et al. 2018). Prokaryotic viruses are known to influence human health by affecting bacterial community structure and function (Reyes et al. 2010). However, the intricate pathways by which this influence is exerted are yet to be fully clarified. Nevertheless, it has been shown that (1) temperate phages are common; (2) bacteriophages vary widely between individual hosts but not within a single subject; and (3) the variety of bacteriophages present increases in adulthood, and the diet affects the composition of phage communities (Minot et al. 2013).

4.3 Virome Functions Within the Human Gut Microbiome

Up to date, there are few clear data about gut virome functions within the gut microbiota ecosystem, although the life cycle of viruses provides indirect information about their possible roles. The great part of the phages found in the human gut shows atypical "temperate" behavior, which justifies the hypothesis that their composition is quite stable during the host's life (Lozupone et al. 2012). Furthermore, several authors have used the terms "stability" and "variability" to define phage behavior. The two kinds of these characteristics, belonging to the bacteriophages in the intestine, are linked as in a "virtuous" cycle. In fact, the stability of the viral genome is responsible for that of other microorganisms, such as the bacteria of the gut microbiota (Hofer 2013). This is proved by the fact that gut virome composition mimics the evolution of the infant bacterial microbiota (Palmer et al. 2007). These are common between phages and bacteria and are implicated in bacterial wall adhesion and immunoglobulin receptor synthesis, contributing to maintaining viral-bacterial immune tolerance in the gut. This allows the persistence of bacterial and viral species in the gut, exerting their effects on enterocytes and on the host (Liu et al. 2002). On the other hand, the presence of one intrinsic variability, typical of the few lytic phages found in the intestine, is an interesting characteristic of these viruses, which allowed the generation of new species in a short time frame and allowed them to escape extinction (Hofer 2013). Indeed, among the genes stably conserved during intestinal viral evolution discovered by metagenomics, there are also those involved in energy harvesting such as for carbohydrate transport and degradation (Markine-Goriaynoff et al. 2004). So these properties are common to diet-derived viruses of plants that can modulate human bacterial microbiota/host metabolism (e.g., carbohydrate synthesis/degradation, protein synthesis) (Lagier et al. 2012). According to "Darwin postulate" on animal species survival, the most common genes mapped by the largest part of sampled individuals were that responsible for DNA replication and repair, namely, a feature of "adaptation for survival" (Waller et al. 2014). As a result, gut virome has a significant impact on our gut microbiome and may potentially play a role in human genome maintenance over the generations. Among the genes encoded in cryptic prophages of *Escherichia coli*, those for resistance to antibiotics and other stress factors have been found (Kumarasamy et al. 2010). These explain the strict interaction between viral and bacterial particles in the intestine, which lead to the classical concept of bacteriophages as "predators" of bacteria. In fact, the transmission of genes between virus and the infected bacteria helps the host to resist oxidative stress and antibiotic use, another proof of the "temperate" lifestyle of the gut virome (Scarpellini et al. 2015).

4.4 Tools for Human Virome Identification

The absence of validated protocol to study virome might be due to the numerous obstacles in virome study such as viral diversity, host contamination, and lack of common conserved sequence within virus genome (Krishnamurthy and Wang 2017). Plaque assay is a culture-based technique and has been used to study and quantify phages and to detect their host range in environmental samples (Hamdi et al. 2017). However, culture-based approaches are not suitable to study viruses in a complex ecosystem such as human gut (Sutton and Hill 2019). Recently, the development of the metagenomics allowed the study of the natural viruses within complex microbial samples (Ye et al. 2019). Metagenomics is a molecular-based technique of non-culturable organisms used to study environmental samples containing complex of microbes by analyzing their genomes based on function and sequence (Riesenfeld et al. 2004). Most studied viromes are DNA and very low RNA viruses; this is because of the high mutation rate of RNA viruses and lack of standard amplification and bioinformatic tool (Marz et al. 2014).

The main steps in metagenomics include sampling, homogenization, filtration, concentration of the purified viral particle, extraction and amplification of viral nucleic acid, genome sequencing, and, finally, assembly and data analysis (Fig. 4.1) (Kumar et al. 2017; Manoussopoulos and Anastassopoulou 2020). Individualization and combination of various steps were used according to sample source and isolated virus to reach high viral titer and low host cell contamination. Different approaches used in metagenomics are discussed in this section.

4.4.1 Sampling

The selection of sample type and site is crucial starting in the studying and isolation of virome. Many common clinical samples yielded low abundance viruses and relative noise background from other microbiome and host cells (Haynes and Rohwer 2011; Rascovan et al. 2016).

The most studied human part containing a huge number of natural viruses is the gut and stool sample (Haynes and Rohwer 2011; Popgeorgiev et al. 2013; Rascovan et al. 2016). Other studied biopsies containing viromes include skin, respiratory

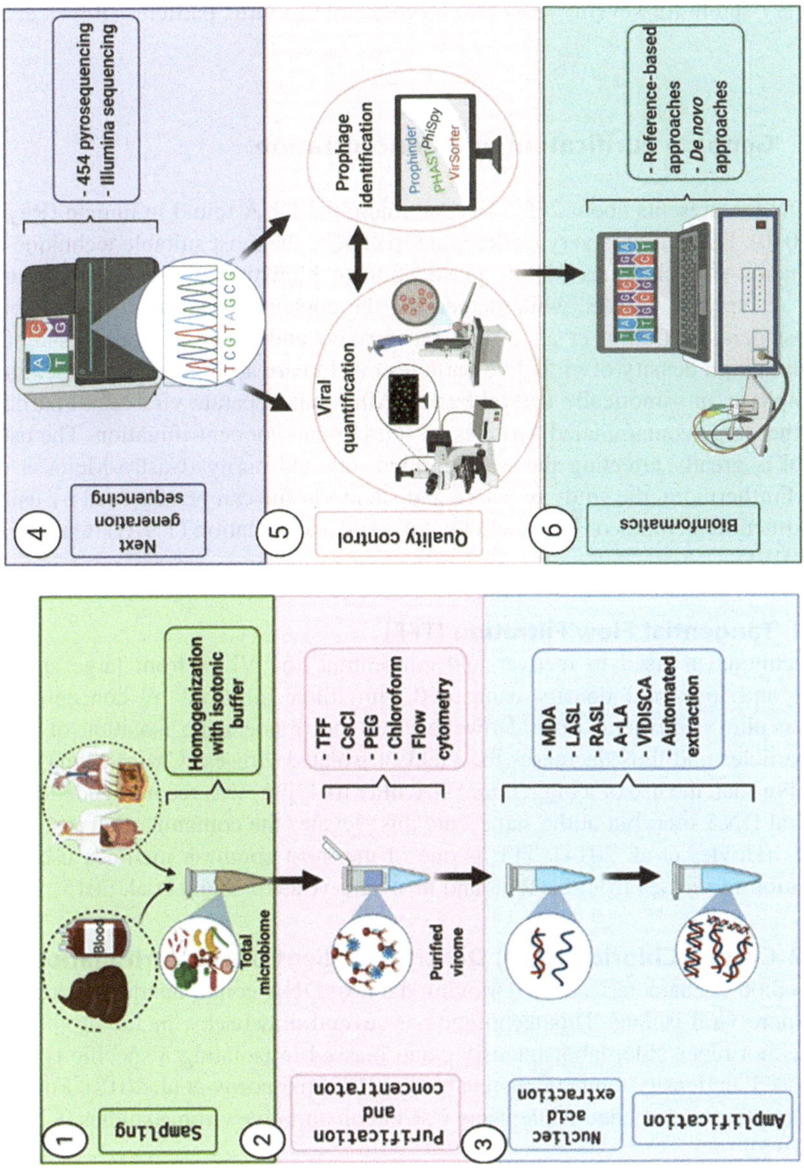

Fig. 4.1 Schematic presentation of the human virome metagenomics study

tract, genitourinary tract, nervous cells, and blood (Popgeorgiev et al. 2013). Collected samples should be homogenized with isotonic buffer and then purified immediately or kept at −80 °C to prevent the change of the ratio between different present microbes (Shkoporov et al. 2018). Fecal material stored at −80 °C was found to be very stable for several years and reproducible to virus particles (Reyes et al. 2010).

4.4.2 Genome Purification and Concentration

Viral DNA represents about 2–5% of total microbial DNA found in human (Reyes et al. 2010). For that, it is very challenging to choose the most suitable technique or combination of techniques for the purification of high titer of viral genome and isolate all types of viruses, while decreasing the contamination of other microbial and host genome (Vibin et al. 2018). Purification and concentration of samples containing high density of virus-like particles (VLPs) are achieved by resuspending the sample in an osmotically neutral buffer. After that, separate viral small particle from other large contaminated particles by filtration and/or centrifugation. The used protocol is greatly affecting the viral purified rate and purity (Castro-Mejía et al. 2015). Furthermore, the study of active and silent virome can be achieved by using both extraction methods of VLPs and total nucleic acid isolation (TNAI) (Garmaeva et al. 2019).

4.4.2.1 Tangential Flow Filtration (TFF)

This technique is used to recover and concentrate the VLPs from large sample volume and low viral density using a 0.2-μm filter followed by concentration using an ultra centrifugal filter. However, this tool renders the isolation of large virus particles and thus decreases the yield of isolated viruses (Castro-Mejía et al. 2015). For that, the use of a bigger pore size filter (0.45μm) was recommended to get high viral DNA titer, but at the same time this elevates the contamination with host particles (Hoyles et al. 2014). TFF is one of the most common methods used for purification of viruses from bacteria and host cells (Castro-Mejía et al. 2015).

4.4.2.2 Cesium Chloride (CsCl) Density Gradient Ultracentrifugation

This method is characterized by removing the host DNA contamination to produce highly pure viral isolate. This technique has several drawbacks, including the fact that it is not repeatable, labor-intensive, and biassed in isolating a specific type of virus based on density range (Callanan et al. 2018; Shkoporov et al. 2018). Furthermore, attention was needed while using CsCl in quantitative virome studies (Kleiner et al. 2015).

4.4.2.3 Precipitation with Polyethylene Glycol

This method is used when large sample volume is present. The polyethylene glycol (PEG) was added to the sample followed by centrifugation, filtration, and dialysis.

Precipitation approach is very useful and more effective than TFF in virome extraction (Castro-Mejía et al. 2015).

4.4.2.4 Chloroform and Nuclease Treatment

This method favors small circular genome than large linear one and is more successful with DNA viruses. Chloroform disrupts the lipid layer of a bacterial cell; then the free bacterial genome is removed by the addition of DNase and RNase to decrease sample contamination (McLaughlin et al. 2006). However, chloroform alters the stability of some enveloped and non-enveloped viruses (Conceição-Neto et al. 2015).

4.4.2.5 Flow-Cytometry-Based Methods

Flow cytometric analysis is used for counting the stained phage particles (Brown et al. 2015) or to fractionate the phage from mixed microbial sample such as human fecal sample according to its size and fluorescence intensity using fluorescence-activated cell sorting (FACS) (Džunková et al. 2015). The main concept of flow cytometry is using a fluorescent dye such as SYBR Green II for labeling of the VLPs and then separation (Roux et al. 2016). This method decreases significantly the host and bacterial contamination and is an added value to neglect the amplification step of genome before sequencing (Shkoporov and Hill 2019). The main limitation here is the low sensitivity in virus detection, thereby leading to great loss of many viruses within the sample (Warwick-Dugdale et al. 2019).

4.4.3 Extraction and Amplification of VLP-Derived DNA

The nucleic acid of viruses was extracted after purification step/s, but with an abundance level under the limit needed for sequencing. To overcome low viral genome abundance, numerous amplification methods are used. Viruses with DNA genome are the most prescribed gut virome in humans. In RNA viruses, the RNA genome is converted to cDNA by reverse-transcription PCR (RT-PCR) with random primers followed by PCR amplification (Emerson et al. 2018).

4.4.3.1 Multiple Displacement Amplification (MDA)

MDA is a fast and sensitive isothermal approach; it amplifies DNA viral genome before sequencing using Phi29 polymerase and random hexamers (Angly et al. 2006). MDA amplifies large amount of the whole viral genome (WVG) of both single-stranded DNA (ssDNA) and dsDNA. The limitation is the overamplification of small circular ssDNA viruses but low presentation of high GC content viruses (Kim et al. 2008). This method favors small circular genome than large linear one and is more successful with DNA viruses (Chen et al. 2014a; Roux et al. 2016).

4.4.3.2 Linker Amplified Shotgun Library (LASL)

LASL is used to amplify DNA based on PCR. The viral DNA template is ligated with dsDNA linker and then amplified using Vent DNA polymerase, ligated again into vector for cloning and finally electroporated. This approach overcomes the problem of bactericidal genes within viral DNA and the use of modified nucleotides. However, the main disadvantages of using PCR are as follows: only double-stranded genome can be amplified, and large DNA concentration is required, which leads to mixed amplification templates (Breitbart et al. 2003; Hindiyeh et al. 2019; Reyes et al. 2012).

4.4.3.3 Random Amplified Shotgun Library (RASL)

RASL method, like LASL, is based on the use of thermal cycler but with the use of random primers to amplify DNA in randomly amplified shotgun libraries. This procedure is rapid, very useful to amplify nanograms of DNA with non-culturable viruses, and suitable for shotgun sequencing (Rohwer et al. 2001).

4.4.3.4 Adaptase-Linker Amplification (A-LA)

A-LA is one of the most used genome amplification tools for both ssDNA and dsDNA templates. This method involves the use of adaptase before the linker amplification (Roux et al. 2016).

4.4.3.5 Virus Discovery cDNA-AFLP (VIDISCA) Technique

This method is a cDNA amplification approach that doesn't require prior genome sequencing (de Vries et al. 2011; van der Heijden et al. 2012). ViSeq, for example, is one of these amplification techniques which is sensitive and give quantitative results that uses adapter-specific primers to document the viral entire genome in humans (Cotten et al. 2014).

4.4.3.6 Fully Automated Virus Extraction

MagNA Pure 96 and eMAG are automated, sensitive, and specific viral genome extraction tools (Hindiyeh et al. 2019).

4.4.4 Sequencing Strategies

Next-generation sequencing method enables the sequencing and identification of unculturable novel viruses. This technique represents the basic step in metagenomic studies (Reyes et al. 2012).

4.4.4.1 Pyrosequencing

This technique is based on the detection of the released pyrophosphate by pyrosequencing. This yields a long genome length of 400 to 500 nucleotides each run (de Vries et al. 2011).

4.4.4.2 Illumina Sequencing

Based on the emitted fluorescent sequence unique to each base, Illumina sequencing simultaneously identifies the DNA bases. The HiSeq2000 platform (Xie et al. 2016) and the MiSeq deep sequencing platform (Cotten et al. 2014) are two examples of Illumina sequencing technique.

4.4.5 Quality Control

Quality control measurements should be applied to overcome the limitations of different approaches and the absence of valid protocol to study virome. The main boundaries during the study of human viromes are the contamination with host genome, abundance of viral genome, and detection of prophages (Sutton and Hill 2019). For that, the quality of sequenced sample, based on the mentioned obstacles, should be checked before the assembly and annotation.

4.4.5.1 Viral Quantification

Epifluorescence Microscopy

It is a rapid, simple, and reproducible tool to count the number of viruses. The sample was stained by SYBR Gold after filtration; then the slides were examined under the microscope, images were captured, and finally, the viral titers were calculated (Patel et al. 2007; Thurber et al. 2009).

Transmission Electron Microscope (TEM)

Tem is an old but labor-intensive method. the samples are fixed and stained using a negative staining technique, and then electron micrographs are taken to estimate the number of virus particles (Hoyles et al. 2014).

4.4.5.2 Prophage Identification Applications

Many automated and computational applications are present to predict prophages in host genome, and their main approach is to detect a sequence similar to known sequence of viral genome. Phage-finder (Fouts 2006), other program was developed based on the detection of prophages by analyzing the difference of dinucleotide relative abundance (Srividhya et al. 2007), Prophinder (Lima-Mendez et al. 2008), PHAge Search Tool (Zhou et al. 2011), PhiSpy application that identifies de novo or known virome based on similarity and composition analysis (Akhter et al. 2012), and VirSorter that can detect prophages in complete and fragmented (meta)genomic databases (Roux et al. 2015).

4.4.6 Computational Approaches for Characterizing Sequenced Viromes

Bioinformatics is the most used method to analyze the produced data. The choice of annotation or assembly software had a great impact on virome analysis outcome more than sequence technology used (Sutton et al. 2019). It could be divided into similarity-dependent and similarity-independent approaches (Haynes and Rohwer

2011). The main advantages of approaches are as follows: they are fast and simple and no special reagents are needed. Furthermore, they are able to distinguish between novel and known viruses by the conserved region on the genome (Delwart 2007).

4.4.6.1 Similarity-Dependent or Reference-Based Approach

This is the original and most used method to analyze the sequence data of microbiome by database search for finding segment similarity. The limitations of this tool in virome study are that they are not applicable in studying de novo viruses and the high diversity of viral genome makes the virus not similar to any known sequence (Brister et al. 2015). For similarity search, BLAST is the most used tool, which is based on the sequence of nucleic acid and amino acid (McGinnis and Madden 2004). In addition, microarray hybridization pattern is used to characterize new viral nucleic acids by comparing the shared sequences (Urisman et al. 2005).

4.4.6.2 Similarity-Independent or De Novo Approaches

In this approach, there is no need for database search, while more focus on the viral dark matter is applied. Phage Communities from Contig Spectrum (PHACCS) is a program for the detection and modeling of de novo viral diversity by using the spectrum of contig from the sequenced data (Angly et al. 2005). Numerous programs are used in the assembly of human virome, while some approaches are used to diagnose particular taxa based on GC % in the genome or dinucleotide combination rate (Sutton et al. 2019; Willner et al. 2009). SPAdes (meta) software (Nurk et al. 2017) showed the best accuracy and genome recovery (Sutton et al. 2019). However, the main limitation of de novo approaches is the finding of the lysogenic or silent virome from host genome (Delwart 2007).

4.5 Virome-Associated Dysbiosis

Imbalances in the makeup of gut microbiome, also termed dysbiosis, are now increasingly linked with a wide spectrum of diseases and disorders (both gut associated and those relating to extra-intestinal organ systems). These range from inflammatory bowel diseases, cancer, to metabolic disorders, obesity, and even autism and Alzheimer's (Lynch and Pedersen 2016). Emphasis is now being placed on delineating whether dysbiosis of the microbiome is a cause or consequence of some of these diseases and how manipulation of the gut microbiome may aid prophylaxis, diagnosis, or treatment (Sommer et al. 2017). The large part of the gut microbiota is composed of viruses and contracts both prokaryotic and eukaryotic cells to form the gut virome (Santiago-Rodriguez and Hollister 2019). As the role of the virome in the gut microbial community has been started to be uncovered, evidences have highlighted that this viral community also reflects the driving diversity and functionality co-evolution of host and microbe within the gut (Koskella

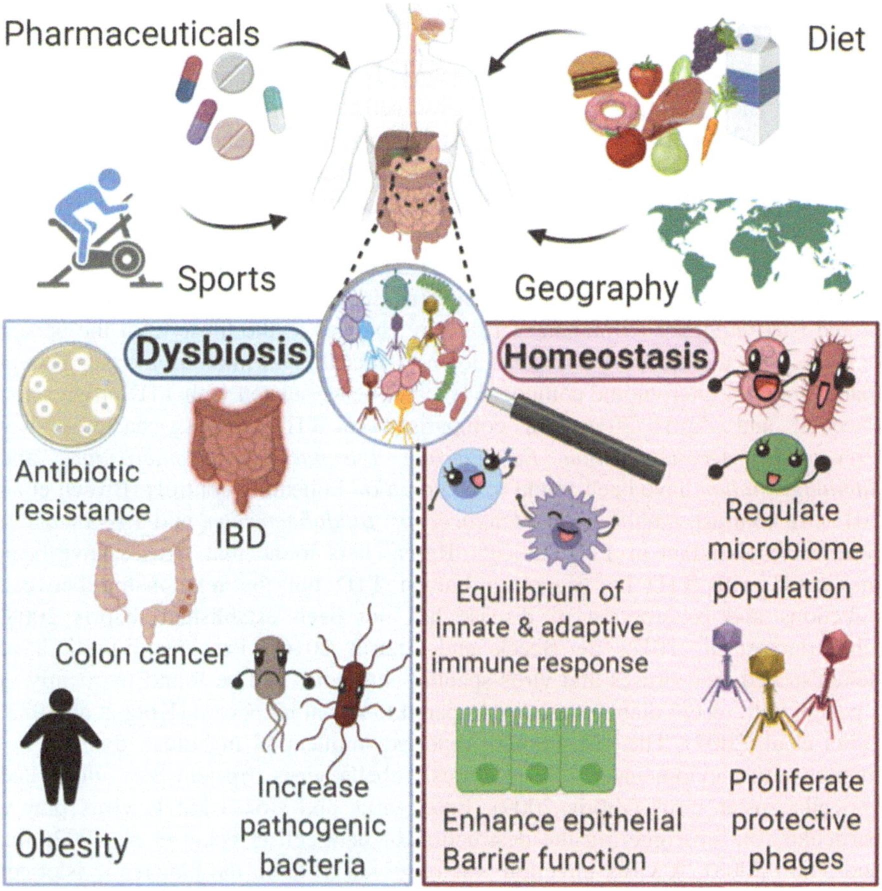

Fig. 4.2 Implications of gut microbiome on human health

and Brockhurst 2014). Recognition of the potential of phage to drive ecological functioning and evolutionary change (Koskella and Lively 2009) has understandably ignited interest in investigating the role of these prokaryotic viruses within the human gut virome and as part of the human gut microbiome. The concept of dysbiosis and the impacts of such perturbations on human health have begun to be considered from the perspective of the virome or phageome. There is a growing consensus that this concept should also be extended to the phage component of the gut ecosystem (Dalmasso et al. 2014). Although translational animal models are used to understand cause-effect relationships, virome analyses of the human gut are providing evidence on the potential role(s) of viruses in maintaining homeostasis or promoting disease (Fig. 4.2).

4.5.1 Type 1 Diabetes (T1D)

Type 1 diabetes (T1D) is a proinflammatory disease that targets beta cells of the pancreatic islet, resulting in a loss of these cells. In this disease, the genetic and environmental factors are incorporated, including complex genetic elements, patient exposures, and the gut microbiome. Viral infections and broader gut dysbioses have been identified as contributing factors or potential causes of T1D. However, human studies have not yet identified microbial functional or compositional triggers that are predictive of T1D or islet autoimmunity. T1D can be also known as the presence of antibodies against beta-cell autoantigens including insulin, zinc transporter 8, and islet antigen 2 (Morahan 2012). Any changes in the lifestyle of the person or changes in the dietary habits are known to alter the gut microbiome, suggesting that altered gut microbiome composition might be associated with T1D progression (Needell and Zipris 2016). In comparison to T1D patients, bacteria such *Prevotella, Faecalibacterium, Eubacterium, Fusobacterium, Anaerostipes*, and *Subdoligranulum* have been found to be common in healthy controls (Brown et al. 2011). In contrast, *Lactobacillus*, *Lactococcus*, *Bifidobacterium*, and *Streptococcus* were more abundant in T1D subjects. Researchers found that viruses have been implicated with T1D by triggering human T1D but the relationship between infections and progression of disease has not been established (Zipris 2008; Ghazarian et al. 2012; de Beeck and Eizirik 2016). Previous reports have demonstrated that viruses and virus-specific antibodies can be found frequently in a person with recent diabetes onset compared to a healthy person (King et al. 1983; Dotta et al. 2007). The viruses with reported implication in human diabetes are mumps virus, cytomegalovirus, rotavirus, rubella virus, Epstein-Barr virus, and varicella zoster virus (Zipris 2009). Enterovirus and Coxsackie B virus play a particular role in triggering the destruction of beta cell (Jaeckel et al. 2002; Jun and Yoon 2003). A virus infection was observed to harm the pancreatic islets of Langerhans in T1D murine models, suggesting a link between the virus and T1D model (Filippi and Herrath 2008). In 1993, an increase in diabetes rate was observed in young children notably after 2 years of released measles epidemic, suggesting that the increased T1D rate might have been associated to the measles outbreak (Lipman et al. 2002). To date, it is unknown how viral infections lead to T1D. It might be that the viral infection leads to diabetes due to a number of mechanisms, including molecular mimicry, bystander activation of T cells, beta-cell damage resulting in autoantigen release and activation of autoreactive T cells, and the induction of stress pathways in beta cells (Ghazarian et al. 2012) (Fig. 4.2).

4.5.2 Type 2 Diabetes Mellitus (T2DM)

Type 2 diabetes mellitus (T2DM) is a metabolic disorder distinguished by a high blood glucose level due to insulin resistance (IS) (Karim et al. 2014). T2D progression is associated with many factors, especially both genetic and environmental factors, in particular dietary habits, and most recently has been associated with change in microbiome composition (Larsen et al. 2010). Continuation studies identified several *Clostridium* spp., as well as *Bacteroides intestinalis*, *Akkermansia muciniphila*, and *E. coli* being enriched in subjects with T2D (Qin et al. 2012). Based on these data, an association was found between specific members of the human gut microbiome and T2D. Additionally, an association was observed between T2D and those viruses belonging to the *Siphoviridae* family, including *Lactobacillus*, *Pseudomonas*, and *Staphylococcus* (temperate and some are strictly lytic) (Ma et al. 2018). Viral infection is one of the important causes of diabetes, liver damage, neurological disorders, and several other diseases worldwide. The most important viruses related to the T2DM are HSV, hepatitis viruses, West Nile virus (WNV), influenza viruses (H1N1, H5N1 serotypes), picornavirus, cytomegalovirus (CMV), enterovirus, and Borna disease virus (BDV) (Cadranel et al. 2008; Carter 2010). Herpes simplex virus type 1 is a member of herpesvirus family. The HSV genome is large, double stranded, containing 74 genes, and encased within an icosahedral capsid protein, which is covered by an envelope (Mettenleiter et al. 2006; McGeoch et al. 2006). Recent clinical and epidemiological studies have reported association between HSV infection and T2DM (Cadranel et al. 2008; Ott 1999; Kroner 2009). Chronic inflammation is closely and early involved in the pathogenesis of T2DM (Pickup 2004; Helmersson et al. 2004). Cytomegalovirus (CMV) belongs to *Herpesviridae* family, and it is extremely a cause of human infections (Sweet 1999; Koichi et al. 2007; Ryan and Ray 2004). Human CMV infects ~40% of the world population, especially elderly people, without causing any specific symptoms (Offermanns and Rosenthal 2008). Reports have showed that CMV harms cells in the pancreas and causes T1D and T2D. Due to T1D, the immune system becomes weak which favors more susceptibility to infection with CMV, and chances of acquiring T2DM become 12 times higher (Rachael 2012; Sarah 2012). Enteroviruses are members of the *Picornaviridae* family and cause infections to humans and other mammals (Li et al. 2005). A study involving a group of patients diagnosed with T2DM were compared to others without T2DM explored the link between enterovirus infection and diabetes. It was observed that 40% of the pancreas of people with T2DM contained the enteroviral protein. The presence of enterovirus in the pancreas of people diagnosed with T2DM was three times higher than that in those without it (Richardson et al. 2009) (Fig. 4.2).

4.5.3 Inflammatory Bowel Disease (IBD)

Inflammatory bowel disease (IBD) is an immune-mediated disease and also defined as a chronic illness of childhood that causes inflammation to the gastrointestinal tract and includes Crohn's disease (CD) and ulcerative colitis (UC). IBD requires medication in some cases, and in extreme cases, surgery is required. Current research suggests that several factors including genetic, environmental, autoimmune, dietary, and microbial play a role in the pathogenesis of IBD (Adamiak et al. 2013; Eszter Muller et al. 2014). Bacteria, fungi, and viruses are enteric microbiome purported that play a role in the pathogenesis of IBD (Wang et al. 2015). Bacteria can differ based on the kind of IBD (CD versus UC) and the operation has been performed (Halfvarson et al. 2017). In general, these bacteria include *Faecalibacterium prausnitzii*, *Roseburia* spp., *Bifidobacterium* spp., and *Lactobacillus* spp., as well as *E. coli*, *Oscillospira*, and unclassified Ruminococcaceae (Morgan et al. 2012; Celiberto et al. 2018). Various studies have investigated the association between the enteric virome and IBD. Early studies to estimate that link showed an increased abundance of phages infecting bacterial orders in subjects with IBD, including Alteromonadales, Clostridiales, and *Clostridium acetobutylicum*, and also elevated viruses from the *Retroviridae* family (Pérez-Brocal et al. 2015). Another study looked at the gut virome of the mucosa of a murine model of colitis and found that the quantity of viruses from the Caudovirales order had increased, but the variety had decreased (Duerkop et al. 2018). Interestingly, phages infecting enterobacteria were safely more represented in mice with colitis. Increased abundance and decreased diversity of phages are in agreement with a reduced number of phage-related functions related with UC (Duerkop et al. 2018). These results are totally opening the possibility of examining therapeutics to target the virome in IBD subtypes. An interesting study reported by Norman et al. detected disease-specific changes. The gut virome was less presented in both CD and UC. The primary difference in the IBD-associated virome was the increased abundance of *Caudovirales* phages on a taxonomic level; however, the exact viruses that may perform such change were different in CD when compared to UC (Norman et al. 2015). All data obtained by Norman et al. was recently reanalyzed by Clooney et al. The results confirmed IBD-specific changes in the virome, loss of the "core phageome," and the stimulated development of phages in patients with CD (Clooney et al. 2019). Moreover, the changes in the structure of virome reflected shifts in bacterial structure. Indeed, both virome and bacteriome alterations were more obvious in patients with CD when compared to those with UC, which reflect the disease severity. Hence, integrating both bacteriome and virome assessment offers higher classification power between healthy and unhealthy states in IBD (Clooney et al. 2019) (Fig. 4.2).

4.5.4 Cancer

Pioneering research studies that linked microorganisms and cancer have led to enormous implications for public health. *Helicobacter pylori* and its relation to gastric cancer helped to classify this bacterium as a class I carcinogen (Santiago-Rodriguez and Hollister 2019). Hepatitis B virus, hepatitis C virus, human papillomavirus (HPV), HIV-1, Epstein-Barr virus (EBV), and human T-cell lymphotropic virus type 1 (HTLV-1) are viruses linked with different types of cancer that are also classified as class I carcinogens (Chen et al. 2014b). Some of these viruses are known as a part of the human virome due to their implications with cancer. However, certain viruses, including HPV, may be known as a part of the human virome as those are considered to be low and high risk, and those with unknown pathogenicity have been identified with no symptoms (Santiago-

Rodriguez and Hollister 2019). Furthermore, only a small percentage of those subjects infected with high-risk HPV develop cancer (Van Dyne et al. 2018). Other viruses can also cause cancer including human polyomavirus (Prado et al. 2018). Knowledge regarding eukaryotic viruses and disease might lead to understand the relationship between disease and gut microbiome. This knowledge provides key information regarding the mechanisms during which viral infection could cause various cancer types, including chronic inflammation, immunodeficiency, and virus oncogenes. Furthermore, a growing number of research investigations have linked microorganisms and certain bacteria members to several malignancies, including melanoma (Matson et al. 2018; Routy et al. 2018; Gopalakrishnan et al. 2018), non-Hodgkin's lymphoma (Montassier et al. 2015), cervical cancer (Lam et al. 2018), acute lymphoblastic leukemia (ALL) (Chua et al. 2017), and colorectal cancer (CRC). CRC is characterized by decreased bacterial variety in feces and mucosal samples (Wong et al. 2017). This decreased diversity is related to the absence of bacteria that may be implicated in preserving a healthy state and the detection of taxa associated with CRC and tumorigenesis, including *Desulfovibrio* spp., *Bilophila wadsworthia*, *Fusobacterium nucleatum*, *Parvimona*, *Alistipes*, and *E. coli* (Rubinstein et al. 2013; Veziant et al. 2016; Maisonneuve et al. 2017; Tilg et al. 2018). These bacteria are being considered as potential candidates for therapeutic approaches (Matson et al. 2018; Routy et al. 2018; Gopalakrishnan et al. 2018) and potential diagnostic. Although the eukaryotic viral infection of the enteric virome has been associated with CRC to a lower extent, phage communities have the potential to be the most effective in this context (Hannigan et al. 2018). Like bacteria, viral infection has been identified in CRC, mostly being bacteriophages from the *Siphoviridae* and *Myoviridae* families (Hannigan et al. 2018). Another study that did not isolate viruses specifically found >20 viral genera that distinguish between subjects with CRC and healthy controls (Nakatsu et al. 2018). *Orthobunyavirus*, *Tunavirus*, *Phikzvirus*, *Betabaculovirus*, and *Zindervirus* are eukaryotic viruses that were remarkably more represented in subjects with CRC (Nakatsu et al. 2018). Other viruses such as *Fromanvirus* seemed to be represented only in the healthy cohort. Also, the most abundant in subjects with CRC include the phageome, *Streptococcus* phage SpSL1, *Streptococcus* phage 5093, *Streptococcus* phage K13, Enterobacteria phage HK544, and *Vibrio* phage pYD38-A (Nakatsu et al. 2018). Alterations in the gut virome represent the opportunity to elucidate bacterial-viral-host interactions in promoting CRC (Fig. 4.2).

4.6 Communication Between Enteric Virome and Human Gut Probiotics: Implication on Gut Health

It is well described that gut microbiomes have a duty in the modulation of the immune system, keeping balance and preventing infections (Domínguez-Díaz et al. 2019). The prokaryotes, eukaryotes, archaea, and viruses (Laforest-Lapointe and Arrieta 2018; Mukhopadhya et al. 2019) living coexist in the intestinal lumen that is an optimal environment for microbial community interactions (Hillman et al. 2017).

Hundreds or thousands of bacteriophage-bacteria pairs could interact at any time, making this hard to study (Sausset et al. 2020). The very recent study unveiling these interactions was reported by Marbouty et al. (2020) who used meta3C proximity ligation to analyze the phage-bacteria interaction in healthy human guts. In this study, they identified 6651 unique host-phage relationships. Half of the detected phages seemed to be lysogenic phages, and one-fourth represented potentially active phages (Marbouty et al. 2020). This outcome matches with previous reports that stated that the predominant state of a bacteriophage in the gut is the lysogenic state instead of the lytic state (Anthenelli et al. 2020).

At the moment, there is a sharp gap in interpreting the gut virome and gut prokaryote interaction, because the current models are not enough to translate this complex relation. However, some predictions have been done in this direction (Beller and Matthijnssens 2019).

4.6.1 Probiotics and Bacteriophages

Probiotics are "live organisms that, when administered in adequate amounts, confer a health benefit to the host." Although traditionally probiotics, mainly *Lactobacillus* and *Bifidobacterium*, have been obtained from the gut and fermented foods, the next generation of probiotics (NGP) are based on commensal bacteria (Martín and Langella 2019). These commensal bacteria are colonizing human mucosal surfaces (Khan et al. 2019). The enteric virome might also be provided the health benefits of probiotics (Łusiak-Szelachowska et al. 2017), and they probably have a stabilizing role in the gut ecosystem (Draper et al. 2018).

The virome can interact with bacterial probiotics. For example, treatment with lytic and lysogenic bacteriophages increases the abundance of *Lactobacillus* and *Bifidobacterium* in mice (Bao et al. 2018). Also, the reduction of *Lactococcus* in Parkinson's disease patients is associated with a higher abundance of *Lactococcus*-lytic phages in comparison to healthy controls (Tetz et al. 2018).

Lactobacillus reuteri is a probiotic that produces antibiotics and can shape the commensal microbiota in the gut (Mu et al. 2018). Nearly all human strains of these bacteria contain active prophages. Studies on an *L. reuteri* harboring two active prophages from the *Siphoviridae* family showed that prophages reduce the fitness of the bacteria during gastrointestinal transit. The phages are induced in the distal intestinal tract rather than in small intestinal regions in an SOS-dependent manner. The phage production provides a competitive advantage by killing a competitor strain (Oh et al. 2019).

4.6.2 Strategies of Interactions Between Prokaryotes and Viruses in the Gut

Bacteriophage predation and lysogenic conversion play an important role in the regulation of bacterial biomass and microbial diversity (Shkoporov and Hill 2019).

However, in the gut, there is no observation of biomass control from phages, and the virus-to-microbe ratio remains low (Shkoporov and Hill 2019). This might be explained by the recently proposed "piggyback-the-winner" strategy. This strategy proposes that phages take advantage of the high microbial abundance and growth rates of their hosts by remaining integrated in them (as a prophage) (Anthenelli et al. 2020; Guo et al. 2020; Silveira and Rohwer 2016). This strategy, observed in the gut and mucosal surfaces, is different to the models explaining the interaction and co-evolution between virus and hosts in other ecosystems, like the ocean where the phage/bacteria ratio is 10:1 (Maurice 2019). These strategies are as follows: first, "the arms race" model - in which the host acquires a mutation that makes it resistant to the virus, but then the virus acquires a mutation that allows it to reinfect the new resistant population (Avrani et al. 2012). Second, the "kill-the-winner" model in which the growth of the most active bacteria population is controlled by a virus and, therefore, there is an increase in the diversity of microbial communities (Maslov and Sneppen 2017; Winter et al. 2010).

Guo et al. (2020) proposed that the interactions between phages and hosts follow the piggyback-the-winner strategy (Guo et al. 2020). Research comparing the gut virome of stunted and non-stunted children demonstrated that non-stunted children had more temperate phages than stunted children. In the latter, microbial interactions could be following the "kill-the-winner" strategy (Mirzaei et al. 2020). An alteration in the abundance of lytic phages in comparison to temperate phages could lead to the development of diseases such as Parkinson's disease (Tetz et al. 2018).

4.6.3 Role of Lytic Phages in the Gut

Bacteriophages might control and regulate the abundance of bacteria in the gut (Bao et al. 2018), but little is known about the dynamics of phage predation in the human gastrointestinal tract (Hsu et al. 2018). Phages in the gut, apart from having an impact (knockdown) on the population of susceptible bacteria, might also have an impact on other microbial populations after a cascade effect caused by inter-bacterial interactions (Hsu et al. 2018). Although Hsu et al. (2018)) worked with phages with a narrow host range, recent studies pointed out that in the ecosystems, there are more bacteriophages with a broad range of hosts than previously thought (de Jonge et al. 2019). In relation to this, Marbouty et al. (2020) found out that most gut phages are specific to their host but nearly one-third of the identified phages showed contact with more than one MAG (Marbouty et al. 2020).

To understand the bacteriophage-host interaction, it is necessary to know what the host target of a bacteriophage. To predict this, several computation analyses have been used such as sequence homology, CRISPR spacers, occurrence profiles (Edwards et al. 2016), and abundance profiles (Stern et al. 2012). Recent research has found out that the abundance of phages is related to the abundance of susceptible bacteria (Oh et al. 2019). Programs such as VirHostMatcher (Ahlgren et al. 2017) and WIsH (Galiez et al. 2017) have shown high accuracy with the predictions.

4.6.4 Horizontal Gene Transfer Between Bacteriophages and Bacteria

The viruses that inhabit the gut are involved in horizontal gene transfer to prokaryotes (Mukhopadhya et al. 2019). This mechanism is determinant to shape complex ecosystems, including the gut, and acquire important genes (Callier 2019; Sutton and Hill 2019). The dynamics of how phage-mediated HGT occurs and regulates the evolution of bacteria in this ecosystem have been recently studied by Frazão et al. (2019)). Their experiment in a mice's gut showed that in the presence of a resident *E. coli*, the colonization of an invading *E. coli* can be successful if it adapts via HGT followed by mutation. The invader *E. coli* showed a rapid evolution, and it acquired two new genomic regions from the resident *E. coli*. They corresponded to two complete prophage regions, named KingRac and Nef. Further studies to characterize the process of inductions of each prophage responsible for the HGT revealed that the prophages of resident and invader *E. coli* can form active phage particles. The lysogenic invaders acquired phage-killing potential and a metabolic advantage related to the uptake of carbon sources (Frazão et al. 2019).

The HGT of antibiotic resistance (AR) genes in the gut is of great interest; however, how it works remains unknown (Kent et al. 2020). Górska et al. (2018) suggested that phages could be involved in the HGT of AR genes. Their pilot research has suggested that the abundance of phages carrying antibiotic resistance genes increases in patients that are under antibiotic treatment. However, further research is needed to know the insights of the integration of the phage in the bacterial genome (Górska et al. 2018). Other genes acquired by horizontal gene transfer are the Shiga genes that codify for the main virulence factors of some *E. coli* strains (Krüger and Lucchesi 2015) and the cholera toxin that was transferred to *Vibrio cholerae* (Faruque and Mekalanos 2012).

4.6.5 Pathogenic Interactions

There is evidence that enteric virome can cause significant diseases and could be utilizied in the gut to promote infection. Apparently, infection with enteric viruses is affected in the absence of gut microbiota because viral-bacteria mixed infections are worse than viral infections alone. How bacteria could facilitate the viral co-infection of cells is an active area of research and has been dug mainly in poliovirus and norovirus (Berger and Mainou 2018; Huang 2020).

Poliovirus is a non-enveloped ssRNA virus, which was a major cause of paralysis until 1950. It has been shown that polysaccharides of bacterial surfaces containing N-acetylglucosamine, like LPS, could bind to poliovirus and stabilize the particles by preventing the premature release of RNA. Also, LPS helps in the attachment of poliovirus to the host cells (Robinson et al. 2014). Bacteria can, as well, enhance the genetic recombination between two different viruses that helps them to increase the viral fitness and drive adaptation (Erickson et al. 2018). Human norovirus (NoV) causes viral gastroenteritis. The presence of commensal bacteria allows for efficient infection with that virus (Baldridge et al. 2015). It has been described that NoV binds to histo-blood group antigens (HBGAs), expressed by several enteric bacteria such as *Enterobacter cloacae*. This moiety is differing between bacterial strains (Almand et al. 2019).

4.6.6 Gaps in Knowledge

The processes by which bacteriophages regulate the microbial community is multi-factorial (Maronek et al. 2020). Current research is being carried out in phage-bacteria interactions seeking to find out what conditions of the human gut can trigger the lytic stage of prophages and how that influences the community. It is known that if phages are released from bacteria, then the competition between them to infect new hosts would increase. Also, the death of many bacteria would set free a niche for many other bacteria to occupy (that could be pathogenic or commensal), and this gives place to new interactions (Maurice 2019; Mirzaei and Maurice 2017). Finally, it has been recently suggested that the interactions between phages and bacteria could be age dependent (Mirzaei et al. 2020). Understanding and learning about these interactions will lead to the potential use of the enteric virome in therapy (Altamura et al. 2020).

4.7 Human Virome Therapeutic Implications and Future Directions

Bacteriophages are the most abundant members of the microbiota and have the ability to shape microbial communities (Garmaeva et al. 2019). It was suggested that they play a role in dysbiosis (Lin and Lin 2019), which is the decline in the diversity of the gut microbiome (DuPont et al. 2020). This microbial imbalance, alongside genetics, immune responses, and functional microbial activity, is associated with many gut diseases (Kostic et al. 2014). The study of the participation of the enteric virome in the development of intestinal diseases could lead to the development of a new diagnostic biomarker or antiviral drugs (Ansari et al. 2020). Also, bacteriophage-based therapies could emerge as a particular treatment option for microbiota-related diseases (Gogokhia et al. 2019).

4.7.1 Importance of the Enteric Virome in Fecal Microbial Transplantation

FMT (fecal microbial transplantation) is the administration of fecal slurry from a donor to the intestinal tract of a recipient, aiming to restore microbiota diversity and composition and provide a health benefit. This treatment has been proven to be effective against *Clostridium difficile* infection (CDI) (Gupta et al. 2016) with cure rates up to 90% (Basson et al. 2020).

The virome participation in the effectiveness of FMT treatment has been poorly studied, but there are signs that bacteriophages might have a role in the FMT outcomes (Broecker et al. 2017). For example, Ott et al. (2017)) showed that sterile fecal filtrate transfer (without living organisms) is effectively eliminating the symptoms of CDI. These results suggested that bacteriophages could be taking part in the FMT results (Ott et al. 2017). Zuo et al. (2018) further studied the differences in the phageome community between CDI patients and healthy controls and their importance in FMT. They found that CDI patients have more abundance but less diversity, evenness, and richness of *Caudovirales* compared to healthy controls. When the *Caudovirales* richness of the donor was higher than that of the recipient, the CDI patients achieved a positive response for FMT. Unlike the FMT non-responders, the FMT responders after the treatment showed more donor-derived *Caudovirales* contigs present in larger fractions in their enteric virome (Zuo et al. 2018). Recent research pointed out that the virome of the recipient after FMT ends up being very similar to the donor and that this can last up to 12 months (Draper et al. 2018). These might indicate that phages could be used as disease treatment (Manrique et al. 2017).

4.7.2 Phage Therapy

Bacteriophages can be used as biocontrol agents for specific hosts (Bao et al. 2018). Based on that, phage therapy is reemerging due to the rising antibiotic resistance problem. However, it is restricted for its use in humans because they can interact with our immune system and evolve inside the body. Additionally, they could influence the composition of the gut microbiome (Divya Ganeshan and Hosseinidoust 2019). Several studies have been done to show their effectiveness and specificity. For example, Hu et al. (2018) did an in vitro study with SalmoFresh™, which is a mixture of six strictly lytic phages of *Salmonella* (Zhang et al. 2019). They revealed that this phage cocktail can be used to target that bacteria in a very specific way. In comparison to the antibiotic azithromycin, phage treatment did not cause perturbation in non-targeted microbial communities (Hu et al. 2018).

Llanos-Chea et al. (2019)) evaluated the effect of the bacteriophage Φ2457T, which is a *Shigella flexneri* 2457 T-specific bacteriophage, in a human intestinal organoid-derived epithelial monolayer model. Their results showed that Φ2457T efficiently killed *Shigella flexneri* 2457 T in a highly specific manner (Llanos-Chea et al. 2019).

Dissanayake et al. (2019)) used "Foodborne Outbreak Pill" (FOP) (a combination of *E. coli* O157:H7, *Salmonella* spp., and *L. monocytogenes*-targeting lytic bacteriophages) to combat pathogenic *E. coli* in mice. FOP was demonstrated to be effective against *E. coli* O157:H7. They found out that FOP was highly specific and was better to maintain the natural richness and diversity of the gut microbiome in comparison to the control treated with ampicillin (Dissanayake et al. 2019). These results were consistent with those of Cieplak et al. (2018)) who used *E. coli* specific bacteriophages against *E. coli* DSM 1058 (Cieplak et al. 2018).

Adherent-invasive *E. coli* (AIEC) may be correlated with sustaining inflammation in the preexisting inflammatory mucosa (Lee et al. 2019) and with the exacerbation of intestinal inflammation (Delmas et al. 2019). Regular treatments with AIEC-bacteriophages isolated from the gut microbiome can reduce and control the outgrowth of targeted bacteria in the intestine and protect from AIEC and invasive bacteria-exacerbated colorectal cancer (Gogokhia et al. 2019). An important challenge to sort out when orally consuming bacteriophages would be to protect the phages from low pH due to the gastric acids. Therefore, encapsulating them with resistant material would make the treatment more effective (Vinner et al. 2019).

4.7.3 Enteric Virome Associated with Gastrointestinal Diseases

4.7.3.1 Inflammatory Bowel Disease (IBD)

Inflammatory bowel disease (IBD) is a convoluted, microbiome-driven immunological inflammatory disorder. There are two subtypes of IBD: Crohn's disease (CD) and ulcerative colitis (UC) (Panaccione 2013). UC and CD patients have different bacteriophage communities in comparison to healthy donors. The order of *Caudovirales* increased in UC and CD patients. Also, it has been suggested that the virome is specific for patients with UC and for patients with CD (Norman et al. 2015). As a therapy, FMT has proven to be not as effective as with CD (Basson et al. 2020). However, mixing donor products has been a way to expand the potential value of FMT when treating chronic diseases (DuPont et al. 2020).

Since the disease phenotypes in adult and pediatric-onset IBD are different, a study in children alone was necessary. Research in kids has shown that minor patterns in gut virome differentiate IBD patients and healthy donors. Among the differences they found was that *Caudovirales* were more abundant in CD (Crohn's disease) and UC (ulcerative colitis) patients than in healthy controls. It was also found that the richness of *Microviridae* is higher in controls than in CD patients (Fernandes et al. 2019). A study of virome transference in FMT between three pediatric UC patients and a healthy donor confirmed that the transference of viruses is mainly associated with temperate-phage replication, specially associated with the group *Siphoviridae*. None of them can replicate in human cells (Chehoud et al. 2016).

4.7.3.2 Celiac Disease Autoimmunity (CDA)

Celiac disease autoimmunity (CDA) is an autoimmune disease that is triggered mainly by genetics and dietary gluten. Recent reports indicated that microbiome disturbance could be behind celiac disease pathogenesis. However, the role or the virome participation in CDA is debatable (Akobeng et al. 2020). Recently, a study with children carrying a genotype for increased risk of celiac disease suggested that *Enterovirus A* and *Enterovirus B* are significantly associated with CDA. This research, using longitudinal birth control analysis, also discarded the possibility that adenovirus is involved with CDA (Kahrs et al. 2019). Also, it seems that there is a cumulative effect of enterovirus and high gluten intake for the development of CDA (Lindfors et al. 2020). On the other hand, previous studies have pointed out that the infection with reovirus could trigger CDA (Bouziat et al. 2017).

4.7.3.3 Enteric Virome Implication in Obesity and Diabetes

Obesity and type 2 diabetes are also related to gut microbiota dysbiosis (Maruvada et al. 2017; Ridaura et al. 2013). Fecal virome transplantation (FVT) – where sterile filtered donor feces that contain enteric viruses but not bacteria are transferred – has proven to reduce weight gain and normalize blood glucose parameters in mice (Rasmussen et al. 2020). In the case of type 1 diabetes (TID), differences between the viromes of patients with TID and controls have also been reported. Zhao et al. (2017)) found that the gut viromes of patients were less diverse than those of healthy controls. For example, healthy patients were richer in the eukaryotic virus *Circoviridae* and the bacteriophages *Microviridae*, *Myoviridae*, and *Podoviridae* (Zhao et al. 2017). Park and Zhao (2018) found differences in the virome population preceding initial signs of T1D (Park and Zhao 2018).

4.7.3.4 Enteric Virome Implication in Parkinson's Disease

Enteric virome has been recently linked to Parkinson's disease (PD) as a possible factor for the development of this neurodegenerative disorder. In PD patients, there is a significant reduction of *Lactococcus* and *Lactobacillus*. *Lactococcus* is involved in the production of microbiota-derived neurochemicals such as dopamine and with gut permeability. Their decrease could be involved with the triggering of PD. Tetz et al. (2018) showed that the decline of *Lactococcus* was not accompanied by the decline of their phages. The parallel decrease would have been normal in case the bacteriophages were integrated as prophage. They correlated that to the higher abundance of lactococcal lytic phages, especially the ones belonging to c-2-like and group 936, in PD patients. This finding could be useful to use the virome composition as a diagnostic tool or target for therapeutic intervention (Santos et al. 2019; Tetz et al. 2018).

4.7.4 Eukaryotic Viruses and Their Implications in Gastrointestinal Diseases

Apart from bacteriophages, the gut virome has eukaryotic viruses. Although they are considered as pathogens, they can live innocuously in the healthy human intestine (Mukhopadhya et al. 2019). The differences in the eukaryotic virome between patients and healthy controls have been reported. For example, Norman et al. (2015) found more presence of sequences from the eukaryotic virus *Anellovirus* in IBD patients (Norman et al. 2015). Also, Ansari et al. (2020) reported fewer abundance and diversity of members of *Megavirales* (Ansari et al. 2020). According to Conceição-Neto et al. (2018), who studied the implication of eukaryotic viruses in FMT treatment, healthy donors have a lower richness of eukaryotic virome than UC patients. Surprisingly, UC patients that responded to the FMT treatment already presented lower viral richness than non-responders. This suggests that eukaryotic virome richness could be important in UC and might lead to an increase in treatment success (Conceição-Neto et al. 2018).

4.7.5 Challenges and Future Directions

There are several challenges that need to be overcome to understand the role of the virome in human health. First is the inefficiency to identify all viruses due to the lack of a universal viral sequence. Second, the detection of DNA viruses has an advantage over RNA viruses because of inadequate sampling strategies that currently focus only on sequencing DNA, leaving aside an important source of diversity. Third is the absence of culture systems to propagate components of the virome and perform the Koch's postulate to know if the virome plays a causative role. Fourth is the need to change from in vitro systems to experimental animal infection models (Wang 2020). There are some gaps in knowledge for using FMT in IBD, such as identifying the optimal route, dose, and frequency of FMT, as well as what makes an optimum donor. Further studies to understand the potential contribution of the virome to the efficacy of the FMT in IBD are needed (Yalchin et al. 2019). Also, even though FMT has proven to be effective, there is a need to develop targeted microbiome therapies (Russell et al. 2018). These will come out as more information is generated on the mechanisms and diversity of human microbiome.

References

Adamiak T, Walkiewicz-Jedrzejczak D, Fish D et al (2013) Incidence, clinical characteristics, and natural history of pediatric IBD in Wisconsin: a population-based epidemiological study. Inflamm Bowel Dis 19:1218–1223

Ahlgren NA, Ren J, Lu YY, Fuhrman JA, Sun F (2017) Alignment-free d_2^* oligonucleotide frequency dissimilarity measure improves prediction of hosts from metagenomically-derived viral sequences. Nucleic Acids Res 45(1):39–53

Ahmed S, Macfarlane GT, Fite A, McBain AJ, Gilbert P, Macfarlane S (2007) Mucosa-associated bacterial diversity in relation to human terminal ileum and colonic biopsy samples. Appl Environ Microbiol 73:7435–7442. https://doi.org/10.1128/AEM.01143-07

Akhter S, Aziz RK, Edwards RA (2012) Nucleic Acids Res 40:1–13

Akobeng AK, Singh P, Kumar M, Al Khodor S (2020) Role of the gut microbiota in the pathogenesis of coeliac disease and potential therapeutic implications. Eur J Nutr 59 (8):3369–3390. https://doi.org/10.1007/s00394-020-02324-y

Allander T (2008) Human bocavirus. J Clin Virol 41(1):29–33

Almand EA, Moore MD, Jaykus LA (2017) Virus bacteria interactions: an emerging topic in human infection. Viruses 9:58

Almand EA, Moore MD, Jaykus L-A (2019) Characterization of human norovirus binding to gut-associated bacterial ligands. BMC Res Notes 12(1):607

Altamura F, Maurice CF, Castagner B (2020) Drugging the gut microbiota: toward rational modulation of bacterial composition in the gut. Curr Opin Chem Biol 56:10–15

Angly F, Rodriguez-Brito B, Bangor D, McNairnie P, Breitbart M, Salamon P, Felts B, Nulton J, Mahaffy J, Rohwer F (2005) PHACCS, an online tool for estimating the structure and diversity of uncultured viral communities using metagenomic information. BMC Bioinform 6:1–9

Angly FE, Felts B, Breitbart M, Salamon P, Edwards RA, Carlson C, Chan AM, Haynes M, Kelley S, Liu H, Mahaffy JM, Mueller JE, Nulton J, Olson R, Parsons R, Rayhawk S, Suttle CA, Rohwer F (2006) PLoS Biol 4:2121–2131

Ansari MH, Ebrahimi M, Fattahi MR, Gardner MG, Safarpour AR, Faghihi MA, Lankarani KB (2020) Viral metagenomic analysis of fecal samples reveals an enteric virome signature in irritable bowel syndrome. BMC Microbiol 20(1):123

Anthenelli M, Jasien E, Edwards R, Bailey B, Felts B, Katira P, Nulton J, Salamon P, Rohwer F, Silveira CB, Luque A (2020) Phage and bacteria diversification through a prophage acquisition ratchet, p 2020.04.08.028340. https://doi.org/10.1101/2020.04.08.028340

Avrani S, Schwartz DA, Lindell D (2012) Virus-host swinging party in the oceans: incorporating biological complexity into paradigms of antagonistic coexistence. Mob Genet Elem 2(2):88–95

Ayukekbong J, Lindh M, Nenonen N, Tah F, Nkuo-Akenji T, Bergström T (2011) Enteric viruses in healthy children in Cameroon: viral load and genotyping of norovirus strains. J Med Virol 83 (12):2135–2142

Bagaglio S, Sitia G, Prati D, Cella D, Hasson H, Novati R, Lazzarin A, Morsica G (2002) Mother-to-child transmission of TT virus: sequence analysis of non-coding region of TT virus in infected mother-infant pairs. Arch Virol 147(4):803–812

Baldridge MT, Nice TJ, McCune BT, Yokoyama CC, Kambal A, Wheadon M, Diamond MS, Ivanova Y, Artyomov M, Virgin HW (2015) Commensal microbes and interferon-λ determine persistence of enteric murine norovirus infection. Science 347(6219):266–269

Banyai K, Jakab F, Reuter G et al (2003) Sequence heterogeneity among human picobirnaviruses detected in a gastroenteritis outbreak. Arch Virol 148(12):2281–2291

Bao H-D, Pang M, Olaniran A, Zhang X-H, Zhang H, Zhou Y, Sun L-C, Schmidt S, Wang R (2018) Alterations in the diversity and composition of mice gut microbiota by lytic or temperate gut phage treatment. Appl Microbiol Biotechnol 102(23):10219–10230

Barr JJ, Auro R, Furlan M et al (2013) Bacteriophage adhering to mucus provide a non-host-derived immunity. Proc Natl Acad Sci U S A 110:10771–10776

Barreira DMPG, Ferreira MSR, Fumian TM, Checon R, de Sadovsky ADI, Leite JPG, Miagostovich MP, Spano LC (2010) Viral load and genotypes of noroviruses in symptomatic and asymptomatic children in southeastern Brazil. J Clin Virol 47(1):60–64

Basson AR, Zhou Y, Seo B, Rodriguez-Palacios A, Cominelli F (2020) Autologous fecal microbiota transplantation for the treatment of inflammatory bowel disease. Transl Res 226:1–11. https://doi.org/10.1016/j.trsl.2020.05.008

Beatty JK, Bhargava A, Buret AG (2014) Post-infectious irritable bowel syndrome: mechanistic insights into chronic disturbances following enteric infection. World J Gastroenterol 20:3976–3985

Beller L, Matthijnssens J (2019) What is (not) known about the dynamics of the human gut virome in health and disease. Curr Opin Virol 37:52–57

Belshaw R, Pereira V, Katzourakis A, Talbot G, Pačes J, Burt A, Tristem M (2004) Long-term reinfection of the human genome by endogenous retroviruses. Proc Natl Acad Sci 101 (14):4894–4899

Berger AK, Mainou BA (2018) Interactions between enteric Bacteria and eukaryotic viruses impact the outcome of infection. Viruses 10(1):19. https://doi.org/10.3390/v10010019

Biagini P, Gallian P, Cantaloube JF, De Micco P, de Lamballerie X (1998) Presence of TT virus in French blood donors and intravenous drug users. J Hepatol 29(4):684–685

Blaise S, de Parseval N, Bénit L, Heidmann T (2003) Genomewide screening for fusogenic human endogenous retrovirus envelopes identifies syncytin 2, a gene conserved on primate evolution. Proc Natl Acad Sci 100(22):13013–13018

Blekhman R, Goodrich JK, Huang K, Sun Q, Bukowski R, Bell JT, Spector TD, Keinan A, Ley RE, Gevers D et al (2015) Host genetic variation impacts microbiome composition across human body sites. Genome Biol 16:191

Bonvicini F, Manaresi E, Gentilomi GA, Di Furio F, Zerbini M, Musiani M, Gallinella G (2011) Evidence of human bocavirus viremia in healthy blood donors. Diagn Microbiol Infect Dis 71 (4):460–462

Bouziat R, Hinterleitner R, Brown JJ, Stencel-Baerenwald JE, Ikizler M, Mayassi T, Meisel M, Kim SM, Discepolo V, Pruijssers AJ, Ernest JD, Iskarpatyoti JA, Costes LMM, Lawrence I, Palanski BA, Varma M, Zurenski MA, Khomandiak S, McAllister N et al (2017) Reovirus infection triggers inflammatory responses to dietary antigens and development of celiac disease. Science 356(6333):44–50

Breitbart M, Rohwer F (2005) Method for discovering novel DNA viruses in blood using viral particle selection and shotgun sequencing. Biotechniques 39(5):729–736

Breitbart M, Hewson I, Felts B, Mahaffy JM, Nulton J, Salamon P, Rohwer F (2003) J Bacteriol 185:6220–6223

Brister JR, Ako-Adjei D, Bao Y, Blinkova O (2015) Nucleic Acids Res 43:571–577

Broecker F, Russo G, Klumpp J, Moelling K (2017) Stable core virome despite variable microbiome after fecal transfer. Gut Microbe 8(3):214–220

Brown CT, Davis-Richardson AG, Giongo A, Gano KA, Crabb DB, Mukherjee N, Casella G, Drew JC, Ilonen J, Knip M et al (2011) Gut microbiome metagenomics analysis suggests a functional model for the development of autoimmunity for type 1 diabetes. PLoS One 6:e25792

Brown MR, Camézuli S, Davenport RJ, Petelenz-Kurdziel E, Øvreås L, Curtis TP (2015) Flow cytometric quantification of viruses in activated sludge. Water Res 68:414–422

Cadranel JF, Di Martino V, Lambrey G et al (2008) Prevalence of hepatitis C infection and risk factors in hospitalized diabetic patients: results of a cross-sectional study. Eur J Gastroenterol Hepatol 20:829–836

Callanan J, Stockdale SR, Shkoporov A, Draper LA, Ross RP, Hill C (2018) Viruses 10:1–17

Callier V (2019) Core concept: gene transfers from bacteria and viruses may be shaping complex organisms. Proc Natl Acad Sci U S A 116(28):13714–13716

Cani PD, Geurts L, Matamoros S et al (2014) Glucose metabolism: focus on gut micro-biota, the endocannabinoid system and beyond. Diabete Metab 40:246–257

Carter CJ (2010) Familial and late-onset Alzheimer's disease: evidence for an auto-immune component triggered by viral, microbial and allergen antigens with homology to beta-amyloid and APP mutant peptides. Nat Preced 1:4662

Castro-Mejía JL, Muhammed MK, Kot W, Neve H, Franz CMAP, Hansen LH, Vogensen FK, Nielsen DS (2015) Optimizing protocols for extraction of bacteriophages prior to metagenomic analyses of phage communities in the human gut. Microbiome 3:64

Celiberto LS, Graef FA, Healey GR, Bosman ES, Jacobson K, Sly LM, Vallance BA (2018) Inflammatory bowel disease and immunonutrition: novel therapeutic approaches through modulation of diet and the gut microbiome. Immunology 155:36–52

Chehoud C, Dryga A, Hwang Y, Nagy-Szakal D, Hollister EB, Luna RA, Versalovic J, Kellermayer R, Bushman FD (2016) Transfer of viral communities between human individuals during fecal microbiota transplantation. MBio 7(2):e00322

Chen Y, Williams V, Filippova M, Filippov V, Duerksen-Hughes P (2014a) Viral carcinogenesis: factors inducing DNA damage and virus integration. Cancers (Basel) 6:2155–2186

Chen M, Song P, Zou D, Hu X, Zhao S, Gao S, Ling F (2014b) PLoS One 9:1–12

Chua LL, Rajasuriar R, Azanan MS, Abdullah NK, Tang MS, Lee SC, Woo YL, Lim YAL, Arian H, Loke P (2017) Reduced microbial diversity in adult survivors of childhood acute lymphoblastic leukemia and microbial associations with increased immune activation. Microbiome 5:35

Cieplak T, Soffer N, Sulakvelidze A, Nielsen DS (2018) A bacteriophage cocktail targeting Escherichia coli reduces E. coli in simulated gut conditions, while preserving a non-targeted representative commensal normal microbiota. Gut Microbe 9(5):391–399

Clemente JC, Pehrsson EC, Blaser MJ, Sandhu K, Gao Z, Wang B, Magris M, Hidalgo G, Contreras M, Noya-Alarcón Ó et al (2015) The microbiome of uncontacted Amerindians. Sci Adv 1:e1500183

Clooney AG, Sutton TDS, Shkoporov AN, Holohan RK, Daly KM, O'Regan O, Ryan FJ, Draper LA, Plevy SE, Ross RP et al (2019) Whole-virome analysis sheds light on viral dark matter in inflammatory bowel disease. Cell Host Microbe 26:764–778

Conceição-Neto N, Zeller M, Lefrère H, De Bruyn P, Beller L, Deboutte W, Yinda CK, Lavigne R, Maes P, Van Ranst M, Heylen E, Matthijnssens J (2015) Sci Rep 5:1–14

Conceição-Neto N, Deboutte W, Dierckx T, Machiels K, Wang J, Yinda KC, Maes P, Van Ranst M, Joossens M, Raes J, Vermeire S, Matthijnssens J (2018) Low eukaryotic viral richness is associated with faecal microbiota transplantation success in patients with UC [Review of Low eukaryotic viral richness is associated with faecal microbiota transplantation success in patients with UC]. Gut 67(8):1558–1559

Cotten M, Oude Munnink B, Canuti M, Deijs M, Watson SJ, Kellam P, Van Der Hoek L (2014) PLoS One 9:e93269

Dalmasso M, Hill C, Ross RP (2014) Exploiting gut bacteriophages for human health. Trends Microbiol 22:399–405

David LA, Maurice CF, Carmody RN, Gootenberg DB, Button JE, Wolfe BE, Ling AV, Devlin AS, Varma Y, Fischbach MA et al (2014) Diet rapidly and reproducibly alters the human gut microbiome. Nature 505:559–563

De Beeck AO, Eizirik DL (2016) Viral infections in type 1 diabetes mellitus—why the [beta] cells? Nat Rev Endocrinol 12(5):263–273. https://doi.org/10.1038/nrendo.30

de Jonge PA, Nobrega FL, Brouns SJJ, Dutilh BE (2019) Molecular and evolutionary determinants of bacteriophage host range. Trends Microbiol 27(1):51–63

de Vries M, Deijs M, Canuti M, van Schaik BDC, Faria NR, van de Garde MDB, Jachimowski LCM, Jebbink MF, Jakobs M, Luyf ACM, Coenjaerts FEJ, Claas ECJ, Molenkamp R, Koekkoek SM, Lammens C, Leus F, Goossens H, Ieven M, Baas F, van der Hoek L (2011) A sensitive assay for virus discovery in respiratory clinical samples. PLoS One 6

Delbue S, Tremolada S, Elia F, Carloni C, Amico S, Tavazzi E, Marchioni E, Novati S, Maserati R, Ferrante P (2010) Lymphotropic polyomavirus is detected in peripheral blood from immunocompromised and healthy subjects. J Clin Virol 47(2):156–160

Delmas J, Gibold L, Faïs T, Batista S, Leremboure M, Sinel C, Vazeille E, Cattoir V, Buisson A, Barnich N, Dalmasso G, Bonnet R (2019) Metabolic adaptation of adherent-invasive Escherichia coli to exposure to bile salts. Sci Rep 9(1):2175

Delwart E (2007) Viral metagenomics. Rev Med Virol 17:115–131

Devaraj S, Hemarajata P, Versalovic J (2013) The human gut microbiome andbody metabolism: implications for obesity and diabetes. Clin Chem 59:617–628

Dinleyici EC, Eren M, Ozen M et al (2012) Effectiveness and safety of Saccharomyces boulardii for acute infectious diarrhea. Expert Opin Biol Ther 12:395–410

Dissanayake U, Ukhanova M, Moye ZD, Sulakvelidze A, Mai V (2019) Bacteriophages reduce pathogenic Escherichia coli counts in mice without distorting gut microbiota. Front Microbiol 10:1984

Divya Ganeshan S, Hosseinidoust Z (2019) Phage therapy with a focus on the human microbiota. Antibiotics (Basel, Switzerland) 8:131. https://doi.org/10.3390/antibiotics8030131

Dominguez-Bello MG, Costello EK, Contreras M, Magris M, Hidalgo G, Fierer N, Knight R (2010) Delivery mode shapes the acquisition and structure of the initial microbiota across multiple body habitats in newborns. Proc Natl Acad Sci U S A 107:11971–11975

Domínguez-Díaz C, García-Orozco A, Riera-Leal A, Padilla-Arellano JR, Fafutis-Morris M (2019) Microbiota and its role on viral evasion: is it with us or against us? Front Cell Infect Microbiol 9:256

Dotta F, Censini S, van Halteren AGS, Marselli L, Masini M, Dionisi S et al (2007) Coxsackie B4 virus infection of beta cells and natural killer cell insulitis in recent-onset type 1 diabetic patients. Proc Natl Acad Sci U S A 104(12):5115–5120

Draper LA, Ryan FJ, Smith MK, Jalanka J, Mattila E, Arkkila PA, Ross RP, Satokari R, Hill C (2018) Long-term colonisation with donor bacteriophages following successful faecal microbial transplantation. Microbiome 6(1):220

Duerkop BA, Kleiner M, Paez-Espino D, ZhuW BB, Hassell B, Winter SE, Kyrpides NC, Hooper LV (2018) Murine colitis reveals a disease associated bacteriophage community. Nat Microbiol 3:11023–11031

DuPont HL, Jiang Z-D, DuPont AW, Utay NS (2020) Abnormal intestinal microbiome in medical disorders and potential reversibility by fecal microbiota transplantation. Dig Dis Sci 65 (3):741–756

Džunková M, D'Auria G, Moya A (2015) Direct sequencing of human gut virome fractions obtained by flow cytometry. Front Microbiol 6:955

Eckardt AJ, Baumgart DC (2011) Viral gastroenteritis in adults. Recent Pat Antiinfect Drug Discov 6(1):54–63

Edwards RA, McNair K, Faust K, Raes J, Dutilh BE (2016) Computational approaches to predict bacteriophage-host relationships. FEMS Microbiol Rev 40(2):258–272

Egli A, Infanti L, Dumoulin A, Buser A, Samaridis J, Stebler C, Gosert R, Hirsch HH (2009) Prevalence of polyomavirus BK and JC infection and replication in 400 healthy blood donors. J Infect Dis 199(6):837–846

Emerson JB, Roux S, Brum JR, Bolduc B, Woodcroft BJ, Jang HB, Singleton CM, Solden LM, Naas AE, Boyd JA, Hodgkins SB, Wilson RM, Trubl G, Li C, Frolking S, Pope PB, Wrighton KC, Crill PM, Chanton JP, Saleska SR, Tyson GW, Rich VI, Sullivan MB (2018) Nat. Microbiology 3:870–880

Erickson AK, Jesudhasan PR, Mayer MJ, Narbad A, Winter SE, Pfeiffer JK (2018) Bacteria facilitate enteric virus co-infection of mammalian cells and promote genetic recombination. Cell Host Microbe 23(1):77–88.e5

Eszter Muller K, Laszlo Lakatos P, Papp M et al (2014) Incidence and Paris classification of pediatric inflammatory bowel disease. Gastroenterol Res Pract 2014:904307

Faruque SM, Mekalanos JJ (2012) Phage-bacterial interactions in the evolution of toxigenic Vibrio cholerae. Virulence 3(7):556–565

Fernandes MA, Verstraete SG, Phan TG, Deng X, Stekol E, LaMere B, Lynch SV, Heyman MB, Delwart E (2019) Enteric virome and bacterial microbiota in children with ulcerative colitis and Crohn disease. J Pediatr Gastroenterol Nutr 68(1):30–36

Fields BN, Knipe DM, Howley PM (2007) Fields' virology, 5th edn. Lippincott-Williams, Philadelphia, PA

Filippi CM, Herrath MG (2008) von viral trigger for type 1 diabetes. Diabetes 57:2863–2871

Fouts DE (2006) Phage_Finder: automated identification and classification of prophage regions in complete bacterial genome sequences. Nucleic Acids Res 34:5839–5851

Foxman EF, Iwasaki A (2011) Genome-virome interactions: examining the role of common viral infections in complex disease. Nat Rev Microbiol 9(4):254–264

Francino MP (2016) Antibiotics and the human gut microbiome: dysbioses and accumulation of resistances. Front Microbiol 6:1543

Frazão N, Sousa A, Lässig M, Gordo I (2019) Horizontal gene transfer overrides mutation in Escherichia coli colonizing the mammalian gut. Proc Natl Acad Sci U S A 116 (36):17906–17915

Friedman SL (2013) Liver fibrosis in 2012: convergent pathways that cause hepatic fibrosis in NASH. Nat Rev Gastroenterol Hepatol 10:71–72

Fryer JF, Delwart E, Hecht FM, Bernardin F, Jones MS, Shah N, Baylis SA (2007) Frequent detection of the parvoviruses, PARV4 and PARV5, in plasma from blood donors and symptomatic individuals. Transfusion 47(6):1054–1061

Gabbay YB, da Luz CRN, Costa IV, Cavalcante-Pepino EL, Sousa MS, Oliveira KK, Wanzeller ALM, Mascarenhas JD, Leite JPG, Linhares AC (2005) Prevalence and genetic diversity of astroviruses in children with and without diarrhea in São Luís, Maranhão, Brazil. Mem Inst Oswaldo Cruz 100(7):709–714

Galiez C, Siebert M, Enault F, Vincent J, Söding J (2017) WIsH: who is the host? Predicting prokaryotic hosts from metagenomic phage contigs. Bioinformatics 33(19):3113–3114
Garmaeva S, Sinha T, Kurilshikov A, Fu J, Wijmenga C, Zhernakova A (2019) BMC Biol 17:1–14
Geurts L, Neyrinck AM, Delzenne NM et al (2014) Gut microbiota controls adipose tissue expansion, gut barrier and glucose metabolism: novel insights into molecular targets and interventions using prebiotics. Benef Microb 5:3–17
Ghazarian L, Diana J, Simoni Y, Beaudoin L, Lehuen A (2012) Prevention or acceleration of type 1 diabetes by viruses. Cell Mol Life Sci 70(2):239–255. https://doi.org/10.1007/s00018-012-1042-1
Girons IS, Bourhy P, Ottone C et al (2000) The LE1 bacteriophage replicates as a plasmid within Leptospira biflexa: construction of an L. biflexa– Escherichia coli shuttle vector. J Bacteriol 182:5700–5705
Gogokhia L, Buhrke K, Bell R, Hoffman B, Brown DG, Hanke-Gogokhia C, Ajami NJ, Wong MC, Ghazaryan A, Valentine JF, Porter N, Martens E, O'Connell R, Jacob V, Scherl E, Crawford C, Stephens WZ, Casjens SR, Longman RS, Round JL (2019) Expansion of bacteriophages is linked to aggravated intestinal inflammation and colitis. Cell Host Microbe 25(2):285–299.e8
Gómara MI, Kang G, Mammen A, Jana AK, Abraham M, Desselberger U, Brown D, Gray J (2004) Characterization of G10P [11] rotaviruses causing acute gastroenteritis in neonates and infants in Vellore, India. J Clin Microbiol 42(6):2541–2547
Gopalakrishnan V, Spencer CN, Nezi L, Reuben A, Andrews MC, Karpinets TV, Prieto PA, Vicente D, Homan K, Wei SC et al (2018) Gut microbiome modulates response to anti-PD-1 immunotherapy in melanoma patients. Science 359:97–103
Górska A, Peter S, Willmann M, Autenrieth I, Schlaberg R, Huson DH (2018) Dynamics of the human gut phageome during antibiotic treatment. Comput Biol Chem 74:420–427
Greninger AL, Runckel C, Chiu CY, Haggerty T, Parsonnet J, Ganem D, DeRisi JL (2009) The complete genome of klassevirus–a novel picornavirus in pediatric stool. Virol J 6(1):82
Guarner C, Soriano G (2005) Bacterial translocation and its consequences in patients with cirrhosis. Eur J Gastroenterol Hepatol 17(1):27–31
Guerin E, Shkoporov A, Stockdale S et al (2018) Biology and taxonomy of crAss-like bacteriophages, the most abundant virus in the human gut. Cell Host Microbe 24:653–664
Guo M, Liu G, Chen J, Ma J, Lin J, Fu Y, Fan G, Lee SM-Y, Zhang L (2020) Dynamics of bacteriophages in gut of giant pandas reveal a potential regulation of dietary intake on bacteriophage composition. Sci Total Environ 734:139424
Gupta S, Allen-Vercoe E, Petrof EO (2016) Fecal microbiota transplantation: in perspective. Therap Adv Gastroenterol 9(2):229–239
Halfvarson J, Brislawn CJ, Lamendella R, Vázquez-Baeza Y, Walters WA, Bramer LM, D'Amato M, Bonfiglio F, McDonald D, Gonzalez A et al (2017) Dynamics of the human gut microbiome in inflammatory bowel disease. Nat Microbiol 2:17004
Hamdi S, Rousseau GM, Labrie SJ, Tremblay DM, Kourda RS, Slama KB, Moineau S (2017) Characterization of two polyvalent phages infecting Enterobacteriaceae. Sci Rep 7:40349
Hannigan GD, Duhaime MB, Run MT, Koumpouras CC, Schloss PD (2018) Diagnostic potential and interactive dynamics of the colorectal cancer virome. MBio 9
Haynes M, Rohwer F (2011) The human virome. Springer, New York, NY, pp 63–77
Helmersson J, Vessby B, Larsson A, Basu S (2004) Association of type 2 diabetes with cyclooxygenase-mediated inflammation and oxidative stress in an elderly population. Circulation 109:1729–1734
Hillman ET, Lu H, Yao T, Nakatsu CH (2017) Microbial ecology along the gastrointestinal tract. Microb Environ 32(4):300–313
Hindiyeh M, Mor O, Pando R, Mannasse B, Kabat A, Assraf-Zarfati H, Mendelson E, Sofer D, Mandelboim M (2019) Comparison of the new fully automated extraction platform eMAG to the MagNA PURE 96 and the well-established easyMAG for detection of common human respiratory viruses. PLoS One 14:e0211079
Hofer U (2013) Variation in the gut virome. Nat Rev Microbiol 11(9):596

Holtz LR, Cao S, Zhao G et al (2014) Geographic variation in the eukaryotic virome of human diarrhea. Virology 468–470:556–564
Hooper LV, Macpherson AJ (2010) Immune adaptations that maintain homeostasis with the intestinal microbiota. Nat Rev Immunol 10(3):159–169
Hoyles L, McCartney AL, Neve H, Gibson GR, Sanderson JD, Heller KJ, van Sinderen D (2014) Characterization of virus-like particles associated with the human faecal and caecal microbiota. Res Microbiol 165:803–812
Hsu BB, Gibson TE, Yeliseyev V, Liu Q, Bry L, Silver PA, Gerber GK (2018) Bacteriophages dynamically modulate the gut microbiota and metabolome, p 454579. https://doi.org/10.1101/454579
Hu YOO, Hugerth LW, Bengtsson C, Alisjahbana A, Seifert M, Kamal A, Sjöling Å, Midtvedt T, Norin E, Du J, Engstrand L (2018) Bacteriophages synergize with the gut microbial community to combat Salmonella. mSystems 3(5):e00119. https://doi.org/10.1128/mSystems.00119-18
Huang PN (2020) Microbiota and enteric viruses infection. Med Microecol 3:100006
Hunter P (2013) The secret garden's gardeners: research increasingly appreciates the crucial role of gut viruses for human health and disease. EMBO Rep 14:683–685
Jaeckel E, Manns M, von Herrath M (2002) Viruses and diabetes. Ann N Y Acad Sci 958:7–25
Jones MS, Kapoor A, Lukashov VV, Simmonds P, Hecht F, Delwart E (2005) New DNA viruses identified in patients with acute viral infection syndrome. J Virol 79(13):8230–8236
Jun HS, Yoon JW (2003) A new look at viruses in type 1 diabetes. Diabetes Metab Res Rev 19 (1):8–31
Kahrs CR, Chuda K, Tapia G, Stene LC, Mårild K, Rasmussen T, Rønningen KS, Lundin KEA, Kramna L, Cinek O, Størdal K (2019) Enterovirus as trigger of coeliac disease: nested case-control study within prospective birth cohort. BMJ 364:l231
Kane M, Case LK, Kopaskie K et al (2011) Successful transmission of a retrovirus depends on the commensal microbiota. Science 334:245–249
Kapusinszky B, Minor P, Delwart E (2012) Nearly constant shedding of diverse enteric viruses by two healthy infants. J Clin Microbiol 50(11):3427–3434
Karim S, Mirza Z, Kamal MA, Abuzenadah AM, Azhar EI, Al-Qahtani MH, Sohrab SS (2014) An association of virus infection with type 2 diabetes and Alzheimer's disease. CNS Neurol Disord Drug Targets 13(3):429–439. https://doi.org/10.2174/18715273113126660164
Kent AG, Vill AC, Shi Q, Satlin MJ, Brito IL (2020) Widespread transfer of mobile antibiotic resistance genes within individual gut microbiomes revealed through bacterial hi-C. Nat Commun 11(1):4379
Khan R, Petersen FC, Shekhar S (2019) Commensal bacteria: an emerging player in defense against respiratory pathogens. Front Immunol 10:1203
Kim KH, Chang HW, Do Nam Y, Roh SW, Kim MS, Sung Y, Jeon CO, Oh HM, Bae JW (2008) Amplification of uncultured single-stranded DNA viruses from rice paddy soil. Appl Environ Microbiol 74:5975–5985
King ML, Shaikh A, Bidwell D, Voller A, Banatvala JE (1983) Coxsackie- B-virus-specific IgM responses in children with insulin-dependent (juvenile-onset; type I) diabetes mellitus. Lancet 1 (8339):1397–1399
Kleiner M, Hooper LV, Duerkop BA (2015) Evaluation of methods to purify virus-like particles for metagenomic sequencing of intestinal viromes. BMC Genomics 16:1–15
Koichi Y, Arvin AM, Gabriella C et al (2007) Human herpesviruses: biology, therapy, and immunoprophylaxis. Cambridge University Press, Cambridge
Koskella B, Brockhurst MA (2014) Bacteria–phage coevolution as a driver of ecological and evolutionary processes in microbial communities. FEMS Microbiol Rev 38(5):916–931. https://doi.org/10.1111/1574-6976.12072
Koskella B, Lively CM (2009) Evidence for negative frequency-dependent selection during experimental coevolution of a freshwater snail and a sterilizing trematode. Evol Int J Organ Evol 63(9):2213–2221. https://doi.org/10.1111/j.1558-5646.2009.00711.x

Kostic AD, Xavier RJ, Gevers D (2014) The microbiome in inflammatory bowel disease: current status and the future ahead. Gastroenterology 146(6):1489–1499. https://doi.org/10.1053/j.gastro.2014.02.009

Krishnamurthy SR, Wang D (2017) Origins and challenges of viral dark matter. Virus Res 239:136–142

Kroner Z (2009) The relationship between Alzheimer's disease and diabetes: type 3 diabetes? Altern Med Rev 14:373–379

Krüger A, Lucchesi PMA (2015) Shiga toxins and stx phages: highly diverse entities. Microbiology 161(Pt 3):451–462

Kumar A, Murthy S, Kapoor A (2017) Evolution of selective-sequencing approaches for virus discovery and virome analysis. Virus Res 239:172–179

Kumarasamy KK, Toleman MA, Walsh TR et al (2010) Emergence of a new antibi-otic resistance mechanism in India, Pakistan, and the UK: a molecular, biological, and epidemiological study. Lancet Infect Dis 10:597–602

Laforest-Lapointe I, Arrieta M-C (2018) Microbial eukaryotes: a missing link in gut microbiome studies. mSystems 3(2):e00201. https://doi.org/10.1128/mSystems.00201-17

Lagier JC, Million M, Hugon P et al (2012) Human gut microbiota: repertoire and variations. Front Cell Infect Microbiol 2:136

Lam KC, Vyshenska D, Hu J, Rodrigues RR, Nilsen A, Zielke RA, Brown NS, Aarnes E-K, Sikora AE, Shulzhenko N et al (2018) Transkingdom network reveals bacterial players associated with cervical cancer gene expression program. Peer J 6:e5590

Larsen N, Vogensen FK, van Den Berg FWJ, Nielsen DS, Andreasen AS, Pedersen BK, Al-Soud WA, Sørensen SJ, Hansen LH, Jakobsen M (2010) Gut microbiota in human adults with type 2 diabetes differs from non-diabetic adults. PLoS One 5:e9085

Lee JG, Han DS, Jo SV, Lee AR, Park CH, Eun CS, Lee Y (2019) Characteristics and pathogenic role of adherent-invasive Escherichia coli in inflammatory bowel disease: potential impact on clinical outcomes. PLoS One 14(4):e0216165

Li L, He Y, Yang H (2005) Genetic characteristics of human Enterovirus 71 and Coxsackievirus A16 circulating from 1999 to 2004 in Shenzhen, People's Republic of China. J Clin Microbiol 43(8):3835–3839

Lim ES, Zhou Y, Zhao G, Bauer IK, Droit L, Ndao IM, Warner BB, Tarr PI, Wang D, Holtz LR (2015) Early life dynamics of the human gut virome and bacterial microbiome in infants. Nat Med 21(10):1228–1234

Lima-Mendez G, Van Helden J, Toussaint A, Leplae R (2008) Prophinder: a computational tool for prophage prediction in prokaryotic genomes. Bioinformatics 24:863–865

Lin DM, Lin HC (2019) A theoretical model of temperate phages as mediators of gut microbiome dysbiosis. F1000Research 8. https://doi.org/10.12688/f1000research.18480.1

Lin R, Liu W, Piao M, Zhu H (2017) A review of the relationship between the gut microbiota and amino acid metabolism. Amino Acids 49:2083–2090

Lindfors K, Lin J, Lee H-S, Hyöty H, Nykter M, Kurppa K, Liu E, Koletzko S, Rewers M, Hagopian W, Toppari J, Ziegler A-G, Akolkar B, Krischer JP, Petrosino JF, Lloyd RE, Agardh D, TEDDY Study Group (2020) Metagenomics of the faecal virome indicate a cumulative effect of enterovirus and gluten amount on the risk of coeliac disease autoimmunity in genetically at risk children: the TEDDY study. Gut 69(8):1416–1422

Lipman TH, Chang Y, Murphy KM (2002) The epidemiology of type 1 diabetes in children in Philadelphia 1990–1994: evidence of an epidemic. Diabetes Care 25(11):1969–1975

Liu M, Deora R, Doulatov SR et al (2002) Reverse transcriptase-mediated tropism switching in Bordetella bacteriophage. Science 295:2091–2094

Llanos-Chea A, Citorik RJ, Nickerson KP, Ingano L, Serena G, Senger S, Lu TK, Fasano A, Faherty CS (2019) Bacteriophage therapy testing against Shigella flexneri in a novel human intestinal organoid-derived infection model. J Pediatr Gastroenterol Nutr 68(4):509–516

Lozupone CA, Stombaugh JI, Gordon JI et al (2012) Diversity, stability and resilience of the human gut microbiota. Nature 489:220–230

Łusiak-Szelachowska M, Weber-Dąbrowska B, Jończyk-Matysiak E, Wojciechowska R, Górski A (2017) Bacteriophages in the gastrointestinal tract and their implications. Gut Pathog 9:44

Lynch SV, Pedersen O (2016) The human intestinal microbiome in health and disease. N Engl J Med 375(24):2369–2379. https://doi.org/10.1056/NEJMra1600266

Ma Y, You X, Mai G, Tokuyasu T, Liu C (2018) A human gut phage catalog correlates the gut phageome with type 2 diabetes. Microbiome 6:24

Mai V, Draganov PV (2009) Recent advances and remaining gaps in our knowledge of associations between gut microbiota and human health. World J Gastroenterol 15:81–85

Maisonneuve C, Irrazabal T, Martin A, Girardin SE, Philpott DJ (2017) The impact of the gut microbiome on colorectal cancer. Annu Rev Cancer Biol 2:229–249

Manoussopoulos YN, Anastassopoulou CG (2020) Virome: the prodigious little cousin of the family. In: Microbiomics, pp 53–73

Manrique P, Dills M, Young M (2017) The human gut phage community and its implications for health and disease. Viruses 9:141. https://doi.org/10.3390/v9060141

Marbouty M, Thierry A, Koszul R (2020) Phages—bacteria interactions network of the healthy human gut. https://doi.org/10.1101/2020.05.13.093716

Marinelli LJ, Fitz-Gibbon S, Hayes C, Bowman C, Inkeles M, Loncaric A, Russell DA, Jacobs-Sera D, Cokus S, Pellegrini M, Kim J (2012) Propionibacterium acnes bacteriophages display limited genetic diversity and broad killing activity against bacterial skin isolates. MBio 3(5): e00279

Markine-Goriaynoff N, Gillet L, Van Etten JL et al (2004) Glycosyltransferases encoded by viruses. J Gen Virol 85:2741–2754

Maronek M, Link R, Ambro L, Gardlik R (2020) Phages and their role in gastrointestinal disease: focus on inflammatory bowel disease. Cell 9(4):1013. https://doi.org/10.3390/cells9041013

Martín R, Langella P (2019) Emerging health concepts in the probiotics field: streamlining the definitions. Front Microbiol 10:1047

Maruvada P, Leone V, Kaplan LM, Chang EB (2017) The human microbiome and obesity: moving beyond associations. Cell Host Microbe 22(5):589–599

Marz M, Beerenwinkel N, Drosten C, Fricke M, Frishman D, Hofacker IL, Hoffmann D, Middendorf M, Rattei T, Stadler PF, Töpfer A (2014) Bioinformatics 30:1793–1799

Maslov S, Sneppen K (2017) Population cycles and species diversity in dynamic kill-the-winner model of microbial ecosystems. Sci Rep 7:39642

Matson V, Fessler J, Bao R, Chongsuwat T, Zha Y, Alegre ML, Luke JJ, Gajewski TF (2018) The commensal microbiome is associated with anti-PD-1 efficacy in metastatic melanoma patients. Science 359:104–108

Maurice CF (2019) Considering the other half of the gut microbiome: bacteriophages. mSystems 4 (3). https://doi.org/10.1128/mSystems.00102-19

McGeoch DJ, Rixon FJ, Davison AJ (2006) Topics in herpesvirus genomics and evolution. Virus Res 117(1):90–104

McGinnis S, Madden TL (2004) BLAST: at the core of a powerful and diverse set of sequence analysis tools. Nucleic Acids Res 32:20–25

McLaughlin MR, Balaa MF, Sims J, King R (2006) Isolation of Salmonella bacteriophages from swine effluent lagoons. J Environ Qual 35(2):522–528

Méndez-Toss M, Griffin DD, Calva J, Contreras JF, Puerto FI, Mota F, Guiscafré H, Cedillo R, Munoz O, Herrera I, López S (2004) Prevalence and genetic diversity of human astroviruses in Mexican children with symptomatic and asymptomatic infections. J Clin Microbiol 42 (1):151–157

Mettenleiter TC, Klupp BG, Granzow H (2006) Herpesvirus assembly: a tale of two membranes. Curr Opin Microbiol 9(4):423–429

Mi S, Lee X, Li XP, Veldman GM, Finnerty H, Racie L, LaVallie E, Tang XY, Edouard P, Howes S, Keith JC (2000) Syncytin is a captive retroviral envelope protein involved in human placental morphogenesis. Nature 403(6771):785–789

Minot S, Sinha R, Chen J et al (2011) The human gut virome: interindividual variation and dynamic response to diet. Genome Res 21(10):1616–1625

Minot S, Bryson A, Chehoud C, Wu GD, Lewis JD, Bushman FD (2013) Rapid evolution of the human gut virome. Proc Natl Acad Sci U S A 110(30):12450–12455

Mirzaei MK, Maurice CF (2017) Ménage à trois in the human gut: interactions between host, bacteria and phages. Nat Rev Microbiol 15(7):397–408

Mirzaei MK, Khan MAA, Ghosh P, Taranu ZE, Taguer M, Ru J, Chowdhury R, Kabir MM, Deng L, Mondal D, Maurice CF (2020) Bacteriophages isolated from stunted children can regulate gut bacterial communities in an age-specific manner. Cell Host Microbe 27 (2):199–212.e5

Moeller AH, Caro-Quintero A, Mjungu D et al (2016) Cospeciation of gut microbiota with hominids. Science 353:380–382

Mokili JL, Rohwer F, Dutilh BE (2012) Metagenomics and future perspectives in virus discovery. Curr Opin Virol 2:63–77

Montassier E, Gastinne T, Vangay P, Al-Ghalith GA, Bruley Des Varannes S, Massart S, Moreau P, Potel G, de La Cochetière MF, Batard E et al (2015) Chemotherapy-driven dysbiosis in the intestinal microbiome. Aliment Pharmacol Ther 42:515–528

Morahan G (2012) Insights into type 1 diabetes provided by genetic analyses. Curr Opin Endocrinol Diabetes Obes 19:263–270

Morgan XC, Tickle TL, Sokol H, Gevers D, Devaney KL, Ward DV, Reyes JA, Shah SA, LeLeiko N, Snapper SB et al (2012) Dysfunction of the intestinal microbiome in inflammatory bowel disease and treatment. Genome Biol 13:R79

Mu Q, Tavella VJ, Luo XM (2018) Role of Lactobacillus reuteri in human health and diseases. Front Microbiol 9:757

Mukhopadhya I, Segal JP, Carding SR, Hart AL, Hold GL (2019) The gut virome: the "missing link" between gut bacteria and host immunity? Therap Adv Gastroenterol 12:1756284819836620

Nagpal R, Newman TM, Wang S, Jain S, Lovato JF, Yadav H (2018) Obesity-linked gut microbiome Dysbiosis associated with derangements in gut permeability and intestinal cellular homeostasis independent of diet. J Diabetes Res 2018:1–9

Nakatsu G, Zhou H, Wu WKK, Wong SH, Coker OO, Dai Z, Li X, Szeto CH, Sugimura N, Lam TYT et al (2018) Alterations in enteric Virome are associated with colorectal Cancer and survival outcomes. Gastroenterology 155:529–541.e5

Needell JC, Zipris D (2016) The role of the intestinal microbiome in type 1 diabetes pathogenesis. Curr Diab Rep 16:89

Nicholson JK, Holmes E, Kinross J, Burcelin R, Gibson G, Jia W, Pettersson S (2012) Host-gut metabolic interactions. Science 336:1262–1268

Nishizawa T, Okamoto H, Konishi K, Yoshizawa H, Miyakawa Y, Mayumi M (1997) A novel DNA virus (TTV) associated with elevated transaminase levels in posttransfusion hepatitis of unknown etiology. Biochem Biophys Res Commun 241(1):92–97

Norman JM, Handley SA, Baldridge MT, Droit L, Liu CY, Keller BC, Kambal A, Monaco CL, Zhao G, Fleshner P et al (2015) Disease-specific alterations in the enteric virome in inflammatory bowel disease. Cell 160:447–460

Nurk S, Meleshko D, Korobeynikov A, Pevzner PA (2017) metaSPAdes: a new versatile metagenomic assembler. Genome Res 27:824–834

Offermanns S, Rosenthal W (2008) Encyclopedia of molecular pharmacology, 2nd edn. Springer, Berlin, pp 437–438

Oh J-H, Lin XB, Zhang S, Tollenaar SL, Özçam M, Dunphy C, Walter J, van Pijkeren J-P (2019) Prophages in Lactobacillus reuteri are associated with fitness trade-offs but can increase competitiveness in the gut ecosystem. Appl Environ Microbiol 86(1):e01922. https://doi.org/10.1128/AEM.01922-19

Okamoto H, Nishizawa T, Takahashi M, Asabe S, Tsuda F, Yoshikawa A (2001) Heterogeneous distribution of TT virus of distinct genotypes in multiple tissues from infected humans. Virology 288(2):358–368

Okamura A, Yoshioka M, Kubota M, Kikuta H, Ishiko H, Kobayashi K (1999) Detection of a novel DNA virus (TTV) sequence in peripheral blood mononuclear cells. J Med Virol 58(2):174–177

Ott A (1999) Diabetes mellitus and the risk of dementia: the Rotterdam study. Neurology 53:1937–1942

Ott SJ, Waetzig GH, Rehman A, Moltzau-Anderson J, Bharti R, Grasis JA, Cassidy L, Tholey A, Fickenscher H, Seegert D, Rosenstiel P, Schreiber S (2017) Efficacy of sterile fecal filtrate transfer for treating patients with Clostridium difficile infection. Gastroenterology 152 (4):799–811.e7

Palmer C, Bik EM, Di Giulio DB et al (2007) Development of the human infant intestinal microbiota. PLoS Biol 5:e177

Panaccione R (2013) Mechanisms of inflammatory bowel disease. Gastroenterol Hepatol 9 (8):529–532

Park A, Zhao G (2018) Mining the virome for insights into type 1 diabetes. DNA Cell Biol 37 (5):422–425

Patel A, Noble RT, Steele JA, Schwalbach MS, Hewson I, Fuhrman JA (2007) Nat Protoc 2:269–276

Pérez-Brocal V, García-López R, Nos P, Beltrán B, Moret I, Moya A (2015) Metagenomic analysis of Crohn's disease patients identifies changes in the virome and microbiome related to disease status and therapy, and detects potential interactions and biomarkers. Inflamm Bowel Dis 21:2515–2532

Pfeiffer JK, Virgin HW (2016) Viral immunity: transkingdom control of viral infection and immunity in the mammalian intestine. Science 351. https://doi.org/10.1126/science.aad5872

Pickup JC (2004) Inflammation and activated innate immunity in the pathogenesis of type 2 diabetes. Diabetes Care 27:813–823

Pollock J, Glendinning L, Wisedchanwet T, Watson M (2018) The madness of microbiome: attempting to find consensus "best practice" for 16S microbiome studies. Appl Environ Microbiol 84:e02627

Popgeorgiev N, Temmam S, Raoult D, Desnues C (2013) Describing the silent human virome with an emphasis on giant viruses. Intervirology 56(6):395–412

Prado J, Monezi T, Amorim A, Lino V, Paladino A, Boccardo E (2018) Human polyomaviruses and cancer: an overview. Clinics 73:e558s

Qin J, Li R, Raes J, Arumugam M, Burgdorf KS, Manichanh C et al (2010) A human gut microbial gene catalogue established by metagenomic sequencing. Nature 464:59–65. https://doi.org/10.1038/nature08821

Qin J, Li Y, Cai Z et al (2012) A metagenome-wide association study of gut microbiota in type 2 diabetes. Nature 490:55–60

Rachael R (2012) Type 2 diabetes linked to common virus. In: My health news daily, p 1

Rascovan N, Duraisamy R, Desnues C (2016) Metagenomics and the human virome in asymptomatic individuals. Annu Rev Microbiol 70:125–141

Rasmussen TS, Mentzel CMJ, Kot W, Castro-Mejía JL, Zuffa S, Swann JR, Hansen LH, Vogensen FK, Hansen AK, Nielsen DS (2020) Faecal virome transplantation decreases symptoms of type 2 diabetes and obesity in a murine model. Gut 69(12):2122–2130. https://doi.org/10.1136/gutjnl-2019-320005

Reyes A, Haynes M, Hanson N, Angly FE, Andrew C, Rohwer F, Gordon JI (2010) Viruses in the faecal microbiota of monozygotic twins and their mothers. Nature 466:334–338

Reyes A, Semenkovich NP, Whiteson K, Rohwer F, Gordon JI (2012) Going viral: next-generation sequencing applied to phage populations in the human gut. Nat Rev Microbiol 10:607–617. https://doi.org/10.1038/nrmicro2853

Richardson SJ, Willcox A, Bone AJ (2009) The prevalence of enteroviral capsid protein vp1 immunostaining in pancreatic islets in human type 1 diabetes. Diabetologia 52(6):1143–1151

Ridaura VK, Faith JJ, Rey FE et al (2013) Gut microbiota from twins discordant for obesity modulate metabolism in mice. Science 341:1241214

Riesenfeld CS, Schloss PD, Handelsman J (2004) Annu Rev Genet 38:525–552

Robinson CM, Pfeiffer JK (2014) Viruses and the microbiota. Annu Rev Virol 1:55–69

Robinson CM, Jesudhasan PR, Pfeiffer JK (2014) Bacterial lipopolysaccharide binding enhances virion stability and promotes environmental fitness of an enteric virus. Cell Host Microbe 15 (1):36–46

Rohwer F, Seguritan V, Choi DH, Segall AM, Azam F (2001) Production of shotgun libraries using random amplification. Biotechniques 31:108–118

Routy B, Le Chatelier E, Derosa L, Duong CPM, Alou MT, Daillère R, Fluckiger A, Messaoudene M, Rauber C, Roberti MP et al (2018) Gut microbiome influences efficacy of PD-1-based immunotherapy against epithelial tumors. Science 359:91–97

Roux S, Krupovic M, Poulet A et al (2012) Evolution and diversity of the Microviridae viral family through a collection of 81 new complete genomes assembled from virome reads. PLoS One 7: e40418

Roux S, Enault F, Hurwitz BL, Sullivan MB (2015) VirSorter: mining viral signal from microbial genomic data. Peer J 2015:1–20

Roux S, Solonenko NE, Dang VT, Poulos BT, Schwenck SM, Goldsmith DB, Coleman ML, Breitbart M, Sullivan MB (2016) Towards quantitative viromics for both double-stranded and single-stranded DNA viruses. Peer J 4:e2777

Rowland I, Gibson G, Heinken A, Scott K, Swann J, Thiele I, Tuohy K (2018) Gut microbiota functions: metabolism of nutrients and other food components. Eur J Nutr 57:1–24

Rubinstein MR, Wang X, Liu W, Hao Y, Cai G, Han YW (2013) Fusobacterium nucleatum promotes colorectal carcinogenesis by modulating E-cadherin/_-catenin signaling via its fad a Adhesin. Cell Host Microbe 14:195–206

Russell L, Monaghan T, Kao D (2018) The need to move away from fecal transplant towards targeted, refined microbiome therapy [Review of The need to move away from fecal transplant towards targeted, refined microbiome therapy]. J Thorac Dis 10(10):E755–E757

Ryan KJ, Ray CG (2004) Sherris medical microbiology, 4th edn. McGraw Hill, New York, pp 566–569

Santiago-Rodriguez TM, Hollister EB (2019) Human Virome and disease: high-throughput sequencing for virus discovery, identification of phage-bacteria dysbiosis and development of therapeutic approaches with emphasis on the human gut. Viruses 11(7):656. https://doi.org/10.3390/v11070656

Santos SF, de Oliveira HL, Yamada ES, Neves BC, Pereira A Jr (2019) The gut and Parkinson's disease-a bidirectional pathway. Front Neurol 10:574

Sarah S (2012) Could a common virus Lead to type 2 diabetes in seniors? In: My health news daily, p 1

Sausset R, Petit MA, Gaboriau-Routhiau V, De Paepe M (2020) New insights into intestinal phages. Mucosal Immunol 13(2):205–215. https://doi.org/10.1038/s41385-019-0250-5

Scarpellini E, Ianirob G, Attili F, Bassanellid C, De Santisd A, Gasbarrinib A (2015) The human gut microbiota and virome: potential therapeutic implications. Dig Liver Dis 47:1007–1012

Shkoporov AN, Hill C (2019) Bacteriophages of the human gut: the "known unknown" of the microbiome. Cell Host Microbe 25(2):195–209

Shkoporov AN, Ryan FJ, Draper LA, Forde A, Stockdale SR, Daly KM, McDonnell SA, Nolan JA, Sutton TDS, Dalmasso M, McCann A, Ross RP, Hill C (2018) Microbiome 6:68

Silveira CB, Rohwer FL (2016) Piggyback-the-winner in host-associated microbial communities. NPJ Biofilms Microbiomes 2:16010

Sommer F, Anderson JM, Bharti R, Raes J, Rosenstiel P (2017) The resilience of the intestinal microbiota influences health and disease. Nat Rev Microbiol 15(10):630–638. https://doi.org/10.1038/nrmicro.2017.58

Srividhya KV, Alaguraj V, Poornima G, Kumar D, Singh GP, Raghavenderan L, Mohan Katta AVSK, Mehta P, Krishnaswamy S (2007) Identification of prophages in bacterial genomes by dinucleotide relative abundance difference. PLoS One 2:e1193

Stern A, Mick E, Tirosh I, Sagy O, Sorek R (2012) CRISPR targeting reveals a reservoir of common phages associated with the human gut microbiome. Genome Res 22(10):1985–1994

Sutton TDS, Hill C (2019) Gut bacteriophage: current understanding and challenges. Front Endocrinol 10:784

Sutton TDS, Clooney AG, Ryan FJ, Ross RP, Hill C (2019) Choice of assembly software has a critical impact on virome characterisation. Microbiome 7:1–15

Sweet C (1999) The pathogenicity of cytomegalovirus. FEMS Microbiol Rev 23(4):457–482

Tetz G, Brown SM, Hao Y, Tetz V (2018) Parkinson's disease and bacteriophages as its overlooked contributors. Sci Rep 8(1):10812

Thurber RV, Haynes M, Breitbart M, Wegley L, Rohwer F (2009) Laboratory procedures to generate viral metagenomes. Nat Protoc 4:470–483

Tilg H, Adolph TE, Gerner RR, Moschen AR (2018) The intestinal microbiota in colorectal cancer. Cancer Cell 33:954–964

Turnbaugh PJ, Hamady M, Yatsunenko T, Cantarel BL, Duncan A, Ley RE, Sogin ML, Jones WJ, Roe BA, Affourtit JP, Egholm M (2009) A core gut microbiome in obese and lean twins. Nature 457(7228):480–484

Urisman A, Fischer KF, Chiu CY, Kistler AL, Beck S, Wang D, DeRisi JL (2005) E-Predict: a computational strategy for species identification based on observed DNA microarray hybridization patterns. Genome Biol 6:R78

van der Heijden M, de Vries M, van Steenbeek FG, Favier RP, Deijs M, Brinkhof B, Rothuizen J, van der Hoek L, Penning LC (2012) Sequence-independent VIDISCA-454 technique to discover new viruses in canine livers. J Virol Methods 185:152–155

Van Dyne EA, Henley SJ, Saraiya M, Thomas CC, Markowitz LE, Benard VB (2018) Trends in human papillomavirus–associated cancers—United States, 1999–2015. MMWR Morb Mortal Wkly Rep 67:918–924

Vanchiere JA, Abudayyeh S, Copeland CM, Lu LB, Graham DY, Butel JS (2009) Polyomavirus shedding in the stool of healthy adults. J Clin Microbiol 47(8):2388–2391

Veziant J, Gagnière J, Jouberton E, Bonnin V, Sauvanet P, Pezet D, Barnich N, Miot-Noirault E, Bonnet M (2016) Association of colorectal cancer with pathogenic Escherichia coli: focus on mechanisms using optical imaging. World J Clin Oncol 7:293

Vibin J, Chamings A, Collier F, Klaassen M, Nelson TM, Alexandersen S (2018) Metagenomics detection and characterisation of viruses in faecal samples from Australian wild birds. Sci Rep 8:1–23

Vinner GK, Richards K, Leppanen M, Sagona AP, Malik DJ (2019) Microencapsulation of enteric bacteriophages in a pH-responsive solid oral dosage formulation using a scalable membrane emulsification process. Pharmaceutics 11(9):475. https://doi.org/10.3390/pharmaceutics11090475

Virgin HW (2014) The virome in mammalian physiology and disease. Cell 157(1):142–150

Virgin HW, Wherry EJ, Ahmed R (2009) Redefining chronic viral infection. Cell 138(1):30–50

Waller AS, Yamada T, Kristensen DM et al (2014) Classification and quantification of bacteriophage taxa in human gut metagenomes. ISME J 8:1391–1402

Wang D (2020) 5 challenges in understanding the role of the virome in health and disease. PLoS Pathog 16(3):e1008318

Wang W, Jovel J, Halloran B et al (2015) Metagenomic analysis of microbiome in colon tissue from subjects with inflammatory bowel diseases reveals interplay of viruses and bacteria. Inflamm Bowel Dis 21:1419–1427

Wang Z, Koonen D, Hofker M, Fu J (2016) Gut microbiome and lipid metabolism: from associations to mechanisms. Curr Opin Lipidol 27:216–224

Warwick-Dugdale J, Solonenko N, Moore K, Chittick L, Gregory AC, Allen MJ, Sullivan MB, Temperton B (2019) Long-read viral metagenomics captures abundant and microdiverse viral populations and their niche-defining genomic islands. Peer J 7:e6800

Wegley L, Edwards R, Rodriguez-Brito B et al (2007) Metagenomic analysis of the microbial community associated with the coral porites astreoides. Environ Microbiol 9:2707–2719

Weinbauer MG (2004) Ecology of prokaryotic viruses. FEMS Microbiol Rev 28:127–181

Willner D, Furlan M, Haynes M, Schmieder R, Angly FE, Silva J, Tammadoni S, Nosrat B, Conrad D, Rohwer F (2009) Metagenomic analysis of respiratory tract DNA viral communities in cystic fibrosis and non-cystic fibrosis individuals. PLoS One 4:e7370

Willner D, Furlan M, Schmieder R, Grasis JA, Pride DT, Relman DA, Angly FE, McDole T, Mariella RP, Rohwer F, Haynes M (2011) Metagenomic detection of phage-encoded platelet-binding factors in the human oral cavity. Proc Natl Acad Sci 108(supplement 1):4547–4553

Winter C, Bouvier T, Weinbauer MG, Thingstad TF (2010) Trade-offs between competition and defense specialists among unicellular planktonic organisms: the "killing the winner" hypothesis revisited. Microbiol Mol Biol Rev 74(1):42–57

Wong SH, Zhao L, Zhang X, Nakatsu G, Han J, XuW XX, Kwong TNY, Tsoi H, Wu WKK et al (2017) Gavage of fecal samples from patients with colorectal cancer promotes intestinal carcinogenesis in germ-free and conventional mice. Gastroenterology 153:1621–1633

Xie H, Guo R, Zhong H, Feng Q, Lan Z, Qin B, Ward KJ, Jackson MA, Xia Y, Chen X, Chen B, Xia H, Xu C, Li F, Xu X, Al-Aama JY, Yang H, Wang J, Kristiansen K, Wang J, Steves CJ, Bell JT, Li J, Spector TD, Jia H (2016) Shotgun metagenomics of 250 adult twins reveals genetic and environmental impacts on the gut microbiome. Cell Syst 3:572–584.e3

Yalchin M, Segal JP, Mullish BH, Quraishi MN, Iqbal TH, Marchesi JR, Hart AL (2019) Gaps in knowledge and future directions for the use of faecal microbiota transplant in the treatment of inflammatory bowel disease. Therap Adv Gastroenterol 12:1756284819891038

Ye SH, Siddle KJ, Park DJ, Sabeti PC (2019) Benchmarking metagenomics tools for taxonomic classification. Cell 178:779–794

Zhang Z, Geng J, Tang X, Fan H, Xu J, Wen X et al (2014) Spatial heterogeneity and co-occurrence patterns of human mucosal-associated intestinal microbiota. ISMEJ 8:881–893. https://doi.org/10.1038/ismej.2013.185

Zhang X, Niu YD, Nan Y, Stanford K, Holley R, McAllister T, Narváez-Bravo C (2019) SalmoFreshTM effectiveness in controlling Salmonella on romaine lettuce, mung bean sprouts and seeds. Int J Food Microbiol 305:108250

Zhao G, Vatanen T, Droit L, Park A, Kostic AD, Poon TW, Vlamakis H, Siljander H, Härkönen T, Hämäläinen A-M, Peet A, Tillmann V, Ilonen J, Wang D, Knip M, Xavier RJ, Virgin HW (2017) Intestinal virome changes precede autoimmunity in type I diabetes-susceptible children. Proc Natl Acad Sci U S A 114(30):E6166–E6175

Zhou Y, Liang Y, Lynch KH, Dennis JJ, Wishart DS (2011) Nucleic Acids Res 39:347–352

Zipris D (2008) Innate immunity and its role in type 1 diabetes. Curr Opin Endocrinol Diabetes Obes 15(4):326–331

Zipris D (2009) Epidemiology of type 1 diabetes and what animal models teach us about the role of viruses in disease mechanisms. Clin Immunol 131(1):11–23

Zoetendal EG, Vaughan EE, de Vos WM (2006) A microbial world within us. Mol Microbiol 59:1639–1650. https://doi.org/10.1111/j.1365-2958.2006.05056.x

Zuo T, Wong SH, Lam K, Lui R, Cheung K, Tang W, Ching JYL, Chan PKS, Chan MCW, Wu JCY, Chan FKL, Yu J, Sung JJY, Ng SC (2018) Bacteriophage transfer during faecal microbiota transplantation in Clostridium difficile infection is associated with treatment outcome. Gut 67 (4):634–643

Interactions of Microbiome for Gut-Brain Axis Health

5

Dibyajit Lahiri, Moupriya Nag, Ankita Dey, R. Z. Sayyed, and Rina Rani Ray

Abstract

The various types of interactions between the gut microbes and the brain have engrossed the interest of researchers in recent times in the context of precision medicines for a variety of diseases. People infected by human immunodeficiency virus (HIV) face neurocognitive fall in comparison to the general population, thus disrupting the persistent composition of the gut microbiome, i.e., dysbiosis. The signaling between the microbiota-gut-brain axis (MGBA) can occur through various types of pathways that involve number of host neurochemical signaling, immune system, direct enteric nervous system routes, vagus nerve, and various types of secondary metabolites. Various neurological and psychiatric disorders often occur due to the alteration in the gut microbial profiles. The cutting-edge research highlights the concept of direct relationship between gut microbiota and psychological status of a person and the role of probiotics on the regulation of human neurocognitive health.

Keywords

Brain · Gut · Gut-brain axis · Mental development · Microbiota · Probiotics

D. Lahiri · M. Nag
Department of Biotechnology, University of Engineering and Management, Kolkata, West Bengal, India

A. Dey · R. R. Ray (✉)
Department of Biotechnology, Maulana Abul Kalam Azad University of Technology, Haringhata, West Bengal, India

R. Z. Sayyed
Department of Microbiology, PSGVP Mandal's Arts, Science and Commerce College, Shahada, Maharashtra, India

R. Z. Sayyed, M. Khan (eds.), *Microbiome-Gut-Brain Axis*,
https://doi.org/10.1007/978-981-16-1626-6_5

5.1 Introduction

The microbiota existing within the gut comprise bacteria, protozoa, archaea, fungi, and virus that live within the human gastrointestinal tract (GIT). It has been observed that the number of bacterial cells found within the human body exceeds the number of human body cells (Sender et al. 2016), and they play an important role in the maintenance of physiological activities within the human being. It has been observed that mitochondria, which are important for the generation of adenosine triphosphate, originate from bacteria and related to Proteobacteria (Roger et al. 2017). The major group of microbial species in the gut belongs to *Bacteroides* and Firmicutes, but there exist a large number of individual microbial communities, and the terms "dysbiosis" and "healthy gut microbiome" remain controversial (Moloney et al. 2014).

Intercommunication between GIT, peripheral and central nervous system (PNS and CNS) and the microorganisms results in the development of MGBA and thereby causes the transmittance and interpretation of information from periphery to the brain and back. This complex system of communication helps in the maintenance of coordination of gastrointestinal functions, which supports the physiological and behavioral processes (Mörkl et al. 2020). The exact mechanism of gut-brain communication helps in the coordination and maintenance of gastrointestinal functions and various physiological processes. It also has its effect on behaviors, mood, and other cognitive functions. The exact mechanism of communication is still under various studies that involve endocrine pathways comprising cortisol and hypothalamic-pituitary-adrenal (HPA) axis, nervous pathways comprising vagus nerve and enteric nervous system, and immune pathways. Psychiatric disorders often result in the alteration in these pathways.

Various types of microbial flora are inherited at the time of birth, and it shows change depending on the dietary habits and various environmental signals (Gomez de Aguero 2016. Koh et al. 2016, Wahlstrom et al. 2016). Changes in the gut microbiome has severe effect on immune signaling, thus resulting in illness associated with the intestine and distal organs comprising inflammatory bowel diseases (IBD), various types of cancer, and autoimmune diseases (Blander et al. 2017; Roy and Trinchieri 2017). Various types of intrinsic and extrinsic determinants play an important role in the maturation and development of the CNS. It has been further observed that germ-free animals or animals' exposure to broad-spectrum antibiotics often has an effect on their CNS physiology and neurochemical signal transduction (Smith 2015).

In recent times, the target for epigenetic modification is the gut microbiota (Gomez de Aguero 2016) that can be used for treating psychiatric disorders to improve symptoms. Administration of probiotics in adequate proportion provides health benefits to the host (Butel 2014). Improvement in the microbiota-gut-brain axis (MGBA) can be achieved by the use of prebiotics comprising diet rich in nondigestible fibers and modified dietary components, antibiotics, synbiotics (a combination of pre- and probiotics), probiotics comprising fermentation products, and transplantation of fecal microbiota (Zmora et al. 2019). These are considered as the potent psychobiotics as they can be used for the purpose of mental health by modifying the microbiota (Dinan et al. 2013; Sarkar et al. 2016). This review highlights the role of brain-gut-microbiota axis on mental makeup and outlines the new definition of psychobiotics, including both pre- and probiotics, which play a pivotal role in influencing bacteria-brain relationships.

5.2 Profile of Gut Microbiota

The bidirectional communication existing between the gut and the brain is an obvious process responsible for controlling safety signals, hunger, and the factors responsible for the intake of food (Konturek et al. 2004). This type of communication is also responsible for maintenance of social behavior, stress response, and fear expressions. Alteration in this behavior is due to illness associated with GI and results in the discomfort associated with GI. It has been further observed that anxiety, acute and chronic stress, and depression also bring about change within the gut microbiota profile (Dinan et al. 2018). The colonization of microbiota occurs within the gut which gets initially seeded from the maternal vagina (Dominguez-Bello et al. 2010). The microbiota of infants delivered via cesarean section (C-section) differs from those born via vaginal delivery. The microbiota of infants delivered via cesarean section comprise of microbes seeded from the skin and delivery suite (Ng 2000) and possess lesser amount of colonization by *Bacteroides*, *Bifidobacterium*, and *Lactobacillus*. The microbiota thriving within the body of infants born via C-section takes nearly 2 years to resemble the microbiota of infants born via vaginal delivery (Hill et al. 2014), but the differential seeding mechanism often results in the relative risk in childhood causing asthma (Metsala et al. 2015) and obesity (Mueller et al. 2015). The microbiota composition in the early life is dependent on various factors comprising the parent use of antibiotics, geography, breastfeeding, and growth in early years of life (Vatanen et al. 2019). Diet plays a vital role in the maintenance of the composition of the gut microbiome, and change in diet alters the microbiota (David et al. 2014) (Fig. 5.1).

5.3 Signaling Mechanism Associated behind MGBA

Various mechanisms are associated with the signaling (Table 5.1) of MGBA (Fig. 5.2). The microbiota associated with the gut is responsible for the production of bioactive peptides comprising branched-chain amino acids, short-chain fatty acids (SCFAs), gut hormones, and neurotransmitters and transformation of the secondary bile acids. The short-chain fatty acids possess the ability to enter to the bloodstream and act as a possible route for the signal to reach up to the brain (Sarkar et al. 2016). Microbes are responsible to bring about metabolism of tryptophan, thus modulating the serotonin signaling (Kennedy et al. 2017). Gut microbes are responsible for the synthesis of acetylcholine, noradrenaline, dopamine, and GABA (Clarke et al. 2014). The walls of the gut comprise enteric nervous system that are mainly

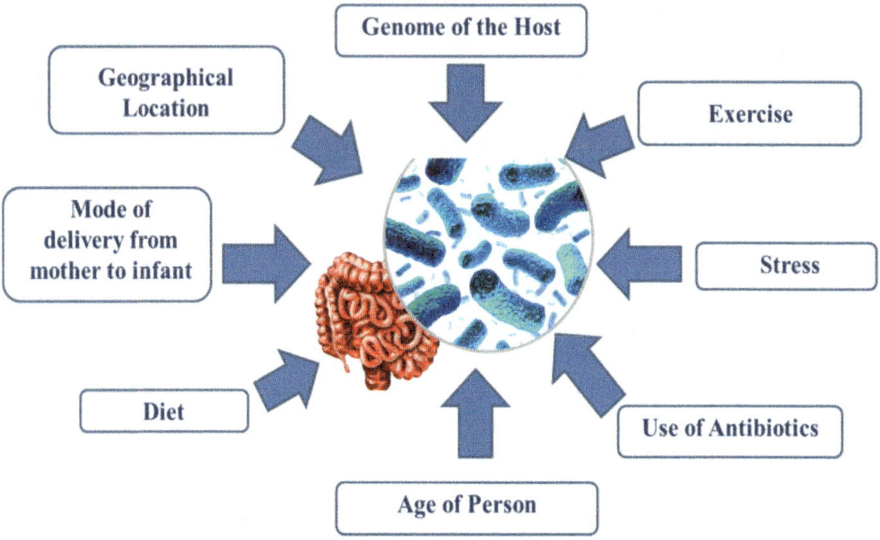

Fig. 5.1 Factors responsible for the maintenance of the gut microbiota

Table 5.1 Different signaling pathways influencing gut microbiota

Type of communication	Mechanism	Reference
Inflammatory signal pathway	Diverse groups of microbial and endogenous signals are responsible for the activation of inflammasome. NLRP6 is an inflammasome signaling that helps in the modulation of the microbiota. Deficiency of NLRP6 results in the distortion of colonization, thereby leading to dysbiosis	Levy (2015)
Type I interferon signaling pathway	Interferon I (IFN-I) plays an important role in the modulation of microbiota. It has been observed that *Lactobacillus acidophilus* possesses the ability to induce antiviral response associated to TLR-2-dependent INF-β. it has been also observed that *Clostridium orbiscindens* helps in the protection of mice from influenza through IFN-I signaling	Weiss (2010) and Steed (2017)
NF-kB signaling pathway	Change in the composition of microbiota results in various inflammatory diseases by the regulation of innate immunity especially by NF-kB signaling. The dysbiosis of the intestine resulting in the killing of *Campylobacter jejuni* causes the activation of NF-kB under the influence of various cytokines that further results in the activation of various immune cells	Masanta (2013)

responsible for the motility of neurotransmitter and short-chain fatty acids (Rea et al. 2016). The gut also comprises immune cells that provide second line of defense against the pathogens after the mucous layer of the gut epithelium, which acts as a

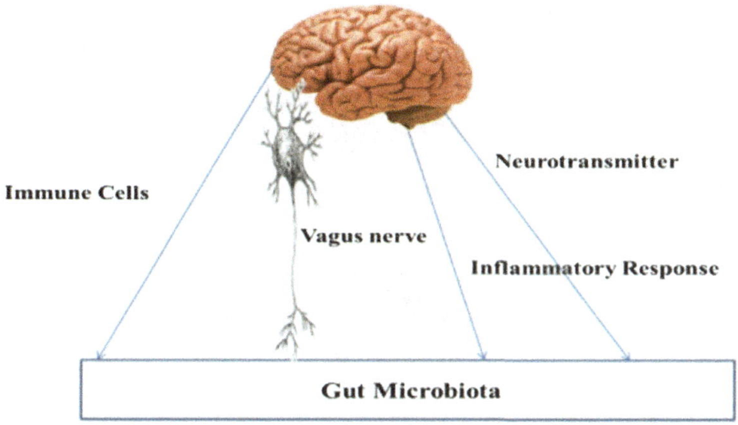

Fig. 5.2 Mechanism of communication between the brain and gut microbiota

physical barrier. Gut microbiota is also responsible for the production of anti- and pro-inflammatory cytokines that act as a signal to the brain via the circulatory system. The permeability of the gut is negatively affected at the time of stress (Vanuytsel et al. 2014). The vagus nerve is regulated by the MGBA signaling (Fig. 5.2), which helps in the maintenance of the communication taking place between the brain and the gut (Fulling et al. 2019). For example, Parkinson's disease can be prevented by the mechanism of vagotomy, which is a surgical method to treat peptic ulcer disease, thereby reducing the possibility of the implication of *Helicobacter pylori* in this disease (Svensson et al. 2015).

5.4 Role of Gut Microbiota in the Development of Brain Behavior

The understanding of the importance of gut microbiota (GM) in the development of the brain and mental state brought about a total paradigm shift in the field of psychology. GM is not only responsible for developing the function of gut-brain but also has its impact on the brain and behavior (Kundu et al. 2017). The disturbance of the microbiota results in the development of mental and brain-associated disorder (Dinan and Cryan 2017). The microbiota present within the infants enhances phylogenetically after birth and resembles the adult form within a 3-year period (Bokulich et al. 2016). The diversification of phylogeny of the microbiota keeps on increasing, but adolescence has a great impact on the composition of the microbiota (Kundu et al. 2017). Gut microbiota plays a vital role in the maintenance of the behavior and mind of the host, but its relevance is often ignored (Vuong et al. 2017). It plays an important role in the perseverance of visceral and peripheral pain response. It has been further observed that supplementation of probiotics after the treatment of antibiotics often results in the suppression of pain sensitivity (Vuong

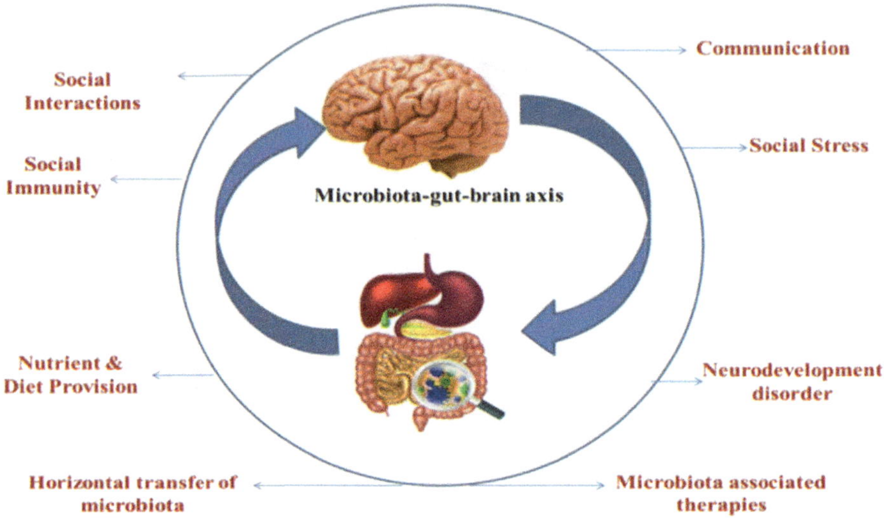

Fig. 5.3 Gut-brain microbiota interaction affecting human activities and management of stress

et al. 2017). Abnormal composition of gut microbiota often results in myalgic encephalomyelitis/chronic fatigue syndrome, and its symptoms can be altered after the supply of probiotics (Rao et al. 2009). Abnormal microbiota is also associated with various pain-associated disorders like migraine, abdominal pain, and chronic back pain (Gawronska et al. 2007). The learning capacity and memory are largely associated to the microbiota being present within the gut (Manderino et al. 2017). It has been often observed that administration of antibiotics often results in the damage of spatial and working memory, which improves with the administration of probiotics (Vuong et al. 2017). It has been often observed that infliction of pathogenic infections results in the development of sickness behaviors with various symptoms comprising social avoidance, fatigue, decreased appetite, and enhancement in anxiety (Gur et al. 2015).

The character and temperament of a person are closely associated with the gut microbiota that possesses the ability to get transplanted from one person to another by the fecal microbiota transplantation (Kim et al. 2017). Management of stress is also associated with the gut microbiota (Fig. 5.3), which plays a vital role in the stress response (Luczynski et al. 2016). It has been observed that psychological stresses help in the activation of neuroendocrine, nervous, and immune systems but also bring about alteration in the mood and gut microbiota (Bharwani et al. 2016). Healthy microbiota helps the host to cope up with stress, whereas alteration in the microbiota enhances the susceptibility of the host to various types of disorders (Vuong et al. 2017).

5.5 Control of Microbiota on the Brain Through Nervous Pathway

Microbiota affects the brain and behavior at a very fast rate via the nervous pathway (Table 5.2). This entire mechanism comprises neural conduction, neurotransmitter, neurogenesis, neurodegeneration, and apoptosis (Thion et al. 2017). Various neuroendocrine pathways are influenced by the gut microenvironment, which is actually maintained by the microbiota.

5.6 Controlling of the Immune System by Gut-Microbiome

The microbiota plays an important role in the maintenance of the immune system. The immunity associated gut mucosa is one of the important parts of the immune system, and immune cells present within the gut-associated lymphoid tissue account for 70–80% of the total immunologically active cells (Tlaskalova-Hogenova et al. 2005). Immune cells bring about regulatory effect upon our body in symbiotic relationship with that of the microbiota. It has been observed that alteration or absence of gut microbiota results in immune deficiency (Gensollen et al. 2016). The gut microbiota helps in the development of adaptive and innate immunity, thus influencing inflammation and neuroimmunity (Freestone et al. 2008).

Table 5.2 Nervous pathways and microbial interaction

Types of nerve pathways	Involvement of microbiota	Reference
Neural conduction	The metabolites produced by the microbiota and its types play a vital role in regulating the activities of the cranial nerves. The microbiota possesses the ability to affect the brain via vagus nerve. The primary afferents possess the ability to first release the impulse, which in turn activates the vagus nerve, thereby sending it to the brain. The microbiota possesses the ability to recognize the signal released by the host and enact promptly	Liang et al. (2018)
Neurotransmitters	Body alone is unable to bring about regulation of neurotransmitter, but microbiota have an important role in the maintenance of neurotransmitters within the body. Gut microbiota possesses the ability to produce a neurotransmitter by altering the metabolism pathways of a neurotransmitter	Liang et al. (2018)
Neurogenesis, neurodegeneration, and apoptosis	The gut microbiota plays an essential role in the maintenance of various physiological activities. The pH concentration of the gut	Liang et al. (2018)

5.6.1 Innate Immunity

GM plays a vital role in the functioning and maturation of the innate immunity. The microbiota regulates the development and functioning of the immune barrier and also helps in the regulation of innate immune cells and pattern recognition receptors (Tlaskalova-Hogenova et al. 2005). The functioning of the blood-brain barrier (BBB) and gut barrier is dependent upon the gut microbiota. Deficiency of the barrier caused by microbiota enhances susceptibility toward various types of diseases (Gensollen et al. 2016). Downregulation of the expression of tight junction, permeability of the BBB, and induction of leaky brain are observed in the absence of microbiota (Kelly et al. 2015). Abnormality in gut microbiota often results in the induction of stress-related disorders and various neurodegenerative diseases (Hoffman et al. 2017).

5.6.2 Adaptive Immunity

The development of adaptive immunity takes place at the time of exposure and combating with microbiota. The process of differentiation and functioning of lymphocytes is dependent on the gut microbiota that further influences the synthesis and release of antibodies (Artis 2008). The immune system possesses the ability to differentiate pathogenic and beneficial group of bacterial cells and possesses the ability to tolerate self-components and harmless materials when exposed to microbiota during early life (Knoop et al. 2017). The gut microbiota helps in regulating the CD4+ T cells and differentiates them to T lymphocytes, which further produce pro-inflammatory responses (Honda and Littman 2016).

5.7 Brain Disorder and Altered Microbiota

Alterations in the gut microbiota result in the development of diseases. Differences in the microbial profiles result in the development of Alzheimer's disease (AD) and various psychological disorders (Table 5.3).

5.8 Therapeutic Target of Gut-Brain Axis

Probiotics are live microorganisms administered in adequate amounts within the host body that have a beneficial effect on human health. The use of probiotics showed its efficacy (Table 5.4) in reducing anxiety-like behavior, depression, and stress within an animal model. Bacterial species like *Bifidobacterium* and *Lactobacillus* are used as probiotics to a large extent. Probiotics do not reside within the gut, but the probiotic formulation requires regular consumption to maintain its positive effect. Probiotics act as components of food that are provided as supplement (O'Toole et al. 2017).

Table 5.3 Microbiota and CNS-associated disorders

Name of the disease	Description	Reference
Multiple sclerosis	It is characterized by immune-associated demyelination of neural axon. Pathogenesis of this disease originates within the immune system and possesses significant contribution of both environmental and genetic factors. Gut microbiota is associated with immune signaling and physiological processes. Thus, it has a control on the pathogenesis at the time of multiple sclerosis	Berer et al. (2011)
Parkinson's disease	It is a neurodegenerative disorder that predominantly occurs due to the malfunction of the motor nerve comprising tremor, muscular rigidity, gait abnormality, and slowness of movement. The composition of bacterial species predominantly regulates the disease. Abundance of Enterobacteriaceae results in postural instability and severity of symptoms. The metabolites produced by the gut microbiota have an essential role in the maintenance of physiological conditions of both the immune system and host	Scheperjans et al. (2015)
Major depressive disorder (MDD)	Alteration in the microbiota of the gut results in the development of major depressive disorder. It has been observed that altered microbiota is observed in patients suffering from MDD	Valles-Colomer et al. (2019) and Dinan and Cryan (2019)
Alzheimer's disease (AD)	This disease is greatly influenced by the presence of the gut microbiome. It has also been observed that the difference in the ratio of Firmicutes/Bacteroidetes forms a parallel link between AD and diabetes mellitus. It has also been observed that the different amounts of microbiota have been observed in the serum of patients suffering from AD	Zhuang et al. (2018) and Arnold et al. (2018)
Schizophrenia	Until now, a limited amount of literature is available on the relationship of microbiota with this disease, but it has been observed that males suffering from schizophrenia possess *Candida albicans* within their gut	Severance et al. (2017)
Autism spectrum disorder (ASD)	Patients suffering from this disease often observed to possess gut-associated comorbidities. Studies have shown that people suffering from ASD show marked alterations in their gut microbiota	Strati et al. (2017) and Coretti et al. (2018)

Table 5.4 Probiotics effective against various diseases and disorders

Types of disorders	Probiotics used	Effect	Reference
MDD	*Lactobacillus acidophilus, L. casei, B. bifidum*	Decrease in depression scores	Akkasheh et al. (2016)
Chronic fatigue syndrome	*L. casei*	Decrease in anxiety syndrome	Rao et al. (2009)
AD	*L. casei, L. acidophilus, L. fermentum, B. bifidum*	Changes in blood lipid profile and also bring about alteration in carbohydrate metabolism	Akbari et al. (2016)
HIV	*S. thermophilus, L. plantarum, B. breve, L. paracasei, L. delbrueckii*	Enhancement in neurocognitive performance	Ceccarelli et al. (2017)
Schizophrenia	*L. rhamnosus, B. animalis, B. breve*	Decrease in anxiety and depression	Okubo et al. (2019)

5.9 Conclusion and Future Aspects

The interaction among microbiome, gut, and brain plays an important role in the maintenance of the physiology and psychology of humans. Although a considerable number of researches are ongoing between gut microbiome and the CNS since the last decade, the question that persists is the relevance of pathophysiology, pathogenesis, and treatment of human brain gut disorders. But in recent times, highly controlled, large-scale, and longitudinal studies need to be performed to analyze dysbiotic gut states and various degrees of psychological illness.

References

Akbari E, Asemi Z, Daneshvar Kakhaki R, Bahmani F, Kouchaki E (2016) Effect of probiotic supplementation on cognitive function and metabolic status in Alzheimer's disease: a randomized, double-blind and controlled trial. Front Aging Neurosci 8:256

Akkasheh G, Kashani-Poor Z, Tajabadi-Ebrahimi M, Jafari P, Akbari H (2016) Clinical and metabolic response to probiotic administration in patients with major depressive disorder: a randomized, double-blind, placebo-controlled trial. Nutrition 32:315–320

Arnold SE, Arvanitakis Z, Macauley-Rambach SL, Koenig AM, Wang HY et al (2018) Brain insulin resistance in type 2 diabetes and Alzheimer disease: concepts and conundrums. Nat Rev Neurol 14:168–181

Artis D (2008) Epithelial-cell recognition of commensal bacteria and maintenance of immune homeostasis in the gut. Nat Rev Immunol 8:411–420

Berer K, Mues M, Koutrolos M, Rasbi ZA, Boziki M, Johner C, Wekerle H, Krishnamoorthy G (2011) Commensal microbiota and myelin autoantigen cooperate to trigger autoimmune demyelination. Nature 479(7374):538–541

Bharwani A, Mian MF, Foster JA, Surette MG, Bienenstock J, Forsythe P (2016) Structural & functional consequences of chronic psychosocial stress on the microbiome & host. Psychoneuroendocrinology 63:2010–2025
Blander JM, Longman RS, Iliev ID, Sonnenberg GF, Artis D (2017) Regulation of inflammation by microbiota interactions with the host. Nat Immunol 18:851–860
Bokulich NA, Chung J, Battaglia T, Henderson N, Jay M, Li H et al (2016) Antibiotics, birth mode, and diet shape microbiome maturation during early life. Sci Transl Med 8:343ra382
Butel MJ (2014) Probiotics, gut microbiota and health. Med Mal Infect 44(1):1–8
Ceccarelli G, Brenchley JM, Cavallari EN, Scheri GC, Fratino M (2017) Impact of high-dose multistrain probiotic supplementation on neurocognitive performance and central nervous system immune activation of HIV-1 infected individuals. Nutrients 9:1269
Clarke G, Stilling RM, Kennedy PJ, Stanton C, Cryan JF, Dinan TG (2014) Minireview: gut microbiota: the neglected endocrine organ. Mol Endocrinol 28:1221–1238
Coretti L, Paparo L, Riccio MP, Amato F, Cuomo M et al (2018) Gut microbiota features in young children with autism spectrum disorders. Front Microbiol 9:3146
David LA, Maurice CF, Carmody RN, Gootenberg DB, Button JE (2014) Diet rapidly and reproducibly alters the human gut microbiome. Nature 505:559–563
Dinan TG, Cryan JF (2017) Gut instincts: microbiota as a key regulator of brain development, ageing and neurodegeneration. J Physiol 595:489–503
Dinan TG, Cryan JF (2019) Gut microbes and depression: still waiting for Godot. Brain Behav Immun 79:1–2
Dinan TG, Stanton C, Cryan JF (2013) Psychobiotics: a novel class of psychotropic. Biol Psychiatry 74(10):720–726
Dinan TG, Cryan JF, Stanton C (2018) Gut microbes and brain development have black box connectivity. Biol Psychiatry 83:97–99
Dominguez-Bello MG, Costello EK, Contreras M, Magris M, Hidalgo G (2010) Delivery mode shapes the acquisition and structure of the initial microbiota across multiple body habitats in newborns. PNAS 107:11971–11975
Freestone PPE, Sandrini SM, Haigh RD, Lyte M (2008) Microbial endocrinology: how stress influences susceptibility to infection. Trends Microbiol 16:55–64
Fulling C, Dinan TG, Cryan JF (2019) Gut microbe to brain signaling: what happens in vagus. Neuron 101:998–1002
Gawronska A, Dziechciarz P, Horvath A, Szajewska H (2007) A randomized double-blind placebo-controlled trial of Lactobacillus GG for abdominal pain disorders in children. Aliment Pharmacol Ther 25:177–184
Gensollen T, Iyer SS, Kasper DL, Blumberg RS (2016) How colonization by microbiota in early life shapes the immune system. Science 352:539–544
Gomez de Aguero M (2016) The maternal microbiota drives early postnatal innate immune development. Science 351:1296–1302
Gur TL, Worly BL, Bailey MT (2015) Stress and the commensal microbiota: importance in parturition and infant neurodevelopment. Front Psych 6:5
Hill C, Guarner F, Reid G, Gibson GR, Merenstein DJ (2014) The International Scientific Association for Probiotics and Prebiotics consensus statement on the scope and appropriate use of the term probiotic. Nat Rev Gastroenterol Hepatol 11:506–514
Hoffman JD, Parikh I, Green SJ, Chlipala G, Mohney RP, Keaton M (2017) Age drives distortion of brain metabolic, vascular and cognitive functions, and the gut microbiome. Front Aging Neurosci 9:298
Honda K, Littman DR (2016) The microbiota in adaptive immune homeostasis and disease. Nature 535:75–84
Kelly JR, Kennedy PJ, Cryan JF, Dinan TG, Clarke G, Hyland NP (2015) Breaking down the barriers: the gut microbiome, intestinal permeability and stress-related psychiatric disorders. Front Cell Neurosci 9:392

Kennedy PJ, Cryan JF, Dinan TG, Clarke G (2017) Kynurenine pathway metabolism and the microbiota-gut-brain axis. Neuropharmacology 112:399–412

Kim HN, Yun Y, Ryu S, Chang Y, Kwon MJ, Cho J (2017) Correlation between gut microbiota and personality in adults: a cross-sectional study. Brain Behav Immun 69:374–385. https://doi.org/10.1016/j.bbi.2017.12.012

Knoop KA, Gustafsson JK, McDonald KG, Kulkarni DH, Coughlin PE, McCrate S et al (2017) Microbial antigen encounter during a preweaning interval is critical for tolerance to gut bacteria. Sci Immunol 2:eaao1314

Koh A, De Vadder F, Kovatcheva-Datchary P, Backhed F (2016) From dietary Fiber to host physiology: short-chain fatty acids as key bacterial metabolites. Cell 165:1332–1345

Konturek SJ, Konturek JW, Pawlik T, Brzozowski T (2004) Brain-gut axis and its role in the control of food intake. J Physiol Pharmacol 55:137–154

Kundu P, Blacher E, Elinav E, Pettersson S (2017) Our gut microbiome: the evolving inner self. Cell 171:1481–1493

Levy M (2015) Microbiota-modulated metabolites shape the intestinal microenvironment by regulating NLRP6 Inflammasome signaling. Cell 163:1428–1443

Liang S, Wu X, Jin F (2018) Gut-brain psychology: rethinking psychology from the microbiota-gut-brain axis. Front Integr Neurosci 12:33

Luczynski P, McVey Neufeld KA, Oriach CS, Clarke G, Dinan TG, Cryan JF (2016) Growing up in a bubble: using germfree animals to assess the influence of the gut microbiota on brain and behavior. Int J Neuropsychopharmacol 19:pyw020

Manderino L, Carroll I, Azcarate-Peril MA, Rochette A, Heinberg L, Peat C (2017) Preliminary evidence for an association between the composition of the gut microbiome and cognitive function in neurologically healthy older adults. J Int Neuropsychol Soc 23:700–705

Masanta WO (2013) Modification of intestinal microbiota and its consequences for innate immune response in the pathogenesis of campylobacteriosis. Clin Dev Immunol 2013:526860

Metsala J, Lundqvist A, Virta LJ, Kaila M, Gissler M, Virtanen SM (2015) Prenatal and post-natal exposure to antibiotics and risk of asthma in childhood. Clin Exp Allergy 45:137–145

Moloney RD, Desbonnet L, Clarke G, Dinan TG, Cryan JF (2014) The microbiome: stress, health and disease. Mamm Genome 25(1–2):49–74

Mörkl S, Butler MI, Holl A, Cryan JF, Dinan TG (2020) Probiotics and the microbiota-gut-brain axis: focus on psychiatry. Curr Nutr Rep 9(3):171–182

Mueller NT, Whyatt R, Hoepner L, Oberfield S, Dominguez-Bello MG (2015) Prenatal exposure to antibiotics, cesarean section and risk of childhood obesity. Int J Obes (Lond) 39:665–670

Ng PC (2000) The fetal and neonatal hypothalamic-pituitary-adrenal axis. Arch Dis Child Fetal Neonatal Ed 82:F250–F254

O'Toole PW, Marchesi JR, Hill C (2017) Next-generation probiotics: the spectrum from probiotics to live biotherapeutics. Nat Microbiol 2:17057

Okubo R, Koga M, Katsumata N, Odamaki T, Matsuyama S (2019) Effect of *bifidobacterium breve* A-*1* on *anxiety* and depressive symptoms in schizophrenia: a proof-of-concept study. J Affect Disord 245:377–385

Rao AV, Bested AC, Beaulne TM, Katzman MA, Iorio C (2009) A randomized, double-blind, placebo-controlled pilot study of a probiotic in emotional symptoms of chronic fatigue syndrome. Gut Pathog 1:6

Rea K, Dinan TG, Cryan JF (2016) The microbiome: a key regulator of stress and neuroinflammation. Neurobiol Stress 4:23–33

Roger AJ, Muñoz-Gómez SA, Kamikawa R (2017) The origin and diversification of mitochondria. Curr Biol 27(21):R1177–R1R92

Roy S, Trinchieri G (2017) Microbiota: a key orchestrator of cancer therapy. Nat Rev Cancer 17:271–285

Sarkar A, Lehto SM, Harty S, Dinan TG, Cryan JF, Burnet PWJ (2016) Psychobiotics and the manipulation of bacteria-gut-brain signals. Trends Neurosci 39:763–781

Scheperjans F, Aho V, Pereira PA, Koskinen K, Paulin L, Pekkonen E, Haapaniemi E, Kaakkola S, Eerola-Rautio J, Pohja M, Kinnunen E, Murros K, Auvinen P (2015) Gut microbiota are related to Parkinson's disease and clinical phenotype. Mov Disord 30(3):350–358

Sender R, Fuchs S, Milo R (2016) Revised estimates for the number of human and bacteria cells in the body. PLoS Biol 14(8):e1002533

Severance EG, Gressitt KL, Stallings CR, Katsafanas E, Schweinfurth LA (2017) Probiotic normalization of Candida albicans in schizophrenia: a randomized, placebo-controlled, longitudinal pilot study. Brain Behav Immun 62:41–45

Smith PA (2015) The tantalizing links between gut microbes and the brain. Nature 526:312–314

Steed AL (2017) The microbial metabolite desaminotyrosine protects from influenza through type I interferon. Science 357:498–502

Strati F, Cavalieri D, Albanese D, De Felice C, Donati C et al (2017) New evidences on the altered gut microbiota in autism spectrum disorders. Microbiome 5:24 Mayer EA, Padua D, Tillisch K. 2014. Altered brain-gut axis in autism: comorbidity or causative mechanisms? Bioessays 36:933–939

Svensson E, Horvath-Puho E, Thomsen RW, Djurhuus JC, Pedersen L (2015) Vagotomy and subsequent risk of Parkinson's disease. Ann Neurol 78:522–529

Thion MS, Low D, Silvin A, Chen J, Grisel P, Schulte-Schrepping J (2017) Microbiome influences prenatal and adult microglia in a sex-specific manner. Cell 172:500.e16–516.e16

Tlaskalova-Hogenova H, Tuckova L, Mestecky J, Kolinska J, Rossmann P, Stepankova R et al (2005) Interaction of mucosal microbiota with the innate immune system. Scand J Immunol 62 (Suppl. 1):106–113. https://doi.org/10.1111/j.1365-3083.2005.01618.x

Valles-Colomer M, Falony G, Darzi Y, Tigchelaar EF, Wang J (2019) The neuroactive potential of the human gut microbiota in quality of life and depression. Nat Microbiol 4:623–632

Vanuytsel T, van Wanrooy S, Vanheel H, Vanormelingen C, Verschueren S (2014) Psychological stress and corticotropin-releasing hormone increase intestinal permeability in humans by a mast cell– dependent mechanism. Gut 63:1293–1299

Vatanen T, Plichta DR, Somani J, Munch PC, Arthur TD (2019) Genomic variation and strain-specific functional adaptation in the human gut microbiome during early life. Nat Microbiol 4:470–479

Vuong HE, Yano JM, Fung TC, Hsiao EY (2017) The microbiome and host behavior. Ann Rev Neurosci 40:21–49

Wahlstrom A, Sayin SI, Marschall HU, Backhed F (2016) Intestinal crosstalk between bile acids and microbiota and its impact on host metabolism. Cell Metab 24:41–50

Weiss G (2010) Lactobacillus acidophilus induces virus immune defence genes in murine dendritic cells by a toll-like receptor-2-dependent mechanism. Immunology 131:268–281

Zhuang ZQ, Shen LL, Li WW, Fu X, Zeng F et al (2018) Gut microbiota is altered in patients with Alzheimer's disease. J Alzheimers Dis 63:1337–1346

Zmora N, Soffer E, Elinav E (2019) Transforming medicine with the microbiome. Sci Transl Med 11(477):eaaw1815

Diet-Gut Microbiota-Brain Axis and IgE-Mediated Food Allergy

6

Mahejibin Khan and Nidhi Sori

Abstract

Food allergy is a common chronic inflammatory disorder caused by abnormal immune reactions of the body to certain food components. Emerging evidence has correlated the prevalence of food allergy to the composition and population of intestinal commensal microbiota. Microbial exposure during infancy and early stages of life plays a major role in shaping the host commensal microbial diversity. Further, nutritional status, diet, and microbial metabolites regulate complex host-microbe interactions enhancing intestinal gut barrier, which contributes to the development of the strong immune system and protection from allergic disorders. This review focuses on the potentials of diet-gut microbiome interaction to restore gut eubiosis for a scope through microbiome-based food toward managing food allergy.

Keywords

Diet · Food allergy · Gut dysbiosis · Gut microbiome · SCFA

6.1 Introduction

Gut-brain axis is defined as the bidirectional interaction between gut microbiota and the central nervous system. Commensal microorganisms of the gut coordinate communication between enteric and central nervous systems and regulate various gastrointestinal (GI) functions such as digestion, nutrient absorption, motility,

M. Khan (✉) · N. Sori
Department of Microbiology and Fermentation Technology, CSIR-Central Food Technological Research Institute, Mysuru, Karnataka, India
e-mail: mahejibin@cftri.res.in

R. Z. Sayyed, M. Khan (eds.), *Microbiome-Gut-Brain Axis*,
https://doi.org/10.1007/978-981-16-1626-6_6

secretion, appetite, energy balance, and metabolism (Tremaroli and Bäckhed 2012; Dockray 2014; Byrne et al. 2016; Cani 2018). These microbes produce a number of neuroactive and immunomodulatory compounds including tryptophan, γ-aminobutyric acid (GABA), serotonin, dopamine, acetylcholine, histamine, and short-chain fatty acids (SCFAs) that are essential for the development of the immune and nervous systems of the human host, thereby playing a crucial role in maintaining intestinal homeostasis and host's health (Macfarlane and Macfarlane 2012; Cryan and Dinan 2012; Aziz et al. 2013; Lin and Zhang 2017). Alterations in gut microbiota that cause imbalance in host-microbe interactions lead to dysbiosis, which not only causes gastrointestinal disorders like irritable bowel syndrome (IBS), inflammatory bowel disease (IBD), and Crohn's disease but also affects the maturation of the immune system leading to inflammatory diseases affecting cognitive functions (Correa-Oliveira et al. 2016; Von Martels et al. 2017; Khan et al. 2020, 2021). Allergy, also referred to as hypersensitivity, is one such common pathological response caused by abnormal immune reactions. When allergic reaction is elicited by a particular food or food component (allergens) in susceptible individuals, it is referred to as food allergy. While food intolerance is a condition that does not involve any immunological mechanism and occur due to lack of certain enzymes in the digestive tract, clinical immunological reactions elicited by IgE-mediated food allergy are induced by food proteins that affect the gastrointestinal tract in response to specific dietary antigens. In such reactions, sensitization and antigen re-exposure through the gut are essential factors (Molloy et al. 2013; Allen and Koplin 2012). Generally, IgE-mediated allergy is characterized by rapid symptoms that range from mild itching to burning sensation of the lips and mouth, swelling, vomiting, skin rashes to severe wheezing, breathing distress, rapid fall in blood pressure, and, in extreme cases, loss of consciousness. In some instances, anaphylactic shocks have also been recorded within minutes of consumption of the food allergen.

Primarily, eosinophils, mast cells, and basophils play a significant role in allergic reactions due to the presence of high-affinity IgE receptor, also known as FcεRI, or Fc epsilon. Crosslinking of the FcεRI via IgE-antigen complexes leads to degranulation of mast cells or basophils and release of inflammatory mediators. In gut mucosa, sensitization of the dendritic cells by the first exposure of the allergen elicits Th2 response, provoking production of pro-inflammatory cytokines IL4, IL5, IL13, and IL9, which trigger B cells to produce IgE antibody and their localization on mast cells. Further exposure to the allergen results in crosslinking of the mast cells' surface IgE leading to degranulation and release of histamine and pro-inflammatory cytokines that together cause allergic reaction (Wambre et al. 2017; Sampson et al. 2018; Schmiechen et al. 2019). Mechanism of food sensitization and allergy is shown in Fig. 6.1.

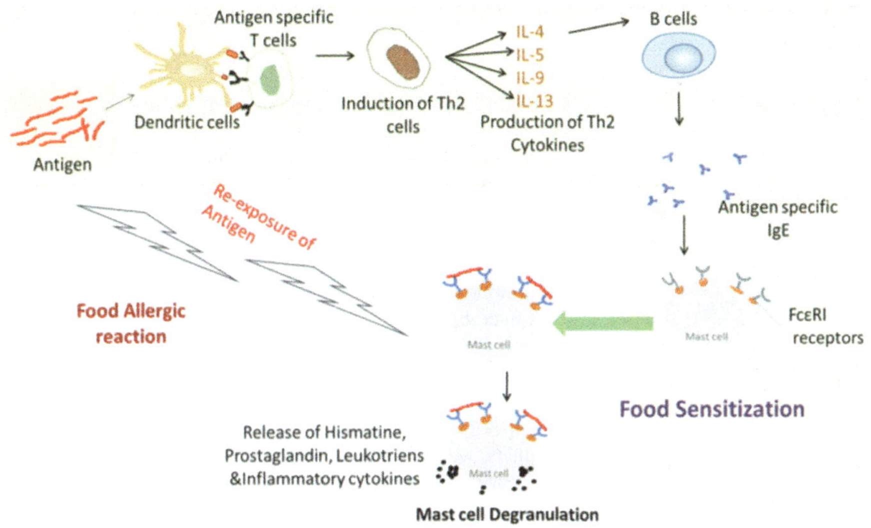

Fig. 6.1 Mechanism of food sensitization and allergy

6.2 Food Allergy

Recent data show that incidences of food allergy, especially in young children, are increasing rapidly in both developed and developing countries (Messina and Venter 2020). Over the last two decades, approximately 50% increase in food allergy incidences and sevenfold increase in hospital admissions for anaphylactic reaction have also been reported in children. Foods such as eggs, cow milk, nuts, peanuts, fish, shellfish, soy, and wheat are among the most common food allergens (Pawankar et al. 2011; Prescott 2013; Brosseau et al. 2019).

There are a number of theories explaining the rapid increase in food allergy (Table 6.1). According to the most common hypothesis, environmental factors, changes in lifestyle, more consumption of high-fat and low-fiber diets, fresh fruits, and other fermentable foods, exposure to antibiotics, and rise in cesarean section deliveries of babies are some of the factors that play a central role in allergic sensitization (Hanski et al. 2012; David et al. 2014; Wypych and Marsland 2018).

The epidemiological data indicate high prevalence of allergic disorders in developed countries and urban population of developing countries. Low incidence of allergic diseases among rural population was reasoned to "hygiene hypothesis" (Strachan 2000) whereby exposure to low doses of infection by nonpathogenic diverse environmental microbial diversity aided in the development of gut commensal community for a robust immune system (Wills-Karp et al. 2001; Blaser 2011; Djuardi et al. 2011; Jie et al. 2013; Mbow et al. 2014).

Table 6.1 Factors affecting the rising prevalence of food allergy

Factors affecting FA	Summary	References
Environmental factors/diet	• Lifestyle changes associated with industrialization and less exposure to farm and other rural areas have altered exposure to commensal and other environmental strains • Consumption of highly processed, high-fat, low-fiber foods causes shifts in microbial communities	Riedler et al. (2001), de Meer et al. (2005), Hanski et al. (2012) and David et al. (2014)
Hygiene hypothesis	• Low or non-exposure to pathogens during childhood due to the use of pasteurized and sterilized milk, food products, water, etc. • Small family size leads to increase in the prevalence of IgE-specific allergic disease	Strachan (2000) and Bloomfield et al. (2006)
Use of antibiotics, cesarean section mode of delivery	• Use of antibiotics during infancy/childhood age influences colonization of gut microbiota and disrupts microbial signaling and immune response • Birth by CS mode of delivery disrupts the colonization of natural microbial flora and leads to an increased risk of developing allergy	Celedón et al. (2004), Noverr et al. (2005), Ly et al. (2006) and Wypych and Marsland (2018)

6.3 Gut Microbiome and Allergy

Gut microbiome or microbiota of the gastrointestinal tract (GIT) predominantly refers to bacterial and archaeal genome even though protozoa, viruses, and fungi also occur in the GIT. Trillions of bacteria reside symbiotically in the gastrointestinal tract. The human body hosts approximately 100 times more microbial genes than self-genes. About 400–450 species of bacteria are present in a healthy human gut (Steinhoff 2005; Qin et al. 2010). Of these, 80–90% belong to Firmicutes and Bacteroidetes, and the remaining 10–20% are Proteobacteria and Actinobacteria (Jandhyala et al. 2015). Gut microbiota co-evolved with the host from the time of birth, and infant gut microbiome symbiosis shaped up commensal microbial community of the GIT. The co-evolution of host-commensal gut microorganisms is regulated by complex interplay of host genetics, mode of delivery of babies, nutritional status during development phase, exposure to antibiotics, and also environmental factors such as hygiene and lifestyle.

During the last few years, several epidemiological studies have associated high prevalence of allergic disorders to the gut microbiota composition and diversity

(Sjogren et al. 2009; Hirata and Kunisawa 2017; Contijoch et al. 2019). Even though the mechanism of association between gut microbiota and food allergy is not completely understood in humans, commensal bacteria inducing mucosal IgA secretion and activating regulatory T cell, Treg, have been demonstrated important for host-microbe homeostasis (Maynard et al. 2012). In this regard, the observations of Stefka et al. (2014) on gut microbiota colonization pattern at neonatal stages, alteration or modification in gut microbiota during infancy stages of life affecting immunity development that subsequently potentiate allergic response and influence the risk of prevalence of allergy are also important. The study suggested a direct link between early-stage intestinal commensal bacteria and the development of allergic disorders. In mice, sensitization to a food allergen enhanced when treated with some antibiotics. In another study, food allergy could be induced in allergy-resistant mice when gut microbiota from allergic mice were transplanted (Yamashita et al. 2012). Very recently, Iweala and Nagler (2019) proposed a mechanism on how gut microbiota contribute to food tolerance and provide protection against food allergy (Fig. 6.2).

Studies conducted at different time points with different populations have reported variations in the gut microbial composition in the infancy stages of children, who developed allergic disease in later growth stages. Penders et al. (2007) showed that the use of antibiotics during infancy stages resulted in decreased population of *Bacteroides* species and increased food allergen sensitization and risk of allergy development later in life. Lower prevalence of *Akkermansia*, *Faecalibacterium*, and *Bifidobacterium* in neonates that could modulate T-cell differentiation and suppress Treg cell activation leading to allergy susceptibility was also reported (Fujimura et al. 2016; Van Den Elsen et al. 2017). Excessive use of antibiotics before and during pregnancy also contributes to increased incidence of allergic diseases. Metsala et al. (2013) explored the association of cow milk allergy with the use of antibiotics in 1 month-old infants and found a positive correlation. The effect of a clinical dose of two different antibiotics, vancomycin and streptomycin, was also analyzed in neonatal and adult mice in relation to gut microbial population and disease severity. Data revealed a little effect of streptomycin on gut microbial diversity and disease development in both neonatal and adult mice. On the other hand, oral administration of vancomycin altered microbial diversity and affected host immunity in neonatal mice. The reaction resulted in exacerbation of airway allergic inflammation and serum IgE without significant changes in adult mice. Vancomycin treatment also reduced *Clostridium* species (clusters IV and XIVa) and *Bacteroides*. Since both the phyla are critical for induction and differentiation of Treg cells, their depletion directly correlated with reduced cellular expression of $CD4^{+}Foxp3^{+}$ Tregs and severity of allergy (Atarashi et al. 2011).

To study the early infantile microbiota and its effect on the development of allergic disorders, Wang et al. (2008) analyzed the gut microbial diversity of 35 infants at the age of 1 week and found significantly lower microbial diversity in the infants who later developed allergy by 18 months of age. Studies of Ismail et al. (2012) on the association of microbial abundance in the early postnatal period and development of allergic disorder in later stages of life, conducted with 98 infants,

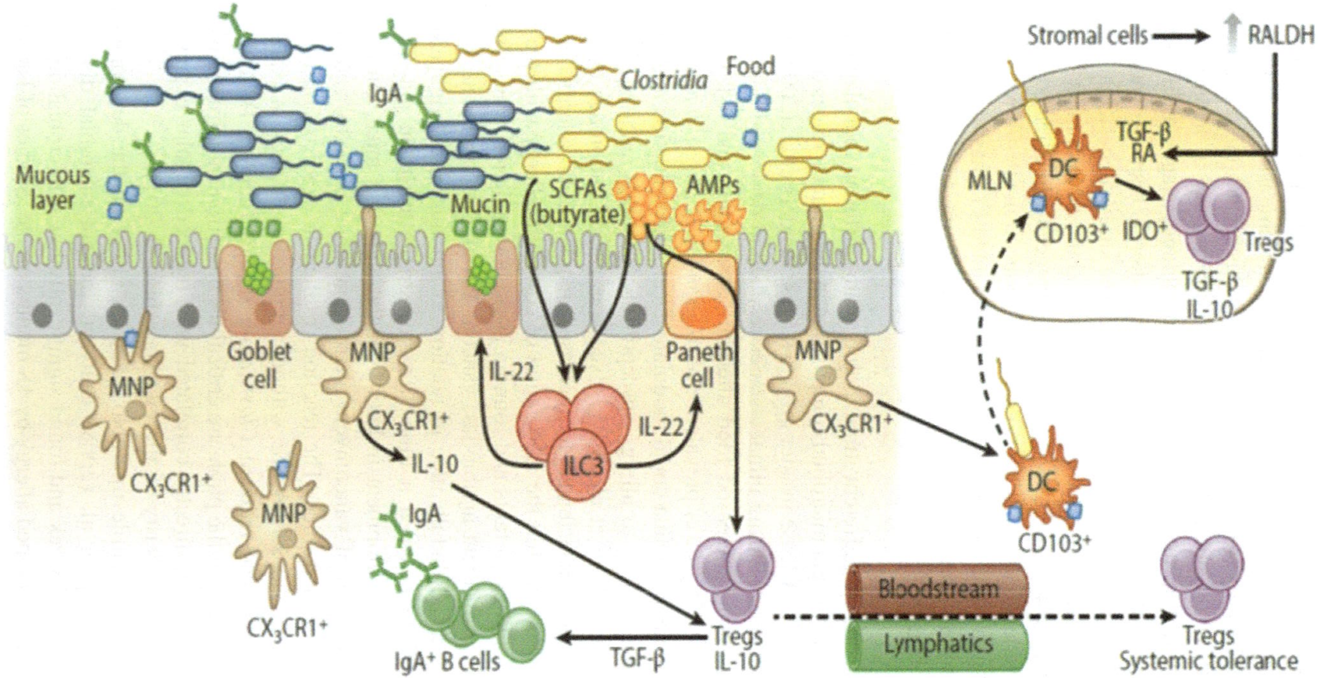

Fig. 6.2 Role of commensal bacteria in protection against allergic sensitization. In healthy individuals, clostridia (and possibly other allergy-protective commensal bacteria) maintain epithelial barrier integrity by stimulating ILC3s to produce IL-22. IL-22 promotes the production of mucus from goblet cells and antimicrobial peptides from Paneth cells and reduces the ability of food allergens to gain access to the systemic circulation. Clostridial metabolites, including SCFAs, directly induce Treg development. Treg-derived TGF-β favors local IgA antibody production, and circulating Tregs promote systemic tolerance (Source: Iweala and Nagler Annu. Rev. Immunol. 2019. 37:377–403)

demonstrated lower microbial diversity in the first week of life that had direct effect with the development of atopic eczema by 1 year of age. In infants, correlation of allergic disorders with lower intestinal microbial diversity was shown by Abrahamsson et al. (2012). It also revealed the importance of *Bacteroides* species in preventing allergic disorders. Ling et al. (2014) analyzed fecal microbial diversity of 34 infants who were diagnosed with IgE and non-IgE-mediated food allergy. Very high counts of Firmicutes, Bacteroidetes, and Proteobacteria were found in allergy groups. Also, Actinobacteria diversity was significantly lower in comparison to healthy groups. The study resulted in the identification of signature microbial groups that could distinguish infants with IgE-mediated and non-IgE-mediated food allergy. Furthermore, higher proportion of *Clostridium* and *Anaerobacter* and decreased levels of *Bacteroides* and *Clostridium XVIII* were correlated to IgE-mediated food allergy. Ho and Bunyavanich (2019) also characterized 17 *clostridia* strains that were capable of reducing the prevalence of allergic inflammation and protecting the host from food allergy. Azad et al. (2015) studied the risk of food sensitization and gut microbiota in a cohort of 166 infants. Higher Enterobacteriaceae/Bacteroidaceae ratio and low gut microbial diversity were associated with a higher risk of food sensitization at the later age of life. Colonization of commensal microbes at the neonatal stage was critical in the development of mucosal immunity and protection against allergic inflammation during later growth stages (Zhuang et al. 2019). Higher accumulation of invariant natural killer T (iNKT) cells was observed in the colonic lamina propria and lungs of germ-free mice. Exposure of gut microbiota prevented accumulation of mucosal iNKT and airway inflammation at neonatal stages but not in adult mice. Early commensal microbial diversity was also linked with serum IgE responses and protection against sensitization to food allergens. Germ-free mice or mice with less microbial exposure at early age expressed more IgE at mucosal sites and induced systemic anaphylaxis (Cahenzli et al. 2013). The American Gut Project of food allergy participants revealed marked difference in microbial richness for alpha and beta diversity between the allergic and non-allergic groups. The study suggested a positive correlation of allergy associated with *Bacteroides fragilis* and a negative correlation with Clostridiales, *Prevotella*, and Ruminococcaceae (Hua et al. 2016). Clostridia, a mucosa-associated commensal, reportedly induced the production of IL-22, important for maintaining epithelial barrier integrity, which plays a pivotal role in preventing food allergen sensitization (Cao et al. 2014; Sabat et al. 2014). To understand the correlation of gut microbiome and exacerbation of food allergy, Feehley et al. (2019) transplanted fecal samples of healthy human infants and infants with cow milk allergy to germ-free mice. Those mice that received samples from the allergic babies became sensitized to the milk protein β-lactoglobulin (BLG) due to which allergic reactions arose upon repeated exposure to the protein. Mice that had received transplants from healthy infants tolerated the dietary antigen without any symptoms of allergy to the milk protein. From the microbiome data, allergic response was correlated to a significant reduction of *Anaerostipes caccae*. The team also showed that transferring this species to germ-free mice was sufficient to protect against an allergic response to cow milk. Thus, appropriate gut microbiota at the early life is critical for the regulation of allergic inflammation in the lung and GI tract.

Cesarean section delivery of babies increased the risk of developing IgE-mediated food allergy (Koplin et al. 2008). Allergic disease was related to imbalance of Th1/Th2 response with enhanced Th2 response. An enhanced Th2 immune response induces IL-4, IL-5, and IL-13 cytokines that contribute to triggering and maintenance of allergic inflammation. During pregnancy, Th1 immune response of a fetus was suppressed to prevent excessive immune response to maternal antigens, and therefore, immune response of the fetus was skewed toward Th2. Immediately after birth, exposure of the infant to maternal gut microbiota shifted Th1 response for immune tolerance and maintained a Th1/Th2 balance. Jakobsson et al. (2014) showed that infants born through CS had reduced microbial diversity due to less microbial exposure at the time of birth. CS infants also displayed poor Th1 response which correlated with lower abundances of *Bacteroidetes* diversity and development of allergic disorder at the later stages of life.

6.4 Diet-Gut Microbiota Interaction and Allergy

The development of molecular techniques during recent years has advanced our knowledge about food-intestinal microbiota interactions and their crucial role in immunity development for establishing healthy host status. Commensals residing in the gut metabolize food and produce a variety of dietary metabolites that serve as signaling molecules and influence the metabolic process of the host (Khan et al. 2013).

Dietary fibers are nondigestible plant polysaccharides. By fermentation, gut bacteria converted these to short-chain fatty acids (SCFAs) such as acetate, propionate, and butyrate. The type and amount of SCFAs produced depended on the composition of microorganisms present in the Gut. SCFAs can shape the mucosal immunological environment and influence the severity of allergic inflammation. SCFAs are either utilized by the gut microbiota for their own metabolism or cross the gut epithelium barrier for release into the lumen, where SCFAs interact directly with the gut-associated lymphoid tissue (GALT) or modify cellular regulation for innate/adaptive immune cells, gene expression, differentiation, proliferation, and apoptosis. Different types of SCFA elicit anti-inflammatory reactions by interacting with "metabolite-sensing" G-protein-coupled receptors (GPCR) that are important for maintaining gut homeostasis and play a vital role in the regulation of Treg cell biology and inflammatory responses. SCFA may activate signaling pathways (Venegas et al. 2019) through GPR41 (free fatty acid receptor 3; FFAR3), GPR43 (free fatty acid receptor 2; FFAR2), and GPR109A (hydroxycarboxylic acid receptor 2; HCAR2). SCFAs also regulate inflammatory cytokine production by inhibiting the transcription factor, the nuclear factor kappa B (NF-κB). Further, SCFAs promote mucous production by intestinal goblet cells that induce secretory IgA and activate inflammasomes resulting in IL-18 secretion, which can recruit different types of effector cells and coordinate the innate immune response. SCFA also helps in maturation of T cells into its subtype and activation of Treg cells that provide oral tolerance and protection against food allergy. Increase in the total number of Treg cells and reduced anaphylaxis also correlated with a higher proportion of $CD103^+$ DCs and retinal dehydrogenase activity in mice model (Tan et al. 2016). Comparative nuclear magnetic resonance–based metabolome analysis disclosed a positive correlation between colonic Treg cell numbers and luminal concentrations of SCFAs (Furusawa et al. 2013).

SCFA levels, specifically butyrate, can serve as metabolic signature to assess the gut microbial diversity and distinguish healthy and diseased states in children with asthma and lower atopic dermatitis or food allergy (Bottcher et al. 2000). Mice fed with high-fiber diet increased the level of circulating SCFAs and were protected against allergic inflammation. It was further observed that mice fed with propionate-supplemented diet displayed higher macrophage count and reduced allergic inflammation through GPR41 receptors since allergic inflammation is exacerbated in GPR41- or GPR43-deficient mice (Trompette et al. 2014). Similar observation was also reported in germ-free mice that were unable to produce SCFA due to lack of commensal microbiota (Maslowski et al. 2009). A distinct microbial diversity and high production of acetate, a type of SCFA, were reported in mice fed with high-fiber diet (Thorburn et al. 2015). Elevated acetate production increased acetylation at the Foxp3 promoter that downregulated certain genes and led to marked suppression of allergic airway disease. It was also shown that high-fiber diet in pregnant mice could suppress robust allergic response in their offspring.

Dysbiosis at the early age of life and shift in microbial composition increased the risk of developing allergic disorders. Arrieta et al. (2015) compared gut microbial diversity of 319 subjects, and higher-risk groups were characterized by reduced level of fecal acetate and significantly lower abundance of *Faecalibacterium*, *Lachnospira*, *Veillonella*, *Rothia*, and some other bacterial genera in their gut. It was further demonstrated that inoculation of specific microbial genera in germ-free mice could ameliorate allergic inflammation in the offspring, which confirmed the role of these bacterial communities and their metabolite in reducing allergy susceptibility. *Faecalibacterium prausnitzii* was identified as a key butyrate producer in different studies.

Cait et al. (2019) also investigated the mechanism of SCFAs to declined allergic inflammation and demonstrated that SCFA could ameliorate allergic disorder by reducing T cells and dendritic cell stimulation. The reaction was reasoned to lessen the levels of circulating immunoglobulin E (IgE) and reduce interleukin-4 (IL4) producing $CD4^+$ T cells. *Bacteroides fragilis* enhanced Treg function through capsular polysaccharide A that promoted tolerance (Round et al. 2011). Similarly, the presence of *Clostridium* species (clusters IV and XIVa) regulated Foxp3+ regulatory3+ T cells via short-chain fatty acid (SCFA) for allergic disease prevention. In an infant cohort of 301 children of ~1 year age, a positive correlation between diet intake, SCFA production, and prevention of food allergy was reported. It was observed that children with low SCFA levels were prone to food sensitization and were at higher risk of developing allergy at later age (Roduit et al. 2019; Costanzo et al. 2020).

Besides SCFA, other metabolites produced by commensal microorganisms such as long-chain fatty acids (LCFAs) including omega-3 and omega-6, glycolipids, and vitamins exert anti-inflammatory and anti-allergic effects. These metabolites lowered Th2 response and reduced the severity of allergic inflammation (Hirata and Kunisawa 2017; Suther et al. 2020).

The finding concludes that diet-gut microbiota interaction shapes mucosal immunity and alleviates allergic inflammation severity through food supplements that include defined microbiota.

6.5 Modulation of Gut Microbiota as a Preventive Measure

Latest research and several epidemiological studies have established the fact that gut microbiota of allergic subjects, both in human and murine models, are distinct from those of healthy controls. Moreover, it is also evident from various studies that alteration of gut microbiota may help in the prevention of allergic diseases (Bunyavanich et al. 2016; Chen et al. 2016). Dietary modifications are recognized as a simple and promising method of modulating gut microbiota. Changes in diet exert a short-term or long-term effect on the intestinal microbial population (Noval Rivas et al. 2015; Nagata et al. 2017). Introduction of high-fiber-rich diet can naturally increase the population of SCFAs producing bacteria in the gut. There are growing evidences to support the protective and anti-inflammatory properties of SCFAs. Underlining the importance of dietary fiber rich diets for the modulation of gut microbiome suggested a correlation between butyrate producing bacterial groups and preventing allergic inflammation. Enhancing the population of SCFAs producing bacteria naturally by intervention through a high-dietary-fiber-rich diet has invoked a potential treatment for allergic diseases (Suther et al. 2020). High-fiber diets used in mice were found to decrease allergic sensitization as well.

During the last few years, pre- and probiotic supplementation has been considered as an attractive and safe option for the modulation of the gut microbiota, which influences the onset of food allergy (Brosseau et al. 2019). Synbiotics are a combination of pre- and probiotics to achieve a synergistic effect (Fox et al. 2019). Prebiotics are nondigestible, small chains oligosaccharides that selectively stimulate the growth and activities of specific groups of bacteria in the colon. Studies suggest a correlation between prevalence of allergic diseases and prebiotics. It has been reported that supplementation of fructo-oligosaccharide (FOS) in food allergy-prone mice group reduced the activation of mast cells and inflammation in the duodenum. In another study, FOS demostrated a regulatory role in activation of naive CD4+ T cells and attenuation of the gut Th2 response (Fujitani et al. 2007; Tsuda et al. 2017). An intervention review on "Prebiotics in infants for prevention of allergy" published meta-analysis of four studies for claims toward significant reduction in eczema (1218 infants, typical risk ratio, 0.68; 95% CI 0.48 to 0.97; typical risk difference, −0.04; 95% CI -0.07 to −0.00; number needed to treat to benefit (NNTB), 25; 95% CI 14 to >100; $P = 0.03$). Pectin-based prebiotics have been shown to stimulate the growth of *Faecalibacterium prausnitzii* and *Eubacterium eligens* DSM3376. These bacterial species are known for their butyrogenic properties and enhance in vitro secretion of the anti-inflammatory cytokine IL-10 (Nagata et al. 2017). In addition, other dietary fibers such as guar gum and cellulose (35% crude fiber) along with vitamin A also protected mice against peanut allergy. High-fiber diet reshaped the gut microbiota and stimulated the growth of SCFAs producing genera such as *Bacteroides*, *Lactobacillus*, and *Bifidobacterium* and inhibited the growth of Firmicutes (Tan et al. 2016).

Dietary intervention with probiotic supplementation effected prevention and treatment of allergy. *Lactobacillus rhamnosus* GG (LGG), a well-characterized probiotic, prevented the occurrence of cow milk allergy through the accumulation of Treg cells in the intestine and other regulatory cytokines like TNF-α, IL6, and IL-10. In addition, *Lactobacillus rhamnosus* GG influenced strain level bacterial diversity in fecal samples of LGG-supplemented infant groups by increasing demethylation of FoxP3 and DNA methylation of IL-4 and IL-5 (Canani et al. 2016, 2017; Paparo et al. 2019). In murine model, *Bifidobacterium longum* subsp. *infantis* LA308 skewed the allergic response toward Th1 through the expression of IL-10. In Zhang et al. (2016), meta-analysis that included 17 trials of 2947 infants, combined probiotics supplementation in prenatal to pregnant mothers and postnatal to infants reduced the risk of food sensitization, RR0.77; 95% CI 0.61–0.98. In a later study by Zhang et al. (2017), oral administration of *Clostridium butyricum* CGMCC0313–1 was found effective in inhibiting β-lactoglobulin (BLG) induced intestinal anaphylaxis. *C. butyricum* stimulated secretion of the IgA and expression of CD4+ CD25+ Foxp3 Treg that reversed the Th1/Th2 imbalance. Canani et al. (2017), based on a large prospective study with randomly selected 220 infants of 5 month median age, with suspected IgE-mediated cow milk allergy, found that the group fed with extensively hydrolyzed casein formula (EHCF) + LGG exhibited less symptoms of allergic manifestation compared to the group fed with only EHCF. Postbiotics (nonviable cells or cell fractions), when administered in adequate amounts, conferred health benefits (Rad et al. 2021), thus showing another strategy for protection against food allergy.

6.6 Conclusion and Future Prospects

Though literature has overwhelmingly supported the use of gut microbiome components to treat the prevailing severity of food allergy, no specific curative medicine or treatment is yet available. Since a vast community of symbionts and commensal microorganisms inhabiting the gastrointestinal tract play a vital role in the development of the immune system, nutrient processing targeting specific gut microbiota in situ appears a strategy for combating food allergy. Current genomic revolution and development of culture-independent methods, like metagenomics, not only offer a scope and opportunity to identify the molecular foundations deciphering gut-microbe relationships but can also decipher the complex interplay between gut microbial compositions toward mucosal microbial diversity, essential for immune homeostasis for microbiome-based diet formulations that stimulate different arms of immune response and prevent allergic disorders. Such formulations should offset imbalances in microbial composition at the early stages of life to shift microbial diversity leading to corrections in complex mucosal systems to overcome allergic disorders in later stages of life. Hence, designing microbiome-based functional foods appears to be a possibility for gut-immunity homeostasis, specifically for protection against IgE-mediated allergic inflammation.

Acknowledgments The authors acknowledge the Indian Council of Medical Research for research funding (ICMR grant no. 2020-9384). NS thanks the Council of Scientific and Industrial Research (CSIR) for Junior Research Fellowship.

References

Abrahamsson TR, Jakobsson HE, Andersson AF, Bjorksten B, Engstrand L, Jenmalm MC (2012) Low diversity of the gut microbiota in infants with atopic eczema. J Allergy Clin Immunol 129 (2):434–440., 440.e1–2. https://doi.org/10.1016/j.jaci.2011.10.025

Allen KJ, Koplin JJ (2012) The epidemiology of IgE-mediated food allergy and anaphylaxis. Immunol Allergy Clin N Am 32:35–50

Arrieta MC, Stiemsma LT, Dimitriu PA, Thorson L, Russell S, Yurist-Doutsch L et al (2015) Early infancy microbial and metabolic alterations affect risk of childhood asthma. Sci Transl Med 7 (307):307ra152. https://doi.org/10.1126/scitranslmed.aab2271

Atarashi K, Tanoue T, Shima T, Imaoka A, Kuwahara T, Momose Y et al (2011) Induction of colonic regulatory T cells by indigenous *Clostridium* species. Science 331(6015):337–341

Azad MB, Konya T, Guttman DS, Field CJ, Sears MR, Hay Glass KT et al (2015) Infant gut microbiota and food sensitization: associations in the first year of life. Clin Exp Allergy 45 (3):632–643

Aziz Q, Doré J, Emmanuel A, Guarner F, Quigley EM (2013) Gut microbiota and gastrointestinal health: current concepts and future directions. Neurogastroenterol Motil 25(1):4–15

Blaser M (2011) Antibiotic overuse: stop the killing of beneficial bacteria. Nature 476 (7361):393–394

Bloomfield SF, Stanwell-Smith R, Crevel RW, Pickup J (2006) Too clean, or not too clean: the hygiene hypothesis and home hygiene. Clin Exp Allergy 36(4):402–425. https://doi.org/10.1111/j.1365-2222.2006.02463.x

Bottcher MF, Nordin EK, Sandin A, Midtvedt T, Bjorksten B (2000) Microflora-associated characteristics in faeces from allergic and nonallergic infants. Clin Exp Allergy 30 (11):1590–1596. https://doi.org/10.1046/j.2222.2000.00982.x

Brosseau C, Selle A, Palmer DJ, Prescott SL, Barbarot S, Bodinier M (2019) Prebiotics: mechanisms and preventive effects in allergy. Nutrients 11:1841

Bunyavanich S, Shen N, Grishin A, Wood R, Burks W, Dawson P et al (2016) Early-life gut microbiome composition and milk allergy resolution. J Allergy Clin Immunol 138 (4):1122–1130

Byrne CS, Chambers ES, Alhabeeb H, Chhina N, Morrison DJ, Preston T et al (2016) Increased colonic propionate reduces anticipatory reward responses in the human striatum to high-energy foods. Am J Clin Nutr 104(1):5–14. https://doi.org/10.3945/ajcn.115.126706

Cahenzli J, Koller Y, Wyss M, Geuking MB, McCoy KD (2013) Intestinal microbial diversity during early-life colonization shapes long-term IgE levels. Cell Host Microbe 14:559–570

Cait A, Cardenas E, Dimitriu PA, Amenyogbe N, Dai D et al (2019) Reduced genetic potential for butyrate fermentation in the gut microbiome of infants who develop allergic sensitization. J Allergy Clin Immunol 144(6):1638–1647. https://doi.org/10.1016/j.jaci.2019.06.029

Canani RB, Sangwan N, Stefka AT (2016) *Lactobacillus rhamnosus* GG supplemented formula expands butyrate-producing bacterial strains in food allergic infants. ISME J 10(3):742–750

Canani RB, Di Costanzo M, Bedogni G (2017) Extensively hydrolysed casein formula containing *Lactobacillus rhamnosus* GG reduces the occurrence of other allergic manifestations in children with cow's milk allergy: 3-year randomized controlled trial. J Allergy Clin Immunol 139 (6):1906–1913

Cani PD (2018) Human gut microbiome: hopes, threats and promises. Gut 67(9):1716–1725. https://doi.org/10.1136/gutjnl-2018-316723

Cao S, Feehley TJ, Nagler CR (2014) The role of commensal bacteria in the regulation of sensitization to food allergens. FEBS Lett 588(22):4258–4266. https://doi.org/10.1016/j.febslet.2014.04.026

Celedón JC, Fuhlbrigge A, Rifas-Shiman S, Weiss ST, Finkelstein JA (2004) Antibiotic use in the first year of life and asthma in early childhood. Clin Exp Allergy 34(7):1011–1016

Chen CC, Chen KJ, Kong MS, Chang HJ, Huang JL (2016) Alterations in the gut microbiotas of children with food sensitization in early life. Pediatr Allergy Immunol 27(3):254–262

Contijoch EJ, Britton GJ, Yang C, Mogno I, Li Z, Ng R et al (2019) Gut microbiota density influences host physiology and is shaped by host and microbial factors. Elife 8:e40553

Correa-Oliveira R, Fachi JL, Vieira A, Sato FT, Vinolo MA (2016) Regulation of immune cell function by short-chain fatty acids. Clin Transl Immunol 5(4):e73. https://doi.org/10.1038/cti.2016.17

Costanzo MD, Carucci L, Canani RB, Biasucci G (2020) Gut microbiome modulation for preventing and treating pediatric food allergies. Int J Mol Sci 21(15):5275. https://doi.org/10.3390/ijms21155275

Cryan JF, Dinan TG (2012) Mind-altering microorganisms: the impact of the gut microbiota on brain and behaviour. Nat Rev Neurosci 13(10):701–712

David LA, Maurice CF, Carmody RN, Gootenberg DB, Button JE, Wolfe BE et al (2014) Diet rapidly and reproducibly alters the human gut microbiome. Nature 505(7484):559–563

de Meer G, Janssen NAH, Brunekreef B (2005) Early childhood environment related to microbial exposure and the occurrence of atopic disease at school age. Allergy 60(5):619–625

Djuardi Y, Wammes LJ, Supali T, Sartono E, Yazdanbakhsh M (2011) Immunological footprint: the development of a child's immune system in environments rich in microorganisms and parasites. Parasitology 138(12):1508–1518

Dockray GJ (2014) Gastrointestinal hormones and the dialogue between gut and brain. J Physiol 592(14):2927–2941. https://doi.org/10.1113/jphysiol.2014.270850

Feehley T, Plunkett CH, Bao R, Hong SMC, Culleen E, Belda-Ferre P et al (2019) Healthy infants harbor intestinal bacteria that protect against food allergy. Nat Med 25(3):448–453. https://doi.org/10.1038/s41591-018-0324-z

Fox A, Bird JA, Fiocchi A, Knol J, Meyer R, Salminen S et al (2019) The potential for pre-, pro- and synbiotics in the management of infants at risk of cow's milk allergy or with cow's milk allergy: an exploration of the rationale, available evidence and remaining questions. World Allergy Org J 12(5):100034. https://doi.org/10.1016/j.waojou.2019.100034

Fujimura KE, Sitarik AR, Havstad S, Lin DL, Levan S, Fadrosh D et al (2016) Neonatal gut microbiota associates with childhood multi sensitized atopy and T cell differentiation. Nat Med 22(10):1187–1191

Fujitani S, Ueno K, Kamiya T, Tsukahara T, Ishihara K, Kitabayashi T et al (2007) Increased number of CCR4-positive cells in the duodenum of ovalbumin-induced food allergy model NC/jic mice and anti-allergic activity of fructooligosaccharides. Allergol Int 56(2):131–138

Furusawa Y, Obata Y, Fukuda S, Endo TA, Nakato G, Takahashi D et al (2013) Commensal microbe-derived butyrate induces differentiation of colonic regulatory T cells. Nature 504 (7480):446–450

Hanski I, Hertzen LV, Fyhrquist N, Koskinen K, Torppa K, Laatikainen T et al (2012) Environmental biodiversity, human microbiota, and allergy are interrelated. PNAS 109(21):8334–8339

Hirata S, Kunisawa J (2017) Gut microbiome, metabolome, and allergic diseases. Allergol Int 66 (4):523–528

Ho HE, Bunyavanich S (2019) Microbial adjuncts for food allergen immunotherapy. Curr Allergy Asthma Rep 19(5):25. https://doi.org/10.1007/s11882-019-0859-1

Hua X, Goedert JJ, Pu A, Yu G, Shi J (2016) Allergy associations with the adult fecal microbiota: analysis of the American gut project. EBio Med 3:172–179

Ismail IH, Oppedisano F, Joseph SJ, Boyle RJ, Licciardi PV, Robins-Browne RM et al (2012) Reduced gut microbial diversity in early life is associated with later development of eczema but not atopy in high-risk infants. Pediatr Allergy Immunol 23(7):674–681

Iweala OI, Nagler CR (2019) The microbiome and food allergy. Annu Rev Immunol 37:377–403. https://doi.org/10.1146/annurev-immunol-042718-041621

Jandhyala SM, Talukdar R, Subramanyam C, Vuyyuru H, Sasikala M, Nageshwar Reddy D (2015) Role of the normal gut microbiota. World J Gastroenterol 21(29):8787–8803

Jakobsson HE, Abrahamsson TR, Jenmalm MC, Harris K, Quince C, Jernberg C, Björkstén B, Engstrand L, Andersson AF (2014) Decreased gut microbiota diversity, delayed Bacteroidetes colonisation and reduced Th1 responses in infants delivered by caesarean section. Gut 63 (4):559–566. https://doi.org/10.1136/gutjnl-2012-303249

Jie Y, Isa ZM, Jie X, Ju ZL, Ismail NH (2013) Urban vs. rural factors that affect adult asthma. Rev Environ Contam Toxicol 226:33–63. https://doi.org/10.1007/978-1-4614-6898-1_2

Khan M, Nakkeeran E, Umesh-Kumar S (2013) Potential application of pectinase in developing functional foods. Annu Rev Food Sci Technol 4:21–34. https://doi.org/10.1146/annurevfood-030212-182525

Khan MS, Ikram M, Park JS, Park TJ, Kim MO (2020) Gut microbiota, its role in induction of Alzheimer's disease pathology, and possible therapeutic interventions: special focus on anthocyanins. Cell 9(4):853. https://doi.org/10.3390/cells9040853

Khan M, Mathew BJ, Gupta P, Garg G, Khadanga S, Vyas AK, Singh AK (2021) Gut dysbiosis and IL-21 response in patients with severe COVID-19. Microorganisms 9(6):1292. https://doi.org/10.3390/microorganisms9061292

Koplin J, Allen K, Gurrin L, Osborne N, Tang ML, Dharmage S (2008) Is caesarean delivery associated with sensitization to food allergens and IgE-mediated food allergy: a systematic review. Pediatr Allergy Immunol 19(8):682–687

Lin L, Zhang J (2017) Role of intestinal microbiota and metabolites on gut homeostasis and human diseases. BMC Immunol 18:2. https://doi.org/10.1186/s12865-016-0187-3

Ling Z, Li Z, Liu X, Cheng Y, Luo Y, Tong X et al (2014) Altered fecal microbiota composition associated with food allergy in infants. Appl Environ Microbiol 80(8):2546–2554

Ly NP, Ruiz-Pérez B, Onderdonk AB, Tzianabos AO, Litonjua AA, Liang C et al (2006) Mode of delivery and cord blood cytokines: a birth cohort study. Clin Mol Allergy 4:13

Macfarlane GT, Macfarlane S (2012) Bacteria, colonic fermentation, and gastrointestinal health. J AOAC Int 95(1):50–60

Maslowski KM, Vieira AT, Ng A, Kranich J, Sierro F, Yu D et al (2009) Regulation of inflammatory responses by gut microbiota and chemoattractant receptor GPR43. Nature 461 (7268):1282–1286

Maynard CL, Elson CO, Hatton RD, Weaver CT (2012) Reciprocal interactions of the intestinal microbiota and immune system. Nature 489:231–241. https://doi.org/10.1038/nature11551

Mbow M, De Jong SE, Meurs L, Mboup S, Dieye TN, Polman K et al (2014) Changes in immunological profile as a function of urbanization and lifestyle. Immunology 143(4):569–577

Messina M, Venter C (2020) Recent surveys on food allergy prevalence. Nutr Today 55(1):22–29. https://doi.org/10.1097/NT.0000000000000389

Metsala J, Lundqvist A, Virta LJ, Kaila M, Gissler M, Virtanen SM (2013) Mother's and offspring's use of antibiotics and infant allergy to cow's milk. Epidemiology 24(2):303–309

Molloy J, Allen K, Collier F, Tang ML, Ward AC, Vuillermin P (2013) The potential link between gut microbiota and IgE-mediated food allergy in early life. Int J Environ Res Public Health 10 (12):7235–7256

Nagata Y, Yamamoto T, Hayashi M, Hayashi S, Kadowaki M (2017) Improvement of therapeutic efficacy of oral immunotherapy in combination with regulatory T cell-inducer kakkonto in a murine food allergy model. PLoS One 12(1):e0170577

Noval Rivas M, Burton OT, Wise P, Charbonnier LM, Georgiev P, Oettgen HC, Rachid R, Chatila TA (2015) Regulatory T cell reprogramming toward a Th2-cell-like lineage impairs oral tolerance and promotes food allergy. Immunity 42(3):512–523

Noverr MC, Falkowski NR, McDonald RA, McKenzie AN, Huffnagle GB (2005) Development of allergic airway disease in mice following antibiotic therapy and fungal microbiota increase: role of host genetics, antigen, and interleukin-13. Infect Immun 73(1):30–38

Paparo L, Nocerino R, Bruno C, Di Scala C, Cosenza L, Bedogni G et al (2019) Publisher correction: randomized controlled trial on the influence of dietary interventionon epigenetic mechanisms in children with cow's milk allergy: the EPICMA study. Sci Rep 9(1):9504

Pawankar R, Canonica GW, Holgate ST, Lockey R (2011) World allergy organization (WAO) white book on allergy. World Allergy Organization, Milwaukee, WI

Penders J, Thijs C, van den Brandt PA, Kummeling I, Snijders B, Stelma F et al (2007) Gut microbiota composition and development of atopic manifestations in infancy: the KOALA birth cohort study. Gut 56(5):661–667

Prescott SL (2013) Early-life environmental determinants of allergic diseases and the wider pandemic of inflammatory non-communicable diseases. J Allergy Clin Immunol 131(1):23–30

Qin J, Li R, Raes J, Arumugam M, Burgdorf HS, Manichanh C et al (2010) A human gut microbial gene catalogue established by metagenomic sequencing. Nature 464(7285):59–65

Rad AH, Maleki LA, Kafil HS, Abbasi A (2021) Postbiotics: a novel strategy in food allergy treatment. Crit Rev Food Sci Nutr 61(3):492–499

Riedler J, Braun-Fährlander C, Eder W, Schreuer M, Waser M, Maisch S et al (2001) Exposure to farming in early life and development of asthma and allergy: a cross-sectional survey. Lancet 358(9288):1129–1133

Roduit C, Frei R, Ferstl R, Loeliger S, Westermann P, Rhyner C et al (2019) High levels of butyrate and propionate in early life are associated with protection against atopy. Allergy 74(4):799–809

Round JL, Lee SM, Li J, Tran G, Jabri B, Chatila TA et al (2011) The toll-like receptor 2 pathway establishes colonization by a commensal of the human microbiota. Science 332(6032):974–977

Sabat R, Ouyang W, Wolk K (2014) Therapeutic opportunities of the IL-22-IL-22R1 system. Nat Rev Drug Discov 13(1):21–38

Sampson HA, O'mahony L, Burks AW, Plaut M, Lack G, Akdis CA (2018) Mechanisms of food allergy. J Allergy Clin Immunol 141(1):11–19

Schmiechen ZC, Weissler KA, Frischmeyer-Guerrerio PA (2019) Recent developments in understanding the mechanisms of food allergy. Curr Opin Pediatr 31(6):807–814. https://doi.org/10.1097/MOP.0000000000000806

Sjogren YM, Jenmalm MC, Bottcher MF, Bjorksten B, Sverremark-Ekstrom E (2009) Altered early infant gut microbiota in children developing allergy up to 5 years of age. Clin Exp Allergy 39 (4):518–526

Stefka AT, Feehley T, Tripathi P, Qiu J, McCoy K, Mazmanian SK et al (2014) Commensal bacteria protect against food allergen sensitization. PNAS 111(36):13145–13150. https://doi.org/10.1073/pnas.1412008111

Steinhoff U (2005) Who controls the crowd? New findings and old questions about the intestinal microflora. Immunol Lett 99(1):12–16

Strachan DP (2000) Family size, infection and atopy: the first decade of the "hygiene hypothesis". Thorax 55(Suppl 1):S2–S10

Suther C, Moore MD, Beigelman A, Zhou Y (2020) The gut microbiome and the big eight. Nutrients 12(12):3728. https://doi.org/10.3390/nu12123728

Tan J, McKenzie C, Vuillermin PJ, Goverse G, Vinuesa CG, Mebius RE et al (2016) Dietary fiber and bacterial SCFA enhance oral tolerance and protect against food allergy through diverse cellular pathways. Cell Rep 15(12):2809–2824

Thorburn AN, McKenzie CI, Shen S, Stanley D, Macia L, Mason LJ et al (2015) Evidence that asthma is a developmental origin disease influenced by maternal diet and bacterial metabolites. Nat Commun 6:7320

Tremaroli V, Bäckhed F (2012) Functional interactions between the gut microbiota and host metabolism. Nature 489(7415):242–249

Trompette A, Gollwitzer ES, Yadava K, Sichelstiel AK, Sprenger N, Ngom-Bru C et al (2014) Gut microbiota metabolism of dietary fiber influences allergic airway disease and hematopoiesis. Nat Med 20(2):159–166

Tsuda M, Arakawa H, Ishii N, Ubukata C, Michimori M, Noda M et al (2017) Dietary Fructo-oligosaccharides attenuate early activation of CD4+ T cells which produce both Th1 and Th2 cytokines in the intestinal lymphoid tissues of a murine food allergy model. Int Arch Allergy Immunol 174:121–132

Van Den Elsen LWJ, Poyntz HC, Weyrich LS, Young W, Forbes-Blom EE (2017) Embracing the gut microbiota: the new frontier for inflammatory and infectious diseases. Clin Transl Immunol 6(1):e125. https://doi.org/10.1038/cti.2016.91

Venegas DP, De la Fuente MK, Landskron G, Gonzalez MJ, Quera R, Dijkstra G et al (2019) Short chain fatty acids (SCFAs)-mediated gut epithelial and immune regulation and its relevance for inflammatory bowel diseases. Front Immunol 10:277

Von Martels JZH, Sadaghian SM, Bourgonje AR, Blokzijl T, Dijkstra G, Faber KN et al (2017) The role of gut microbiota in health and disease: in vitro modeling of host-microbe interactions at the aerobe-anaerobe interphase of the human gut. Anaerobe 44:3–12. https://doi.org/10.1016/j.anaerobe.2017.01.001

Wambre E, Bajzik V, DeLong JH, O'Brien K, Nguyen QA, Speake C et al (2017) A phenotypically and functionally distinct human TH2 cell subpopulation is associated with allergic disorders. Sci Transl Med 9(401):eaam9171

Wang M, Karlsson C, Olsson C, Adlerberth I, Wold AE, Strachan DP et al (2008) Reduced diversity in the early fecal microbiota of infants with atopic eczema. J Allergy Clin Immunol 121(1):129–134

Wills-Karp M, Santeliz J, Karp CL (2001) The germless theory of allergic disease: revisiting the hygiene hypothesis. Nat Rev Immunol 1(1):69–75

Wypych TP, Marsland BJ (2018) Antibiotics as instigators of microbial dysbiosis: implications for asthma and allergy. Trends Immunol 39(9):697–711

Yamashita H, Takahashi K, Tanaka H, Nagai H, Inagaki N (2012) Overcoming food allergy through acquired tolerance conferred by transfer of tregs in a murine model. Allergy 67 (2):201–209

Zhang GQ, Hu HJ, Liu CY, Zhang Q, Shakya S, Li ZY (2016) Probiotics for prevention of atopy and food hypersensitivity in early childhood: a PRISMA compliant systematic review and meta-analysis of randomized controlled trials. Medicine (Baltimore) 95(8):e2562

Zhang J, Su H, Li Q, Wu H, Liu M, Huang J et al (2017) Oral administration of *Clostridium butyricum* CGMCC0313-1 inhibits β-lactoglobulin-induced intestinal anaphylaxis in a mouse model of food allergy. Gut Pathog 9:11. https://doi.org/10.1186/s13099-017-0160-6

Zhuang L, Chen H, Zhang S, Zhuang J, Li Q, Feng Z (2019) Intestinal microbiota in early life and its implications on childhood health. Genomics Proteomics Bioinformatics 17(1):13–25. https://doi.org/10.1016/j.gpb.2018.10.002

Gut Microbiome Composition as the Key Factor for Immunomodulation in the Host

7

Mohd Iqbal Bhat, Ankita Kumari, Suman Kapila, and Rajeev Kapila

Abstract

Gut microbiota is an intricate assortment of microbes that naturally thrive in the digestive tract of humans and other animals. These microbes are very critical for host development, immunity and nutrition. Ample scientific evidences establish the role of gut microbes in human health and diseases. Especially with respect to host immunity, the interaction is highly interlinked with microbiota influencing the induction, training and function of host immune cells, thereby regulating immune homeostasis. In turn, the immune system has a central role in shaping composition, diversity and distribution of host gut microbiota. When immune system–microbiota alliance is operating optimally, a myriad of health benefits are rendered to the host including protection against pathogens, intact intestinal barrier integrity, immunohomeostasis and others. Any disturbance in this intricate association is strongly associated with immunological dysregulation with aberrant immune responses that result in inflammation and tissue injury and subsequently can cause autoimmunity, allergy and cancer. This clearly reflects interdependence of host immune system and their gut microbiota as well the critical role of immune system–microbiota cross talk in the host health and disease.

Keywords

Microbiota · Innate immunity · Pathogen · Immune system · Immune homeostasis

M. I. Bhat
Government Degree College (Boys) Pulwama, Pulwama, Jammu and Kashmir, India

A. Kumari · S. Kapila · R. Kapila (✉)
Animal Biochemistry Division, ICAR-National Dairy Research Institute, Karnal, Haryana, India

R. Z. Sayyed, M. Khan (eds.), *Microbiome-Gut-Brain Axis*,
https://doi.org/10.1007/978-981-16-1626-6_7

7.1 Introduction

The gut microbiome is an intricate multifaceted network of about 100 trillion microbes that constitute about 1000 different species (Gerard 2016). These microbes naturally grow within length and breadth of different parts of the gastrointestinal tract with their immaculate ability to get replenished within a short duration of time. These gut microbes develop very quickly right from birth and just after 2–3 years, a stable gut microbiota composition is established. The composition of these microbes is dependent on age, environmental factors (diet, physical activity, etc.) and pathological conditions. In healthy conditions, the microbial diversity and richness enhance with age attaining the highest complexity during adulthood. Ample investigations have demonstrated the role of gut microbiome in human health, nutrition, disease as well as antibiotic resistance (Kau et al. 2011; Ley 2010; Sommer et al. 2009). Due to the critical association between dysbiosed microbiome and the development and progression of various ailments affecting the gut and other visceral organs, microbiota has become a hot research area, especially to biomedical research teams.

Based on various in vitro and in vivo animal models, it has been conclusively revealed that a balanced microbiota composition is very critical for proper functioning of the individual (Pearce et al. 2018). The composition of gut microbiota does change continuously throughout the growth and development of an individual even under normal circumstance. Importantly, this gut microbial population can be manipulated in a desired manner so as to have a better lifelong health benefits. These microbes include archaea, bacteria, viruses, fungi and others that make an intricate ecosystem along the gastrointestinal tract. The total collection of these microbes along with their respective genomes constitutes the microbiome (Bhat and Kapila 2017). Not only the gastrointestinal tract, microorganisms also colonize other anatomical parts such as oral cavity, urogenital tract, mammary gland, skin, mucosa and respiratory tract, but major contribution is from the microbes present within the gut that harbours the densest microbes collectively called as gut microbiota (Dekaboruah et al. 2020). The total number of these microbes is far more than the total number of host cells reflecting that we are much more bacteria as we are human (Sender et al. 2016). Both the microbe number and diversity are very critical in determining healthy as well as diseased conditions of the body. However, the composition of this microbial community is very much host specific that evolves throughout an individual's lifetime with modulations both due to external and internal factors. Further, each bacterial strain harbours thousands of genes, and hence, the collective bacterial genome is about 100 times more than what the host possesses and therefore has a massive effect on host development and functions. These microbes co-evolve and continue to live in the host. The microbes themselves or their metabolites act as critical environmental stimuli and thus affect host functions. Because of this intricacy and interdependence, gut microbiota is now regarded as a virtual organ of the human body for its role in host health and disease (Bhat and Kapila 2017).

With the development of new tools and technique in recent years, the gut microbiome composition, diversity and individual variations have been extensively explored to know dynamic operations of commensal microbiota, though this research area is still in infancy and comprehensive research efforts are still required to know this critical gut microbiota–host interdependence. Further, the main focus is to determine the different microbial species that are predominant during healthy and unhealthy conditions as well as those species that are present during early and elderly stages of development. The gut microbiota has enormous variations consisting of the abundant Firmicutes, Bacteroidetes, Fusobacteria, Actinobacteria, Proteobacteria and Verrucomicrobia with Firmicutes and Bacteroidetes together constituting about 90% of gut microbes (Rinninella et al. 2019). The phylum Firmicutes alone includes more than 200 different genera out of which 95% of total Firmicutes is contributed by *Bacillus*, *Clostridium*, *Lactobacillus*, *Enterococcus* and *Ruminococcus*. The highest microbial density lies in the colon (10^{10}–10^{12}/g) with round about 400 bacterial species. In the lower part of the intestine, mostly anaerobic microbes are found that include *Bacteroides*, bifidobacteria, Fusobacteria and peptostreptococci compared to aerobes and facultative aerobes that include enterobacteria and lactobacilli (Illiano et al. 2020) However, the microbiome composition is never the same, and this complex microbe aggregation changes both due to genetic and environmental factors that may include diet, place of living and exposure to environmental pollution such as toxin, carcinogens and antibiotics (Hasan and Yang 2019). In fact, the microbial composition is used to indicate the healthy or unhealthy status of an individual.

Besides, the gut microbiota is closely involved with numerous aspects of host physiology that includes nutritional status, mental and behavioural patterns, responses to variable stress factors as well as health and disease status of the host (Kho and Lal 2018). The proper execution of various gastrointestinal tract functions that include digestion, absorption as well as protection against colonization by pathogens is largely dependent upon these microbes (Bhat and Kapila 2017). Gut microbes are also essential reservoirs of vitamins like K and B, short-chain fatty acids (SCFAs), cholesterol metabolism as well as digestion of dietary polysaccharides that otherwise will remain un-metabolized. Importantly, gut microbiota is closely linked with the development of the host immune system right from the infancy to the elderly stage. Not only the intestinal mucosal immunity but also the systemic immune systems heavily depend on the host microbiota for their proper development and maturation (Zheng et al. 2020). In fact, various immunological pathologies, notably inflammatory bowel disease, celiac disease (multisystemic autoimmune disorder), inflammatory bowel disease, psoriatic arthritis, atopic eczema and others occur due to changes in gut microbial bacterial diversity and functions (Valdes et al. 2018). Studies have shown the close intimacy between the gut microbiota composition and immune system maturation and development along with associated health complications that could develop due to mismatches in the interactions of these two important components of an individual.

7.2 Interdependence of Gut Microbiota and Host Immune Functions

The gut microbiota and host immune system are two critical host components that largely decide the overall functionality of an individual. Importantly both gut microbiota and the immune system affect the functioning as well as the development of each other. The host coexists with these microbes but simultaneously mount a strong and rapid response to the pathogenic microbes (Pickard et al. 2017). The gut microbial communities co-exist in dynamic relationships with the host through intricate networks of interactions and signals. Investigation has shown the cell surface or cytosolic pattern recognition receptors such as Toll-like receptors (TLR), C-type lectin receptors (CLRs), AIM2-like receptors (ALRs), OAS-like receptors (OLRs) and NOD-like receptors (NLR) mediated interactions between gastrointestinal innate immune system and commensal microbes that are responsible for effective defence mechanisms against pathogenic and non-pathogenic dangers (Strowig et al. 2018). Gut microbiota maintains a proper balance between host self-defence and immune tolerance that ultimately results in homoeostatic conditions within the gut. In addition, a number of metabolites, short-chain fatty acids, polyamines, polyphenols and vitamins are released in the gut with the help of gut microbiota, which also influence the host immune functions (Table 7.1). An imbalanced communication between host immune cells and gut microbiota could therefore have many ill effects on host functions, especially with respect to gut immune functions.

7.2.1 Gut Microbiota-Dependent Immune System Development and Maturation

The human gastrointestinal tract (GIT) is home to a vast and most vivid microbial community called microbiota that has a key role in shaping the integrity of host gut immune functions. In fact, the gut microbiota is closely associated with many developmental aspects of the adaptive immune system as well as innate immunity. As reported previously, the early colonization of host's mucosal surfaces in mammals has a key role in the maturation of host gut immune system (Gensollen et al. 2016). In a recent investigation, it was shown that pregnant women with IBD and their offspring had lower bacterial diversity and altered bacterial composition (Torres et al. 2020). When this altered microbiota was transferred to germ-free mice, they showed immature intestinal immune system maturation with fewer class-switched memory B cells and regulatory T cells in the colon as compared to control women and their babies. The role of microbiota in immune system development can be explored especially using germ-free (GF) and gnotobiotic animals. Using specific receptors such as Toll-like receptors (TLRs), nucleotide-binding oligomerization (NOD)-like receptors (NLRs), C-type lectin receptors and others, immune cell system could sense these microorganisms (Thaiss et al. 2016). This receptor-based interaction with microbial components, like LPS, flagellin, bacterial DNA, etc.,

Table 7.1 Microbiota-dependent host immunomodulation

Microbes/ metabolites/ components	Role in regulation	Consequences	References
Faecalibacterium prausnitzii	Induces high levels of IL-10 and reduces the levels of IL-12 and IFN-γ; therefore, it has anti-inflammatory roles	Lower levels of *F. prausnitzii* were observed in Crohn's disease while levels increased during psoriasis	Codoñer et al. (2018)
Polysaccharide A from *Bacteroides fragilis*	Induces regulatory T cells (Tregs) to produce IL-10, suppresses Th17 cell activity and also protects the host from *Helicobacter hepaticus*-induced colitis	Loss of *Bacteroides* was observed during IBD	Round et al. (2011), Chiu et al. (2014)
Butyrate	Exhibits anti-inflammatory activity by inducing intestinal microbiota to release IL-10	Decrease in butyrate observed in IBD	Singh et al. (2014)
Bacteroides thetaiotaomicron	Reduces the levels of inflammatory cytokines by increasing the nuclear export of the RelA subunit of NF-κB, which is responsible for expression of inflammatory genes	High levels of inflammatory cytokines in inflammatory diseases	Kelly et al. (2004)
Bacterial flagellin	Stimulates ILC3 cells to produce IL-22, which provides defence against various pathogens via inducing the production of anti-microbial proteins	Irregular levels of IL-22 observed in human intestinal mucosa in helminth infection	Leung (2013)
Clostridia	Releases butyrate, stimulates differentiation of colonic T regulatory cells (Treg), which have a role in the suppression of inflammatory and allergic responses	Reduced in IBD in association with low amount of butyrate	Furusawa et al. (2013)

results in a cascade of signalling networks including the activation of transcription factor NF-κB, which results in the release of various chemokines, cytokines and anti-microbial proteins (Francino 2014). Members of the microbial community promote pro- and anti-inflammatory responses in the host, which are critical mediators in the maintenance of immune homeostasis (McDermott and Huffnagle 2014). In a recent investigation, five NF-κB suppressive strains were identified belonging to *Clostridium* clusters IV, XIVa and XV that independently suppressed the secretion of the chemokine IL-8 from blood mononuclear cells and gut epithelial organoids (Giri et al. 2019). These NF-κB suppressive microbes suppressed the cytokine-driven inflammatory responses and endoplasmic reticulum stress in gut epithelial organoids that was responsible for immunomodulatory effects, suggesting the extrinsic regulator role of microbiome in host immunity.

Investigation has shown that gut microbiota plays an important role in the differentiation of T cells into different types of cells including helper T cells (Th1, Th2 and Th17) or regulatory T cells (Tregs) (Belkaid and Hand 2014). For example, segmented filamentous bacteria promote the development of Th17 cells in the intestine, which secrete IL-17 and IL-22, which increased inflammatory response in the host against pathogenic bacteria such as *C. rodentium* (Ivanov et al. 2009). Several studies explained the role of clostridia in the development of Foxp3^{+} regulatory cells in the intestine, which are anti-inflammatory in nature (Atarashi et al. 2013). Other members such as *Escherichia*, *Akkermansia*, *Bacteroides*, *Clostridium*, *Lactobacillus* and *Streptococcus* were also found to induce these regulatory cells (Geva-Zatorsky et al. 2017). The microbial composition is also able to regulate the generation of CD4^{+} and CD8^{+} T cells, which are activated in viral infections and enhanced anti-tumour immunity (Ichinohe et al. 2011; Tanoue et al. 2019). In addition to the regulation of T cell functioning, gut bacteria also stimulate the migration of macrophages and neutrophils in the intestinal tissues for providing protection against pathogens (Kamada and Núñez 2014). Microbes of the gut regularly stimulate the macrophages for IL-10 production, which can further induce Tregs that control the unregulated development of Th17 cells (Rivollier et al. 2012). Recently identified innate lymphoid cells (ILCs) are mainly dependent on the colonization of microbiota for their proper functioning (Kim and Kim 2016). ILCs comprised cytotoxic and non-cytotoxic cells (ILC1, 2 and 3), and most of the studies defined the role of ILC3 in host–microbiota interactions. These lymphocytes restrict the response of T cells to commensal bacteria and thus promote their colonization. Gut microbes also stimulate innate lymphoid cells 3 (ILC3) that subsequently releases IL-22 which acts as activating factor for the enzyme fucosyltransferase 2 (galactoside 2-α-L-fucosyltransferase 2) that protects from enteric pathogens (Thaiss et al. 2016). The development and function of neutrophils are largely dependent locally as well as systemically upon gastrointestinal microbiota. Further, the gastrointestinal tract microbiota affects the differentiation of T cell populations either into the different helper cells that include Th1, Th2 and Th17 or into regulatory T cells (Tregs) (Francino 2014). Conclusively, it can be said that all branches of the immune system are influenced by the microbiota, reflecting the immense role of these microbes in shaping the host immune system.

7.2.2 Role of the Immune System in Shaping Gut Microbiota Complexity

Just like gut microbes influence the immune system functions, the immune system in turn has a key role in deciding the composition and diversity of gut microbiota. In fact, the major proportion of the immune system, approximately up to 70%, resides in the intestine. The host gut defence system includes a diverse array of mechanisms including the multilayered mucus layer and secreted immunoglobulin (sIgA) along with the release of a number of anti-microbial peptides that on one side provide the host defence but at the same time keep the microbiota in check and maintain a mutual beneficial relationship with them (Dolle et al. 2016). At the same time, the mucosal immunity fights out the potential danger that could result from microbiota-derived antigens through the production of specific antibodies. The secretory immunoglobulin (sIgA) in particular is known to play a vital role in deciding the microbiota diversity and composition (Pabst and Slack 2020).

The immune system consists of lymphoid organs and immune cells such as macrophages, dendritic cells, neutrophils and natural killer cells. In addition to immune cells, epithelial cells of the gastrointestinal (GI) tract also have an important role in maintaining the integrity of gut functionality (Zhang et al. 2015). They act as a strong physical barrier to pathogens and toxins and also work along with other components of the immune system in defence mechanisms. Underneath the epithelial layer, antigen-presenting cells (APCs) and lymphocytes present in the lamina propria and gut-associated lymphoid tissue (GALT), respectively, are components of the gut immune system that respond in an antigen-specific manner (Takiishi et al. 2017). All components function synergistically to combat the pathogen invasion in mucosal tissue. The GI tract is the first one to interact with external stimuli; the epithelium of the GI tract is regularly exposed to several types of antigens like food components, commensal bacteria, pathogens and toxins. Due to these continuous exposures, it can distinguish between commensal bacteria and pathogens and opposed the colonization of pathogenic organisms in the gut. In addition to its role in defence against pathogenic microorganisms, the immune system also plays an important role in shaping the commensal bacteria, which is beneficial for host health. Bilateral interactions of gut microbiota with the immune system generate a number of immune responses, and reversibly the immune system could sense and differentiate commensal microorganisms and pathogens, thus developing tolerance (Zheng et al. 2020).

A number of interactions between the two are characterized over the years. In particular, the mucus layer in the intestine forms a double layer, which acts as a primary barrier to the host's defence. The outer layer of mucus supports the colonization of microorganisms and provides nutrition to them (Kashyap et al. 2013). Some of the cytokines such as TGF-β and IL-10 released from host immune cells are known to maintain mucosal tolerance and also support the colonization of commensal bacteria through stimulation of secretory IgA and thus contribute to intestinal homeostasis (Lazar et al. 2018). Intestinal cells have a pivotal role in maintaining intestinal homeostasis as these cells express a range of immune receptors on their surface. Previouly, it has beeen reported that NOD1 of epithelial cells is necessary for the secretion of C-C motif chemokine 20 (CCL20) that has key role in the developmemt of isolated lymphoid follicles (IFLs) responsible for the production of antigen-specific intestinal IgA immunoglobulins (Bouskra et al. 2008; Fenton et al. 2020). NRLP6 in epithelial cells encourages the inflammasome-mediated IL-18 production as well as the secretion of mucus by goblet cells, which contribute to homeostatic regulation of host–microbiota interface (Wlodarska et al. 2014). It has also an important role in the regulation of anti-viral immunity (Wang et al. 2015).

The role of the innate immune system in the development of the community of gut microorganisms can be best studied in mice models having immune deficiencies. The host's innate immune system might promote the growth of microbiota during dysbiosis. For example, in the case of intestinal infection, fucosylated proteins of intestinal cells induced by ILC3 provide energy to commensal bacteria (Pickard et al. 2014). Signalling via TLR1 during *Yersinia enterocolitica* infection also has a role in the maintenance of intestinal ecological homeostasis (Kamdar et al. 2016).

7.3 Microbiota Released Metabolite and Immune System Modulation

Not only the gut microbes but their derived metabolites such as short-chain fatty acids, polyamines, polyphenols and others have a significant outcome on the host immune functions as shown in Fig. 7.1. Microbial metabolites interact with the host's immune system by interacting with stromal and epithelial cells. In the case of microbial metabolites, short-chain fatty acids (SCFAs), like butyrate, acetate, propionate, succinate and lactate, are the most studied (Morrison and Preston 2016). These metabolites are produced through the action of gut microbiota by fermenting non-digestible carbohydrates like dietary fibres and resistant starch (Bhat and Kapila 2017). These metabolites can get incorporated in intestinal epithelial cells or can diffuse across the epithelium into the underlying intestinal lamina propria, thus influencing different host's immune system. For example, microbiota-produced butyrate regulates transepithelial fluid transport along with reduction of mucosal inflammation. Butyrate is considered an essential secondary metabolite that has a key role in the development and functioning of several immune cell lineages (Man et al. 2020). Previously, gut microbiota-derived butyrate was observed to impart anti-inflammatory effects in the colon through increased histone acetylation of the Foxp3 (forkhead box P3) locus in naive $CD4^+$ T cells, which subsequently increased Foxp3 expression that stimulates the differentiation of Treg cells (Furusawa et al. 2013). Similarly, butyrate-dependent colonic Treg differentiation was reported in myeloid cells through histone deacetylase inhibition (Arpaia et al. 2013). Further, SCFAs derived from gut commensal bacteria increased the naive $CD4^+$ T cells, Tregs and other immune cell populations.

In addition, microbial metabolism of dietary foods in the gut also produces biologically active polyphenolic compounds and polyamines (Bhat and Kapila 2017). Polyphenolic compounds are transformed into various derivatives of aromatic SCFAs such as phenylacetate and phenylbutyrate through the action of microbes such as *Bacteroides* species, *Clostridium* species, *Eubacterium limosum* and *Eggerthella lenta* and subsequently bring out various health benefits. For example, polyphenol fisetin is reported to modulate immune functions when incubated with human monocytic THP-1 cells through the epigenetic inhibition of the expression of NF-jB genes, IL-6 and TNF-α (Kim et al. 2012). Similarly, polyamines have been reported to exert regulatory functions on immune cells possibly by regulating transcription, protein translation, stress protein responses

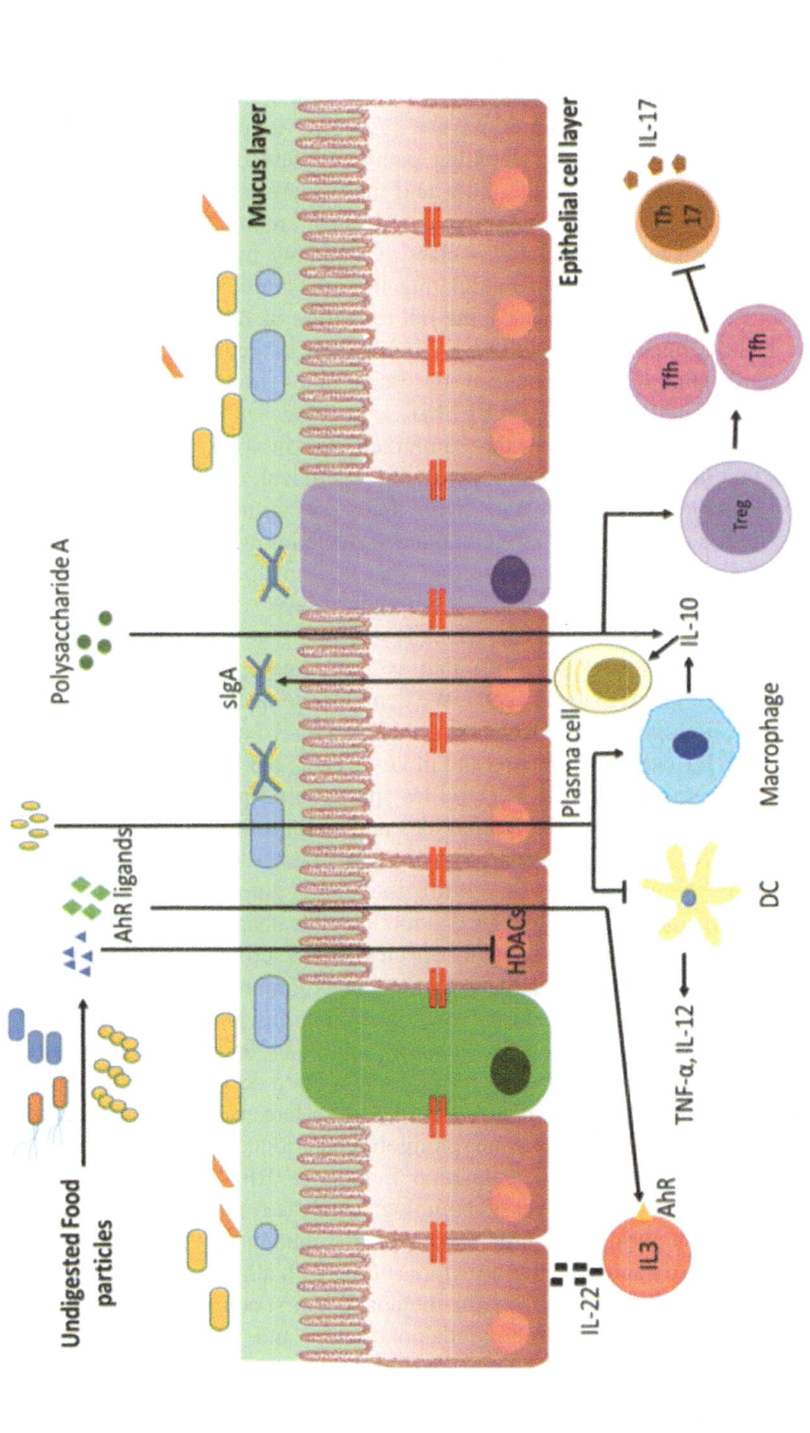

Fig. 7.1 Interdependence of gut microbiota and host immune functions. An array of different bacteria reside in the gut that release a variety of metabolites such as SCFAs (blue, triangle), tryptophan catabolites (green, diamond) and secondary bile acids (yellow, ellipse), which affect the different components of the gut immune system

and cellular metabolism. Polyamines exert anti-inflammatory effects by suppressing inflammatory T cells along with production of cytokines and nitric oxide (NO), thereby having an immunomodulatory effect (Keough et al. 2011).

While the human gut microbiota are suspected to produce diffusible small molecules that modulate host signaling pathways, few of these molecules have been identified. Species of Bacteroides and their relatives, which often comprise. Fifty percent of the gut community, are unusual among bacteria in that their membrane is rich in sphingolipids, a class of signaling molecules that play a key role in inducing apoptosis and modulating the host immune response. Although known for more than three decades, the full repertoire of Bacteroides sphingolipids has not been defined. Here, we use a combination of genetics and chemistry to identify the sphingolipids produced by Bacteroides fragilis NCTC 9343. We constructed a deletion mutant of BF2461, a putative serine palmitoyltransferase whose yeast homolog catalyzes the committed step in sphingolipid biosynthesis. We show that the D2461 mutant is sphingolipid deficient, enabling us to purify and solve the structures of three alkaline-stable lipids present in the wild-type strain but absent from the mutant. The first compound was the known sphingolipid ceramide phosphorylethanolamine, and the second was its corresponding dihydroceramide base. Unexpectedly, the third compound was the glycosphingolipid a-galactosylceramide (a-GalCerBf), which is structurally related to a sponge-derived sphingolipid (a-GalCer, KRN7000) that is the prototypical agonist of CD1d-restricted natural killer T (iNKT) cells. We demonstrate that a-GalCer Bf has similar immunological properties to KRN7000: it binds to CD1d and activates both mouse and human iNKT cells both in vitro and in vivo. Thus, our study reveals BF2461 as the first known member of the Bacteroides sphingolipid pathway, and it indicates that the committed steps of the Bacteroides and eukaryotic sphingolipid pathways are identical. Moreover, our data suggest that some Bacteroides sphingolipids might influence host immune homeostasis. While the human gut microbiota are suspected to produce diffusible small molecules that modulate host signaling pathways, few of these molecules have been identified. Species of Bacteroides and their relatives, which often comprise. Fifty percent of the gut community, are unusual among bacteria in that their membrane is rich in sphingolipids, a class of signaling molecules that play a key role in inducing apoptosis and modulating the host immune response. Although known for more than three decades, the full repertoire of Bacteroides sphingolipids has not been defined. Here, we use a combination of genetics and chemistry to identify the sphingolipids produced by Bacteroides fragilis NCTC 9343. We constructed a deletion mutant of BF2461, a putative serine palmitoyltransferase whose yeast homolog catalyzes the committed step in sphingolipid biosynthesis. We show that the D2461 mutant is sphingolipid deficient, enabling us to purify and solve the structures of three alkaline-stable lipids present in the wild-type strain but absent from the mutant. The first compound was the known sphingolipid ceramide phosphorylethanolamine, and the second was its corresponding dihydroceramide base. Unexpectedly, the third compound was the glycosphingolipid a-galactosylceramide (a-GalCer Bf), which is structurally related to a sponge-derived sphingolipid (a-GalCer, KRN7000) that is the prototypical agonist of CD1d-restricted natural killer T (iNKT) cells. We demonstrate that a-GalCer Bf has similar immunological properties to KRN7000: it binds to CD1d and activates both mouse and human iNKT cells both in vitro and in vivo. Thus, our study reveals BF2461 as the first known member of the Bacteroides sphingolipid pathway, and it indicates that the committed steps of the Bacteroides and eukaryotic sphingolipid pathways are identical. Moreover, our data suggest that some Bacteroides sphingolipids might influence host immune homeostasis.

The glycosphingolipid α-galactosylceramide (αGalCer) derived from human gut microbe *Bacteroides* and their relatives serves as an important class of signalling molecules that have a key role in inducing cellular apoptosis and modulating the host immune response (Von Gerichten et al. 2019). It was reported that α-galactosylceramide (αGalCer) binds to CD1d and activates both mouse and human invariant natural killer T cells both in vitro and in vivo, suggesting the role of *Bacteroides* sphingolipids in influencing host immune homeostasis. Further, it was noticed that when α-galactosylceramide was presented to cluster of differentiation 1d (CD1d) receptors on antigen-presenting cells, it was observed to efficiently modulate immune responses against tumours, microbial and viral infections and autoimmune diseases. Interestingly, decreased αGalCer production was observed in mice when gut microbiota composition was altered due to colitis and influenza A virus infection. Previously, α-galactosylceramide was also found to diminish inflammation of the intestine in mice colitis model, which subsequently maintained intestinal homeostasis (An et al. 2014). Collectively, these studies demonstrated the critical role of microbiota-derived glycosphingolipid αGalCer in maintaining gut homeostasis, thereby having a key role in mediating local and systemic immune responses.

The microbial community is also a source of secondary bile acids having anti-inflammatory properties and is found to repress the production of tumour necrosis factor (TNF)-α and IL-12 from dendritic cells in addition to increasing the production of IL-10 from macrophages (Fiorucci et al. 2018). One of the tryptophan metabolites, indole-3-aldehyde, acts as ligands for the host receptors such as aryl hydrocarbon receptor (AhR), which enhance the transcription of IL-22, which plays a critical *role* in antibacterial immunity and mediates host defence through the mucosal barrier (Zelante et al. 2013). Other than these metabolites, cell wall components such as polysaccharide A (PSA) from *Bacteroides* species are responsible for the induction of Tregs and production of anti-inflammatory interleukin, IL-10, which are known as major contributors to the maintenance of immune homeostasis (Round and Mazmanian 2010). Similarly, the vitamin A lipid metabolite retinoic acid has been reported to maintain the balance between pro-inflammatory and anti-inflammatory immune responses. Further, it was found that retinoic acid deficiency was found to affect both the composition of the microbiota and immune system function that subsequently resulted in the decreased number of T helper 17 (T_H17) cells (Cha et al. 2010).

7.4 Mechanistic View of Host Innate Immune System and Microbiota Interaction

After understanding the intimate associations between gut microbes and host immune functions, it is very exciting to know about these cellular and molecular mechanisms responsible for these complex pathways. Though the mechanisms involved in the interaction of gut microbiota and host immune cells are not fully explored yet, it believed that these interactions are regulated both at transcriptional and epigenetic levels. Regarding molecular studies, several genes involved in the absorption of nutrients, gut barrier functionality, intestinal immunity and metabolism of xenobiotics are studied over the years to exploit the interactions of the immune system with commensals. The transcription programming of these genes mainly depends upon the sensing of microbial components by intestinal cells (Sommer et al. 2015). These components regulate the expression of the above genes through the regulation of ubiquitin signalling and translocation of p65 transcriptional factor to activate NF-κB inflammatory pathway and also via vesicular trafficking (Thaiss et al. 2016).

Besides transcriptional regulation, host–microbiota interactions are also studied by means of epigenetic modifications. Interactions of cells with microbes could bring changes in the chromatin structure, which further affects the chromatin accessibility to transcriptional factors. These epigenetic events might affect the transcriptional programming of the cells. Takiishi et al. (2017) studied the role of commensal bacteria on the host's innate immune system via the epigenetic pathways. High levels of methylation on the promoter region of TLR4 in colonized mice suggested that commensal bacteria regulated the immune system by suppressing the PRRs. Deletion of histone deacetylase (HDAC3) from intestinal cells resulted in damage to the integrity of the intestinal barrier (Alenghat et al. 2013). Similarly, gut microbiota-dependent epigenetic regulations have been reported to regulate the development of various types of immune cells including $CD4^+$ T cells, Tregs and other immune cells (Alenghat and Artis 2014). Microbiota-derived metabolites such as short-chain fatty acids, polyamines and polyphenols affect the host gut functions including immune functions through epigenetic modulations involving DNA methylation and demethylation, histone acetylation and deacetylation as well as RNA interference (Bhat and Kapila 2017).

7.5 Consequences of Mismatched Interaction Between Gut Microbiota and Immune Cells

From investigation, it is quite obvious that balanced gut microbiota composition is very critical for the overall functioning of an individual. Alterations in interactions of the immune system and gut microbiota because of perturbation in gut microbiota composition might result in faulty functions of the immune system, which could further result in autoimmune and inflammatory diseases. The most common inflammatory diseases due to dysregulation of the microbiota-immune system are inflammatory bowel disorders including inflammatory bowel disease and ulcerative colitis (UC), while autoimmune diseases include type 1 diabetes, multiple sclerosis (MS) and rheumatoid arthritis (Gianchecchi and Fierabracci 2019).

In a recent investigation, it was reported that differential microbiota composition exists in the small intestine of healthy and unhealthy children with inflammatory bowel disease (IBD) symptoms (Rapozo et al. 2017). Microbial investigation revealed that the children with IBD had decreased total microbial counts of *Collinsella*, *Lactobacillus*, *Bacillus*, Firmicutes, Actinobacteria and Bacteroidetes (Krogius-Kurikka et al. 2009). The dysbiosed intestinal microbe composition makes these children more susceptible to malabsorption of micronutrient resulting in the depletion of essential nutrients within their body, reflecting the interdependence of gut microbiota and host functions. In fact, dysbiosed gut microbiota is associated with the progression of metabolic disorders like obesity, type 2 diabetes mellitus (T2DM), cardiovascular diseases and cancer (Li et al. 2019). Previous investigation has showed the difference in the composition of gut microbiome even between twins, which was later found to be related with the development of obese conditions within one individual than the other (Harley and Karp 2012).

During the development of inflammatory bowel disease (IBD), a constant decrease of microbes belonging to *Faecalibacterium*, *Clostridium* and *Eubacterium* species in contrast to a consistent increase in members of Enterobacteriaceae, *Ruminococcus gnavus* and *Fusobacterium nucleatum* was observed (Brown et al. 2019). In addition to these microbes, loss of PSA and sphingolipids producing *Bacteroides* during IBD further hampers the immune homeostasis (Chiu et al. 2014). In the case of autoimmune disease, MS and a low count of *Bacteroides* and *Faecalibacterium* are also observed (Miyake et al. 2015). Further, it has been found that development of type 1 diabetes is correlated with high abundance of intestinal bacteroids and lower numbers of Clostridiales (Giongo et al. 2011). Alterations in microbiota in RA are mostly associated with an increase in *Prevotella* species, which subsequently enhance the sensitivity for chemically induced colitis and can further contribute to inflammatory diseases (Maeda et al. 2016). In diseased conditions, activities of both innate and adaptive immune cells are debited from their normal functions, especially T cells. T helper (Th) and natural killer T (NKT) cells are known to further contribute to maintaining the inflammation.

7.6 Combating Mismatched Gut Microbiota and Immune System Interactions

Gastrointestinal disorders are mainly associated with alterations in gut microbiota. Therefore, restoration of normal gut microbiota could be one critical step in combating the problems associated with the gut or other visceral organs. At present, faecal microbial transplantation (FMT), that is, a process of transferring faeces from a healthy person to the intestine of the person having gut disorders, is very popular to mediate intestinal homeostasis. The consumption of probiotics, which are defined as live microorganisms, when consumed in adequate amounts, confers benefits to the host and could be one way to restore these mismatched interactions. These probiotic microbes have been reported to have ample health benefits, especially with reference to intestinal homeostasis in addition to their prophylactic and therapeutic effects in various disease models. However, probiotics have strain-specific effects, and hence, it is very imperative to decide the specific microbes for combating microbiota host mismatches. The combination of these proved to be more effective in the treatment of diseases in comparison to a single strain (Kim et al. 2017).

7.7 Future Perspectives

To know the precise mechanism involved in microbiota, dependent host immune modulation needs to be explored. Epigenetic mechanisms appeared to have a critical effect in mediation of the host gut microbiota immune system modulation but are still in infancy and need to be elucidated further. Detailed investigations involved in host–microbiome interactions could become a suitable target for investigators to ponder upon for coming years.

References

Alenghat T, Artis D (2014) Epigenomic regulation of host–microbiota interactions. Trends Immunol 35(11):518–525

Alenghat T, Osborne LC, Saenz SA, Kobuley D, Ziegler CG, Mullican SE, Choi I, Grunberg S, Sinha R, Wynosky-Dolfi M, Snyder A (2013) Histone deacetylase 3 coordinates commensal-bacteria-dependent intestinal homeostasis. Nature 504(7478):153–157

An D, Oh SF, Olszak T, Neves JF, Avci FY, Erturk-Hasdemir D, Lu X, Zeissig S, Blumberg RS, Kasper DL (2014) Sphingolipids from a symbiotic microbe regulate homeostasis of host intestinal natural killer T cells. Cell 156(1–2):123–133

Arpaia N, Campbell C, Fan X, Dikiy S, Van Der Veeken J, Deroos P, Liu H, Cross JR, Pfeffer K, Coffer PJ, Rudensky AY (2013) Metabolites produced by commensal bacteria promote peripheral regulatory T-cell generation. Nature 504(7480):451–455

Atarashi K, Tanoue T, Oshima K, Suda W, Nagano Y, Nishikawa H, Fukuda S, Saito T, Narushima S, Hase K, Kim S (2013) T reg induction by a rationally selected mixture of Clostridia strains from the human microbiota. Nature 500(7461):232–236

Belkaid Y, Hand TW (2014) Role of the microbiota in immunity and inflammation. Cell 157 (1):121–141

Bhat MI, Kapila R (2017) Dietary metabolites derived from gut microbiota: critical modulators of epigenetic changes in mammals. Nutr Rev 75(5):374–389

Bouskra D, Brézillon C, Bérard M, Werts C, Varona R, Boneca IG, Eberl G (2008) Lymphoid tissue genesis induced by commensals through NOD1 regulates intestinal homeostasis. Nature 456(7221):507–510

Brown EM, Kenny DJ, Xavier RJ (2019) Gut microbiota regulation of T cells during inflammation and autoimmunity. Annu Rev Immunol 37:599–624

Cha HR, Chang SY, Chang JH, Kim JO, Yang JY, Kim CH, Kweon MN (2010) Downregulation of Th17 cells in the small intestine by disruption of gut flora in the absence of retinoic acid. J Immunol 184(12):6799–6806

Chiu CC, Ching YH, Wang YC, Liu JY, Li YP, Huang YT, Chuang HL (2014) Monocolonization of germ-free mice with Bacteroides fragilis protects against dextran sulfate sodium-induced acute colitis. Biomed Res Int 2014:675786

Codoñer FM, Ramírez-Bosca A, Climent E, Carrión-Gutierrez M, Guerrero M, Pérez-Orquín JM, De La Parte JH, Genovés S, Ramón D, Navarro-López V, Chenoll E (2018) Gut microbial composition in patients with psoriasis. Sci Rep 8(1):1–7

Dekaboruah E, Suryavanshi MV, Chettri D, Verma AK (2020) Human microbiome: an academic update on human body site specific surveillance and its possible role. Arch Microbiol 202 (8):2147–2167

Dolle L, Tran HQ, Etienne-Mesmin L, Chassaing B (2016) Policing of gut microbiota by the adaptive immune system. BMC Med 14(1):1–4

Fenton TM, Jørgensen PB, Niss K, Rubin SJ, Mörbe UM, Riis LB, Da Silva C, Plumb A, Vandamme J, Jakobsen HL, Brunak S (2020) Immune profiling of human gut-associated lymphoid tissue identifies a role for isolated lymphoid follicles in priming of region-specific immunity. Immunity 52(3):557–570

Fiorucci S, Biagioli M, Zampella A, Distrutti E (2018) Bile acids activated receptors regulate innate immunity. Front Immunol 9:1853

Francino MP (2014) Early development of the gut microbiota and immune health. Pathogens 3 (3):769–790

Furusawa Y, Obata Y, Fukuda S, Endo TA, Nakato G, Takahashi D, Nakanishi Y, Uetake C, Kato K, Kato T, Takahashi M (2013) Commensal microbe-derived butyrate induces the differentiation of colonic regulatory T cells. Nature 504(7480):446–450

Gensollen T, Iyer SS, Kasper DL, Blumberg RS (2016) How colonization by microbiota in early life shapes the immune system. Science 352(6285):539–544

Gerard P (2016) Gut microbiota and obesity. Cell Mol Life Sci 73(1):147–162

Geva-Zatorsky N, Sefik E, Kua L, Pasman L, Tan TG, Ortiz-Lopez A, Yanortsang TB, Yang L, Jupp R, Mathis D, Benoist C (2017) Mining the human gut microbiota for immunomodulatory organisms. Cell 168(5):928–943

Gianchecchi E, Fierabracci A (2019) Recent advances on microbiota involvement in the pathogenesis of autoimmunity. Int J Mol Sci 20(2):283

Giongo A, Gano KA, Crabb DB, Mukherjee N, Novelo LL, Casella G, Drew JC, Ilonen J, Knip M, Hyöty H, Veijola R (2011) Toward defining the autoimmune microbiome for type 1 diabetes. ISME J 5(1):82–91

Giri R, Hoedt EC, Shamsunnahar K, McGuckin MA, Morrison M, Capon RJ, Begun J, Cuív PÓ (2019) Secreted microbial metabolites modulate gut immunity and inflammatory tone. BioRxiv

Harley IT, Karp CL (2012) Obesity and the gut microbiome: striving for causality. Mol Metab 1 (1–2):21–31

Hasan N, Yang H (2019) Factors affecting the composition of the gut microbiota, and its modulation. PeerJ 7:e7502

Ichinohe T, Pang IK, Kumamoto Y, Peaper DR, Ho JH, Murray TS, Iwasaki A (2011) Microbiota regulates immune defense against respiratory tract influenza A virus infection. Proc Natl Acad Sci 108(13):5354–5359

Illiano P, Brambilla R, Parolini C (2020) The mutual interplay of gut microbiota, diet and human disease. FEBS J 287(5):833–55

Ivanov II, Atarashi K, Manel N, Brodie EL, Shima T, Karaoz U, Wei D, Goldfarb KC, Santee CA, Lynch SV, Tanoue T (2009) Induction of intestinal Th17 cells by segmented filamentous bacteria. Cell 139(3):485–498

Kamada N, Núñez G (2014) Regulation of the immune system by the resident intestinal bacteria. Gastroenterology 146(6):1477–1488

Kamdar K, Khakpour S, Chen J, Leone V, Brulc J, Mangatu T, Antonopoulos DA, Chang EB, Kahn SA, Kirschner BS, Young G (2016) Genetic and metabolic signals during acute enteric bacterial infection alter the microbiota and drive progression to chronic inflammatory disease. Cell Host Microbe 19(1):21–31

Kashyap PC, Marcobal A, Ursell LK, Smits SA, Sonnenburg ED, Costello EK, Higginbottom SK, Domino SE, Holmes SP, Relman DA, Knight R (2013) Genetically dictated change in host mucus carbohydrate landscape exerts a diet-dependent effect on the gut microbiota. Proc Natl Acad Sci 110(42):17059–17064

Kau AL, Ahern PP, Griffin NW, Goodman AL, Gordon JI (2011) Human nutrition, the gut microbiome and the immune system. Nature 474(7351):327–336

Kelly D, Campbell JI, King TP, Grant G, Jansson EA, Coutts AG, Pettersson S, Conway S (2004) Commensal anaerobic gut bacteria attenuate inflammation by regulating nuclear-cytoplasmic shuttling of PPAR-γ and RelA. Nat Immunol 5(1):104–112

Keough MP, Hayes CS, DeFeo K, Gilmour SK (2011) Elevated epidermal ornithine decarboxylase activity suppresses contact hypersensitivity. J Investig Dermatol 131(1):158–166

Kho ZY, Lal SK (2018) The human gut microbiome–a potential controller of wellness and disease. Front Microbiol 9:1835

Kim M, Kim CH (2016) Colonization and effector functions of innate lymphoid cells in mucosal tissues. Microbes Infect 18(10):604–614

Kim HJ, Kim SH, Yun JM (2012) Fisetin inhibits hyperglycemia-induced proinflammatory cytokine production by epigenetic mechanisms. Evid Based Complement Alternat Med 2012:639469

Kim D, Zeng MY, Núñez G (2017) The interplay between host immune cells and gut microbiota in chronic inflammatory diseases. Exp Mol Med 49(5):e339

Krogius-Kurikka L, Lyra A, Malinen E, Aarnikunnas J, Tuimala J, Paulin L, Mäkivuokko H, Kajander K, Palva A (2009) Microbial community analysis reveals high level phylogenetic alterations in the overall gastrointestinal microbiota of diarrhoea-predominant irritable bowel syndrome sufferers. BMC Gastroenterol 9(1):1

Lazar V, Ditu LM, Pircalabioru GG, Gheorghe I, Curutiu C, Holban AM, Picu A, Petcu L, Chifiriuc MC (2018) Aspects of gut microbiota and immune system interactions in infectious diseases, immunopathology, and cancer. Front Immunol 9:1830

Leung JM (2013) A role for IL-22 in the relationship between intestinal helminths, gut microbiota and mucosal immunity. Int J Parasitol 43(3–4):253–257

Ley RE (2010) Obesity and the human microbiome. Curr Opin Gastroenterol 26(1):5–11

Li BY, Xu XY, Gan RY, Sun QC, Meng JM, Shang A, Mao QQ, Li HB (2019) Targeting gut microbiota for the prevention and management of diabetes mellitus by dietary natural products. Foods 8(10):440

Maeda Y, Kurakawa T, Umemoto E, Motooka D, Ito Y, Gotoh K, Hirota K, Matsushita M, Furuta Y, Narazaki M, Sakaguchi N (2016) Dysbiosis contributes to arthritis development via activation of autoreactive T cells in the intestine. Arthritis Rheumatol 68(11):2646–2661

Man AW, Zhou Y, Xia N, Li H (2020) Involvement of gut microbiota, microbial metabolites and interaction with polyphenol in host immunometabolism. Nutrients 12(10):3054

McDermott AJ, Huffnagle GB (2014) The microbiome and regulation of mucosal immunity. Immunology 142(1):24–31

Miyake S, Kim S, Suda W, Oshima K, Nakamura M, Matsuoka T, Chihara N, Tomita A, Sato W, Kim SW, Morita H (2015) Dysbiosis in the gut microbiota of patients with multiple sclerosis, with a striking depletion of species belonging to clostridia XIVa and IV clusters. PLoS One 10 (9):e0137429

Morrison DJ, Preston T (2016) Formation of short chain fatty acids by the gut microbiota and their impact on human metabolism. Gut Microbes 7:189–200

Pabst O, Slack E (2020) IgA and the intestinal microbiota: the importance of being specific. Mucosal Immunol 13(1):12–21

Pearce SC, Coia HG, Karl JP, Pantoja-Feliciano IG, Zachos NC, Racicot K (2018) Intestinal in vitro and ex vivo models to study host-microbiome interactions and acute stressors. Front Physiol 9:1584

Pickard JM, Maurice CF, Kinnebrew MA, Abt MC, Schenten D, Golovkina TV, Bogatyrev SR, Ismagilov RF, Pamer EG, Turnbaugh PJ, Chervonsky AV (2014) Rapid fucosylation of intestinal epithelium sustains host–commensal symbiosis in sickness. Nature 514 (7524):638–641

Pickard JM, Zeng MY, Caruso R, Núñez G (2017) Gut microbiota: role in pathogen colonization, immune responses, and inflammatory disease. Immunol Rev 279(1):70–89

Rapozo DC, Bernardazzi C, de Souza HS (2017) Diet and microbiota in inflammatory bowel disease: The gut in disharmony. World J Gastroenterol 23(12):2124

Rinninella E, Raoul P, Cintoni M, Franceschi F, Miggiano GA, Gasbarrini A, Mele MC (2019) What is the healthy gut microbiota composition? A changing ecosystem across age, environment, diet, and diseases. Microorganisms 7(1):14

Rivollier A, He J, Kole A, Valatas V, Kelsall BL (2012) Inflammation switches the differentiation program of Ly6Chi monocytes from antiinflammatory macrophages to inflammatory dendritic cells in the colon. J Exp Med 209(1):139–155

Round JL, Mazmanian SK (2010) Inducible Foxp3+ regulatory T-cell development by a commensal bacterium of the intestinal microbiota. Proc Natl Acad Sci 107(27):12204–12209

Round JL, Lee SM, Li J, Tran G, Jabri B, Chatila TA, Mazmanian SK (2011) The Toll-like receptor 2 pathway establishes colonization by a commensal of the human microbiota. Science 332 (6032):974–977

Sender R, Fuchs S, Milo R (2016) Revised estimates for the number of human and bacteria cells in the body. PLoS Biol 14(8):e1002533

Singh N, Gurav A, Sivaprakasam S, Brady E, Padia R, Shi H, Thangaraju M, Prasad PD, Manicassamy S, Munn DH, Lee JR (2014) Activation of Gpr109a, receptor for niacin and the commensal metabolite butyrate, suppresses colonic inflammation and carcinogenesis. Immunity 40(1):128–139

Sommer MOA, Dantas G, Church GM (2009) Functional characterization of the antibiotic resistance reservoir in the human microflora. Science 325:112821131

Sommer F, Nookaew I, Sommer N, Fogelstrand P, Bäckhed F (2015) Site-specific programming of the host epithelial transcriptome by the gut microbiota. Genome Biol 16(1):1–5

Strowig T, Thiemann S, Diefenbach A (2018) Microbiome and gut immunity: innate immune cells. In: The gut microbiome in health and disease. Springer, Cham, pp 103–118

Takiishi T, Fenero CI, Câmara NO (2017) Intestinal barrier and gut microbiota: shaping our immune responses throughout life. Tissue Barriers 5(4):e1373208

Tanoue T, Morita S, Plichta DR, Skelly AN, Suda W, Sugiura Y, Narushima S, Vlamakis H, Motoo I, Sugita K, Shiota A (2019) A defined commensal consortium elicits CD8 T cells and anti-cancer immunity. Nature 565(7741):600–605

Thaiss CA, Zmora N, Levy M, Elinav E (2016) The microbiome and innate immunity. Nature 535 (7610):65–74

Torres J, Hu J, Seki A, Eisele C, Nair N, Huang R, Tarassishin L, Jharap B, Cote-Daigneault J, Mao Q, Mogno I (2020) Infants born to mothers with IBD present with altered gut microbiome that transfers abnormalities of the adaptive immune system to germ-free mice. Gut 69(1):42–51

Valdes AM, Walter J, Segal E, Spector TD (2018) Role of the gut microbiota in nutrition and health. BMJ 13:361

Von Gerichten J, Lamprecht D, Opálka L, Soulard D, Marsching C, Pilz R, Sencio V, Herzer S, Galy B, Nordström V, Hopf C (2019) Bacterial immunogenic α-galactosylceramide identified in the murine large intestine: dependency on diet and inflammation. J Lipid Res 60(11):1892–1904

Wang P, Zhu S, Yang L, Cui S, Pan W, Jackson R, Zheng Y, Rongvaux A, Sun Q, Yang G, Gao S (2015) Nlrp6 regulates intestinal antiviral innate immunity. Science 350(6262):826–830

Wlodarska M, Thaiss CA, Nowarski R, Henao-Mejia J, Zhang JP, Brown EM, Frankel G, Levy M, Katz MN, Philbrick WM, Elinav E (2014) NLRP6 inflammasome orchestrates the colonic host-microbial interface by regulating goblet cell mucus secretion. Cell 156(5):1045–1059

Zelante T, Iannitti RG, Cunha C, De Luca A, Giovannini G, Pieraccini G, Zecchi R, D'Angelo C, Massi-Benedetti C, Fallarino F, Carvalho A (2013) Tryptophan catabolites from microbiota engage aryl hydrocarbon receptor and balance mucosal reactivity via interleukin-22. Immunity 39(2):372–385

Zhang K, Hornef MW, Dupont A (2015) The intestinal epithelium as guardian of gut barrier integrity. Cell Microbiol 17(11):1561–1569

Zheng D, Liwinski T, Elinav E (2020) Interaction between microbiota and immunity in health and disease. Cell Res 30(6):492–506

Gut–Brain Axis: Role of the Gut Microbiome on Human Health

8

Lipi Poojara, Dhaval K. Acharya, Jalpa Patel, and Rakesh M. Rawal

Abstract

The gut microbiome is considered as an organ that contributes to the regulation of host metabolism. Mammals possess an 'extended genome' of millions of microbial genomes located in the intestine: the microbiome. To date, there is rapidly booming evidence for host–microbe interaction at virtually all levels of complexity, ranging from direct cell-to-cell communication to comprehensive systemic signalling and engaging various organs and organ systems, including the central nervous system. As such, the disclosure of differential microbial composition is associated with alterations in behaviour, and cognition has consequently subsidized to establish the microbiota–gut–brain axis as an extension of the well-accepted gut–brain axis concept. Numerous exertions have been focused on demarcating a role for this axis in health and disease, ranging from stress-associated conditions such as depression, anxiety and irritable bowel syndrome (IBS) to neurodevelopmental conditions such as autism. Besides this, the gut–brain axis is also reported to influence brain disorders, e.g. Alzheimer's disease, Parkinson's disease and schizophrenia. There is bidirectional communication network that links the enteric and central nervous systems. This network is not merely anatomical, but it encompasses endocrine, humoral, metabolic and immune routes of intercommunication as well. The autonomic nervous system, hypothalamic–pituitary–adrenal (HPA) axis and nerves within the gastrointestinal tract all link the gut and the brain, allowing the brain to influence intestinal

L. Poojara · J. Patel · R. M. Rawal (✉)
Department of Biochemistry & Forensic Science, University School of Sciences, Gujarat University, Ahmedabad, Gujarat, India

D. K. Acharya
Department of Microbiology, B N Patel Institute of Paramedical and Sciences, Anand, Gujarat, India

R. Z. Sayyed, M. Khan (eds.), *Microbiome-Gut-Brain Axis*,
https://doi.org/10.1007/978-981-16-1626-6_8

activities, including activity of functional immune effector cells, and the gut to influence mood and behaviour, cognition and mental and reproductive health. In this chapter, we have focused on how gut microbiomes influence physical and mental health.

Keywords

Autonomic nervous system · Depression · Gut microbiome · Gut–brain axis · Health

8.1 Introduction: Gut Microbiome and Brain Broadcast

8.1.1 Composition and Dynamics of Healthy Adult Microbiota

Earlier it was depicted that the gut microbiota is comprised of 500–1000 species of microbes (Ramakrishna and Krishnan 2007), but a large-scale study in 2007 has estimated that the collective human gut microflora is made up of more than 35,000 bacterial species (Frank et al. 2007). Additionally, if well specified from a standpoint of total bacterial genes, Human Microbiome Project and Metagenomics of the Human Intestinal Tract studies reveal that there is a presence of more than 10 million non-redundant genes in the human microbiome.

Considering the human body as an environment, human microbiota is the entire assemblage of microorganisms living at the surface and inside of our body (Dewhirst et al. 2010; Grice et al. 2006; González et al. 2014; Arumugam et al. 2011). These communities of microorganisms are vital for many more important aspects of human physiology, digestion, detoxification and immune system development. Some of the microbes that live in the gut encode proteins that are essential for the host's health, such as enzymes that are required for the breakdown of indigestible food components and vitamin production (Flint et al. 2012; Qin et al. 2010). So we humans are having two genomes, one inherited from our parents and the other one is acquired, i.e. 'the microbiome'. This concept is the foundation for the characterization of humans as 'superorganisms' (Walsh et al. 2014). The most significant difference between these two genomes is that the inherited genome remains nearly stable during our entire lifetime, but the genome acquired from microbiome is extremely dynamic and can be affected by numerous factors like age (Gajer et al. 2012), diet (David et al. 2014; Wu et al. 2011), hormonal cycles (Koren et al. 2012), travel (Yatsunenko et al. 2012), therapies, treatments (Perez-Cobas et al. 2013) and illness (Perez-Cobas et al. 2013).

8.1.2 Formation of Gut Microbiota During the Early Stages of Life

Infants who are fully term, vaginally delivered, breastfed, and not antibiotic-treated have the best chance of developing a healthy gut flora (Alex et al. 2013). In these newborns, facultative anaerobes like *enterobacteria*, *staphylococci* and *streptococci* are the most primitive microbes starting to colonize and further taking advantage of the redox potential and available oxygen in the newborn gut. These initial colonizers consume available oxygen in the gut; by this way, it creates an anaerobic ecosystem and permits the proliferation of the strict anaerobes, *Clostridium*, *Bacteroides* and *bifidobacteria*; after that, *bifidobacteria* become dominant and more numerous than all other bacterial groups and species within the first few weeks of human life. The newborn microbiota is extremely dynamic, and it is exemplified by low stability and low variety. By the end of first year of life, newborns develop a microbial profile different for each infant and attains the characteristic microbiota of an adult gut microbiome, and by age of 2.5 years, the microbiota completely resembles that of an adult in terms of composition (Lobo et al. 2014).

8.1.3 What Consists Gut Microbiota?

The adult microbiota has been reported to be relatively stable over time in addition to being more complex than that of the neonate (Hamady and Knight 2009). Healthy gut microbiota is mainly composed of phyla Firmicutes and Bacteroidetes. followed by phyla Actinobacteria and Verrucomicrobia. Yet this general profile remains persistent; gut microbiota displays both temporal and spatial differences in distribution at the genus level and beyond. There is a notable variation in the range and quantity of bacteria from the oesophagus distally to the rectum, ranging from 10^1 per gram of contents in the oesophagus and stomach to about 10^{12} per gram of insides in the colon and distal gut (O'Hara and Shanahan 2006). Figure 8.1 shows the time-based diversity of the gut microbiota from oesophagus distally to the colon. *Streptococcus* seems to be the leading genus in the distal oesophagus, duodenum and also jejunum (Pei et al. 2004; Justesen et al. 1984).

Helicobacter is the regulatory genus present in the stomach and regulates the entire microbial population of the gastric flora; that is, when *Helicobacter pylori* (*H. pylori*) populates in the stomach as a commensal, at that time, the gut attains a rich diversity with another dominant genus like *Streptococcus* (most dominant), *Prevotella*, *Veillonella* and *Rothia* (Blaser 1999; Andersson et al. 2008). This range of microbes gets disturbed when *H. pylori* acquires a pathogenic phenotype. The large intestine comprises more than 70% of all microbes that reside in our body. The main phyla that inhabit in the large intestine are Firmicutes and Bacteroidetes. Eventually, Firmicutes/Bacteroidetes ratio has been obtained in predisposition to disease states (Ley et al. 2006).

The remarkable variability even in healthy persons that has been noticed in the current studies makes the implication of this ratio controversial. Additionally, from Firmicutes and Bacteroidetes, the human colon is similarly having primary pathogens like *Campylobacter jejuni*, *Salmonella enterica*, *Vibrio cholera*, *Escherichia coli* (*E. coli*) and *Bacteroides fragilis*, but with very less abundance (0.1% or less of the entire gut microbiome) (Human Microbiome Project Consortium 2012; Gillespie et al. 2011). The phylum Proteobacteria is markedly low, and its

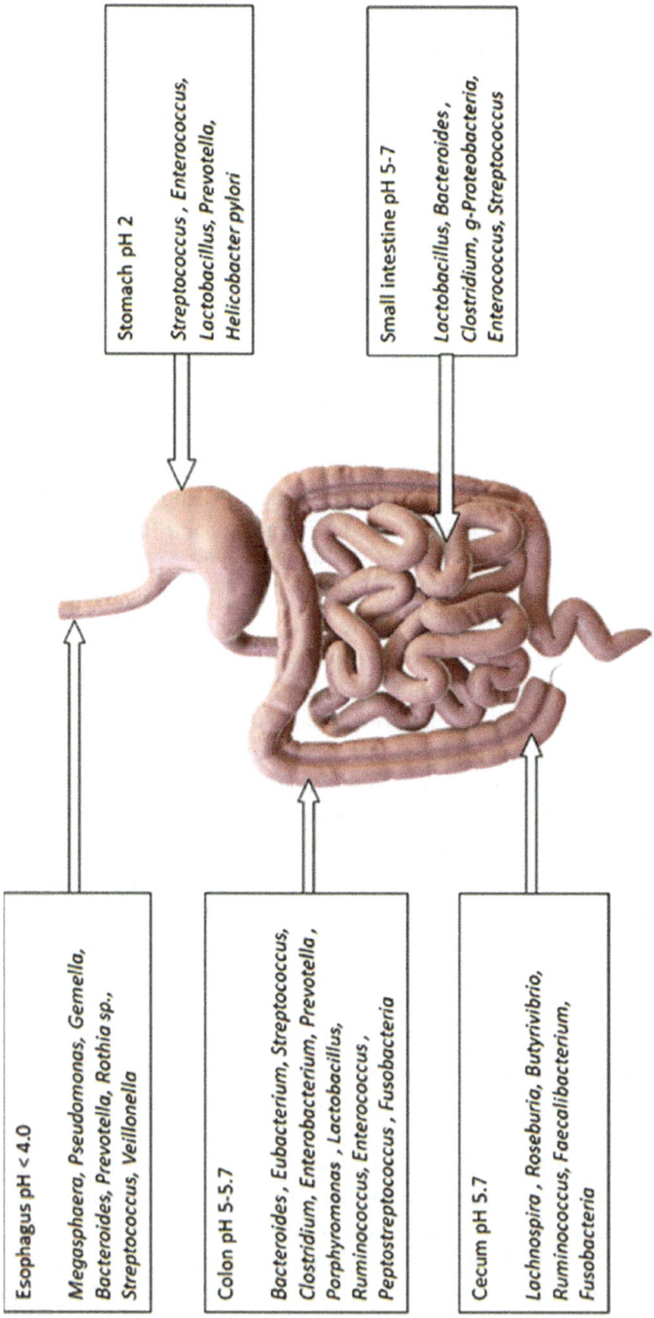

Fig. 8.1 Distribution of the normal human gut flora

deficiency along with high abundance of genera *Bacteroides*, *Prevotella* and *Ruminococcus* suggests a healthy gut microbiota (Hollister et al. 2014). Moreover, this longitudinal divergence, we do have axial discrepancy from the lumen to the mucosal surface of the intestine. Although Enterobacteriaceae, *Enterococcus*, *Clostridium*, *Lactobacillus*, *Bacteroides*, *Bifidobacterium*, *Streptococcus* and *Ruminococcus* are the predominant luminal microbial genera (can be recognized from stool analysis), solitary *Clostridium*, *Lactobacillus*, *Enterococcus* and *Akkermansia* are the principal mucosa and mucus-linked genera (which can be detected in the mucus layer and epithelial crypts of the small intestine) (Swidsinski et al. 2005). These intestinal microbiotas are known to play a key role in several metabolic, nutritional, physiological and immunological processes (O'Hara and Shanahan 2006).

Throughout human life, the healthy gut microbiota composition increases in both variety and richness (Scholtens et al. 2012) and gets maximum complexity in the human adult, with several hundred species-level phylotypes dominated by the phyla Bacteroidetes and Firmicutes (Rajilic-Stojanovic et al. 2012). Each human individual reaches a homeostatic climax composition, which likely remains relatively stable during most of a healthy adult's life. Although the individual microbial composition has an 'individual core' that varies at the bacterial phylotype level and depends on the lifestyle of that individual (Zoetendal et al. 2008; Jalanka-Tuovinen et al. 2011), at the late stages of life, the microbiota composition becomes again less diverse and more dynamic, characterized by a higher *Bacteroides* to Firmicutes ratio, increase in Proteobacteria and decrease in *Bifidobacterium* (Biagi et al. 2010).

Establishment of the gut microbiota population in early life plays a key role in the microbial makeup and disease predisposition throughout the entire life span (Scholtens et al. 2012). Sometimes, a dissimilar microbiota composition is linked with chronic intestinal disorders and the severity of distress during disease and subsequent use of antibiotic (Sekirov et al. 2010). An additional important factor in microbiota composition improvement is diet. In early life, diet already has an effect on the gut microbiome. Breastfed babies has a microbiota that is more heterogeneous than that of formula-fed babies and has a better taxonomic variety (Schwiertz et al. 2010). In addition, food habits till the age of 3 years can also impact gut microbiota composition; in a malnourished child, there is lower abundance of Bacteroidetes; those are proven to be specific in breaking down of carbohydrates from energy-rich western diet foods. Briefly, human gut microbiota is having a symbiotic relationship with the gut mucosa and reveals significant nutrient metabolic, xenobiotic and drug metabolism, antimicrobial protection and immunomodulation and gut protecting jobs in the healthy person. And it obtains its beneficial nutrients from host dietary components and shed epithelial cells. As a result, it is an organ with wide metabolic competence and significant signals from the brain that can affect the motor, sensory, and secretory systems of the gut and functional smoothness.

8.1.4 Gut–Brain Axis

An estimated 90 percent of cells found in the human body are not belongs to the human after all but of mostly prokaryotic origin, derived from at least 40,000 bacterial strains in 1800 genera (Forsythe and Kunze 2013; Frank and Pace 2008; Luckey 1972). Though considerably smaller in size, these approximately 100 trillion cells add up to a mass of almost 1–2 kg in an adult individual (Forsythe and Kunze 2013)—approximately the weight of a full-grown human brain (1.5 kg).

The discovery that differential microbial composition is associated with alterations in behaviour and cognition has significantly contributed to establish the 'microbiota–gut–brain axis' as an extension of the well-accepted 'gut–brain axis' concept. This concept is used to describe the bidirectional communication between the central nervous system (CNS) and intestinal organs and was first introduced in terms of 'peripheral regulation of emotions' by William James and Carl Lange in the 1880s and further challenged and refined by Walter Cannon in the 1920s as 'primacy of the brain in regulating gastrointestinal function'. So gut–brain axis is a bidirectional interaction network that links both enteric and central nervous systems. This correlation is not only anatomical but also extends to incorporate endocrine, humoral, metabolic and immune paths of transmission as well. Furthermore, autonomic nervous system, hypothalamic–pituitary–adrenal (HPA) axis and nerves within the gastrointestinal (GI) tract all link the gut and the brain together, allowing the brain to influence intestinal activities and activities of functional immune effector cells; moreover, they influence mood, cognition and mental health. This host–microbe interaction is present at all levels of complexity, ranging from direct cell-to-cell communication to extensive systemic signalling and involving various organs and organ systems.

Signals coming from the brain can affect motor, sensory and secretory sensory systems of the gut, and on the other hand, visceral messages from the gut can impact brain functioning with the help of this gut–brain bidirectional transmission network (Grenham et al. 2011; Montiel-Castro et al. 2013). This correlation of brain functioning with enteric gut microbiota is less extensively studied but increasingly accepted and appreciated (Khanna and Tosh 2014). Gut microbiota predominantly consists of bacteria but also contains archaea, protozoa, fungi and viruses, all of which have co-evolved with the human host. Our colon harbours the largest numbers of microorganisms in the gut; most of these native microbes are strict anaerobes in nature (Eckburg et al. 2005). Synthesis and role of these intestinal microbiota have constantly been the subject of intense study; primarily it was analysed using culture-based microbiological methods (Grenham et al. 2011), and right now, culture-independent 16S rRNA gene sequence-based techniques are in use, and these techniques allow better understanding of microbial structure and assortment of this complex study (Arboleya et al. 2008; Qin et al. 2010). With evolving improvements in metagenomic technologies, we are able to disclose the composition of the human gut microbiota from early childhood (Palmer et al. 2007) to elderly (Claesson et al. 2012). Although lesser is known regarding the physiological impact of these microbiota on host health, comprising that of the brain, understanding the stimulus of gut microbiota on the host well-being has been portrayed as one of the most exciting areas in entire medicine (Shanahan 2012).

At the time of birth, our brain is extremely under-developed, and gut is generally interpreted as completely sterile. As described in Sect. 8.1.2, preliminary colonization is influenced by mother's microbe environment and the environment of the hospital. This colonization plays an important role in brain development in the early post-natal period. The subsequent microbial arrangement of the newborn gut is affected by several factors including diet, use of antibiotics, mode of delivery, surrounding environment and the main maternal microbiota (Koenig et al. 2011; Marques et al. 2010; Dominguez-Bello et al. 2010).

These properties of intestinal microbiota identified in healthy full-term infants are distressed in preterm infants (Dennison 1976) that are commonly delivered via caesarean section, take antibiotics and are sometimes not fed properly (Hoy et al. 2000). Moreover, preterm infants are having functionally immature or not properly developed gut which has low levels of acidity in the stomach, because they are lacking in gastric acid secretion and they need to be fed more frequently (Hoy et al. 2000; Sondheimer and Clark 1985; Sondheimer et al. 1985), and it leads to an increase in the incidence of potentially pathogenic bacteria in the gastrointestinal (GI) tract, and preterm infants have a smaller amount of microbial variety than full-term infants (Arboleya et al. 2008; Chang et al. 2011; Jacquot et al. 2011). And these characteristics, which have been linked to the development of cerebral palsy and autism, have been the focus of research and ongoing controversy (Mangiola et al. 2016).

In the case of the elderly, when these microbiota compositions of elderly people in nursing homes are compared with those living in the community with their families, large-scale alterations were noticed. Those admitted in nursing homes have a far less varied microbiota, and this can be a result of less diverse diet (Claesson et al. 2012). It is also thinkable, sometimes, that pathological factors lead to admission into nursing homes, likewise worsening cognitive functionality and declining physical activity, might be having an important role in the reduced microbial richness and not a less diverse diet. Current studies should explain this issue, and this can be a challenge for the food industry to discover diets for the elderly to help them sustain their microbial variety. What we can justify here is that a dysregulated gut microbiota either in early childhood or in an elderly population meaningfully increases the possibility of brain dysfunction.

8.1.5 How Gut Microbiota Communicates with the Brain?

There are various possible direct and indirect communication routes through which the gut microbiota can communicate with the brain including neuroendocrine, neuroanatomical immune and through neurotransmitters.

8.1.5.1 Neuroanatomical Pathway

Human gut can interrelate with the brain with the help of two neuroanatomical pathways. One is mutual information interchange straight between the gut and the brain by autonomic nervous system (ANS) and vagus nerve (VN) in the spinal cord, and the other one is a bidirectional signalling between the gut and the brain through communication between enteric nervous system (ENS) within the gut and ANS and VN; inside the spinal cord, information from the heart, lungs, liver, pancreas, stomach and intestines is conveyed to the brain via sensory fibres in the vagus nerve (Travagli et al. 2003). Sensory vagal inputs reach the nucleus of the solitary tract (NTS) and are thence conveyed to extensive zones of the CNS and also the cerebral cortex and medulla oblongata. Preclinical studies have implicated the vagus nerve as a key route of neural communication between microbes of the gut and centrally mediated behavioural effects, as confirmed with the elimination of central *Lactobacillus rhamnosus* after vagotomy (Bravo et al. 2011), and those who underwent vagotomy at an early age have a reduced risk of certain neurologic disorders (Svensson et al. 2015)

8.1.5.2 Neuroendocrine-HPA Axis

Neuroendocrine-HPA axis provides the principal control of the stress reaction and can have a considerable impact on the brain–gut–microbiota axis (Wang and Kasper 2014; Tillisch 2014; Scott et al. 2013; Moloney et al. 2014; O'Mahony et al. 2009, 2011, 2017). It is fair enough and maybe of significance in several pathologic conditions psychological or physical stress can considerably dysregulate the HPA axis and in result the brain–gut microbiota axis, e.g. in IBS (Dinan et al. 2006). Human brain recruits these same methods to control the composition of the gut microbiota, for example, in conditions of stress. The hypothalamic–pituitary–adrenal (HPA) axis controls cortisol secretion, and cortisol can in turn impact immune cells (including cytokine secretion) locally within gut as well as systemically in body. This cortisol level can also alter gut permeability and barrier function and can in turn alter gut microbiota composition. Additionally, the gut microbiota and probiotic agents can modify the levels of circulating cytokines, and this can be effective on brain functioning.

Stress and HPA axis can also affect the formation of the gut microbiome. Initial stress and separation of the mother may possibly lead to a long-term change of HPA and had an extended effect on the microbial population (Desbonnet et al. 2008; Barouei et al. 2012). When it is evaluated with rats not separated from the mother, an assortment of 16S ribosomal RNA in adult rats, who have been through mother separation for around 3 h/day starting from day 2 to day 12 after birth, unveiled that stress extremely altered microbiome detected from faeces (O'Mahony et al. 2009). A mouse that is exposed to a long-term stress microbiome configuration was comparably different from a non-stressed mouse (Bendtsen et al. 2012). Recently, with the use of the above theories, it can be concluded that repeated social interaction and stress can diminish the number of *Bacteroides* in the caecum and augment the number of *Clostridium*. Stress can also upsurge interleukin-6 (IL 6) and monocyte chemoattractant protein 1 (MCP-1) levels in blood. MCP-1 was significantly related to the variations of three different kinds of stress-inducing bacterial strains, namely, *Enterococcus faecalis*, *Pseudobutyrivibrio* and aerogenic bacteria *Dorea*.

8.1.5.3 Immunological Pathway

The development of gut immune system is dependent on the gut microbiota (Furusawa et al. 2013; Mayer et al. 2014). Germ-free mice nearly had no immune activity, but they were able to generate immunity when fed with certain microbiota. For instance, the segmented filamentous bacterium in the gut can re-establish its full functions of gut B and T lymphocytes (Umesaki et al. 1995, 1999; Talham et al. 1999). These bacteria can communicate with the host through a variety of routes, and Toll-like receptors (TLRs) of a host cell play an important role in the broadcast between bacteria and host. Currently, ten different types of TLRs are in the human innate immune system; all of these have been identified as pattern recognition receptors (Takeuchi and Akira 2010). And they are part of the innate immune system, performs the initial step in the production of cytokine response, also widely distributed on neurons (McKernan et al. 2011). Thus, neurons likewise respond to bacterial and viral components. Thus, neurons likewise respond to bacterial and viral components. Intestinal epithelial cells are able to transfer microbial composition or metabolites in the internal environment and also with the nervous system (O'Brien et al. 2004). The equilibrium of gut microbiota may alter the regulation of inflammatory response, and this method may also engage in the control of emotion and behaviour.

Immune signalling from the gut to the brain facilitated by cytokine molecules is an additional documented route of communication (El Aidy et al. 2014). Cytokines produced at the level of the gut can penetrate bloodstream to the brain. Under normal physiologic conditions, it is unlikely that they cross the blood–brain barrier (BBB), but growing evidence implies a capacity to signal across the BBB and to affect brain areas like hypothalamus, where the BBB is lacking. It is through the latter mechanism the cytokines interleukin (IL)-1 and IL-6 activate the hypothalamic–pituitary–adrenal (HPA) axis, bringing about the release of cortisol. This is the most potent activator of the stress system.

8.1.5.4 Neurotransmitters Regulating Gut–Brain Axis

Gut microbiota likewise regulates important central neurotransmitters, such as serotonin, with varying levels of precursors; for example, *Bifidobacterium infantis* has shown to raise plasma tryptophan levels, and so it influences central serotonin (5HT) transmission (O'Brien et al. 2004). Interestingly, some bacteria associated in the synthesis and release of neurotransmitters have been already reported. *Lactobacillus* and *Bifidobacterium* spp. can synthesize g-aminobutyric acid (GABA); *Escherichia*, *Bacillus* and *Saccharomyces* spp. are able to produce noradrenaline; *Candida*, *Streptococcus*, *Escherichia* and *Enterococcus* spp. have been synthesizing serotonin; *Bacillus* can produce dopamine; likewise *Lactobacillus* can generate acetylcholine (Lyte 2013, 2014). These neurotransmitters of microbial origin are able to penetrate into the mucosal layer of the intestine, even though it is extremely improbable that these bacterial species can directly affect brain function. Even if they enter into the bloodstream, which is by no means sure, they will be capable of crossing the blood–brain barrier (BBB). That is why their effect on brain function is almost indirect, by acting on the enteric nervous system (ENS). SCFAs (short-chain fatty acids), which include butyrate, propionate and acetate, are indispensable metabolic end products of gut microbial activity and may apply central effects through G-protein–coupled receptors, even though such receptors are sparsely concentrated in the brain. It is more obvious that they act as epigenetic modulators through histone deacetylases (Stilling et al. 2014) SCFAs are also engaged in energy balance and metabolism and able to regulate adipose tissue, liver tissue and skeletal muscle and function (Canfora et al. 2015). Therefore, a lot of essential neurotransmitters in the body are formed by the gut microbiota, employing impact on the human body including the brain. Therefore, a lot of essential neurotransmitters in the body are formed by the gut microbiota, employing impact on the human body including the brain from which several neurotransmitters produced by gut microbiota are defined as critical molecules.

8.2 Gut–Microbiome–Brain Implications on Physical Health

Generally, the intestinal microbiota composition of healthy individuals is comparatively stable; however, alterations in the microbiota community may lead to a permanent imbalance known as dysbiosis (Lynch and Pedersen 2016). Numerous factors such as antibiotics, diet (comprising specific probiotic and prebiotic intake), the host immune system and acidic environment have been seen to influence the microbiota composition of the gut. Perturbation to the gut microbiota ecosystem resulting in dysbiosis can lead to gastrointestinal diseases. With current research advising dysbiosis of gut microbiota is having potential implication not only in IBS, but also in other disorders such as obesity (Turnbaugh and Gordon 2009), diabetes (Qin et al. 2012), metabolic syndrome (D'Aversa et al. 2013), cardiovascular disease and IBD as well as on reproductive health.

8.2.1 Irritable Bowel Syndrome (IBS)

Irritable bowel syndrome (IBS) is a common gastrointestinal (GI) disorder categorized by persistent abdominal pain allied with alterations in bowel habits. Aspects associated to IBS symptom development comprise history of enteric infection, deviations in the gut microbiota, immunomodulation, alterations in brain–gut processing and vagaries in visceral sensation and motility (Ford et al. 2017). IBS can be clinically subtyped into IBS with constipation (IBS-C), IBS with diarrhoea (IBS-D) and mixed IBS (IBS-M). In addition, IBS patients seemed to have a higher degree of psychosocial stress, a poorer quality of life and inferior levels of work productivity. Alterations in the normal gut microbiota have been proposed as etiologic factors in the development of functional gastrointestinal disorders such as IBS and functional dyspepsia and shared GI disorders of unknown aetiology

(Upadhyay et al. 2018). The pathogenesis and pathophysiology of IBS are incompletely unstated, but abnormal GI motility, visceral hypersensitivity, altered brain–gut function, low-grade inflammation and psychosocial factors are considered to subsidize. IBS has been significantly associated with small intestinal bacterial overgrowth (SIBO) (4–78%) (Ghoshal and Ghoshal 2017) and prior GI infection (5–32%), suggesting that enteric dysbiosis (i.e. disrupted microbial homeostasis) is a potential pathogenic mechanism of IBS. In recent years, many research groups have engrossed on recognizing the gut microbiota composition of the large intestine of IBS patients, using modern culture-independent techniques. Next-generation sequencing has revealed that IBS patients, compared with healthy controls, show significantly lower abundance in enteric *Lactobacillus*, *Bifidobacterium* and *Faecalibacterium prausnitzii* (O'Mahony et al. 2005). During periods of dysbiosis, the gut microbiome influences inflammation metabolism inside the GI tract, primarily through the production of cytokines (such as interleukin [IL]-10 and IL-4) and other cellular communication mediators, such as interferon-gamma. In irritable bowel syndrome (IBS), atypical microbiota populations stimulate mucosal innate immune responses, which increases gut epithelial permeability, triggers gut pain sensory pathways and dysregulates the enteric nervous system (Mayer et al. 2014); both brain–gut and gut–brain dysfunctions arise, the prior being dominant. Obstruction in the gut–brain axis affects intestinal motility and secretion, confers to visceral hypersensitivity and leads to cellular alterations of the entero-endocrine and immune systems (Kennedy et al. 2014).

8.2.2 Metabolic Diseases

The human gut microbiota has been studied for more than a century. Examination that the gut microbiota, as an environmental factor, donates to adiposity and has further increased curiosity in the field. The human microbiota can be altered by diet, and macronutrients work as substrates for various microbial metabolites, such as short-chain fatty acids (SCFA) and bile acids, and are able to modulate host metabolism. Obesity predisposes towards type 2 diabetes and cardiovascular disease. The gut microbiota shows a significant role in the regulation of the host's metabolism and the extraction of energy from ingested food. Gut microbiotas have not only beneficial roles for the host but also have pathophysiological relations with the host, particularly in the case of obesity and related metabolic disorders. Recent studies have revealed that changes in the gut microbiota may be associated in the pathogenesis of obesity and diabetes. Obesity is the outcome of a long-term positive imbalance between energy intake and expenditure, which is controlled by multiple pathways comprising metabolites, hormones and neuropeptides (Upadhyay et al. 2018) Gut hormones seem to interconnect information from the gastrointestinal tract to the regulatory appetite centres within the central nervous system (CNS) via the so-called gut–brain axis. Such messages may be transferred to the CNS either via vagal or non-vagal afferent nerve signalling or directly via blood circulation (Bueter et al. 2009). Complex neural networks, distributed throughout the forebrain and brainstem, are in control of feeding and energy homeostasis (Schwartz et al. 2000). Novel research shows that the gut microbiota is involved in obesity and metabolic disorders, revealing that obese animal and human subjects have alterations in the composition of the gut microbiota compared to their lean counterparts. Moreover, transplantation of the microbiota of either obese or lean mice influences body weight in the germ-free recipient mice, suggesting that the gut ecosystem is a significant target for weight management (Harakeh et al. 2016). Native gut microbes may regulate body weight by inducing the host's metabolic, neuroendocrine and immune functions. The intestinal microbiota, as a whole, offers supplementary metabolic functions and regulates the host's gene expression, improving the ability to extract and store energy from the diet and contributing to body-weight gain (Ley et al. 2005). Inequalities in the gut microbiota and increasing plasma lipopolysaccharide can also act as inflammatory factors linked to the growth of atherosclerosis, insulin resistance and weight gain.

Onset of diabetes has increased rapidly and became a major public health concern worldwide. Type 1 diabetes (T1D) is an autoimmune disease characterized by insufficient insulin production because of T-cell-mediated destruction of insulin-secreting pancreatic beta cells, while type 2 diabetes (T2D) is a condition in which the body does not produce or use insulin well. Various factors are associated with the development of diabetes, such as diet, genome and intestinal microbiota. Changes in the gut microbiota can influence the levels of gut hormones involved in the regulation of satiety and glycaemic control, such as glucagon-like peptide-1 (GLP-1), which stimulates insulin secretion from the pancreas (Baggio and Drucker 2007; Tolhurst et al. 2012)

In obese individuals and patients with metabolic syndrome, an increase in insulin sensitivity is noted after 6 weeks of allogeneic or autologous faecal microbiota transplantation from normal individuals (Vrieze et al. 2012). Same results were observed earlier in mice as well (Bäckhed et al. 2004), which has become the trigger point for the researchers to study gut microbiota in diabetes (Gravitz 2012).

Carbohydrates are an essential nutritional factor for all mammals and their gut microbiota. These bacteria greatly influence glycaemic control. Undigested polysaccharides and partially digested carbohydrates reach the gut microbiota in the distal gut, where they are metabolized by bacterial enzymes (Musso et al. 2011). It has been investigated that the genera *Ruminococcus*, *Fusobacterium* and *Blautia* are positively associated with T2D, whereas the genera *Bifidobacterium*, *Bacteroides*, *Faecalibacterium*, *Akkermansia* and *Roseburia* are negatively associated with T2D (Mangiola et al. 2016).

The disrupted GDM (gestational diabetes gut microbiota) is very similar to gut microbiota in individual patients with type 2 diabetes and associated intermediary metabolic dysfunctions. Eight months postpartum, previous GDM women have different gut microbiota than the woman with normal pregnancy. This microbial dysbiosis may increase the risk of T2D, which needs to be investigated. Feeding of probiotic dahi containing *Lactobacillus acidophilus* NCDC14 and *L. casei* NCDC19 has been tested to substantially reduce STZ-induced oxidative damage in pancreatic tissues. Thus, the modulation of the intestinal microbiota by probiotics may be effective towards prevention and management of T1D and T2D. Supplements of prebiotics improve *Bifidobacterium* abundance, which alters microbial dysbiosis and improves glucose tolerance in mice (Cani et al. 2007).

There is much reasonable curiosity in the interplay of drugs and intestinal microbiota. It is well known that anti-diabetic drugs can modulate microbiota and improve diabetes. Improvements in fasting blood glucose, glucose tolerance and insulin resistance were observed with the combined therapy of a prebiotic and metformin in diabetic mice (Zheng et al. 2018). Multivariate research found that there are significant discrepancies in gut composition between T2DM (Type 2 Diabetes Mellitus) and in metformin-untreated participants significant increases was observed in *Escherichia* species and decreases in *Intestinibacter* following metformin therapy (Harsch and Konturek 2018). But there is still some uncertainty in this emerging field. Whether microbiota causes diabetes or diabetes affects intestinal microbiota is not yet quite simple. Investigation of altered gut microbiota can help in the early detection of diabetes even before serological tests (Nair et al. 2018).

8.2.3 Reproductive Health

As of today, researchers understand that residents in human gut form a symbiotic relationship with the host and offer several benefits to the host. For example, commensal microbes consistently provide a set of services to the host such as modulation of the immune system, inhibition of pathogen colonization and releasing nutrients from food (Kim et al. 2020). Reportedly, dysbiosis of gut microbiota has been implicated in many disease states, including diabetes, obesity and cardiovascular disease (Razavi et al. 2019). Recently, a novel theory of 'microgenderome' associated to the potential bidirectional interaction roles between the sex hormones and gut microbiota has emerged (Aguilera et al. 2020). It has been reported that the composition of commensal microbes of male and female animals deviated at the time of puberty, which imbedded that sex hormone levels put forth particular influences on the composition of the microbiota. Abstraction of gut microbiota increased the testosterone concentration in female mice but decreased the concentration in male mice. Thus, the commensal gut microbiota also had effects on the production of male sex hormone (Yuan et al. 2020).

Despite the advances in assisted reproductive technology (ART) in women as well as in men, approximately 8–12% of the global population willing to conceive is unable to do so. Available evidence advises that vaginal and uterine microbiota have a close relationship with female infertility (Moreno et al. 2016). In fact, microbiota analysis using the 16S rRNA amplicon sequencing of cervical swabs revealed significant differences regarding the relative read count of the genus *Gardnerella* between females diagnosed with infectious infertility and fertile controls (Benner et al. 2018). Several mechanisms have been proposed to suggest that dysbiosis of gut microbiota can be involved in the development of polycystic ovary syndrome (PCOS). However, the data obtained from cross-sectional studies are insufficient to reveal the causality of the relationship (Zhao et al. 2020; Yurtdaş and Akdevelioğlu 2020).

8.3 Gut–Microbiome–Brain Implications on Mental Health

The recently emerged concept of the bidirectional communication of the gut–microbiota–brain emphasizes the relevance to study associations between neurodegenerative diseases and the gut microbiota. There exists growing evidence that gut microbiota may affect the central nervous system through communication via the vagus nerve, signalling mediators of the immune system, enteric hormones and gut microbiota-derived products (Sherwin et al. 2016). Gut bacteria produce neuroactive compounds and can modulate neuronal function, plasticity and behaviour. Furthermore, intestinal microorganisms impact the host's metabolism and immune status which in turn affect neuronal pathways in the enteric and central nervous systems. Communication pathways between gut microbiota and the central nervous system could include autonomic, neuroendocrine, enteric and immune systems, with pathology resulting in disruption to neurotransmitter balance, increases in chronic inflammation or exacerbated hypothalamic–pituitary–adrenal axis activity.

8.3.1 Stress/Depression

Depression is a major form of mood disorder characterized by depressed mood and/or recurrent thoughts of death and/or loss of interest or pleasure in life activities present over a period of at least 2 weeks. It results from neuro-psychiatric disturbance, immunological deregulation, genetic factors and environmental influences; nevertheless, a correlation with gut microbiota is emerging (Mangiola et al. 2016). Growing evidence links gut microbiome to the development and maturation of the central nervous system, which are regulated by microbiota potentially through stress response, neurotransmitter, neuroimmune, and endocrine pathways. The dysfunction of such microbiota–gut–brain axis is implicated in neuropsychiatric disorders, depression and other stress-related conditions (Kuo and Chung 2019). Bipolar disorder and major depression are associated with substantial disability, morbidity and reduced life expectancy. People with mood maladies have shown higher ratios of unhealthy lifestyle choices, including poor diet quality and suboptimal nutrition (Balanza-Martinez et al. 2020). Coello et al. (2019) found that gut microbiota community association differed between patients with newly diagnosed bipolar disorder and healthy individuals. Having a newly diagnosed bipolar illness was related with the prevalence of *Flavonifractor*, even after controlling for age, gender, physical activity, and waist size, and was mitigated by smoking status. The presence of *Flavonifractor* may possibly influence oxidative stress and inflammation in its host and could possibly link gut microbiota with illness pathology of bipolar disorder (Coello et al. 2019).

Sudo et al. (2004) demonstrated that the presence of gut microbiota modulated the long-range hypothalamic–pituitary–adrenal reaction to stress. These experiments showed that germ-free mice (mice raised in a sterile condition and lacking gut bacteria) exhibited a higher stress response as measured by an increased adrenocorticotropic hormone and corticosterone release compared to control mice with gut microbiota. This exaggerated hypothalamic–pituitary–adrenal response was reversed by the introduction of *Bifidobacterium infantis* and was somewhat reversed with stool from orthodoxly raised mice. Germ-free mice also exhibit reduced anxiety-like behaviour in addition to altered levels of brain-derived neurotrophic factors and other neurotransmitters. In 2017, Meson et al. investigated that certain gut bacteria were connected to mood symptoms in a clinical cohort of major depressive disorder patients. In this study, species richness, or the total number of detected gut bacteria, was predictive of insomnia and depression, while abundance of Enterobacteriaceae was predictive of anxiety. In the same investigation, *Lactobacillus* and *Enterococcus* abundance was also positively related to psychomotor agitation. In 2015, Luna and Foster suggested particular administration of *Lactobacillus* sp., *Bifidobacterium* sp., *L. helveticus*, *B. longum*, *L. rhamnosus* and *Lactobacillus farciminis* in murine sample led to an improvement of depression and anxiety symptoms.

8.3.2 Autism

Autism spectrum disorder (ASD) is a prevalent neurodevelopmental condition with no known aetiology or cure. Several possible contributing factors, both genetic and environmental, are being actively investigated. Amongst these, maternal immune dysregulation has been identified as potentially involved in promoting ASD in the offspring. An important role of gut microbiota in the maintenance of physiological state into the gastrointestinal system is supported by several studies that have shown a qualitative and quantitative alteration of the intestinal flora in a number of gastrointestinal and extra-gastrointestinal diseases. Approximately 30–50% of children and adults with autism spectrum disorders have chronic gastrointestinal symptoms, typically constipation, diarrhoea and alternating constipation and diarrhoea (Adams et al. 2019); many of them also show abnormal behavioural patterns such as aggression, anxiety and tendency to self-injure (Afroz and Alvina 2019). It has been demonstrated that a large amount of species under the genus *Clostridium* (ten times more) characterized the qualitative composition of faecal samples of autistic children. The composition of microbiota has been characterized, showing an imbalance of the phyla Bacteroidetes and Firmicutes (Mangiola et al. 2016). Some of the microbial products, e.g. various metabolites of aromatic amino acids, have the potential to be neuroactive and affect the functions of the enteric and central nervous systems.

Moreover, ASD patients have significantly higher intestinal permeability which causes leakage of lymphocytes and pro-inflammatory cytokines into the circulatory system. Those inflammatory molecules eventually reach the brain and cause immune activation there (Alexeev et al. 2018; Ashwood et al. 2011). As gut dysbiosis is responsible for the increased permeability of the intestinal epithelial cells, this evidence supports the idea that there is an important effect of gut dysbiosis on immune dysregulation and possibly on ASD (Afroz and Alvina 2019; Quigley 2016). Averina et al. (2020) using a whole metagenome sequencing approach found that significant differences with decreases in average abundance in the microbiota of ASD children were found for the genera *Barnesiella* and *Parabacteroides* and species *Alistipes putredinis*, *B. caccae*, *Bacteroides intestinihominis*, *Eubacterium rectale*, *Parabacteroides distasonis* and *Ruminococcus lactaris*. They also noted decreases in the abundance of genes linked to production of GABA, melatonin and butyric acid in the ASD metagenomes. In a recent research with a mouse model of autism, *Sutterella* correlated with a low performance in social and obsessive-compulsive disorder (marble burying) tests and TNF-α levels (Coretti et al. 2017).

8.3.3 Parkinson's Disease

Although Parkinson's disease (PD) has been the most intensively studied, the microbiome is of interest across a range of neurodegenerative disorders. PD presently is conceptualized as a protein aggregation disease in which pathology involves both the enteric and the central nervous system, possibly spreading from one to another via the vagus nerves. PD may be of particular relevance, given the high prevalence of gastrointestinal disturbances that often precede the more well-recognized motor symptoms. An overstimulation of the innate immune system due to gut dysbiosis and/or small intestinal bacterial overgrowth, together with higher intestinal barrier permeability, may provoke local and systemic inflammation as well as enteric neuroglial activation, ultimately triggering the development of alpha-synuclein pathology. The gut microbiota and its relevant metabolites interact with the host via a series of biochemical and functional inputs, thereby affecting host homeostasis and health. Indeed, a dysregulated microbiota–gut–brain axis in PD might lie at the basis of gastrointestinal dysfunctions (Caputi and Giron 2018).

Although findings have been varied, there are some clear trends evident in the microbiome composition of patients with PD. Several studies showed an increase of *Lactobacillus*, *Bifidobacterium*, *Akkermansia* and Verrucomicrobiaceae in PD, while *Faecalibacterium*, *Coprococcus*, *Blautia* and *Prevotella* appear to be under-represented (Quigley 2017; Butler et al. 2019). Similarly, Scheperjans et al. (2015) and Unger et al. (2016) found that PD patients showed a different gut microbiota than healthy controls, which was characterized by lower abundance of Prevotellaceae, Lactobacillaceae and the butyrate producer *Faecalibacterium prausnitzii*, whereas Enterobacteriaceae and *Bifidobacterium* spp. were more abundant.

Although most of the differences were associated with disease duration, lower abundance in Lachnospiraceae was the only difference between de novo PD patient and healthy control (remaining lower across almost all PD duration strata).

Decreased Lachnospiraceae and increased Lactobacillaceae and Christensenellaceae were associated with a worse clinical profile, including higher frequencies of cognitive impairment, gait disturbances and postural instability. Gut microbiota may be an environmental modulator of the pathogenesis of PD and may contribute to the interindividual variability of clinical features (Barichella et al. 2019). Trace amines and their primary receptor, trace amine-associated receptor-1 (TAAR1), are widely studied for their involvement in the pathogenesis of neuropsychiatric disorders despite being found in the gastrointestinal tract at physiological levels. A therapeutic benefit of TAAR1 compounds in clinical trials is thoughtful manipulation of the brain–gut–microbiome axis to modulate symptoms of neuropsychiatric disease (Bugda Gwilt et al. 2020).

8.3.4 Alzheimer's Disease

Alzheimer's disease (AD) is the most common form of dementia and one of the major causes of disability and dependency in older people. The diversity of the gut microbiota declines in the elderly and in patients with Alzheimer's disease (AD). Restoring the diversity with probiotic treatment alleviates the psychiatric and histopathological findings. The three different linkages between the present gut microbiome hypothesis and the other major theories for the pathogenesis of AD are as follows: bacterial metabolites and amyloids can trigger central nervous system inflammation and cerebrovascular degeneration; impaired gut microbiome flora inhibits the autophagy-mediated protein clearance process; and gut microbiomes can change the neurotransmitter levels in the brain through the vagal afferent fibres (Bostanciklioglu 2019).

Moreover, impaired memory and learning involve the dysfunction neurotransmission of glutamate, the agonist of the N-methyl-D-aspartate receptor and a major excitatory neurotransmitter in the brain. Gut microbiota including *Bacteroides vulgatus* and *Campylobacter jejuni* affect glutamate metabolism and decrease the glutamate metabolite 2-keto-glutaramic acid. Meanwhile, gut bacteria with glutamate racemase including *Corynebacterium glutamicum*, *Brevibacterium lactofermentum* and *Brevibacterium avium* can convert L-glutamate to D-glutamate. N-methyl-D-aspartate receptor (NMDAR)-enhancing agents have been found to potentially improve cognition in AD or Parkinson's disease patients. These findings suggest that D-glutamate (D-form glutamate) metabolized by the gut bacteria may influence the glutamate NMDAR and cognitive function in dementia patients (Chang et al. 2020). Through metabolic activity of non-pathological microorganisms and secretion of functional by-products that increase the permeability of the intestinal mucosa, the gut microbiota influences both the production and absorption of neurotransmitters (e.g. serotonin and GABA), increasing their bioavailability to the CNS. It has been further shown some components of the gut microbiota—predominantly bacteria—synthesize and release amyloid peptides and lipopolysaccharides, which in turn activate inflammatory signalling through the release of cytokines, with potential effects on the pathophysiological cascade of Alzheimer's disease (Vanessa et al. 2018). Depleting intestinal microbiota in AD animal models reduces amyloid-beta (Abeta) plaque deposition. Age-related changes in the microbiota contribute to immunologic and physiologic decline. Translationally relevant dietary manipulations may be an effective approach to slow microbiota changes during aging.

8.4 Conclusion

Due to the rapid pace of microbial science discovery, many additional functions of the microbiome are likely to be discovered. Researchers are increasingly aware that the gut and the brain communicate and are looking to leverage actions of healthy gut microbiota to treat psychological conditions. Diversity in the gut microbiota is vital not only for gut health but also for normal physiologic functioning in other organs, especially the brain. Sometimes, an altered gut microbiota in the form of dysbiosis at the extremes of life, both in the neonate and in the elderly, can have a profound impact on brain functioning. The brain is reliant on gut microbes for essential metabolic outcomes; it is not surprising that a dysbiosis can have serious negative consequences for brain function both from neurologic and mental health perspectives. However, the microbiome is a complex and dynamic ecosystem, and understanding its role in host illness and its potential for the treatment of neurological disorder will ultimately require more study.

References

Adams JB, Borody TJ, Dae-Wook K, Alexander K, Rosa K, Michael JS (2019) Microbiota transplant therapy and autism: lessons for the clinic. Expert Rev Gastroenterol Hepatol 13 (11):1033–1037

Afroz KF, Alvina K (2019) Maternal elevated salt consumption and the development of autism spectrum disorder in the offspring. J Neuroinflammation 16(1):265

Aguilera M, Gálvez-Ontiveros Y, Rivas A (2020) Endobolome, a new concept for determining the influence of microbiota disrupting chemicals (MDC) in relation to specific endocrine pathogenesis. Front Microbiol 11:578007

Alex S, Lange K, Amolo T, Grinstead JS, Haakonsson AK, Szalowska E, Koppen A, Mudde K, Haenen D, Lahham S, Roelofsen H, Houtman R, Burg B, Mandrup BA, Kalkhoven E, Müller M, Hooiveld G, Kersten S (2013) Short chain fatty acids stimulate angiopoietin-like 4 synthesis in human colon adenocarcinoma cells by activating peroxisome proliferator-activated receptor gamma. Mol Cell Biol 33:1303–1316

Alexeev EE, Lanis JM, Kao DJ, Campbell EL, Kelly CJ, Battista KD, Gerich ME, Jenkins BR, Walk ST, Kominsky DJ, Colgan SP (2018) Microbiota-derived indole metabolites promote human and murine intestinal homeostasis through regulation of interleukin-10 receptor. Am J Pathol 188(5):1183–1194

Andersson AF, Lindberg M, Jakobsson H, Bäckhed F, Nyrén P (2008) Comparative analysis of human gut microbiota by barcoded pyrosequencing. PLoS One 3:e2836

Arboleya S, Binetti A, Salazar N, Fernández N, Solís G, Hernandez-Barranco A, Armougom F, Raoult D (2008) Use of pyrosequencing and DNA barcodes to monitor variations in Firmicutes and Bacteroidetes communities in the gut microbiota of obese humans. BMC Genomics 9:576

Arumugam M, Raes J, Pelletier E, Arumugam M, Raes J, Pelletier E, Paslier D, Takuji Y, Daniel RM, Gabriel RF, Julien T, Thomas B, Jean MB, Marcelo B, Natalia B, Francesc C, Leyden F, Laurent G, Torben H, Masahira H, Tetsuya H, Michiel K, Ken K, Marion L, Florence L, Chaysavanh M, Nielsen H, Trine N, Nicolas P, Julie P, Junjie Q, Thomas S, Sebastian T, David T, Edgardo U, Erwin GZ, Junwang FG, Oluf P, Willem M, Joel D, Metahit C, Jean W, Ehrlich SD (2011) Enterotypes of the human gut microbiome. Nature 473:174–180

Ashwood P, Krakowiak P, Hertz-Picciotto I, Hansen R, Pessah I, Van de Water J (2011) Elevated plasma cytokines in autism spectrum disorders provide evidence of immune dysfunction and are associated with impaired behavioral outcome. Brain Behav Immun 25(1):40–45

Averina OV, Kovtun AS, Polyakova SI, Savilova AM, Rebrikov DV, Danilenko VN (2020) The bacterial neurometabolic signature of the gut microbiota of young children with autism spectrum disorders. J Med Microbiol 69(4):558–571

Bäckhed F, Ding H, Wang T, Hooper LV, Koh GY, Nagy A, Clay FS, Gordon JI (2004) The gut microbiota as an environmental factor that regulates fat storage. Proc Natl Acad Sci 101 (44):15718–15723

Baggio LL, Drucker DJ (2007) Biology of incretins: GLP-1 and GIP. Gastroenterology 132 (6):2131–2157

Balanza-Martinez V, Shansis FM, Tatay-Manteiga A, López-García P (2020) Diet and neurocognition in mood disorders - an overview of the overlooked. Curr Pharm Des 26 (20):2353–2362

Barichella M, Severgnini M, Roberto C, Erica C, Carlotta B, Serena C, Valentina F, Raffaella C, Camilla C, Samanta F, Giovanna P, Gianluca B, Luigi Z, Emanuele C, Clarissa C, Gianni P (2019) Unraveling gut microbiota in Parkinson's disease and atypical parkinsonism. Mov Disord 34(3):396–405

Barouei J, Moussavi M, Hodgson DM (2012) Effect of maternal probiotic intervention on HPA axis, immunity and gut microbiota in a rat model of irritable bowel syndrome. PLoS One 7(10): e46051

Bendtsen KMB, Krych L, Sørensen DB, Pang W, Nielsen DS, Josefsen K, Lars HH, Søren JS, Hansen AK (2012) Gut microbiota composition is correlated to grid floor induced stress and behavior in the BALB/c mouse. PLoS One 7(10):e46231

Benner M, Ferwerda G, Joosten I, Van der Molen RG (2018) How uterine microbiota might be responsible for a receptive, fertile endometrium. Hum Reprod Update 24(4):393–415

Biagi E, Nylund L, Candela M, Ostan R, Bucci L, Pini E, Nikkïla J, Monti D, Satokari R, Franceschi C, Brigidi P (2010) Through ageing, and beyond: gut microbiota and inflammatory status in seniors and centenarians. PLoS One 5(5):e10667

Blaser MJ (1999) Hypothesis: the changing relationships of Helicobacter pylori and humans: implications for health and disease. J Infect Dis 179:1523–1530

Bostanciklioglu M (2019) The role of gut microbiota in pathogenesis of Alzheimer's disease. J Appl Microbiol 127(4):954–967

Bravo JA, Forsythe P, Chew MV, Emily E, Hélène M, Timothy GD, John B, John FC (2011) Ingestion of Lactobacillus strain regulates emotional behavior and central GABA receptor expression in a mouse via the vagus nerve. Proc Natl Acad Sci U S A 108(38):16050–16055

Bueter M, Ashrafian H, le Roux CW (2009) Mechanisms of weight loss after gastric bypass and gastric banding. Obes Facts 2(5):325–331

Bugda Gwilt K, Gonzalez DP, Olliffe N, Oller H, Hoffing R, Puzan M, El Aidy S, Miller GM (2020) Actions of trace amines in the brain-gut-microbiome axis via trace amine-associated receptor-1 (TAAR1). Cell Mol Neurobiol 40(2):191–201

Butler MI, Sabrina M, Kiran VS, John FC, Timothy GD (2019) The gut microbiome and mental health: what should we tell our patients. Can J Psychiatr 64(11):747–760

Canfora EE, Jocken JW, Blaak EE (2015) Short-chain fatty acids in control of body weight and insulin sensitivity. Nat Rev Endocrinol 11(10):577

Cani PD, Amar J, Iglesias MA, Poggi M, Knauf C, Bastelica D, Burcelin R, Audrey MN, Francesca F, Kieran MT, Chantal C, Aurélie W, Evelyne D, Béatrice C, Thierry S,

Bernard C, Jean F, Jean-François T, Glenn RG, Louis C, Nathalie MD, Marie CA, Rémy B (2007) Metabolic endotoxemia initiates obesity and insulin resistance. Diabetes 56 (7):1761–1772

Caputi V, Giron MC (2018) Microbiome-gut-brain axis and toll-like receptors in Parkinson's disease. Int J Mol Sci 19:6

Chang JY, Shin SM, Chun J, Lee JH, Seo JK (2011) Pyrosequencing-based molecular monitoring of the intestinal bacterial colonization in preterm infants. J Pediatr Gastroenterol Nutr 53 (5):512–519

Chang CH, Lin CH, Lane HY (2020) d-glutamate and gut microbiota in Alzheimer's disease. Int J Mol Sci 21:8

Claesson MJ, Jeffery IB, Conde S, Power SE, O'Connor EM, Cusack S, Susan EP, Eibhlís MO, Siobhán C, Hugh MBH, Mairead C, Bhuvaneswari L, Orla O, Gerald FF, Jennifer D, Michael O, Norma H, Kieran O, Denis O, Douwe S, Martina W, Lorraine B, Catherine S, Julian RM, Anthony PF (2012) Gut microbiota composition correlates with diet and health in the elderly. Nature 488:178

Coello K, Hansen TH, Sørensen N, Munkholm K, Kessing LV, Pedersen O, Vinberg M (2019) Gut microbiota composition in patients with newly diagnosed bipolar disorder and their unaffected first-degree relatives. Brain Behav Immun 75:112–118

Coretti L, Cristiano C, Florio E, Scala G, Lama A, Keller S, Cuomo M, Russo R, Pero R, Paciello O, Raso GM, Meli R, Cocozza S, Calignano A, Chiariotti L, Lembo F (2017) Sex-related alterations of gut microbiota composition in the BTBR mouse model of autism spectrum disorder. Sci Rep 7:45356

D'Aversa F, Tortora A, Ianiro G, Ponziani FR, Annicchiarico BE, Gasbarrini A (2013) Gut microbiota and metabolic syndrome. Intern Emerg Med 8(1):11–15

David LA, Maurice CF, Carmody RN, David BG, Julie EB, Benjamin EW, Alisha VL, Sloan D, Yug V, Michael AF, Sudha BB, Rachel JD, Peter JT (2014) Diet rapidly and reproducibly alters the human gut microbiome. Nature 505:559–563

Dennison B (1976) Definition of preterm delivery. Br Med J 2(6049):1449

Desbonnet L, Garrett L, Clarke G, Bienenstock J, Dinan TG (2008) The probiotic Bifidobacteria infantis: an assessment of potential antidepressant properties in the rat. J Psychiatr Res 43 (2):164–174

Dewhirst FE, Chen T, Izard J, Bruce J, Anne CRT, Wen-Han Y, Abirami L, William GW (2010) The human oral microbiome. J Bacteriol 192:5002–5017

Dinan TG, Quigley EM, Ahmed SM, Scully P, O'Brien S, O'Mahony L, Siobhan OM, Fergus S, Keeling P (2006) Hypothalamic-pituitary-gut axis dysregulation in irritable bowel syndrome: plasma cytokines as a potential biomarker. Gastroenterology 130(2):304–311

Dominguez-Bello MG, Costello EK, Contreras M, Magris M, Hidalgo G, Fierer N, Rob K (2010) Delivery mode shapes the acquisition and structure of the initial microbiota across multiple body habitats in newborns. Proc Natl Acad Sci U S A 107:11971

Eckburg PB, Bik EM, Bernstein CN, Purdom E, Dethlefsen L, Sargent M, Steven RG, Keran EN, David AR (2005) Diversity of the human intestinal microbial flora. Science 308:1635

El Aidy S, Dinan TG, Cryan JF (2014) Immune modulation of the brain-gut-microbe axis. Front Microbiol 5:146

Flint HJ, Scott KP, Duncan SH, Louis P, Forano E (2012) Microbial degradation of complex carbohydrates in the gut. Gut Microbes 3:289–306

Ford AC, Luthra P, Tack J, Boeckxstaens GE, Moayyedi P, Talley NJ (2017) Efficacy of psychotropic drugs in functional dyspepsia: systematic review and meta-analysis. Gut 66 (3):411–420

Forsythe P, Kunze WA (2013) Voices from within: gut microbes and the CNS. Cell Mol Life Sci 70 (1):55–69

Frank DN, Pace NR (2008) Gastrointestinal microbiology enters the metagenomics era. Curr Opin Gastroenterol 24(1):4–10

Frank DN, St Amand AL, Feldman RA, Boedeker EC, Harpaz N, Pace NR (2007) Molecular-phylogenetic characterization of microbial community imbalances in human inflammatory bowel diseases. Proc Natl Acad Sci U S A 104:13780–13785

Furusawa Y, Obata Y, Fukuda S, Endo TA, Nakato G, Takahashi D, Yumiko N, Chikako U, Keiko K, Tamotsu K, Masumi T, Noriko NF, Shinnosuke M, Eiji M, Shingo H, Koji A, Satoshi O, Yumiko F, Trevor L, Julie MC, David LT, Masaru T, Shohei H, Osamu O, Tatsuya M, Haruhiko K, Jun K, Kenya H, Koji H, Ohno H (2013) Commensal microbe-derived butyrate induces the differentiation of colonic regulatory T cells. Nature 504 (7480):446–450

Gajer P, Brotman RM, Bai G, Joyce S, Urcel MES, Xue Z, Sara SKK, Li F, Zhanshan M, Xia Z, Zaid A, Larry JF, Jacques R (2012) Temporal dynamics of the human vaginal microbiota. Sci Transl Med 4:132–152

Ghoshal UC, Ghoshal U (2017) Small intestinal bacterial overgrowth and other intestinal disorders. Gastroenterol Clin 46(1):103–120

Gillespie JJ, Wattam AR, Cammer SA, Gabbard JL, Shukla MP, Dalay O, Driscoll T, Hix D, Mane SP, Mao C, Nordberg EK, Scott M, Schulman JR, Snyder EE, Sullivan DE, Wang C, Warren A, Williams KP, Xue T, Yoo HS, Zhang C, Zhang Y, Will R, Kenyon RW, Sobral BW (2011) PATRIC: the comprehensive bacterial bioinformatics resource with a focus on human pathogenic species. Infect Immun 79:4286–4298

González A, Vázquez-Baeza Y, Knight R (2014) SnapShot: the human microbiome. Cell 158:690–690

Gravitz L (2012) Microbiome: the critters within. Nature 485(7398):S12–S13

Grenham S, Clarke G, Cryan JF, Dinan TG (2011) Brain-gut-microbe communication in health and disease. Front Physiol 2:94

Grice EA, Kong HH, Conlan S, Clayton BD, Joie D, Alice CY, Gerard GB, Robert WB, Patrick RM, Eric DG, Maria LT, Julia S (2006) Topographical and temporal diversity of the human skin microbiome. Science 324:1190–1192

Hamady M, Knight R (2009) Microbial community profiling for human microbiome projects: tools, techniques, and challenges. Genome Res 19(7):1141–1152

Harakeh SM, Khan I, Kumosani T, Barbour E, Almasaudi SB, Bahijri SM, Azhar EI (2016) Gut microbiota: a contributing factor to obesity. Front Cell Infect Microbiol 6:95

Harsch IA, Konturek PC (2018) The role of gut microbiota in obesity and type 2 and type 1 diabetes mellitus: new insights into "old" diseases. Med Sci 6(2):32

Hollister EB, Gao C, Versalovic J (2014) Compositional and functional features of the gastrointestinal microbiome and their effects on human health. Gastroenterology 146:1449–1458

Hoy CM, Wood CM, Hawkey PM, Puntis JW (2000) Duodenal microflora in very-low-birth-weight neonates and relation to necrotizing enterocolitis. J Clin Microbiol 38(12):4539–4547

Human Microbiome Project Consortium (2012) Structure, function and diversity of the healthy human microbiome. Nature 486:22699609

Jacquot A, Neveu D, Aujoulat F, Mercier G, Marchandin H, Jumas-Bilak E, Picaud JC (2011) Dynamics and clinical evolution of bacterial gut microflora in extremely premature patients. J Pediatr 158(3):390–396

Jalanka-Tuovinen J, Salonen A, Nikkila J, Immonen O, Kekkonen R, Lahti L, Palva A, DeVos WM (2011) Intestinal microbiota in healthy adults: temporal analysis reveals individual and common core and relation to intestinal symptoms. PLoS ONE 6:e23035. https://doi.org/10.1371/journal.pone.0023035

Justesen T, Nielsen OH, Jacobsen IE, Lave J, Rasmussen SN (1984) The normal cultivable microflora in upper jejunal fluid in healthy adults. Scand J Gastroenterol 19:279–282

Kennedy PJ, Cryan JF, Dinan TG, Clarke G (2014) Irritable bowel syndrome: a microbiome-gut-brain axis disorder. World J Gastroenterol 20(39):14105

Khanna S, Tosh PK (2014) A clinician's primer on the role of the microbiome in human health and disease. Mayo Clin Proc 89:107–114

Kim YS, Unno T, Kim BY, Park MS (2020) Sex differences in gut microbiota. World J Men's Health 38(1):48

Koenig JE, Spor A, Scalfone N, Fricker AD, Stombaugh J, Knight R, Largus T (2011) Succession of microbial consortia in the developing infant gut microbiome. Proc Natl Acad Sci U S A 108 (Suppl. 1):4578–4585

Koren O, Goodrich JK, Cullender TC, Spor A, Laitinen K, HeleneKling B, Gonzalez A, Jeffrey JW, Largus TA, Rob K, Fredrik B, Erika I, Seppo S, Ruth EL (2012) Host remodeling of the gut microbiome and metabolic changes during pregnancy. Cell 150:470–480

Kuo PH, Chung YE (2019) Moody microbiome: challenges and chances. J Formos Med Assoc 118 (Suppl 1):S42–S54

Ley RE, Bäckhed F, Turnbaugh P, Lozupone CA, Knight RD, Gordon JI (2005) Obesity alters gut microbial ecology. Proc Natl Acad Sci 102(31):11070–11075

Ley RE, Turnbaugh PJ, Klein S, Gordon JI (2006) Microbial ecology: human gut microbes associated with obesity. Nature 444:1022–1023

Lobo C, Moreno-Ventas X, Tapia-Paniagua S, Rodriguez C, Morinigo MA, De La Banda IG (2014) Dietary probiotic supplementation (Shewanella putrefaciens Pdp11) modulates gut microbiota and promotes growth and condition in Senegalese sole larviculture. Fish Physiol Biochem 40:295–309

Luckey TD (1972) Introduction to intestinal microecology. Am J Clin Nutr 25:12

Lynch SV, Pedersen O (2016) The human intestinal microbiome in health and disease. N Engl J Med 375(24):2369–2379

Lyte M (2013) Microbial endocrinology in the microbiome-gut-brain axis: how bacterial production and utilization of neurochemicals influence behavior. PLoS Pathog 9(11):e1003726

Lyte M (2014) Microbial endocrinology and the microbiota-gut-brain axis. Microb Endocrinol 2014:3–24

Mangiola F, Ianiro G, Franceschi F, Fagiuoli S, Gasbarrini G, Gasbarrini A (2016) Gut microbiota in autism and mood disorders. World J Gastroenterol 22(1):361

Marques TM, Wall R, Ross RP, Fitzgerald GF, Ryan CA, Stanton C (2010) Programming infant gut microbiota: influence of dietary and environmental factors. Curr Opin Biotechnol 21:149–156

Mayer EA, Knight R, Mazmanian SK, Cryan JF, Tillisch K (2014) Gut microbes and the brain: paradigm shift in neuroscience. J Neurosci 34(46):15490–15496

McKernan DP, Dennison U, Gaszner G, Cryan JF, Dinan TG (2011) Enhanced peripheral toll-like receptor responses in psychosis: further evidence of a pro-inflammatory phenotype. Transl Psychiatry 1(8):e36

Moloney RD, Desbonnet L, Clarke G, Dinan TG, Cryan JF (2014) The microbiome: stress, health and disease. Mamm Genome 25(1):49–74

Montiel-Castro AJ, Gonzalez-Cervantes RM, Bravo-Ruiseco G, Pacheco-Lopez G (2013) The microbiota-gut-brain axis: neurobehavioral correlates, health and sociality. Front Integr Neurosci 7:70

Moreno I, Codoñer FM, Vilella F, Valbuena D, Martinez-Blanch JF, Jimenez-Almazán J, Simon C (2016) Evidence that the endometrial microbiota has an effect on implantation success or failure. Am J Obstet Gynecol 215(6):684–703

Musso G, Gambino R, Cassader M, Pagano G (2011) Meta-analysis: natural history of non-alcoholic fatty liver disease (NAFLD) and diagnostic accuracy of non-invasive tests for liver disease severity. Ann Med 43(8):617–649

Nair AT, Ramachandran V, Joghee NM, Antony S, Ramalingam G (2018) Gut microbiota dysfunction as reliable non-invasive early diagnostic biomarkers in the pathophysiology of Parkinson's disease: a critical review. J Neurogastroenterol Motil 24(1):30

O'Brien SM, Scott LV, Dinan TG (2004) Cytokines: abnormalities in major depression and implications for pharmacological treatment. Hum Psychopharmacol 19(6):397–403

O'Hara AM, Shanahan F (2006) The gut flora as a forgotten organ. EMBO Rep 7:688–693

O'Mahony L, McCarthy J, Kelly P, Hurley G, Luo F, Chen K, Quigley EM (2005) Lactobacillus and bifidobacterium in irritable bowel syndrome: symptom responses and relationship to cytokine profiles. Gastroenterology 128(3):541–551

O'Mahony SM, Marchesi JR, Scully P, Codling C, Ceolho AM, Quigley EM, John FC, Dinan TG (2009) Early life stress alters behavior, immunity, and microbiota in rats: implications for irritable bowel syndrome and psychiatric illnesses. Biol Psychiatry 65(3):263–267

O'Mahony SM, Hyland NP, Dinan TG, Cryan JF (2011) Maternal separation as a model of brain–gut axis dysfunction. Psychopharmacology 214(1):71–88

O'Mahony SM, Clarke G, Dinan TG, Cryan JF (2017) Early-life adversity and brain development: is the microbiome a missing piece of the puzzle? Neuroscience 342:37–54

Palmer C, Bik EM, Digiulio DB, Relman DA, Brown PO (2007) Development of the human infant intestinal microbiota. PLoS Biol 5:e177

Pei Z, Bini EJ, Yang L, Zhou M, Francois F, Blaser MJ (2004) Bacterial biota in the human distal esophagus. Proc Natl Acad Sci U S A 101:4250–4255

Perez-Cobas AE, Gosalbes MJ, Friedrichs A, Henrik K, Alenjandro A, Kathleen E, Wolfgng O, David R, Rafael B, Martin VB, Sven CN, Carolin D, Femke AH, Amparo L, Coral B, Jana S, MDS V, Stephan JO, Manuel F, Andres M (2013) Gut microbiota disturbance during antibiotic therapy: a multi-omic approach. Gut 62:1591–1601

Qin J, Li R, Raes J et al (2010) A human gut microbial gene catalogue established by metagenomic sequencing. Nature 464:59–65

Qin J, Li Y, Cai Z, Li S, Zhu J, Zhang F, Suisha L, Wenwei Z, Yuanlin G, Dongqian S, Yangqing P, Dongya Z, Zhuye J, Wenxian W, Youwen Q, Wenbin X, Junhua L, Lingchuan H, Donghui L, Peixian W, Yali D, Xiaojuan S, Zesong L, Aifa T, Shilong Z, Xiaoping L, Weineng C, Ran X, Mingbang W, Qiang F, Meihua G, Jing Y, Yanyan Z, Ming Z, Torben H, Gaston S, Jeroen R, Gwen F, Shujiro O, Mathieu A, Emmanuelle L, Pierre R, Nicolas P, Jean-Michel B, Zhaoxi Z, Hua C, Ruifu Y, Weimou Z, Songgang L, Huanming Y, Jian W, Dusko E, Rasmus N, Oluf P, Karsten K, Wang J (2012) A metagenome-wide association study of gut microbiota in type 2 diabetes. Nature 490(7418):55–60

Quigley EM (2016) Leaky gut - concept or clinical entity? Curr Opin Gastroenterol 32(2):74–79

Quigley EMM (2017) Microbiota-brain-gut axis and neurodegenerative diseases. Curr Neurol Neurosci Rep 17(12):94

Rajilic-Stojanovic M, Shanahan F, Guarner F, deVos WM (2012) Phylogenetic analysis of dysbiosis in ulcerative colitis in relapse and remission. Gut 19(3):481–488

Ramakrishna B, Krishnan S (2007) The normal bacterial flora of the human intestine and its regulation. J Clin Gastroenterol 41:S2–S6

Razavi AC, Potts KS, Kelly TN, Bazzano LA (2019) Sex, gut microbiome, and cardiovascular disease risk. Biol Sex Differ 10(1):1–14

Scheperjans F, Aho V, Pereira PA, Koskinen K, Paulin L, Pekkonen E, Haapaniemi E, Kaakkola S, Eerola-Rautio J, Pohja M, Esko K, Kari M, Petri A (2015) Gut microbiota are related to Parkinson's disease and clinical phenotype. Mov Disord 30:350–358

Scholtens PA, Oozeer R, Martin R, Amor KB, Knol J (2012) The early settlers: intestinal microbiology in early life. Annu Rev Food Sci Technol 3:425–447

Schwartz MW, Woods SC, Porte D, Seeley RJ, Baskin DG (2000) Central nervous system control of food intake. Nature 404(6778):661–671

Schwiertz A, Taras D, Schäfer K, Beijer S, Bos NA, Donus C, Hardt PD (2010) Microbiota and SCFA in lean and overweight healthy subjects. Obesity 18(1):190–195

Scott LV, Clarke G, Dinan TG (2013) The brain-gut axis: a target for treating stress-related disorders. Inf Psychiatr 28:90–99

Sekirov I, Russell SL, Antunes LCM, Finlay BB (2010) Gut microbiota in health and disease. Physiol Rev 90(3):859–904

Shanahan F (2012) The gut microbiota-a clinical perspective on lessons learned. Nat Rev Gastroenterol Hepatol 9:609–614

Sherwin E, Sandhu KV, Dinan TG, Cryan JF (2016) May the force be with you: the light and dark sides of the microbiota-gut-brain axis in neuropsychiatry. CNS Drugs 30(11):1019–1041

Sondheimer JM, Clark DA (1985) Gastric pH in healthy preterm infants-effect of age and feeding type. In: Gastroenterology. WB Saunders Co., Philadelphia

Sondheimer JM, Clark DA, Gervaise EP (1985) Continuous gastric pH measurement in young and older healthy preterm infants receiving formula and clear liquid feedings. J Pediatr Gastroenterol Nutr 4(3):352–355

Stilling RM, Dinan TG, Cryan JF (2014) Microbial genes, brain & behaviour–epigenetic regulation of the gut–brain axis. Genes Brain Behav 13(1):69–86

Sudo N, Chida Y, Aiba Y, Sonoda J, Oyama N, Yu XN et al (2004) Postnatal microbial colonization programs the hypothalamic–pituitary–adrenal system for stress response in mice. J Physiol 558 (1):263–275

Svensson E, Horváth-Puhó E, Thomsen RW, Djurhuus JC, Pedersen L, Borghammer P, Sørensen HT (2015) Vagotomy and subsequent risk of Parkinson's disease. Ann Neurol 78(4):522–529

Swidsinski A, Loening-Baucke V, Lochs H, Hale LP (2005) Spatial organization of bacterial flora in normal and inflamed intestine: a fluorescence in situ hybridization study in mice. World J Gastroenterol 11:1131–1140

Takeuchi O, Akira S (2010) Pattern recognition receptors and inflammation. Cell 140(6):805–820

Talham GL, Jiang HQ, Bos NA, Cebra JJ (1999) Segmented filamentous bacteria are potent stimuli of a physiologically normal state of the murine gut mucosal immune system. Infect Immun 67 (4):1992–2000

Tillisch K (2014) The effects of gut microbiota on CNS function in humans. Gut Microbes 5 (3):404–410

Tolhurst G, Heffron H, Lam YS, Parker HE, Habib AM, Diakogiannaki E, Jennifer C, Johannes G, Frank R, Gribble FM (2012) Short-chain fatty acids stimulate glucagon-like peptide-1 secretion via the G-protein–coupled receptor FFAR2. Diabetes 61(2):364–371

Travagli RA, Hermann GE, Browning KN, Rogers RC (2003) Musings on the wanderer: what's new in our understanding of vago-vagal reflexes: III. Activity-dependent plasticity in vago-vagal reflexes controlling the stomach. Am J Physiol 284(2):G180

Turnbaugh PJ, Gordon JI (2009) The core gut microbiome, energy balance and obesity. J Physiol 587(17):4153–4158

Umesaki Y, Okada Y, Matsumoto S, Imaoka A, Setoyama H (1995) Segmented filamentous bacteria are indigenous intestinal bacteria that activate intraepithelial lymphocytes and induce MHC class II molecules and fucosyl asialo GM1 glycolipids on the small intestinal epithelial cells in the ex-germ-free mouse. Microbiol Immunol 39(8):555–562

Umesaki Y, Setoyama H, Matsumoto S, Imaoka A, Itoh K (1999) Differential roles of segmented filamentous bacteria and clostridia in development of the intestinal immune system. Infect Immun 67(7):3504–3511

Unger MM, Spiegel J, Dillmann K-U, Grundmann D, Philippeit H, Bürmann J, Klaus F, Andreas S, Karl-Herbert S (2016) Short chain fatty acids and gut microbiota differ between patients with Parkinson's disease and age-matched controls. Parkinsonism Relat Disord 32:66–72

Upadhyay J, Farr O, Perakakis N, Ghaly W, Mantzoros C (2018) Obesity as a disease. Med Clin 102(1):13–33

Vanessa JRD, Paula V, Forlenza AS (2018) Relevance of gut microbiota in cognition, behaviour and Alzheimer's disease. Pharmacol Res 136:29–34

Vrieze A, Van Nood E, Holleman F, Salojärvi J, Kootte RS, Bartelsman JF, Geesje MD, Mariette TA, Mireille JS, Raish O, Muriel D, Anne D, Johan ET, Van HV, Vincent WB, Albert KG, Hans GH, Erwin GZ, Erik SS, Willem M, Joost BLH, Max N, Nieuwdorp M (2012) Transfer of intestinal microbiota from lean donors increases insulin sensitivity in individuals with metabolic syndrome. Gastroenterology 143(4):913–916

Walsh CJ, Guinane CM, O'Toole PW, Cotter PD (2014) Beneficial modulation of the gut microbiota. FEBS Lett 588(22):4120–4130

Wang Y, Kasper LH (2014) The role of microbiome in central nervous system disorders. Brain Behav Immun 38:1–12

Wu GD, Chen J, Hoffmann C, Bittinger K, Chen YY, Keilbaugh SA, Bewtra M, Knights D, Walters WA, Knight R, Sinha R (2011) Linking long-term dietary patterns with gut microbial enterotypes. Science 334(6052):105–108

Yatsunenko T, Rey FE, Manary MJ, Indi T, Maria GD, Monica C, Magda M, Glida H, Robert NB, Andrey PA, Andrew C, Barbara W, Jens R, Justin KJ, Gregory C, Catherine AL, Christian L, Jose CC, Dan K, Rob K, Jeffrey IG (2012) Human gut microbiome viewed across age and geography. Nature 486:222–227

Yuan X, Chen R, Zhang Y, Lin X, Yang X (2020) Sexual dimorphism of gut microbiota at different pubertal status. Microb Cell Factories 19(1):1–16

Yurtdaş G, Akdevelioğlu Y (2020) A new approach to polycystic ovary syndrome: the gut microbiota. J Am Coll Nutr 39(4):371–382

Zhao X, Jiang Y, Xi H, Chen L, Feng X (2020) Exploration of the relationship between gut microbiota and polycystic ovary syndrome (PCOS): a review. Geburtshilfe Frauenheilkd 80 (2):161

Zheng J, Li H, Zhang X, Jiang M, Luo C, Lu Z, Xu Z, Shi J (2018) Prebiotic mannan-oligosaccharides augment the hypoglycemic effects of metformin in correlation with modulating gut microbiota. J Agric Food Chem 66(23):5821–5831

Zoetendal EG, Rajilic-Stojanovic M, De Vos WM (2008) High-throughput diversity and functionality analysis of the gastrointestinal tract microbiota. Gut 57:1605–1615

Part II

Modulation of Gut Brain Axis Through Pro- and prebiotics

Functional Role of Prebiotic Supplement in Brain Signalling

9

Sreeranjini Sukumaran Rajamma, Venkateshwaran Krishnaswami, and Ruckmani Kandasamy

Abstract

Prebiotics are a class of nutrients, typically non-digestible fibre compounds observed in food which may induce the growth of beneficial microorganisms with greater influence in the gastrointestinal tract and gut microbiota. Prebiotics may pass through the upper part of the gastrointestinal tract and stimulate advantageous bacterial growth. The function and composition of gut microbiota can be altered using prebiotics. Gut bacteria are involved in the physiological processes including immunomodulation, adiposity, energy balance and electrophysiological activity of the central nervous system. Prebiotics not only possess its activity against infectious agents but also have actions on brain-derived neurotrophic factors, neurotransmitters and synaptic proteins. Prebiotics hold greater influence on cognition and psychiatric disorders. Prebiotics like wheat fibre can be considered as a treatment option for autism. In this regimen, the proposed book chapter will focus towards gut microbiota, probiotics, role of prebiotics and its supplements in supporting gut microbial growth, brain signalling, abnormalities in brain signalling, clinical and preclinical findings related to psychiatric changes and in overcoming abnormal psychiatric changes.

This chapter is published in the memory of Dr. K. Akilandeswari, Assistant Professor, Department of Pharmaceutical Technology, University College of Engineering, Anna University, BIT Campus, Tiruchirappalli.

S. S. Rajamma · V. Krishnaswami · R. Kandasamy (✉)
Centre for Excellence in Nanobio Translational Research (CENTRE), Department of Pharmaceutical Technology, University College of Engineering, Anna University, Tiruchirappalli, Tamil Nadu, India
e-mail: ruckmani@aubit.edu.in

R. Z. Sayyed, M. Khan (eds.), *Microbiome-Gut-Brain Axis*,
https://doi.org/10.1007/978-981-16-1626-6_9

Keywords

Prebiotics · Brain · Signalling · Gut · Microbiota

9.1 Introduction

The majority of microorganisms in the gastrointestinal tract are collectively known as human gut microbiota (Xifra et al. 2016). Gut microbiota function as basic physiological processes and at the same time alter the host susceptibility to diseases (Yang et al. 2020). About three million microorganisms and thousands of bacterial phylotypes are involved in a different function of host metabolism. The gut microbiota consists of more than 150 genes. About 80–90% of the bacterial phylotypes include both gram-negative bacteria (Bacteroidetes, Proteobacteria) and gram-positive bacteria (Actinobacteria, Proteobacteria) (Xifra et al. 2016).

The gut microbiome comprises all microorganisms and their genome present in the intestinal tract and contributes towards the development of the hypothalamic-pituitary-adrenal (HPA) axis. Gut bacteria are greatly involved in the regulation of various physiological processes such as immunomodulation, adiposity, energy balance and electrophysiological activity of the enteric nervous system (ENS) (Sarkar et al. 2016). Gut microbiota composition can be altered in patients with both metabolic and neuropsychiatric symptoms. Studies reported that the gut microbiota has a major role in regulating the functions of the gut-brain axis. The functions of the gut-brain axis include metabolism, inflammation, brain function and behaviour (Fernandez et al. 2017).

A core gut microbiota was seen among family members, but there are inter-individual variations in the presence of gut microbiota in each patient (Xifra et al. 2016). High-throughput and low-cost sequencing methods are an effective technique to find the composition and structure of gut microbiota, and hypervariable regions (V1–V9) of the ribosomal RNA present in bacteria may help to find the species easily. It is reported that the microbiota of each individual is varying after birth because the gastrointestinal (GI) tract is getting colonized rapidly and changes in the microbiota occur due to disease, diet and drugs such as antibiotics. The presence of *lactobacilli* is higher in the microbiota of vaginally delivered infants as compared to caesarean section which slowed down and diminished the colonization of the *Bacteroides* genus but colonized by facultative anaerobes like *Clostridium* species. The development of microbiota in its initial stage is controlled by two phyla, such as Actinobacteria and Proteobacteria, with low microbial diversity. The microbiota of infants resembles that of adults, and the composition of the microbiota will be stable in adulthood. The changes in microbiota are expected due to various life events. In geriatric cases (age over 65), the presence of Bacteroidetes phyla and *Clostridium* cluster IV is reported as compared to that of younger subjects. In elderly patients, the metabolic process such as short-chain fatty acid (SCFA) production and amylolysis is reduced, and thereby the proteolytic activity gets increased (Elizabeth and Nathalie 2017). The intestinal mucosa is vital for health management since it is important in the growth and maintenance of the physiological system. Altered intestinal mucosa significantly affects the development and functions of the brain. Infections or paralytic ileus, prolonged hospitalization and death post stroke are the major clinical complications reported for brain injury (Houlden et al. 2016).

The major beneficial gut microbiome are fructo-oligosaccharides (FOS) and galacto-oligosaccharides (GOS) (Burokas et al. 2017). Some of the reported factors which regulate the microbiota are prebiotics, probiotics, antibiotics, bacterial infection, genetically modified bacteria and faecal microbiota transplantation (Burokas et al. 2015). Antimicrobials, prebiotics, probiotics and diet are some of the modulators of the gut microbiome which function according to the type and phase of the disease (Kao et al. 2019).

9.2 Probiotics

Probiotics are living microorganisms that support humans as well as animals when administered adequately. The most commonly used probiotics are different species of *lactobacilli* and *bifidobacteria*. They provide beneficial effects on CNS dysfunction during neurological disorders by increasing microbiota diversity and beneficial bacteria compositions (Bagheri et al. 2019). Probiotics with various gut microbiota such as strains of *bifidobacteria* and *lactobacilli* have shown anxiolytic and procognitive effects in rodents and humans; therefore, this can be used for the treatment of brain disorders (Alexandra et al. 2018). Probiotics are reported to exhibit preventive activity against the neurodegenerative disease such as Alzheimer's disease. The probiotic mixture of *L. acidophilus*, *B. bifidum* and *B. longum* and another combination of *L. acidophilus*, *L. casei*, *B. bifidum* and *L. fermentum* has positive effects on the treatment of Alzheimer's disease. A mixture of probiotics (*L. acidophilus*, *L. fermentum*, *B. lactis and B. longum*) has shown effects on modulating gut microbiota and improves memory deficits and oxidative stress in β-amyloid (1–42)-injected rats (Yang et al. 2020). The abnormal reactions were reversible through probiotic-induced bacterial recolonization (Sarkar et al. 2016). Thus, the probiotics can be used as an agent for regulating the gut microbiome and thereby modifying the health (Kazemi et al. 2019a, b).

9.3 Prebiotics

Prebiotics are selectively fermented ingredients first defined in the mid-1990s. Prebiotics help to improve the health of an organism by specific alterations in the composition and/or activity of the gastrointestinal microbiota (Alexandra et al. 2018; Brownawell et al. 2012). Glenn Gibson and Marcel Roberfroid in 1995 proposed the concepts of prebiotics initially. Prebiotics were described as "a non-digestible food ingredient that beneficially affects the host by selectively stimulating the growth

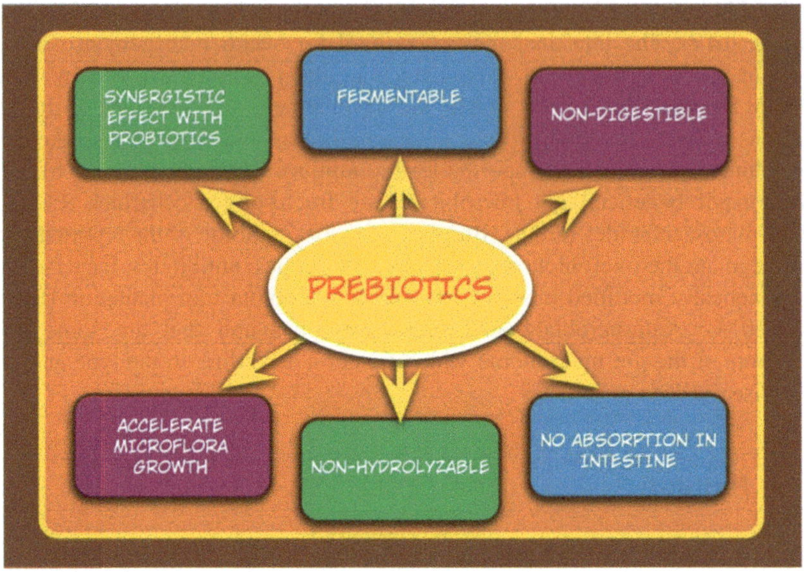

Fig. 9.1 Salient features of prebiotics

and/or activity of one or a limited number of bacteria in the colon and thus improving the host's health". Short- and long-chain fructans (FOS, inulin), lactulose and GOS are some of the prebiotics classified under this definition. The sixth meeting of International Scientific Association of Probiotics and Prebiotics held in 2008 had defined "dietary prebiotics" as "a selectively fermented ingredient that results in specific changes in the composition and/or activity of the gastrointestinal microbiota, thus conferring benefit(s) upon host's health". The various types of prebiotics are fructans (inulin and fructo-oligosaccharide or oligofructose), galacto-oligosaccharides, starch, glucose-derived oligosaccharides, non-carbohydrate oligosaccharides and other oligosaccharides. Small amounts of FOS and GOS are present in foods. An ideal prebiotic should not be degraded in stomach pH and enzymes, absorbed in the gastrointestinal tract and can also be fermented by the gut microbiota (Davani-Davari et al. 2019). The salient features of prebiotics are shown in Fig. 9.1.

9.4 Prebiotics Role in Supporting Gut Microbial Growth

Prebiotics are a group of beneficial nutrients which stimulate some of the bacterial species in the gut microbiota to produce beneficial effects on the host. Its degradation products (short-chain fatty acids) are released into the systemic circulation which affects the functions of the gastrointestinal tract and other organs (Kazemi et al. 2019a, b). The carbohydrates which are not digested in the small intestine may undergo fermentation in the large intestine by the gut microbiota and produce short-chain fatty acids and lactic acids which may stimulate the bacteria (*bifidobacteria* and *lactobacilli*). These bacteria are beneficial to improve the health (Alexandra et al. 2018). Several studies reported the effective role of prebiotics in reducing the risk and severity of GI infections and inflammations such as diarrhoea, inflammatory bowel disease, ulcerative colitis and bowel function disorders like irritable bowel syndrome. Prebiotics promote mineral absorption and lower the risk of obesity (Brownawell et al. 2012).

Prebiotics support the growth and activity of probiotics (Davani-Davari et al. 2019). The use of prebiotics is a better choice for the maintenance of brain health and adjunctive treatment for neuropsychiatric disorders. The central expression of brain-derived neurotrophic factor (BDNF) and *N*-methyl-D-aspartate receptor (NMDAR) subunits is reduced in the absence of gut bacteria. The oral probiotics increase the brain-derived neurotrophic factor (BDNF) and impart significant anxiolytic effects. Prebiotics did not alter glutamate, glutamine, L-serine, L-alanine or D-alanine levels in the brain. The effect of galacto-oligosaccharides on the components of central NMDAR signalling was greater than fructo-oligosaccharides and reflects the proliferative potency of galacto-oligosaccharides on microbiota (Savignac et al. 2013). Some disruptions in the normal gut microbiota may cause depression (Kazemi et al. 2019a, b). Little amounts of prebiotics are usually present in our daily diet, and they have an extensive role in improving health. FOS and GOS are the main source for manufacturing prebiotics (Davani-Davari et al. 2019).

9.5 Brain Signalling (Microbiome-Gut-Brain Axis)

There are considerable interactions between the gut microbiota and the CNS through the gut-brain axis which maintains the health of an organism (Tarr et al. 2015; Sarkar et al. 2016). The mechanisms which are proposed for the effects of microbiota on the gut-brain axis are the regulation of the functions of the autonomic nervous system, the neuroendocrine system and the immune system (Yang et al. 2020). The abnormalities associated with brain signalling are shown in Fig. 9.2.

Usually, the gut microbiome and the CNS get matured during the early period of life; hence, this period is influential for the growth and development of normal physiology of an individual. The gut microbiome helps to improve the neurodevelopmental process through the regulation of neuronal, hormonal and immunological pathways. Interruptions to the gut microbiome may cause abnormalities in the responses of the HPA axis and brain-derived neurotrophic factors which may crucially affect the normal behaviour of an individual (Loughman et al. 2020). The gut microbiome has extensive effects on the functioning of the brain such as psychological processing, behaviour, neurodevelopmental process and during changes in gene expression in particular brain regions (Burokas et al. 2017).

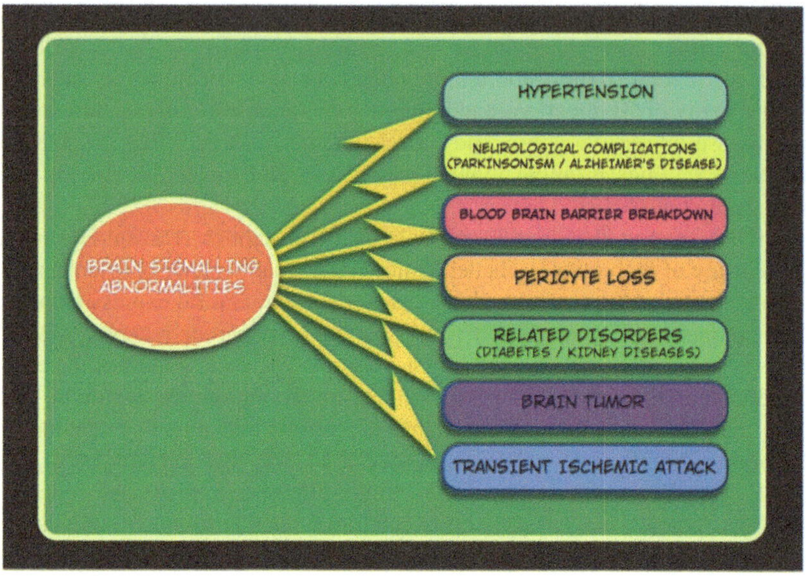

Fig. 9.2 Brain signalling abnormalities

9.5.1 Neural Pathway

The enteric nervous system (ENS) has a significant role in controlling gut functions to maintain the general homeostasis by regulating the neurotransmitters and thereby acts as the "brain of the gut". The active participation of the major parasympathetic nerve (vagus nerve) is essential. With the help of the vagus nerve (a major nerve in the parasympathetic system), dorsal root ganglia and somatosensory afferents, the afferent signals from the GI tract are transmitted to the brain stem (periaqueductal grey) and finally allow the distribution to the brain areas such as the hypothalamus, thalamus, limbic system and somatosensory cortex. The efferent signals are sent back to the enteric nervous system (ENS) through the spinal or vagal efferents (Burokas et al. 2015). The mode of communication of the gut to the brain is through vagus nerve innervation, immune system activation by cytokine/chemokine activity and release of neuropeptides, hormone, cortisol secretion and microbial metabolites. Studies have shown that the vagus nerve has effects on the bidirectional communication of the microbiota-gut-brain axis (Davani-Davari et al. 2019).

9.5.2 Hypothalamic-Pituitary-Adrenal Axis

The hypothalamic-pituitary-adrenal (HPA) axis is a group of structures which include the paraventricular nucleus (PVN) of the hypothalamus, the anterior lobe of the pituitary gland and the adrenal gland. The HPA axis plays a vital role in the regulation of stress responses (Smith and Vale 2006). Alterations in the gut microbiota activate the HPA axis by the release of various mediators (pro-inflammatory cytokines, small bioactive molecules, prostaglandins, microbial antigens, ileal corticosterone and short-chain fatty acids). Some bacterial species interact with the vagus nerve and produce vagal signals which activate the HPA axis that may lead to gut dysbiosis and altered permeability (Misiak et al. 2020). Noradrenaline is the main neurotransmitter in the sympathetic nervous system which can alter the goblet cell function and influences the gut microbiota indirectly. The sympathetic nervous system is activated when nor-adrenaline is released and it circulates in stroke conditions (Houlden et al. 2016).

9.5.3 Tryptophan and 5-Hydroxytryptamine Metabolism

Tryptophan is an essential amino acid usually present in protein-rich diet. It gets absorbed from the gut to the systemic circulation and exists as free and albumin-bound fractions through large amino acid transport, and it can cross the BBB easily for participating in the biosynthesis of 5-hydroxytryptamine (5-HT). The mucosal 5-HT can activate the peristaltic movement of the gastrointestinal tract. *Clostridium* bacteria have been reported for its active role in the activation of 5-HT synthesis and the regulation of peristaltic movement of the GI tract (Golubeva et al. 2017). Both probiotics and prebiotics or its combination can be used for curing GI symptoms and autism spectrum disorder (ASD)-related symptoms. An increased level of *Clostridium* bacterial species with its corresponding increase in plasma 5-hydroxytryptamine was reported in children affected with ASD, and the treatment with a combination of probiotics and fructo-oligosaccharides (FOS) has shown marked reduction in the 5-HT level (Wang et al. 2020). The hippocampal monoamine neurotransmitter gene plays an important role in regulating brain activities such as behaviour and its functions. Derangements in the functions of monoamine neurotransmitter during childhood alter the development of brain functions which leads to the development of brain-related disorders such as neuropsychiatric disorders, pyramidal and extra-pyramidal motor disorders, epilepsy, etc. A study with anxiety and depression-induced germ-free mice models administered with commensal microbiota reported a significant increase in the level of serotonin and dopamine in the striatum which reveals the positive impact of probiotic role in modulating behavioural changes (Pan et al. 2019). The presence of microbiota associated tryptophan metabolites such as 5-hydroxy indoleacetate, melatonin, N-acetyl tryptophan, tryptamine, indol-3-acetate, methyl indole-3-acetate, and methyl indole-3-propionate in urine has also been linked to a direct role of the human metabolome in gut microbial metabolism (Pavlova et al. 2017). An isoquinoline alkaloid, palmitin, is reported for its activity against ulcerative colitis, protecting from gut microbiota dysbiosis and modulating tryptophan catabolism (Zhang et al. 2018).

9.5.4 Immune System

The development and functions of both innate and adaptive immune systems can be regulated by the gut microbiota (Dhar and Mohanty 2020). After injury, a rapid tissue reaction was seen, including the generation of reactive oxygen species, purine metabolites, mitochondrial and polysaccharide components. These factors then bind to pattern recognition receptors (PPRs) such as toll-like receptors (TLRs) and nucleotide binding receptors (NODs), presenting innate cells with gut microbial ligands that help defend against secondary infections (Sabin and Echeverri 2020). TLRs can recognize the microorganism-associated molecular patterns and pathogen-associated molecular patterns and activate the immune responses. Microorganisms such as *Bacteroides*, *Lactobacillus* and *Bifidobacterium* may bind to the innate cell receptors and secrete some metabolites such as short-chain fatty acids (butyrate, acetate, propionate and secondary bile acids) (Dhar and Mohanty 2020). The mood and cognition get altered directly (toll-like receptors) and indirectly (immune activation) through mechanisms involved in some bacterial products such as gram-negative endotoxins (Beilharz et al. 2016). Brain-derived neurotrophic factor (BDNF) which is widely distributed in the nervous system has effect on pro-inflammatory cytokines and regulates the neuroplasticity and inflammation and inhibits the cell apoptosis during gut inflammation. The reduced expression of BDNF is related to inflammation and stress. Studies reported that gut microbiota have effects on BDNF levels and behaviours in mice (Li et al. 2018). Lipopolysaccharides are toll-like receptor ligands that activate the nuclear factor-k (NF-k) pathway, resulting in dysbacteriosis of the gut microbiota, associated low grade chronic systemic inflammation during ageing, and cognitive impairment (Yang et al. 2020). Toll-like receptors, NOD-like receptors (NLRs) and RIG-I-like receptors are some of the pattern recognition receptors which can identify the cell components of gut microbiota such as lipopolysaccharides, peptidoglycan and flagellin and can activate cytokines, hormones and some molecular signals to the CNS (Davani-Davari et al. 2019).

9.5.5 Gut Hormonal Response

The pancreatic hormone GLP-1 is classified as an incretin which can bind to its receptor GLP-1R. in the enteric nervous system and stimulates the vagus nerve to activate the gut-brain-periphery axis. As a result, insulin secretion is stimulated while glucagon release is inhibited. GLP-1 has effects on both the peripheral and central nervous systems. Dysbacteriosis of gut microbiota may lead to degeneration of nerves and muscles in the digestive system and also shows that GLP-1 resistance G proteins, adenylate cyclase, cyclic AMP (cAMP), protein kinase C (PKC), cAMP response element binding protein (CREB), nitric oxide (NO) and NO synthase (NOS) are the main signalling molecules responsible for GLP-1 intracellular action for neurons and B cells (Grasset et al. 2017).

9.6 Factors Affecting the Microbiome-Gut-Brain Axis

The factors affecting the microbiome-gut-brain axis are diet, age, sex and some drugs such as antibiotics. Any changes in the structure of gut microbiota may affect the normal functioning of the host (An et al. 2020).

9.6.1 Role of Diet in the Microbiome-Gut-Brain Axis

The composition of microbiota varies according to the daily diet (Fulling et al. 2020). Intake of high levels of saturated fats and processed sugars leads to obesity, and its short-term exposure impairs memory and causes neurological diseases. Varying diet is independent of hippocampal, hypothalamic and neuroplasticity markers and brain-derived neurotrophic factors. However, it changes the microbial composition in a variety of ways, including memory, inflammation-related hippocampus genes, and the gut microbiome (Burokas et al. 2015). The microbiota-colon-brain axis plays an important role in the regulation of energy metabolism. A study has also reported the importance of gut microbiota in the development of obesity and the memory loss associated with the diet containing high saturated fatty acids (Zhang et al. 2019). Feed deprivation in fish showed dangerous effects in their behaviour and stress physiology, which consequently leads to disease outbreak. Functional ingredients in diet affect the physiology and stress responses of host organism. Feed deprivation in some cases influenced anxiety-like behaviours (Forsatkar et al. 2017). The role of macronutrients in the complete diet is greater and can be considered as a better treatment option for diet-induced memory deficits (Beilharz et al. 2016).

9.6.2 Role of Age in the Microbiome-Gut-Brain Axis

Any exposure to environmental factors causes permanent impact on brain function during adolescence and early adulthood. Mental uneasiness due to the lack of gut bacteria affects gut-brain communication and brain development which may cause psychiatric disorders. A study has reported the effects of gut bacterial depletion from weaning onwards on adult cognitive, social and emotional behaviours. Any depletion in the gut microbiota affects the adult brain by reduced anxiety-induced cognitive deficits, changes in the tryptophan metabolic pathway and reduction in brain-derived neurotrophic factor (BDNF), oxytocin and vasopressin expression (Desbonnet et al. 2015). Adolescence is a critical period of growth that is marked not only by changes in behaviour and the neuroimmune system, and also based on the development of gut microbiota (Fulling et al. 2020). Traumatic brain injury (TBI) is highly reported in children and adolescents. The combination of resveratrol, prebiotic fibre and omega-3 fatty acids can be used for the treatment of TBI which may support to prevent injury-related deficits in medial prefrontal cortex (mPFC) spine density (Salberg et al. 2017). Ageing may alter the composition of gut microbiota which may lead to inflammation and dementia through the suppression of TLR4- and RIG-I-mediated NF-κβ signalling. Activation of microglia and neuro-inflammation occurs not only due to neuronal loss and oxidative changes but also due to age-related dementia (Yang et al. 2020). An increase in pathogenic bacteria was observed in the gut microbiota of aged persons than beneficial microbiota. An impaired microbiota may cause a chronic inflammation and enhance the upregulation of neurotrophic factors such as neurotrophins and neuropoietic cytokines. Both prebiotic and probiotic supplements can modulate gut microbiota and improve the physiological state and thereby serve as a best therapeutic tool for age-related cognitive impairment (Romo-Araiza and Ibarra 2020).

9.6.3 Role of Sex in the Microbiome-Gut-Brain Axis

Sex differences should be examined in neurogastroenterology and psychiatric research with significant changes in behaviour due to changes in gut microbiota in sex differences. Gut microbiota has a great influence in the expression of genes in the medial prefrontal cortex. Any changes in the digestive system during childhood may affect the composition of the gut microbiome, resulting in microbial imbalance, improper functioning of the intestinal barrier and disruptions in brain development owing to dysfunction of gut-brain axis(Rincel et al. 2019).

9.6.4 Role of Drugs in the Microbiome-Gut-Brain Axis

Antibiotics may alter the structure and composition of the gut microbiota (Desbonnet et al. 2015). Triptolide, a plant constituent used for the treatment of autoimmune disease, can produce major hepatic toxicity by destroying the Firmicutes and reducing the short-chain fatty acids. The triptolide toxicity can be reduced using the intake of prebiotics, probiotics and short-chain fatty acids such as propionate (Huang et al. 2020).

9.7 Abnormalities in Brain Signalling

Both the central nervous system and the digestive system are interconnected with neuroendocrine and humoral pathways (MacLaren et al. 2019). Studies reported that stressor exposure and acid suppression significantly alter gut microbiota community (Tarr et al. 2015). Early life stress is a widely reported risk factor for the development of psychiatric disorders and may cause enduring changes in the gut microbiota that lead to the development of abnormal neuronal and endocrine functions. The gut microbiota influence the brain development and function by affecting inflammatory mediators, the hypothalamic-pituitary-adrenal axis and neurotransmission. Early life stress from social isolation can lead to alterations in gut microbiota, anxiety, learning/memory impairment, low levels of hippocampal IL-6, IL-10, and neurogenesis (Doherty et al. 2017).

Gut bacteria can influence appetite by participating in the hunger pathway both locally and centrally through the use of molecular derivatives generated during its various phases of growth. A combination of altered social and feeding behaviours is common in children with autism spectrum disorder (ASD). The α-melanocyte-stimulating hormone (α-MSH) is a specific anorexigenic neuropeptide in the brain acting on melanocortin receptor type 4 (MC4R) which is also involved in the feeding behaviour. Oxytocin is another neuropeptide which is critically involved in the social behaviour. The brain-derived neurotrophic factor (BDNF) signalling produced by the neurons has been observed in the pathophysiology of ASD and contributes to the anorexigenic effects. It has been reported that the release of β-endorphin independent of α-MSH in response to endocannabinoids blocks the proopiomelanocortin (POMC) neurons. Leptin, obtained from adipose tissue and stomach, can directly and indirectly trigger POMC neurons by controlling the energy homeostasis and blocking the inhibitory γ-aminobutyric acid from adjacent neuropeptide Y of the arcuate nucleus (Fetissov et al. 2019).

Gut microorganisms have a major role in the biotransformation of phospholipids. In the presence of the enzyme phospholipase D, certain bacteria in the gut may hydrolyze phosphatidylcholine to choline, and the released choline will be converted in to trimethylamine (Chittim et al. 2019). Trimethylamine (TMA) found in gut microbiota may be oxidised to Timethylamine N-oxide, which can affect the inflammatory process and induce cardiovascular disorders. Choline, L-carnitine and ergothioneine are the main precursors of TMA. Choline is present in many foods in its free and combined forms such as phosphatidylcholine, phosphocholine and sphingomyelin (Janerio et al. 2018). The neural excitation and inhibition balance can be regulated by some neurotransmitters such as γ-aminobutyric acid (GABA) and glutamate (Sarkar et al. 2016).

9.8 Disease Occurring Due to Abnormalities in Brain Signalling

Disorders affecting the microbiota-gut-brain axis are metabolic disorders (obesity, diabetes), functional gastrointestinal disorders (irritable bowel syndrome), stress, anxiety, depression, neurodegenerative disorders (Alzheimer's disease, multiple sclerosis, Parkinson's disease), neurodevelopmental disorders (autism, schizophrenia) and also addiction (alcohol dependence) (Burokas et al. 2015). A drastic change in the social and feeding behaviours of children having autism spectrum disorder was reported. Some neuropeptides such as α-MSH and oxytocin have a pivotal role in controlling the social and feeding behaviour (Fetissov et al. 2019). The human infant gut microbiota has enduring significance in the neurodevelopmental process. The cross-sectional connections between behaviour and gut microbiota in 77 human babies aged 18–27 months revealed a link between phylogenetic diversity and temperamental issues, especially in boys (Loughman et al. 2020). The correlation between prebiotics role and brain signalling is shown in Fig. 9.3.

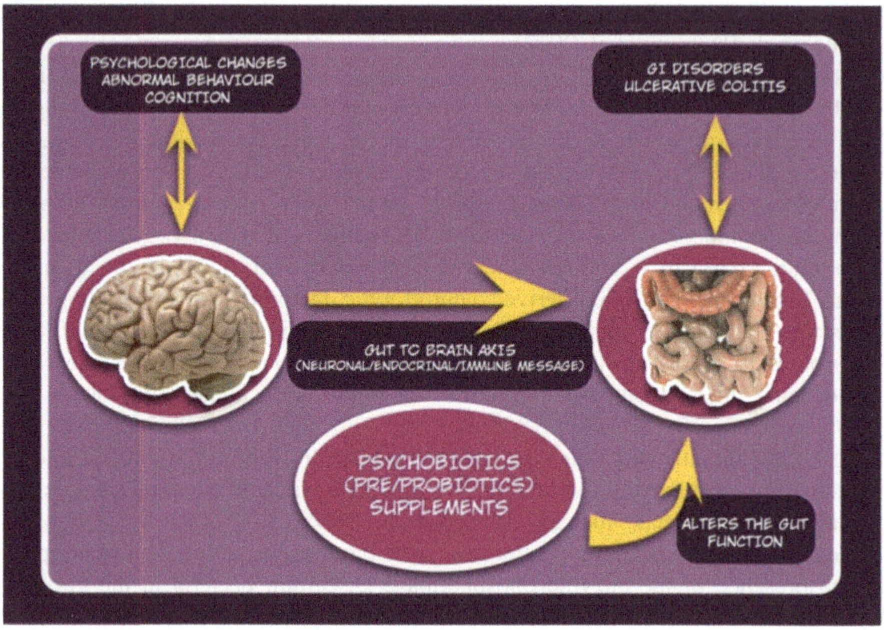

Fig. 9.3 Prebiotics role and brain signalling

Chronic Periodontitis (CP) is an infectious disease caused by inflammation in periodontal tissues, which is followed by the invasion of bacteria, endotoxins, and cytokines into the bloodstream with response to inflammation. Studies reported that CP can alter the normal oral and gut microbiota through the generation of plaque microorganisms and this may lead to the formation of various diseases in both the digestive system and the central nervous system (Xue et al. 2020).

The microbiota-gut-brain axis is extensively involved in the pathology of major depressive disorder (MDD) and regulates the functions of the brain. An increased expression of immune-modulating microbiota such as *Clostridia* was reported in experimental mice exposed to social defeat stress. An increased level of *Lactobacillus*, *Clostridium cluster III*, *Anaerofustis* and *Corynebacterium* was also reported in rats exposed to uncontrolled stress. The increased plasma adrenocorticotrophin and corticosterone levels in response to restraint stress in germ-free mice establish the direct link between the hypothalamic-pituitary-adrenal (HPA) axis and the microbiota (Sarkar et al. 2016).

Traumatic brain injury (TBI) is highly reported in civilians, military personnel and veterans and further aggravates to the symptoms of co-morbid post-traumatic stress disorder (Brenner et al. 2017). Probiotics can be used to treat human allergic diseases such as asthma and atopic diseases. Probiotics use has been shown to minimise the need of antibiotics, improve immune-related illnesses such as inflammatory bowel disease, metabolic syndrome, and diabetes, and to have positive health benefits on anxiety, depression, and gut discomfort (Davani-Davari et al. 2019).

The role of prebiotics in the management of metabolic and central nervous system-related diseases is reported in many studies (Ahmadi et al. 2019). Gut microbiota dysbiosis might be implicated in the pathophysiology of depression. Hence, probiotics, prebiotics and synbiotics have been administered in clinical trials with an attempt to relieve depressive symptoms (Vaghef-Mehrabany et al. 2020). Recent studies reported the significance of gut microbiota in the management of tumour growth (Li et al. 2020).

9.9 Clinical and Preclinical Findings Related to Psychiatric Changes

The role of the microbiota-gut-brain axis in health and diseases such as neuropsychiatric disorders is gaining much importance. The understanding on the composition of prebiotics and the microbiota-derived metabolites acting on signalling of cellular pathways is much interesting (Neri-Numa and Pastore 2020). An altered composition of caecal microbiota with specific changes in Peptococcaceae and Prevotellaceae was observed in experimental stroke models, and the changes in the gut microbiota were observed in traumatic brain injury. To treat individuals with brain damage, altering the gut mucosa is a better alternative (Houlden et al. 2016). In a preclinical experiment, the behavioural, physiological and caecal microbiota profile of aged male mice was studied and reported the role of gut microbiota in increasing gut permeability and peripheral inflammation that cause the deterioration of behavioural, affective and cognitive functions in ageing (Scott et al. 2017).

Altered Schaedler flora (ASF) mice represent a unique model to elucidate mechanisms governing microbiota-gut-brain communication. ASF mice displayed marked anxiogenic behaviour as compared to conventionally reared mice (Lyte et al. 2019). A preclinical study reported the effects of stress and acid suppression on the distribution of gastrointestinal microbiota. The cognitive functions are regulated through several key biological processes in the hippocampus. Neurocognition may alter due to dysbiosis caused by acid suppression during stress (MacLaren et al. 2019).

The cerebrovascular disease like cerebral ischaemic stroke can alter the normal functions of gut microbiota including *Bacteroides*, *Escherichia*, *Shigella*, *Haemophilus*, *Eubacterium nodatum*, *Collinsella*, *Enterococcus*, *Proteus*, *Alistipes*, *Klebsiella*, *Shuttleworthia* and *Faecalibacterium*. The combination therapy for the treatment of cerebrovascular diseases with *Puerariae Lobatae Radix* (PLR) and *Chuanxiong Rhizoma* (CXR) has strong effects on the gut microbiome and cured the cerebral infarction and modified the nerve functions (Chen et al. 2019). It has been reported that child compound Endothelium corneum (CCEC) is effective for the therapy of functional dyspepsia (FD). Further, CCEC significantly enhanced gastric emptying and small intestinal transit of FD-affected rats and prominently suppressed gastrointestinal microinflammation. CCEC supressed over-activated POMC/Stat3/Akt pathway in the hypothalamus. CCEC enhanced gastrointestinal motility probably through rebalancing the homeostasis of the brain-gut-microbiota axis (He et al. 2019).

Clinical investigations comparing standard enteral formula (SEF) against enteric formula with prebiotic content (EFPC) in terms of nutrition treatment-related outcomes in neurocritical care patients revealed the relevance of nutrition therapy in preventing protein debt. The use of EFPC compared to SEF was associated with significant higher total energy, carbohydrate, protein, lipid, enteral volume and fluid intake during each day of nutrition therapy (Tuncay et al. 2018).

A double-blind clinical trial was conducted to compare the effect of probiotic (*Lactobacillus helveticus* and *Bifidobacterium longum*) and prebiotic (galacto-oligosaccharide) supplementation, and an improvement in the BDI score, a decline in the kynurenine/tryptophan ratio and a rise in the tryptophan/isoleucine ratio were observed in MDD subjects who are supplemented with probiotics for 8 weeks as compared to the placebo who are not supplemented with probiotics (Kazemi et al. 2019a, b).

A double-blind placebo-controlled trial reported the effect of prebiotic and probiotic on serum inflammatory cytokines (TNF-α, IL-1ß, IL-6 and IL-10). A reduced BDI score in the prebiotic-treated group, a reduced cortisol level and an equilibrium in the cytokine levels were observed in both groups which are treated with prebiotics and probiotics. Probiotics have marked effects on improving the symptoms of depression (Kazemi et al. 2019a, b).

9.10 Prebiotics Role in Overcoming the Abnormal Psychiatric Changes

Prebiotics possess a greater role in overcoming the brain-related disease associated with the gut microbiota. Sialyllactose is a prebiotic which can alter the colonic and gut microbiota composition and reduce the stress and anxiety by regulating the immune and endocrine functions. The prebiotic combination of FOS and GOS can reduce the level of corticosterone and L-tryptophan in stress conditions (Burokas et al. 2017). ProBiotic-4 is a probiotics product made up of Bifidobacterium lactis, Lactobacillus casei, Bifidobacterium bifidum, and Lactobacillus acidophilus, which have effects on the microbiota-gut-brain axis. A study reported that ProBiotic-4 can be used for the management of cognitive impairments in senile mice having impaired gut microbiota and can also reduce the age-related dysfunction of the intestinal and blood-brain barrier (Yang et al. 2020).

It has been reported that there is a relationship between the effect of prebiotic (oligofructose) treatment on dentate gyrus neurons and spatial memory but prebiotic administration did not improve behavioural alterations and associated reduction of hippocampal neurogenesis. Prebiotic administration improved excessive food intake and glycaemic dysregulations (glucose tolerance and insulin resistance) (Fernandez et al. 2017). Preclinical studies reported the beneficial effect of galacto-oligosaccharide for improving learning and memory deficits. Probiotics can raise the activity of anorexigenic gut hormones (peptide tyrosine tyrosine, glucagon-like peptide 1 and leptin) and can decline the level of orexigenic hormones (ghrelin). Prebiotics are a better option for the treatment of schizophrenia (Kao et al. 2018). Both prebiotics and probiotics can be used for the management of the enteric nervous system (Davani-Davari et al. 2019).

9.11 Prebiotic Supplement in Overcoming the Psychiatric Changes

The lactic acid bacteria present in yogurts can modify the memory deficits during ageing (Scott et al. 2017). Resveratrol ameliorates the hepatic steatosis by modulating the gut microbiota. It increases the gene expression of fasting-induced adipose factor, decreases the expression of lipogenesis-related genes and proteins (SREBP-1, FAS and ACC) and reverses high-fat diet (HFD)-induced gut microbiota dysbiosis, with an increase in the relative abundance of Bacteroidetes and a decrease in that of Firmicutes and Proteobacteria (Xiaohan et al. 2020). A dihydroquinoline analog of agomelatine, *N*-(2-(7-methoxy-3,4-dihydroisoquinolin-1-yl)ethyl) acetamide hydrochloride (NMDEA), has reduced the depression in chronic unpredictable mild stress-induced mice models. NMDEA is involved in regulating the neuro-inflammatory markers (IL-1β, IL-6 and iNOS) and gut microbiota by acting on the microbiota-inflammasome-brain axis and can be used for the treatment of dysbiosis of bacterial species (An et al. 2020). Milk oligosaccharides such as 3′ Sialyllactose and 6′ Sialyllactose support normal microbial communities and behavioural responses during exposure to stress, potentially through effects on the microbiota-gut-brain axis. Milk oligosaccharides helped to maintain normal behaviour on tests of anxiety-like behaviour and normal numbers of DCX+ immature neurons (Tarr et al. 2015). The various prebiotic supplements in overcoming psychiatric changes are shown in Table 9.1.

The role of chronic prebiotic (combination of fructo-oligosaccharides and galacto-oligosaccharides) treatment (for 3 weeks) in anxiety, depression, cognition, stress response and social behaviour using C57BL/6J male mice was evaluated in plasma corticosterone, microbiota composition and caecal short-chain fatty acids and reported for its efficient antidepressant and anxiolytic actions. Prebiotics can regulate the gene expression in the hippocampus and the hypothalamus (Burokas et al. 2017). Bimuno™ galacto-oligosaccharide (B-GOS) is a prebiotic that can activate *N*-methyl-D-aspartate (NMDA) and promote the growth of useful gut bacteria (Gronier et al. 2018). Bimuno™ galacto-oligosaccharide (B-GOS) can reduce the weight gained through olanzapine intake and modify the memory (Kao et al. 2019). Nutrient supplements are essential for the proper development of the brain. Prebiotics and bioactive milk fractions are some of the agents which can act on the microbiota-gut-brain axis and can supplement nutrients for the brain to enhance the memory and maintain the emotional behaviours (Mika et al. 2018). Prebiotics which are supplemented through food undergo biotransformation in the presence of colonic microorganisms and release some metabolites such as short-chain fatty acids into the lumen of the gastrointestinal tract and alter the composition of host (Neri-Numa and

Table 9.1 Prebiotic supplements in overcoming psychiatric changes

Supplement name	Mode of action	Therapeutic benefit	References
Prebiotics inulin or mucin	Mucin fails to inhibit tumour growth in germ-free mice	For colon cancer	Li et al. (2020)
	Inulin enhances the efficacy of a MEK inhibitor against melanoma		
Resveratrol	Increases the gene expression of fasting-induced adipose factor	Ameliorates the hepatic steatosis	Xiaohan et al. (2020)
N-(2-(7-methoxy-3,4-dihydroisoquinolin-1-yl)ethyl)acetamide hydrochloride	Regulating the neuro-inflammatory markers (IL-1β, IL-6 and iNOS)	Antidepressant action	An et al. (2020)
Soybean peptides Maillard reaction products (SMRPs)	Modulating gut microbiota to alleviate ageing-related disorders in D-galactose-induced ICR mice	Flavour enhancer and potential prebiotic which retard the ageing process	Zhang et al. (2020)
Ethanol-precipitated glycans from the softwood hemicellulose autohydrolysate	Stimulate in vitro growth of *Bifidobacterium adolescentis*	Cardioprotective	Deloule et al. (2020)
Bimuno™ galacto-oligosaccharides (B-GOS)	Activate *N*-methyl-D-aspartate (NMDA) and promote the growth of useful gut bacteria	Memory enhancer, treatment option for neuro-inflammation	Gronier et al. (2018), Kao et al. (2019)
Water-soluble, non-digestible polysaccharides isolated from sago and acorn	Reduce high-fat diet-induced defects	Type 2 diabetes mellitus	Ahmadi et al. (2019)
Phenylethanoid glycosides (magnoloside A)	Modulate the composition of gut microbiota	Used for abdominal bloating, pain and indigestion	Xue et al. (2019)
Prebiotics and bioactive milk fractions	Enhance memory and maintain the emotional behaviours	Nutrient supplements for brain development	Mika et al. (2018)
Combination of fructo-oligosaccharides and galacto-oligosaccharides	Regulate the gene expression in the hippocampus and the hypothalamus	Antidepressant and anxiolytic	Burokas et al. (2017)
Lactic acid bacteria present in yogurts	–	Modify the memory deficits during ageing	Scott et al. (2017)
Psychobiotics	Beneficial bacteria have greater influence on the gut	Better anxiolytic and antidepressant agents	Sarkar et al. (2016)

(continued)

Table 9.1 (continued)

Supplement name	Mode of action	Therapeutic benefit	References
	microbiome and the immune system		
Milk oligosaccharides (3′ Sialyllactose and 6′ Sialyllactose)	Support normal microbial communities and behavioural responses during exposure to stress	Maintain normal behaviour on tests of anxiety-like behaviour and normal numbers of DCX+ immature neurons	Tarr et al. (2015)

Pastore 2020). Water-soluble, non-digestible polysaccharides isolated from sago and acorn have shown positive effects to reduce the high-fat diet-induced defects in the biotransformation of glucose through the effect on the microbiota-gut-brain axis and can be used as a better therapeutic tool against type 2 diabetes mellitus (Ahmadi et al. 2019). Soybean peptides Maillard reaction products (SMRPs) as a flavour enhancer were reported as a potential prebiotic on modulating gut microbiota to alleviate ageing-related disorders in D-galactose-induced ICR mice. SMRPs have been reported to elevate the diversity of gut microbiota and ameliorate microbial community structure (Zhang et al. 2020).

The ethanol-precipitated glycans from the softwood hemicellulose autohydrolysate were able to stimulate in vitro growth of *Bifidobacterium adolescentis*, but to a much lesser extent than that of adherent-invasive *E. coli*, *B. adolescentis* was the best producer of SCFA. When mice were fed with the ethanol-precipitated fraction, the relative abundance of Bacteroidetes raised while that of Proteobacteria diminished, suggesting change towards a less obesogenic microbiome. Following treatment, lipid analysis showed a decrease in cholesterol, bile acids and free fatty acids, indicating a potential cardioprotective role (Deloule et al. 2020). The importance of gut microbiota in anti-tumour immunity and the potential therapeutic role for prebiotics were studied by the addition of the prebiotics inulin or mucin to the diet of C57BL/6 mice. Mucin fails to inhibit tumour growth in germ-free mice, indicating that the gut microbiota is required for the activation of the anti-tumour immune response. Inulin limits tumour growth in syngeneic mouse models of colon cancer and NRAS mutant melanoma and enhances the efficacy of a MEK inhibitor against melanoma (Li et al. 2020). Magnoloside A, a phenylethanoid glycoside, was isolated from the traditional Chinese medication 'Hou Po,' which is frequently used to alleviate stomach bloating, discomfort, and indigetion. Magnoloside A can modulate the composition of gut microbiota. Pre-clinical studies reported that magnoloside A has activity against functional dyspepsia through the activation of peptide hormones such as gastrin, motilin and calcitonin and reducing the rate of 5-hydroxytryptamine and nitric oxide synthase (Xue et al. 2019).

Psychobiotics are also beneficial bacteria that have greater influence on the gut microbiome, the brain, the enteric nervous system and the immune system and can be used as better anxiolytic and antidepressant agents (Sarkar et al. 2016).

9.12 Conclusion

Prebiotics are selectively fermented non-digestible food ingredient that helps to maintain the host's health by stimulating the growth or activity of the gastrointestinal microbiota. The gut microbiota has strongly interacted with the central nervous system (CNS) through the gut-brain axis and maintains the health of an organism. Prebiotics play a major role in overcoming the abnormal psychiatric changes such as anxiety-like behaviour, impairment in learning and memory, depression, cognition, stress response, social behaviour, synaptic injuries and neurodegeneration. Non-digestible oligosaccharides with prebiotic properties are extensively used for stimulating the beneficial bacteria such as *bifidobacteria* and *lactobacilli*. Several studies have proven the effect of prebiotics in the central nervous system and its effective role in the management of psychiatric disorders in connection with the gut microbiota. Human microbiota is a potential diagnostic and therapeutic tool for many disorders affecting the gut-brain axis, and the function of prebiotics in the treatment of such diseases is now the most intensive area of study. The future works should be required for understanding the mechanisms involved in microbiota and brain functions, development of novel prebiotics for the regulation of altered gut microbiota and treatment of neurological diseases.

Acknowledgements We gratefully acknowledge the National Facility for Drug Development for Academia, Pharmaceutical and Allied Industries (NFDD) (Ref. No. VI-D&P/349/10-11/TDT/1, Date: 21.10.2010) supported by the Department of Science and Technology (Government of India), New Delhi. Ms. Sreeranjini Sukumaran Rajamma gratefully acknowledges the financial support (Senior Research Fellowship) received from the Indian Council of Medical Research, New Delhi (Ref. No. 45/41/2019-PHA/BMS, Date: 28.06.2019).

References

Ahmadi S, Nagpal R, Wang S, Gagliano J, Kitzman DW, Soleimanian-Zad S, Sheikh-Zeinoddin M, Read R, Yadav H (2019) Prebiotics from acorn and sago prevent high-fat diet-induced insulin resistance via microbiome-gut-brain axis modulation. J Nutr Biochem 67:1–13

Alexandra CR, Morgan M, Arlene JG, Lamiyah FM, Evan FF, Elara MR, Andrew JJ, Andrew TG, Benoit C, Nancy GF (2018) The microbiota influences cell death and microglial colonization in the perinatal mouse brain. Brain Behav Immun 67:218–229

An Q, Li C, Chen Y, Yang Y, Song R, Zhou LX, Tong A, Luo Y, Li J (2020) Scaffold hopping of agomelatine leads to enhanced antidepressant effects by modulation of gut microbiota and host immune responses. Pharmacol Biochem Behav 192:172910

Bagheri S, Heydari A, Alinaghipour A, Salami M (2019) Effect of probiotic supplementation on seizure activity and cognitive performance in PTZ-induced chemical kindling. Epilepsy Behav 95:43–50

Beilharz JE, Kaakoush NO, Maniam J, Morris MJ (2016) The effect of short-term exposure to energy-matched diets enriched in fat or sugar on memory, gut microbiota and markers of brain inflammation and plasticity. Brain Behav Immun 57:304–313

Brenner LA, Stearns-Yoder KA, Hoffberg AS, Penzenik ME, Starosta AJ, Hernandez TD, Hadidi DA, Lowry CA (2017) Growing literature but limited evidence: a systematic review regarding prebiotic and probiotic interventions for those with traumatic brain injury and/or posttraumatic stress disorder. Brain Behav Immun 65:57–67

Brownawell AM, Caers W, Gibson GR, Kendall CWC, Ringel Y, Slavin JL, Lewis KD (2012) Prebiotics and the health benefits of fiber: current regulatory status, future research and goals. J Nutr 142:962–974

Burokas A, Moloney RD, Dinanand TG, Cryan JF (2015) Microbiota regulation of the mammalian gut-brain axis. Adv Appl Microbiol 91:1–62

Burokas A, Arboleya S, Moloney RD, Peterson VL, Murphy K, Clarke G, Stanton C, Dinan TG, Cryan JF (2017) Targeting the microbiota-gut-brain axis: prebiotics have anxiolytic and antidepressant like effects and reverse the impact of chronic stress in mice. Biol Psychiatry 82:472–487

Chen R, Wu P, Cai Z, Fang Y, Zhou H, Lasanajak Y, Tang L, Ye L, Hou C, Zhao J (2019) Puerariae Lobatae Radix with chuanxiong Rhizoma for treatment of cerebral ischemic stroke by remodeling gut microbiota to regulate the brain-gut barriers. J Nutr Biochem 65:101–114

Chittim CL, Campo AMD, Balskus EP (2019) Gut bacterial phospholipase Ds support disease-associated metabolism by generating choline. Nature Microbiol 4(1):155–163

Davani-Davari D, Negahdaripour M, Karimzadeh I, Seifan M, Mohkam M, Masoumi SJ, Berenjian A, Ghasemi Y (2019) Prebiotics: definition, types, sources, mechanisms and clinical applications. Foods 8(3):92

Deloule V, Boisset C, Hannani D, Suau A, Gouellec AL, Chroboczek J, Botte C, Yamaryo-Botte Y, Chirat C, Toussaint B (2020) Prebiotic role of softwood hemicellulose in healthy mice model. J Funct Foods 64:103688

Desbonnet L, Clarke G, Traplin A, Sullivan OO, Crispie F, Moloney RD, Cotter PD, Dinan TG, Cryan JF (2015) Gut microbiota depletion from early adolescence in mice: implications for brain and behaviour. Brain Behav Immun 48:165–173

Dhar D, Mohanty A (2020) Gut microbiota and Covid-19 possible link and implications. Virus Res 285:198018

Doherty FD, Mahony SMO, Peterson VL, Sullivan OO, Crispie F, Cotter PD, Wigmore P, King MV, Cryan JF, Fone KCF (2017) Post-weaning social isolation of rats leads to long-term disruption of the gut microbiota-immune-brain axis. Brain Behav Immun 68:261–273

Elizabeth T, Nathalie J (2017) Introduction to the human gut microbiota. Biochem J 474:1823–1836

Fernandez DCL, Celia F, Julie S, Amandine E, Lucile C, Patrice DC, Sophie L, Nathalie C (2017) Impact of prebiotics on metabolic and behavioral alterations in a mouse model of metabolic syndrome. Brain Behav Immun 64:33–49

Fetissov SO, Averina OV, Danilenko VN (2019) Neuropeptides in the microbiota-brain axis and feeding behavior in autism spectrum disorder. Nutrition 61:43–48

Forsatkar MN, Nematollahi MA, Rafiee G, Farahmand H, Lawrence C (2017) Effects of the prebiotic mannan-oligosaccharide on the stress response of feed deprived zebrafish (Danio rerio). Physiol Behav 180:70–77

Fulling C, Lach G, Bastiaanssen TFS, Fouhy F, Donovan ANO, Ventura-Silva AP, Stanton C, Dinan TG, Cryan JF (2020) Adolescent dietary manipulations differentially affect gut microbiota composition and amygdala neuroimmune gene expression in male mice in adulthood. Brain Behav Immun 87:666–678

Golubeva AV, Joyce SA, Moloney G, Burokas A, Arboleya S, Flynn I, Khochanskiy D, Moya-Perez A, Sherwin E, Peterson V, Murphy K, Makarova O, Buravkov S, Hyland NP, Stanton C, Clarke G, Gahan CGM, Dinan TG, Cryan JF, Rea K (2017) Microbiota-related changes in bile acid & tryptophan metabolism are associated with gastrointestinal dysfunction in a mouse model of autism. EBioMedicine 24:166–178

Grasset E, Puel RBA, Charpentier J, Collet X, Christensen JE, Terce F (2017) A specific gut microbiota dysbiosis of type 2 diabetic mice induces GLP-1 resistance through an enteric NO-dependent and gut-brain axis mechanism. Cell Metab 25:1075–1090

Gronier B, Savignac HM, Miceli MD, Idriss SM, Tzortzis G, Anthony D, Burnet PWJ (2018) Increased cortical neuronal responses to NMDA and improved attentional set-shifting performance in rats following prebiotic (B-GOSs) ingestion. Eur Neuropsychopharmacol 28:211–224

He Y, Yang C, Wang P, Yang L, Wu H, Liu H, Guo Z, Li J, Shi H, Wu X, Hu Z, Qi M (2019) Child compound endothelium corneum attenuates gastrointestinal dysmotility through regulating the homeostasis of brain-gut-microbiota axis in functional dyspepsia rats. J Ethanopharmacol 240:111953

Houlden A, Goldrick M, Brough D, Vizi ES, Enart N, Martinecz B, Roberts IS, Denes A (2016) Brain injury induces specific changes in the caecal microbiota of mice via altered autonomic activity and mucoprotein production. Brain Behav Immun 57:10–20

Huang JF, Zhao Q, Dai MY, Xiao XR, Zhang T, Zhu WF, Li F (2020) Gut microbiota protects from triptolide-induced hepatotoxicity: key role of propionate and its downstream signalling events. Pharmacol Res 155:104752

Janerio MH, Ramirez MJ, Milagro FI, Martinez JA, Solas M (2018) Implication of trimethylamine N-oxide (TMAO) in disease: potential biomarker or new therapeutic target. Nutrients 10 (10):1398

Kao ACC, Burnet PWJ, Lennox B (2018) Can prebiotics assist in the management of cognition and weight gain in schizophrenia? Psychoneuroendocrinology 95:179–185

Kao ACC, Chan KW, Anthony DC, Lennox BR, Burnet PWJ (2019) Prebiotic reduction of brain histone deacetylase (HDAC) activity and olanzapine-mediated weight gain in rats are acetate independent. Neuropharmacology 150:184–191

Kazemi A, Noorbala AA, Azam K, Djafarian K (2019a) Effect of prebiotic and probiotic supplementation on circulating proinflammatory cytokines and urinary cortisol levels in patients with major depressive disorder: a double-blind, placebo-controlled randomized clinical trial. J Funct Foods 52:596–602

Kazemi A, Noorbala AA, Azam K, Eskandari MH, Djafarian K (2019b) Effect of probiotic and prebiotic vs placebo on psychological outcomes in patients with major depressive disorder: a randomized clinical trial. Clin Nutr 38(2):522–528

Li C, Cai YY, Yan ZX (2018) Brain-derived neurotrophic factor preserves intestinal mucosal barrier function and alters gut microbiota in mice. Kaohsiung J Med Sci 34:134–141

Li Y, Elmen L, Schmaltz R, Segota I, Bradley LM, Xian Y, Tinoco R, Ramer-Tait A, Feng Y, Zarecki R, Fujita Y, Ronai ZA, Long T, Munoz RRS, Peterson SN (2020) Prebiotic induced anti-tumor immunity attenuates tumor growth. Cell Rep 30:1753–1766

Loughman A, Ponsonby AL, Hely MO, Symeonides C, Collier F, Tang MLK, Carlin J, Ranganathan S, Allen K, Pezic A, Saffery R, Jacka F, Harrison LC, Sly PD, Vuillermin P (2020) Gut microbiota composition during infancy and subsequent behavioural outcomes. EBioMedicine 52:102640

Lyte JM, Proctor A, Phillips GJ, Lyte M, Wannemuehler M (2019) Altered Schaedler flora mice: an old defined microbiota animal model to study the new science of microbiota-gut-brain axis. Behav Brain Res 356:221–226

MacLaren R, Radcliffe RA, Van Matre ET, Robertson CE, Ir D, Frank DNF (2019) The acute influence of acid suppression with esomeprazole on gastrointestinal microbiota and brain gene expression profiles in a murine model of restraint stress. Neuroscience 3981:206–217

Mika A, Gaffney M, Roller R, Hills R, Bouchet AC, Hulen KA, Thompson RS, Chichlowski M, Berg BM, Fleshner M (2018) Feeding the developing brain: juvenile rats fed diet rich in prebiotics and bioactive milk fractions exhibit reduced anxiety related behavior and modified gene expression in emotion circuits. Neurosci Lett 677:103–109

Misiak B, Loniewski I, Marlicz W, Frydecka D, Rudzki L, Samochowiec J, Szulc A (2020) The HPA axis dysregulation in severe mental illness: can we shift the blame to gut microbiota. Prog Neuro-Psychopharmacol Biol Psychiatry 102:109951

Neri-Numa IA, Pastore GM (2020) Novel insights into prebiotic properties on human health: a review. Food Res Int 131:108973

Pan JX, Deng FL, Zeng BH, Zheng P, Liang WW, Yin B, Wu J, Dong MX, Luo YY, Wang HY, Wei H, Xie P (2019) Absence of gut microbiota during early life affects anxiolytic behaviors and monoamine neurotransmitters system in the hippocampal of mice. J Neurol Sci 400:160–168

Pavlova T, Vidova V, Bienertova-Vasku J, Almasi M, Klanova J, Spacil Z, Janku P (2017) Urinary intermediates of tryptophan as indicators of the gut microbial metabolism. Anal Chim Acta 987:72–80

Rincel M, Aubert P, Chevalier J, Grohard P, Oliveira CM, Helbling JC, Levy E, Chevalier G, Leboyer M, Eberl G, Laye S, Basso L, Capuron L, Vergnolle N, Neunlist M, Boudin H, Lepage P, Darnaudery M (2019) Multi-hit early life adversity affects gut microbiota, brain and behavior in a sex-dependent manner. Brain Behav Immun 80:179–192

Romo-Araiza A, Ibarra A (2020) Prebiotics and probiotics as potential therapy for cognitive impairment. Med Hypotheses 134:109410

Sabin KZ, Echeverri K (2020) The role of the immune system during regeneration of the central nervous system. J Immunol Regen Med 7:100023

Salberg S, Yamakawa G, Christensen J, Kolb B, Mychasiuk R (2017) Assessment of a nutritional supplement containing resveratrol, prebiotic fiber, and omega-3 fatty acids for the prevention and treatment of mild traumatic brain injury in rats. Neuroscience 365:146–157

Sarkar A, Dinan TG, Lehto SM, Cryan JF, Harty S, Burnet PWJ (2016) Psychobiotics and the manipulation of bacteria-gut-brain signals. Trends Neurosci 39(11):763–781

Savignac HM, Corona G, Mills H, Chen L, Spencer JPE, Burnet PWJ, Tzortzis G (2013) Prebiotic feeding elevates central brain derived neurotrophic factor, N-methyl-D aspartate receptor subunits and D-serine. Neurochem Int 63:756–764

Scott KA, Ida M, Peterson VL, Prenderville JA, Moloney GM, Izumo T, Murphy K, Murphy A, Ross RP, Stanton C, Dinan TG, Cryan JF (2017) Revisiting Metchnikoff: age-related alterations in microbiota-gut-brain axis in the mouse. Brain Behav Immun 65:20–32

Smith SM, Vale W (2006) The role of the hypothalamic-pituitary-adrenal axis in neuroendocrine responses to stress. Dialogues Clin Neurosci 8(4):383–395

Tarr AJ, Galley JD, Fisher SE, Chichlowski M, Berg BM, Bailey MT (2015) The prebiotics 3′ sialyllactose and 6′ sialyllactose diminish stressor-induced anxiety-like behavior and colonic microbiota alterations: evidence for effects on the gut-brain axis. Brain Behav Immun 50:166–177

Tuncay P, Arpaci F, Doganay M, Erdem D, Ergun H, Atabey D, Sahna A (2018) Use of standard enteral formula versus enteric formula with prebiotic content in nutrition therapy: a randomized controlled study among neuro-critical care patients. Clin Nutr 25:26–36

Vaghef-Mehrabany E, Maleki V, Behrooz M, Ranjbar F, Ebrahimi-Mameghani M (2020) Can psychobiotics “mood” ify gut? An update systematic review of randomized controlled trials in healthy and clinical subjects on antidepressant effects of probiotics, prebiotics and synbiotics. Clin Nutr 39(5):1395–1410

Wang Y, Li N, Yang JJ, Zhao DM, Chen B, Zhang GQ, Chen S, Cao RF, Yu H, Zhao CY, Zhao L, Ge YS, Liu Y, Zhang LH, Hu W, Zhang L, Gai ZT (2020) Probiotics and fructo-oligosaccharide intervention modulate the microbiota-gut brain axis to improve autism spectrum reducing also the hyper-serotonergic state and the dopamine metabolism disorder. Pharmacol Res 157:104784

Xiaohan Y, Weiyao L, Qingrong L, Hongmin Z, Zihui L, Xinjie Z, Lin Z, Xiang F (2020) Interactions between resveratrol and gut microbiota affect the development of hepatic steatosis: a fecal microbiota transplantation study in high-fat diet mice. J Funct Foods 67:103883

Xifra G, Esteve E, Ricart W, Fernandez-Real JM (2016) Influence of Dietary Factors on Gut Microbiota: The Role on Insulin Resistance and Diabetes Mellitus. In: Didac Mauricio (ed) Molecular Nutrition and Diabetes, A Volume in the Molecular Nutritional Series. Academic Press Publishing, pp 148–154

Xue Z, Wu C, Wei J, Xian M, Wang T, Chen M, Yang B (2019) An orally administered magnoloside A ameliorates functional dyspepsia by modulating brain-gut peptides and gut microbiota. Life Sci 233:116749

Xue L, Zou X, Yang XQ, Peng F, Yu DK, Du JR (2020) Chronic periodontitis induces microbiota-gut-brain axis disorders and cognitive impairment in mice. Exp Neurol 326:113176

Yang X, Yu D, Xue I, Li H, Du J (2020) Probiotics modulates the microbiota-gut-brain axis and improves memory deficits in aged SAMP8 mice. Acta Pharm Sin B 10:475–487

Zhang XJ, Yuan ZW, Qu C, Yu XT, Huang T, Chen PV, Su ZR, Dou YX, Wu JZ, Zeng HF, Xie Y, Chen JN (2018) Palmatine ameliorated murine colitis by suppressing tryptophan metabolism and regulating gut microbiota. Pharmacol Res 137:34–46

Zhang P, Yu Y, Qin Y, Zhou Y, Tang R, Wang Q, Wang H, Weston-Green K, Huang XF, Zheng K, Li X (2019) Alterations to the microbiota-colon-brain axis in high-fat-diet induced obese mice compared to diet-resistant mice. J Nutr Biochem 65:54–65

Zhang Z, He S, Cao X, Ye Y, Yang L, Liu H, Sun H, Wang J (2020) Potential prebiotic activities of soybean peptides Maillard reaction products on modulating gut microbiota to alleviate aging-related disorders in D-galactose-induced ICR mice. J Funct Foods 65:103729

Probiotic Mechanism to Modulate the Gut-Brain Axis (GBA)

10

Mrinal Samtiya, Tejpal Dhewa, and Anil Kumar Puniya

Abstract

Gut-brain axis (GBA) forms the complex network which connect gut-brain bidirectionally (two-way communication). The gut bacteria regulate the brain functions by modulating the gut microflora, which may advance the brain health. Several probiotics are used to improve various health issues of the brain, like stress, anxiety, and depression. "Psychobiotics" is the term used for those probiotics which affect the brain functions. The gut and the brain are linked through various biochemical signaling pathways between the enteric nervous system, which is present in the intestinal region, and the central nervous system, which includes the brain. The vagus nerve establish the primary connection between the gut and the brain; it is also the longest nerve present in the body. The ENS forms various neurotransmitters like the brain, i.e., dopamine, gamma-aminobutyric acid (GABA), and serotonin, which all have functions in regulating the mood. The bidirectional communication between the microbiota and the GBA mainly regulates through the signaling between the gut and the nervous system and from the nervous system to the gut via the nervous, endocrine, and immune systems. Several successful results were observed in brain-related disorders through using probiotics; however, the ultimate process of probiotic-aided improvement of neural health is not completely described, although many studies elucidated that metabolites produced by the different strains of probiotic like neurotransmitter could be the potential moderator which controls the GBA by regulating different signaling pathways of the neural system, the endocrine

M. Samtiya · T. Dhewa (✉)
Department of Nutrition Biology, Central University of Haryana, Mahendergarh, Haryana, India

A. K. Puniya
Department of Dairy Microbiology, National Dairy Research Institute, Karnal, Haryana, India

R. Z. Sayyed, M. Khan (eds.), *Microbiome-Gut-Brain Axis*,
https://doi.org/10.1007/978-981-16-1626-6_10

system, and the immune system. This chapter summarizes the possible probiotic mechanisms concerning the GBA.

Keywords

Probiotics · Gut-brain axis · Psychobiotics · Neurotransmitters · Immunomodulation · Gut microbiota

10.1 Introduction

Probiotics are defined as the "living microorganisms that confer a benefit to the host health when taken into an appropriate amount" according to the FAO/WHO (2002) guidelines. The common genera of the probiotic strains are: *Leuconostoc*, *Lactobacillus* (*L. casei*, *L. johnsonii*, *L. paracasei*, *L. acidophilus*, *L. fermentum*, *L. rhamnosus*, *L. brevis*, *L. plantarum*, and *L. delbrueckii*), *Bifidobacterium* (*B. longum*, *B. animalis*, *B. bifidum*, *B. boulardii*, *B. breve*, *B. adolescentis*, *B. infantis*), *Pediococcus*, *Saccharomyces bayanus*, *Streptococcus*, *Enterococcus*, and *Bacillus* (Fijan 2014). Probiotics have shown various attributes or properties, which make them superior to other microorganisms, such as digestive improvement, absorption of nutrients, maintaining microbiota, immunomodulatory effect, and complement for metabolic conditions. However, several studies conclude that mental health and brain functions of the host also improve through supplementation of traditional functional foods that contain probiotics (Messaoudi et al. 2011; Gareau et al. 2011). Probiotics modulate the physiology of the host, but the exact mechanism of modulation is not fully explained; they may regulate the functions through altering the immune system of the host (Bermudez-Brito et al. 2012). Previous evidence specified that Kimchi isolates *Weissella cibaria* WIKIM28 improves the atopic dermatitis signs by modifying the functions of regulatory T cell and suppressing Th2 (allergic) action (Lim et al. 2017). This GBA concept is not so new; bidirectional interaction between the gut and the brain has already been recognized. Bidirectional concept concludes that the brain regulates the gastrointestinal tract (GI) by modulating secretion, blood flow, absorption, and motility; synchronously, the gastrointestinal tract also affects the brain functions (Grenham et al. 2011). Probiotic bacteria create a healthy environment in the gut by maintaining the microbiota of the gastrointestinal tract and also regulate the host immune system. Probiotic consumption restores the beneficial bacterial concentration by controlling the microflora composition in the gastrointestinal tract (Choi et al. 2015; Mountzouris et al. 2007). Desired action and homeostasis of the body are controlled by the brain. Significant changes and defects in the brain functions lead to various kinds of emotional and physiological damage (Dantzer et al. 2008; Qureshi and Mehler 2013). The mental loss is not only linked with degeneration in brain functions but also associated with the improper functioning of the immune system as well as variations in the microbiota. Microorganisms that reside in the human gastrointestinal region have been denoted as intestinal microflora or gut microbiota.

The symbiotic microorganisms have a multifaceted connective network which is closely connected with the host system and influences their physical functionality. Probiotics are gaining much attention recently from the perspective of nervous function and mental health because of the reason that they modulate the GBA by altering the gut microbiota toward the beneficial state (Bravo et al. 2012). The framework of the GBA comprises the enteric nervous system, neuroendocrine system, autonomic nervous system, central nervous system, immune system, and gastrointestinal tract (Kim et al. 2018). Gut microbiota producing neurotransmitter metabolites regulates the various functions of the brain (Dinan et al. 2015). A previous study by Gareau et al. (2011) reported that gut microbiome and supplementation of probiotics improve the memory or brain-related impairment. The infection-mediated stimulation of cytokines changed the function of the brain and directs the growth of several irregularities which are associated with the behavior (Deverman and Patterson 2009). Numerous previous researches have discovered the association of probiotics in neuroscience and cognition (Bravo et al. 2011; Selhub et al. 2014). This chapter contained the information related to the role of intestinal microbiota in the bidirectional interaction between the gut and the brain as well as considers the recent discoveries related to the modulation of brain functions through the use of probiotics.

10.2 Probiotics and GBA

A complex communication system revealed the gut-brain crosstalk, which is primarily responsible for balancing the maintenance of the gastrointestinal tract and regulating emotional and cognitive functioning. Even though in the last decades, the gut and brain axis has been one of the foremost focus of the research and however, regulation of the signaling pathway by the effective functioning of microbiota is quiet at its initial stages. Table 10.1 shows the possible mechanism(s) of bidirectional communication between GBA. Bravo and colleagues conducted a study that is one of the most prominent studies in this area, which exposed that the tenth cranial nerve (vagus) is unique among the key pathways of the regulation amid the GBA. A study conducted in vagotomized mice shown that continuous administration of *Lactobacillus rhamnosus* JB-1 exert benefits by decreasing anxiety and

Table 10.1 Possible mechanism(s) to modulate the GBA (Carabotti et al. 2015)

Gut microbiome to brain	Brain to gut microbiome
Bacterial metabolites and mucosal immune regulation	Modification in motility
Enteric sensory regulations	Immune function regulation
Expression of neurotransmitters such as serotonin and gamma-aminobutyric acid (GABA) and production of neurotrophic factor (BDNF)	Regulation of biofilm and mucus production
Intestinal epithelial barrier protection and regulation of tight junctional integrity	Modification of intestinal permeability

depressive behaviors in these mouse models. In vagotomized mice, probiotics change the GABA receptors expression due to reducing nervousness and depression in particular regions that entailed in pathogenesis (Bravo et al. 2011). Likewise, the supplementation of *Bifidobacterium longum* in the dextran sodium sulfate-induced chronic colitis animal model was ineffective to decrease the anxiety behavior in mice that experienced vagotomy before colitis initiation (Bercik et al. 2011). A key query in the research area is whether stress-associated ailments can be restored by directing the gut-mind axis. To this end, numerous reports have proven that food that modifies the microbiota, prebiotics, and probiotics can diminish stress-associated behavior and HPA activation (Davis et al. 2017). Probiotics works mainly through modifying intestinal microbiota composition, maintaining intestinal epithelial barrier integrity, inhibiting bacterial translocation and controlling local inflammatory response through immune system regulation that presents in the gastrointestinal tract (Rios et al. 2017). Schnorr and Bachner (2016) reported that food comprising of probiotics, at the third week could increase the general time of sleep, which could lower down the chronic anxiety pressure, in the second week it can reduce Beck Anxiety Inventory score, and suitable variation in microbiota after 2 to 3 weeks, which increase in Lactobacillales and Bacteroides members, and decrease the population of *Clostridium* and Actinomycetes. All the outcomes signify that probiotic consumption could recover and alter the microbiota species, which may get rid of anxiety-like activities (Cepeda et al. 2017). Previous studies confirmed that gut microbiota could be altered through the consumption of probiotics and/or antibiotics, which also supported that microbiota could affect the GBA. All these investigations verify that microbiota influences mind neurochemistry by regulating anxiety and HPA machine (Saulnier et al. 2013). Similarly, probiotics decrease the stress-stimulated release of cortisol, anxiety, and depression-like behavior (Bravo et al. 2011). Moreover, probiotics associated VSL#3 alter the microbiota composition, resulting in an increase in brain-derived neurotrophic factor BDNF expression and reduction of age-linked adjustments in the hippocampus (Distrutti et al. 2014). Evidence illustrates that microbiota includes the vagus nerve during bidirectional communication with the brain, which transfers information from the luminal milieu to the central nervous system. Indeed, vagotomized mice no longer show neurochemical and behavioral effects, figuring out that vagus nerve acts as a crucial modulatory pathway of communication between the microbiota and the mind (Bravo et al. 2011). An earlier evidence stated that *L. farciminis* regulates various activities like hindering permeability, endotoxemia, functioning of HPA axis, stress-induced neural inflammation, and additionally producing a beneficial impact on the mucosal barrier (Kelly et al. 2015). Microbiota may additionally cooperate with the GBA via distinctive functioning mechanisms, the important one possibly modulating the epithelial barrier of the intestine, whose perturbation could affect the entire essential sections. Previously, a water avoidance stressed mice model study reported that probiotic pretreatment improve the stress and also associated with reestablishment of intestinal epithelial barrier and protection of tight junctions (Ait-Belgnaoui et al. 2014).

Additionally, the immune system is the other competent pathway through which the intestine and the brain are regulated to each other. Several probiotic strains which belong to the *Bifidobacterium* and *Lactobacillus* groups revealed the proficiency to re-establish the plasma concentration of adrenocorticotropic hormone (ACTH) and corticosterone along with the reduction of hypothalamic corticotropin-releasing factor (CRF), tumor necrosis factor (TNF)-α levels, and interleukin (IL)-6 which were changed both by chronic stresses and subsequently by increased colonic permeability (Laval et al. 2015; Ait-Belgnaoui et al. 2012; Gareau et al. 2007; Smith et al. 2014). Furthermore, previous evidence signifies that probiotics can have interaction with the GBA through the modulation of neurotransmitter signaling. Indeed, it has been confirmed that *Bifidobacterium infantis* regulate the serotonin (5-hydroxytryptamine) level by increasing the concentration of tryptophan in plasma (Desbonnet et al. 2010). Metabolites produced from bacterial origin confirm the additional pathway of interaction which joins the brain and the gut. Short-chain fatty acids (SCFAs) are one of the main derivatives of bacterial metabolism along with acetic, propionic, and butyric acids. It has been confirmed in rodent's models that SCFAs regulate the histone deacetylases which critically influence memory and learning (Levenson et al. 2004; Ferrante et al. 2003; Silva et al. 2020). Even though intensive understanding of signaling brought by way of SCFAs remains missing, it is previously documented that SCFAs link to G-protein-coupled receptors (GPCR). GPR43 and GPR41 have largely elucidated receptors of SCFA; in addition to hydrocarboxylic acid receptor GPR109a/HCAR2 and GPR164, expressed in a huge collection of cells from the intestinal mucosal layer to the nervous and immune systems (Mohajeri et al. 2018; Bolognini et al. 2016). These receptors' activation can be altered meaningfully according to the cells on which these are expressed, such as SCFAs regulates the stimulation of the glucagon-like peptide 1 and peptide YY secretion by connecting through their receptors which presented on enteroendocrine cells (Cherbut et al. 1998; Greiner and Bäckhed 2016). SCFAs regulate the metabolic homeostasis as well as control the energy expenditure (body) by binding particularly to GPR41 (G-protein-coupled receptors); it has been revealed that probiotics also alter the levels of SCFAs in healthy adults (Kimura et al. 2011; Ferrario et al. 2014). Besides, it has been previously revealed that numerous strains of bacteria can alter the neurotransmitter precursor concentration in the gastrointestinal tract and also individually produced several neurotransmitters, including gamma-aminobutyric acid, serotonin (5-HT), noradrenaline (NA), and dopamine (Sherwin et al. 2018; Fung et al. 2017; Calvani et al. 2018). These neurotransmitters can potentially impact several cerebral functions and microglial activation (Abdel-Haq et al. 2019). Figure 10.1 Depicting bidirectional communications between GBA.

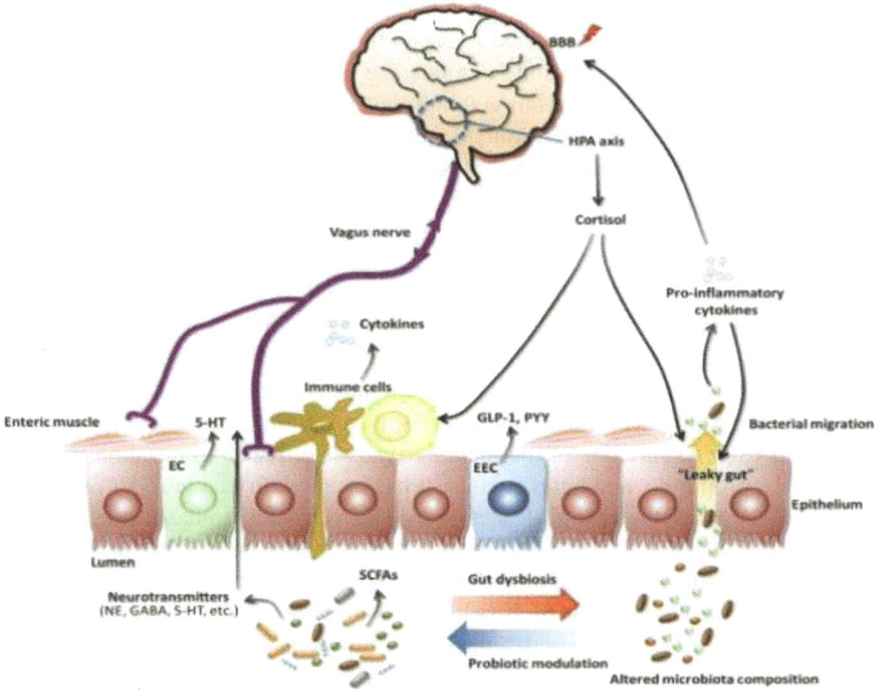

Fig. 10.1 Bidirectional communications between GBA (Kim et al. 2018)

10.3 Bidirectional Communications Within the GBA

The gut and the brain bidirectionally control each other via a number of mechanisms and pathway including neural, neurotransmitters, neuroendocrine and metabolites, and immunological markers. Here, various interaction pathways are highlighted, which are useful for the critical functioning of the GBA.

10.3.1 Possible Mechanisms Through Which Probiotics Can Modulate the GBA

10.3.1.1 Neural Regulatory Interactions Between the Gut and the Brain

The enteric nervous system (ENS) is the primary center of interaction, which connects the intestinal microbiota of the host and the nervous system. The ENS concluded as "the second brain" due to its exceptional capability and neuronal complexity on the same level as the brain and has a distinct unit to control the linked intestinal movements and the immune system's reactions (Furness 2012; Breit et al. 2018). Gut microbiota exchanges signal bidirectionally with the brain through the ENS by using humoral and neural paths (Luan et al. 2019). Neural information transmits to the brain from the interior organs as well as from the gut through the parasympathetic vagus afferents (Breit et al. 2018). Intestinal microflora controls the thresholds of the electrophysiological neurons in the ENS (Sarkar et al. 2016). For instance, calcium-dependent potassium channels in neurons are stimulated by *Lactobacillus reuteri* strain within the colon myenteric plexus of the rat model (Kunze et al. 2009). *B. longum* (NCC3001) produced metabolic compounds that provoked the action potential reduction within the myenteric neurons, followed by electrical stimulation (Bercik et al. 2011). Another neuronal pathway that is connected by the vagus (tenth cranial nerve) nerve starts from the brainstem and ends up within the abdominal region. It forms a bidirectional complex between the gut and the brain; the vagus nerve plays both efferent and afferent functions. About 80% of sensory vagal fibers are present, which transmit the information between the body's organs and the central nervous system (CNS) (Cryan and Dinan 2012; Thayer and Sternberg 2009). The vagus nerve regulates numerous essential features, consisting of bronchial contraction, motility of the gastrointestinal tract, and coronary heart rate. Several recognized consequences have shown that different strains of probiotics and gut microbiome depend on the vagus nerve activity (Goehler et al. 2008; Bercik et al. 2011; Bravo et al. 2011). Studies confirmed that the vagus nerve works as a critical modulator which regulates the *L. rhamnosus* signaling pathways. Due to the *L. rhamnosus* intrusion, enteric neurons and vagus nerve function amplified in the gastrointestinal tract of mice (Perez-Burgos et al. 2013, 2014). Furthermore, based on experimentations, the vagus nerve becomes recognized as a primary modulator within the gut-brain interaction (bidirectional) pathway in vagotomized mice (Bravo et al. 2011).

10.3.1.2 Immune-Facilitated Communication Between the Gut and the Brain

Microbiota residing in the gastrointestinal tract can impact the immune response; this forms an indirect connection which implies the link within the gut microbiome and the brain (Macpherson and Uhr 2004; Bengmark 2013). Lymphoid tissues present in the human gastrointestinal tract form the most important organ of the immune system; it entailed nearly 70% of the entire immune system of the body (Vighi et al. 2008). When pro-inflammatory cytokines are peripherally administrated in rodents, it could stimulate a range of anxiety behaviors, including unstable sleep, loss of appetite, and exploratory behavior suppression, it jointly denoted as sickness behavior (Bilbo and Schwarz 2012). Probiotic microbes endorse that immunoregulatory consequences arise due to regulatory T cells and anti-inflammatory cytokine (IL-10) introduction (Dinan et al. 2013). A previous study suggested that oral supplementation of *B. infantis* (35624) improved the expression of IL-10 in the peripheral blood of human beings (Bilbo and Schwarz 2012). Numerous gastrointestinal and GALT cells, consisting of macrophages, T cells, and dendritic cells, can easily move through the blood-brain barrier and influence the brain neurons and glial cells (Diamond et al. 2011). Moreover, earlier research suggested that gut microbiota regulates the immune response which controls the microglial activation and homeostasis in the CNS of germ-free mice (Erny et al. 2015). Systemic circulation of chemokines, immune factors, and cytokines regulates the functioning of the brain through the use of circumventricular organs and vagus nerve (Hosoi et al. 2002). Inside the nervous system, pro-inflammatory cytokines can generate neural inflammation, by this means affecting the BBB permeability (McCusker and Kelley 2013). Breaching of BBB permeability leads to the infusion of immune cells, reactive gliosis, and deterioration of inflammatory reactions, which ultimately cause neurodegeneration (Obermeier et al. 2013). A research study by Palomar et al. (2014) exhibited that *L. casei* strengthened the immune response of chronic stressed adult mice by enhancing the production of IgA cells, CD4+ cells in the small intestine and luminal secretion of IgA, and decreasing the levels of IFN-γ (Palomar et al. 2014).

10.3.1.3 GBA Regulation by Neuroactive Compounds and Metabolites

Microflora of intestinal origin has the capacity of making a variety of neurotransmitters, neuroactive compounds, and metabolites. GABA, serotonin, catecholamine, and acetylcholine are the neurotransmitters and neuroactive compounds which are produced by gut microbiota. Without a doubt, ENS neurons and enterochromaffin cells of the gut mucosa produce nearly 95% of 5-HT within the body. Besides the point, 5-HT is also intricate in the gastrointestinal secretion, contraction, and relaxation of smooth muscle as well as pain sensitivity (Costedio et al. 2007; McLean et al. 2007), while existing evidence suggested that probiotics communicate with the gut-brain axis through the instruction of neurotransmitter signaling, although signaling pathways (5-HT) are associated with the regulation of the cognition and mood-related behavior. Especially, it has been already proved that *B. infantis* control the serotonin (5-HT) level through elevating tryptophan concentration in plasma (Desbonnet et al. 2010). Gamma-aminobutyric acid can be formed by numerous strains of *Lactobacillus* and *Bifidobacterium*; similarly, norepinephrine can be produced by several species of *Bacillus*, *Escherichia*, and *Saccharomyces*; likewise, dopamine can be formed by *Serratia*, *Bacillus*, *Lactobacillus*, *Escherichia*, and *Lactococcus*. Indirectly gut microbiome controls the manufacturing of neurotransmitters via modulating available neuro-active chemicals precursors or by stimulating the neuroendocrine and enteroendocrine cells of the host (Desbonnet et al. 2010; Yano et al. 2015). Probiotics synthesize neurotransmitters as well as excite the host cells to produce these neurochemicals which could be used as delivery vehicles for transferring neuroactive compounds (Lyte 2011). Furthermore, bacteria produced short-chain fatty acid (SCFA) neuroactive metabolites such as propionate, butyrate, acetate, and lactate (Horn and Klein 2013; Overduin et al. 2013). SCFAs communicated with the nervous system directly, thereby stimulating sympathetic neurons, and can also cross via the blood-brain barrier (BBB), in this manner stimulating behavior and neural signaling (Kimura et al. 2013; Frost et al. 2014; Ríos-Covián et al. 2016). Butyrate provokes the hypothalamic-pituitary-adrenal (HPA) axis and also acts as an active epigenetic modifier by hindering the activity of histone deacetylases (HDACs) (Gagliano et al. 2014; Stilling et al. 2014). Especially, BDNF acts like neurotrophic constituents which are associated with mood-related concerns; as a result, increasing BDNF concentration can be a proficient intervention. Several research studies address this problem; for example, BDNF level increases in the hippocampus after administration of *Bifidobacterium*; equally, *Lactobacillus* residing in the gastrointestinal tract can increase the BDNF concentration (Rios et al. 2017). Previous studies established that certain strains of gut microbiota, for instance, *L. acidophilus*, *Streptococcus*, *Candida*, and *B. infantis*, secrete several neurotransmitters (catecholamine, GABA, glycine, 5-HT) which have well-known therapeutic attributes in regulating psychological illness or controlling endocannabinoid expression. Gut microbiota produces neuroactive molecules that can influence the nerve signals and affect the neuropsychiatric parameters along with sleep, urge for food, mood, and cognition (Kali 2016).

10.3.1.4 Central Nervous System Regulates Gut Microbiota

The central nervous system can transform the composition of the gut microbiota as well as overall biomass by regulating satiety. Moreover, the nervous system plays a significant role in regulating the working of the gut; for example, gastrointestinal movement as well as acid, bicarbonate, and mucus secretion all have a prominent role in biofilm and mucous layer protection (Rhee et al. 2009). Under the regulation of the brain, the direct impact is aided by the secretion, neuron signaling molecules, enterocromaffin, and immune cells, affecting the microbiota. Communication between the microorganisms and the CNS is mainly based on the receptors of neurotransmitters which are present in bacteria. Numerous studies stated that bacteria contain the binding sites for the host enteric neurotransmitters and can stimulate the role of microbiota components, promoting the susceptibility to inflammation and infection stimuli (Hughes and Sperandio 2008). The brain may affect the composition and characteristics of microbiota by changing the intestinal permeability, due to which bacterial antigens pass through the breached epithelium and excite an immune reaction in the mucosal part of the intestine. As a result of acute stress, colonic paracellular penetrability is increased, which comprises interferon-γ overproduction and reduction in ZO-2 (zona occludens) as well as occludin mRNA expression (Demaude et al. 2006). Host neurons cause direct impact by producing neurotransmitters and neurotransmitter receptors, which are displayed on intestinal microbiota. Neurotransmitters which are prepared by hosts affect the function of microbial components through binding, thereby stimulating susceptibility for infection and inflammation stimuli (Carabotti et al. 2015). The central nervous system also directly or indirectly controls the release of antimicrobial proteins, cytokines, and signaling molecules into the intestinal lumen by the enteroendocrine cells, Paneth cells, neurons, and immune cells which are responsible for the secretion of α-defensin. Microbial persistence and their surrounding milieu are influenced by these secreted products (Carabotti et al. 2015). Gut microbiota composition is also controlled by the nervous system by altering the penetrability of the epithelial barrier, thereby permitting the diffusion of bacteria and enabling host and microbiome communication in the mucosa. Under anxiety condition, the HPA axis discharges a stress hormone known as cortisol; by this means, it regulates the intestinal motility and immune responses through cells, secretory immunoglobulin A, and cytokines (Kim et al. 2018).

10.4 Modulation of the GBA by Probiotics

The gut microbiota-brain axis can interact through signaling which followed several different mechanisms. Gut microbiota can form bioactive peptides, comprising neurotransmitters, branched-chain amino acids, several intestinal hormones, and SCFAs, and also cause secondary bile acid alteration. Lactate, butyrate, acetate, and propionate are the SCFAs that are easily entered into the circulatory system, and this may be the possible route through which it would be interacting with the brain (Sarkar et al. 2016). Several health benefits were claimed by probiotics related to psychological and physical disorders. Probiotics positively modulate the gut microbiota as well as promote the nourishment of the commensal microbiota composition (Chaiyasut and Sivamaruthi 2018). In vivo and clinical studies are mentioned below which show that probiotic interaction and possible mechanisms are followed to maintain the GBA.

10.4.1 In Vivo Studies

Previous studies reported that depressive behaviors decreased when *L. rhamnosus* was orally supplemented in healthy mouse (Bravo et al. 2011) as well as chronically stressed mouse models (McVey Neufeld et al. 2018). A study reported that *L. rhamnosus* improved the behavior and physiology of vagotomized rats (Bravo et al. 2011). This report authenticates that the vagus nerve is the key mediator in the *L. rhamnosus* signaling pathway. In the gastrointestinal tract of mice, *L. rhamnosus* intensified the rate of enteric neurons and vagus nerve firing (Perez-Burgos et al. 2013, 2014). Earlier studies indicated that *L. rhamnosus* interacted with the brain through the signaling which followed neural pathways, and it might also manifest an antidepressant effect via impacting the HPA axis and the central GABAergic system. In a mouse model, intake of *L. rhamnosus* changed the mRNA expression of GABA-A and GABA-B receptors and however decreased the depression- and anxiety-like behaviors. Additionally, these properties were reliant on the undamaged vagus nerve (Bravo et al. 2011). In rats, the consumption of *L. casei* reduced both actions as well as the amount of corticotropin-releasing factor (CRF-expressing) cells in the paraventricular nucleus (PVN) while stimulating the vagus afferents (Takada et al. 2016). Intragastric administration of *L. casei* downregulated the activity of the sympathetic efferent in the adrenal glands and liver, and this effect did not occur after vagotomy (Tanida et al. 2014). Another study reported that *L. brevis*-fermented milk enhanced the GABA concentration when given to depressed rats and also revealed an antidepressant potency on the same level with fluoxetine (Ko et al. 2013). In addition to this, the consumption of *L. brevis* enhanced sleep duration in mice by producing GABA content (Han et al. 2017). *L. reuteri* treatment improved the behaviors of depression in chronic stress (Marin et al. 2017) and immobilization stress models of mice (Jang et al. 2019). Antidepressant effects are attenuated by the administration of KYN, which specifies that *L. reuteri* improves depression by decreasing the plasma KYN levels. In chronically stressed mice, supplementation of *L. plantarum* reduced depressive-like behavior (Liu et al. 2016; Dhaliwal et al. 2018). Additionally, earlier research explained the mechanism of *L. reuteri* involved in the regulation of indoleamine 2,3-dioxygenase, a rate-limiting enzyme present in immune cells that catabolizes tryptophan to kynurenine (Réus et al. 2015). When *L. rhamnosus* JB-1 treatment was given to mice, it enhanced the glutamate, gamma-aminobutyric acid, and *N*-acetyl aspartate levels in the brain, demonstrating that brain activity could be controlled by probiotics through regulating the metabolic pathways of probiotics, and it also put forward a probiotic approach into remedies for nervous ailments (Janik et al. 2016). Probiotics regulate the inflammation by improving the blood-brain barrier (BBB) integrity and by reducing the neuroinflammation in patients with neural problems (Felger and Lotrich 2013; Miller et al. 2013). Furthermore, expression of *N*-methyl-D-aspartate (NMDA) receptor and brain-derived neurotrophic factor (BDNF) enhanced in germ-free mice which were supplemented with *B. infantis*; in the hippocampus and cortex region of the brain, these molecules perform an essential role in regulating the learning and memory functions and reduced germ-free mouse expressions (Sudo et al. 2004). A probiotic strain, *L. acidophilus*, which is isolated from a normal human gastrointestinal tract when orally supplemented in the DSS-induced mouse model of colitis inhibited the colitis-allied reaction of the IL-23/T17 axis and also decreased the secretion of cytokines (pro-inflammatory) (Chen et al. 2015). When female wistar rats were administered with *L. farciminis* (10^{11} CFU/day) for 15 days before partial restraint stress. Results showed that *L. farciminis* supplementation inhibited the stress, decreased the permeability in the colon, and phosphorylation of colonocyte myosin light chain compared to the control group (Ait-Belgnaoui et al. 2006). *B. longum* (NCC3001) affects the vagal integrity when administered in the DSS-induced colitis model of mice, without altering the immune responses as well as brain-derived neurotrophic factors. The status of histopathology and functions of myeloperoxidase had been no longer affected by the probiotic intervention (Bercik et al. 2011). In rats' model, anxiolytic-like activity was investigated for probiotics *L. helveticus* (R0052) and *B. longum* (R0175) (PF). The conditioned defensive burying test confirmed that 14 days of probiotic formulation (PF) supplementation reduced anxiety-like behavior in rats (Messaoudi et al. 2011). *Lactobacillus* metabolites (LM) (0.5–1.0%) (such as microelements, lactate, polypeptides, enzymes, organic and amino acids) were administered in rats and also exposed toward ratiometric Ca^{2+} imaging. The outcomes concluded that Ca^{2+} discharge and absorption were improved when LM was supplemented continuously, due to which brain intracellular signaling was stimulated, and these cognitive and psychological functions were improved (Sobol and Belostotskaya 2016). In Table 10.2, recent in vivo studies of probiotic strains which regulated the brain-related functions are mentioned.

Table 10.2 Probiotic strains which regulated the brain-related functions in in vivo trials

Probiotics	Model	Effects	References
Bifidobacterium infantis	Specific pathogen-free (SPF) and gnotobiotic mice	Normalized stress response	Sudo et al. (2004)
	Male adult C57BL/6J mice	Antidepressant effect, increased 5-HT and 5-HTP levels, decreased anxiety, increased BDNF levels	Tian et al. (2019)
Bifidobacterium longum	Male AKR mice	Normalized anxiety behavior and brain-derived neurotrophic factor (BDNF) mRNA	Bercik et al. (2010)
Faecalibacterium prausnitzii (ATCC 27766)	Sprague-Dawley male rats	Decreased anxiety and depression, increased short-chain fatty acids (SCFAs)	Hao et al. (2019)
Lactobacilli (PP)	Rat	Increased Ca^{2+}, stimulating the intracellular signaling, improvement of psychological parameters and cognitive functions of the brain	Sobol and Belostotskaya (2016)
Lactobacillus brevis	Depressed Sprague-Dawley male rats	Decreased depression	Ko et al. (2013)
Lactobacillus farciminis	Female Wistar rats	Prevented stress-induced hypersensitivity	Ait-Belgnaoui et al. (2006)
Lactobacillus helveticus MCC1848	C57BL/6J male mice	Improved anxiety- or depressive-like behaviors	Maehata et al. (2019)
Lactobacillus helveticus R0052 and *Bifidobacterium longum* R0175	Wistar rats	Reduced anxiety-like behavior	Messaoudi et al. (2011)
Lactobacillus helveticus R0052	Mice	Decreased anxiety-like behavior	Ohland et al. (2013)
Lactobacillus kefiranofaciens CGMCC2809 (ZW3)	Mice	Regulation of immune system-mediated biochemical disorders in the hypothalamic-pituitary-adrenal axis and tryptophan metabolism caused by stress, improved depression-like behavior	Sun et al. (2019)
Lactobacillus paracasei K71	Senescence-accelerated female SAMP8 mice	Increased protein expression of BDNF, decreased 5-HT-degrading enzymes, and increased 5-HT levels in brain tissues and serum	Corpuz et al. (2018)
Lactobacillus paracasei PS23	C57BL/6J mice	Reversed chronic corticosterone-induced anxiety- and depression-like behaviors	Wei et al. (2019)

(continued)

Table 10.2 (continued)

Probiotics	Model	Effects	References
Lactobacillus paracasei PS23	Senescence-accelerated male and female SAMP8 mice	Increased TNF-α, decreased IL-10, delayed age-related cognitive decline	Huang et al. (2018)
Lactobacillus plantarum MTCC 9510	Swiss albino male mice	Improved gut and blood-brain barrier integrity, prevented stress	Dhaliwal et al. (2018)
Lactobacillus plantarum PS128	Mice	Ameliorated anxiety- and depression-like behaviors and modulated neurochemicals related to affective disorders	Liu et al. (2016)
Lactobacillus reuteri	Male BALB/cJ, C57BL/6N, and C57BL/6J mice	Reduction in stress-induced increased abdominal IDO expression, decreased stress-stimulated increase in KYN levels, decreased depression	Marin et al. (2017)
Lactobacillus reuteri NK33 and *Bifidobacterium adolescentis* NK98	C57BL/6 male mice	Suppressed anxiety/depression, NF-κB activation was suppressed	Jang et al. (2019)
Lactobacillus rhamnosus	BALB/c male mice	GABAergic system modulation, depression and anxiety reduction	Bravo et al. (2011)
	BALB/c male mice	Reduced depressive behavior	McVey Neufeld et al. (2018)

10.4.2 Clinical Studies

During and after the supplementation of probiotic *L. casei* Shirota (6.5 × 10^9 CFU) containing yogurt, healthy human volunteers' cognition behavior and mood were estimated at the reference line. The obtained results of the study directed that probiotic administration regulates the stress, tension, and depressive-like behaviors of the volunteers. In general, probiotic yogurt advanced good behavior (Benton et al. 2007). During clinical cases, *L. casei* administration reduced the salivary cortisol levels, stress feelings, and occurrence of intestinal and flu-related signs in stressed peoples (Kato-Kataoka et al. 2016; Takada et al. 2016). These research studies indicate that *L. casei* inhibits the overactivity of the HPA axis through the regulation of the vagus nerve and successively decreases the stress-related feelings and ailments; in addition to this, in vitro, GABA is also produced (Oleskin et al. 2014). In patients with major depressive disorder (MDD), Probiotic Sticks comprising *L. helveticus*, along with *B. longum*, decreased the depressive-like behavior in addition to clinical depression (Kazemi et al. 2019). Previous reports indicated that *L. helveticus* may improve cognition by regulating the activity of the central nervous system in addition to the HPA axis and also decrease the depression-like behavior by modulating BDNF expression and 5-HT system (Liang et al. 2015). Mixed species of probiotics that also contained *L. casei* when given to MDD patients can reduce clinical depression symptoms and depressive-like signs (Akkasheh et al. 2016). Probiotic supplementation containing 3 billion CFU of *L. helveticus* R0052 and *B. longum* R0175 (PP) improved the psychological depression when given to human volunteers; it is recorded by measuring the concentration of urinary free cortisol along with Hospital Anxiety and Depression Scale and Hopkins Symptom Checklist. During the study duration, no adverse effects were recorded (Messaoudi et al. 2011). Acute psychological stress is certainly linked with cold or flu. The study had found that *L. helveticus* (R0052), *B. bifidum* (R0071) and *B. infantis* (R0033) significantly decreased the symptoms of cold or flu in academically stressed healthy students when supplemented for six weeks. However, students who consumed *Bifidobacterium* spp. showed higher protective results than other groups of students (Langkamp-Henken et al. 2015). Random groups of healthy petrochemical employees were administered for 6 weeks, with yogurt containing *B. lactis* (BB12) and *L. acidophilus* LA5 or capsule containing probiotics (*B. breve, L. Rhamnosus, L. casei, Streptococcus thermophiles, L. bulgaricus L. acidophilus, B. longum*), or conventional yogurt or control. Furthermore, individual's mental health was recorded through a general health questionnaire, stress scale scores, and depression anxiety. This study concluded that petrochemical workers' psychological state was improved after probiotic-containing yogurt and multispecies capsule administration. In contrast, no health stimulating role was recorded in workers who consume conventional yogurt (Mohammadi et al. 2016). After supplementation of *L. casei* strain (10^9 CFU/day) Shirota for eight weeks, improved the cortisol level, and stress level in academically stressed students (Takada et al. 2016). In Table 10.3, clinical studies in which brain-related functions were regulated by probiotics are listed.

10.5 Conclusions

Nowadays, the GBA concepts have been enthusiastically explored, and many research studies have established that gut microbiota modification and probiotics administration have overcome several brain-related disorders. The disproportion of the gut microflora composition can result in several illnesses. Probiotics can regulate the gut microbiota composition, which could advance the bacterial population, intestine epithelium barrier characteristic, and cytokine production. By examining previous studies, it would be concluded that the gut microbiota and the brain regulated bidirectionally to form the GBA using these links, i.e., microbial metabolites, neuroendocrine and neurotransmitters, and neural immunomodulation. The exact connection between gut microbiota, probiotics, and brain illnesses is not entirely understood in clinical research. According to previous research, it is found that supplementation of probiotics can positively control the gut microflora and brain functions and also regulate the host immune system. In addition, probiotics maintain the microbiota-GBA by secretion of neurotransmitters, metabolites (SCFAs, Acetate, etc.), and promote the growth of beneficial commensal microorganisms. In this

Table 10.3 Probiotic strains which modulated the brain-related functions in clinical trials.

Probiotic species	Model	Effects	References
Bifidobacterium breve A1	Elderly humans with mild cognitive impairment (mean age ≈ 83 years)	Improving cognitive function and maintaining the quality of life of the elderly	Kobayashi et al. (2019)
	Patients with schizophrenia	Significantly improved anxiety/depression score, IL-22 and TRANCE expression was significantly increased	Okubo et al. (2019)
Bifidobacterium longum 1714	Healthy human males (mean age ≈ 25 years)	Decreased salivary cortisol output and anxiety scores, improvements in hippocampus-dependent visuospatial memory performance, reduced stress, and improved memory	Allen et al. (2016)
Clostridium butyricum MIYAIRI 588	Treatment-resistant major depressive disorder (TRD) patients	Provided significant improvement in depression	Miyaoka et al. (2018)
Lactobacillus acidophilus LA5 and *Bifidobacterium lactis* BB12; multispecies probiotic capsule	Seventy petrochemical workers	Multispecies probiotic capsule had beneficial effects on mental health parameters	Mohammadi et al. (2016)
Lactobacillus casei Shirota	Healthy medical students (mean age ≈ 23 years)	Increase in salivary cortisol levels, relieved stress-associated responses of abdominal dysfunction	Kato-Kataoka et al. (2016)
	One hundred and thirty-two healthy members	Improved the mood of those whose mood was initially poor	Benton et al. (2007)
	Healthy medical students	Prevented hypersecretion of cortisol and physical symptoms under stressful conditions, possibly through vagal afferent signaling to the brain and reduced stress	Takada et al. (2016)
Lactobacillus helveticus	Healthy middle-aged humans (mean age ≈ 58 years)	Increased memory and attention	Ohsawa et al. (2018)
Lactobacillus helveticus R0052, *Bifidobacterium longum* ssp. *infantis*	Academically stressed students	Prevented the onset of stress-related cold/flu	Langkamp-Henken et al. (2015)

(continued)

Table 10.3 (continued)

Probiotic species	Model	Effects	References
R0033, *Bifidobacterium bifidum* R0071			
Lactobacillus helveticus R0052 and *Bifidobacterium longum* R0175	Healthy humans	Reduced the cortisol level and improved the anxiety and depression	Messaoudi et al. (2011)
Lactobacillus paracasei MCC1849	Healthy females (mean age ≈ 21 years)	Maintained a desirable mood state, even under mental stress conditions	Murata et al. (2018)
Lactobacillus plantarum P8	Stressed human adults with mild levels of depression (mean age ≈ 31 years)	Decreased plasma IFN-γ and TNF-α levels, decreased anxiety and stress	Lew et al. (2019)
Lactobacillus plantarum 299v	Patients with major depressive disorder (MDD)	Decrease in KYN concentration, improvement of cognitive functions, no significant changes of TNF-α, IL-6 and IL-1b, and cortisol	Rudzki et al. (2019)
Lactobacillus rhamnosus CGMCC1.3724	Obese men and women	Decrease in the Beck depression inventory score	Sanchez et al. (2017)
Lactobacillus rhamnosus HN001	Pregnant women (14–16 weeks)	Significantly lower depression and anxiety scores	Slykerman et al. (2017)

chapter, authors attempted to explain briefly psychobiotics possible mechanisms and also focus on the presently known routes of interaction with the GBA.

References

Abdel-Haq R, Schlachetzki JC, Glass CK, Mazmanian SK (2019) Microbiome–microglia connections via the gut–brain axis. J Exp Med 216(1):41–59

Ait-Belgnaoui A, Han W, Lamine F, Eutamene H, Fioramonti J, Bueno L, Theodorou V (2006) Lactobacillus farciminis treatment suppresses stress induced visceral hypersensitivity: a possible action through interaction with epithelial cell cytoskeleton contraction. Gut 55 (8):1090–1094

Ait-Belgnaoui A, Durand H, Cartier C, Chaumaz G, Eutamene H, Ferrier L, Houdeau E, Fioramonti J, Bueno L, Theodorou V (2012) Prevention of gut leakiness by a probiotic treatment leads to attenuated HPA response to an acute psychological stress in rats. Psychoneuroendocrinology 37(11):1885–1895

Ait-Belgnaoui A, Colom A, Braniste V, Ramalho L, Marrot A, Cartier C, Houdeau E, Theodorou V, Tompkins T (2014) Probiotic gut effect prevents the chronic psychological stress-induced brain activity abnormality in mice. Neurogastroenterol Motil 26(4):510–520

Akkasheh G, Kashani-Poor Z, Tajabadi-Ebrahimi M, Jafari P, Akbari H, Taghizadeh M, Memarzadeh MR, Asemi Z, Esmaillzadeh A (2016) Clinical and metabolic response to

probiotic administration in patients with major depressive disorder: a randomized, double-blind, placebo-controlled trial. Nutrition 32(3):315–320
Allen AP, Hutch W, Borre YE, Kennedy PJ, Temko A, Boylan G, Murphy E, Cryan JF, Dinan TG, Clarke G (2016) *Bifidobacterium longum* 1714 as a translational psychobiotic: modulation of stress, electrophysiology and neurocognition in healthy volunteers. Transl Psychiatry 6(11): e939
Bengmark S (2013) Gut microbiota, immune development and function. Pharmacol Res 69 (1):87–113
Benton D, Williams C, Brown A (2007) Impact of consuming a milk drink containing a probiotic on mood and cognition. Eur J Clin Nutr 61(3):355–361
Bercik P, Verdu EF, Foster JA, Macri J, Potter M, Huang X, Malinowski P, Jackson W, Blennerhassett P, Neufeld KA, Lu J (2010) Chronic gastrointestinal inflammation induces anxiety-like behavior and alters central nervous system biochemistry in mice. Gastroenterology 139(6):2102–2112
Bercik P, Park AJ, Sinclair D, Khoshdel A, Lu J, Huang X, Deng Y, Blennerhassett PA, Fahnestock M, Moine D, Berger B (2011) The anxiolytic effect of Bifidobacterium longum NCC3001 involves vagal pathways for gut–brain communication. Neurogastroenterol Motil 23 (12):1132–1139
Bermudez-Brito M, Plaza-Díaz J, Munoz-Quezada S, Gomez-Llorente C, Gil A (2012) Probiotic mechanisms of action. Ann Nutr Metab 61(2):160–174
Bilbo SD, Schwarz JM (2012) The immune system and developmental programming of brain and behavior. Front Neuroendocrinol 33(3):267–286
Bolognini D, Tobin AB, Milligan G, Moss CE (2016) The pharmacology and function of receptors for short-chain fatty acids. Mol Pharmacol 89(3):388–398
Bravo JA, Forsythe P, Chew MV, Escaravage E, Savignac HM, Dinan TG, Bienenstock J, Cryan JF (2011) Ingestion of Lactobacillus strain regulates emotional behavior and central GABA receptor expression in a mouse via the vagus nerve. Proc Natl Acad Sci 108(38):16050–16055
Bravo JA, Julio-Pieper M, Forsythe P, Kunze W, Dinan TG, Bienenstock J, Cryan JF (2012) Communication between gastrointestinal bacteria and the nervous system. Curr Opin Pharmacol 12(6):667–672
Breit S, Kupferberg A, Rogler G, Hasler G (2018) Vagus nerve as modulator of the brain–gut axis in psychiatric and inflammatory disorders. Front Psych 9:44
Calvani R, Picca A, Lo Monaco MR, Landi F, Bernabei R, Marzetti E (2018) Of microbes and minds: a narrative review on the second brain aging. Front Med 5:53
Carabotti M, Scirocco A, Maselli MA, Severi C (2015) The gut-brain axis: interactions between enteric microbiota, central and enteric nervous systems. Ann Gastroenterol 28(2):203
Cepeda MS, Katz EG, Blacketer C (2017) Microbiome-gut-brain axis: probiotics and their association with depression. J Neuropsychiatry Clin Neurosci 29(1):39–44
Chaiyasut C, Sivamaruthi BS (2018) Influence of probiotic supplementation on brain function: involvement of gut microbiome, inflammation, and stress pathway. In: Gut microbiota-brain axis. IntechOpen, London
Chen L, Zou Y, Peng J, Lu F, Yin Y, Li F, Yang J (2015) Lactobacillus acidophilus suppresses colitis-associated activation of the IL-23/Th17 axis. J Immunol Res 2015:909514
Cherbut C, Ferrier L, Rozé C, Anini Y, Blottière H, Lecannu G, Galmiche JP (1998) Short-chain fatty acids modify colonic motility through nerves and polypeptide YY release in the rat. Am J Physiol Gastrointest Liver Physiol 275(6):G1415–G1422
Choi HJ, Lee NK, Paik HD (2015) Health benefits of lactic acid bacteria isolated from kimchi, with respect to immunomodulatory effects. Food Sci Biotechnol 24(3):783–789
Corpuz HM, Ichikawa S, Arimura M, Mihara T, Kumagai T, Mitani T, Nakamura S, Katayama S (2018) Long-term diet supplementation with Lactobacillus paracasei K71 prevents age-related cognitive decline in senescence-accelerated mouse prone 8. Nutrients 10(6):762
Costedio MM, Hyman N, Mawe GM (2007) Serotonin and its role in colonic function and in gastrointestinal disorders. Dis Colon Rectum 50(3):376–388

Cryan JF, Dinan TG (2012) Mind-altering microorganisms: the impact of the gut microbiota on brain and behaviour. Nat Rev Neurosci 13(10):701–712

Dantzer R, O'Connor JC, Freund GG, Johnson RW, Kelley KW (2008) From inflammation to sickness and depression: when the immune system subjugates the brain. Nat Rev Neurosci 9 (1):46–56

Davis DJ, Hecht PM, Jasarevic E, Beversdorf DQ, Will MJ, Fritsche K, Gillespie CH (2017) Sex-specific effects of docosahexaenoic acid (DHA) on the microbiome and behavior of socially-isolated mice. Brain Behav Immun 59:38–48

Demaude J, Salvador-Cartier C, Fioramonti J, Ferrier L, Bueno L (2006) Phenotypic changes in colonocytes following acute stress or activation of mast cells in mice: implications for delayed epithelial barrier dysfunction. Gut 55(5):655–661

Desbonnet L, Garrett L, Clarke G, Kiely B, Cryan JF, Dinan TG (2010) Effects of the probiotic Bifidobacterium infantis in the maternal separation model of depression. Neuroscience 170 (4):1179–1188

Deverman BE, Patterson PH (2009) Cytokines and CNS development. Neuron 64(1):61–78

Dhaliwal J, Singh DP, Singh S, Pinnaka AK, Boparai RK, Bishnoi M, Kondepudi KK, Chopra K (2018) Lactobacillus plantarum MTCC 9510 supplementation protects from chronic unpredictable and sleep deprivation-induced behaviour, biochemical and selected gut microbial aberrations in mice. J Appl Microbiol 125(1):257–269

Diamond B, Huerta PT, Tracey K, Volpe BT (2011) It takes guts to grow a brain: increasing evidence of the important role of the intestinal microflora in neuro-and immune-modulatory functions during development and adulthood. BioEssays 33(8):588–591

Dinan TG, Stanton C, Cryan JF (2013) Psychobiotics: a novel class of psychotropic. Biol Psychiatry 74(10):720–726

Dinan TG, Stilling RM, Stanton C, Cryan JF (2015) Collective unconscious: how gut microbes shape human behavior. J Psychiatr Res 63:1–9

Distrutti E, O'Reilly JA, McDonald C, Cipriani S, Renga B, Lynch MA, Fiorucci S (2014) Modulation of intestinal microbiota by the probiotic VSL#3 resets brain gene expression and ameliorates the age-related deficit in LTP. PLoS One 9(9):e106503

Erny D, de Angelis ALH, Jaitin D, Wieghofer P, Staszewski O, David E, Keren-Shaul H, Mahlakoiv T, Jakobshagen K, Buch T, Schwierzeck V (2015) Host microbiota constantly control maturation and function of microglia in the CNS. Nat Neurosci 18(7):965

Felger JC, Lotrich FE (2013) Inflammatory cytokines in depression: neurobiological mechanisms and therapeutic implications. Neuroscience 246:199–229

Ferrante RJ, Kubilus JK, Lee J, Ryu H, Beesen A, Zucker B, Smith K, Kowall NW, Ratan RR, Luthi-Carter R, Hersch SM (2003) Histone deacetylase inhibition by sodium butyrate chemotherapy ameliorates the neurodegenerative phenotype in Huntington's disease mice. J Neurosci 23(28):9418–9427

Ferrario C, Taverniti V, Milani C, Fiore W, Laureati M, De Noni I, Stuknyte M, Chouaia B, Riso P, Guglielmetti S (2014) Modulation of fecal Clostridiales bacteria and butyrate by probiotic intervention with Lactobacillus paracasei DG varies among healthy adults. J Nutr 144 (11):1787–1796

Fijan S (2014) Microorganisms with claimed probiotic properties: an overview of recent literature. Int J Environ Res Public Health 11(5):4745–4767

Food and Agriculture Organization of the United Nations/World Health Organization FAO/WHO (2002) Guidelines for the evaluation of probiotics in food. Joint FAO/WHO working group on drafting guidelines for the evaluation of probiotics in food; London, Ontario, Canada

Frost G, Sleeth ML, Sahuri-Arisoylu M, Lizarbe B, Cerdan S, Brody L, Anastasovska J, Ghourab S, Hankir M, Zhang S, Carling D (2014) The short-chain fatty acid acetate reduces appetite via a central homeostatic mechanism. Nat Commun 5(1):1–11

Fung TC, Olson CA, Hsiao EY (2017) Interactions between the microbiota, immune and nervous systems in health and disease. Nat Neurosci 20(2):145

Furness JB (2012) The enteric nervous system and neurogastroenterology. Nat Rev Gastroenterol Hepatol 9(5):286

Gagliano H, Delgado-Morales R, Sanz-Garcia A, Armario A (2014) High doses of the histone deacetylase inhibitor sodium butyrate trigger a stress-like response. Neuropharmacology 79:75–82

Gareau MG, Jury J, MacQueen G, Sherman PM, Perdue MH (2007) Probiotic treatment of rat pups normalises corticosterone release and ameliorates colonic dysfunction induced by maternal separation. Gut 56(11):1522–1528

Gareau MG, Wine E, Rodrigues DM, Cho JH, Whary MT, Philpott DJ, MacQueen G, Sherman PM (2011) Bacterial infection causes stress-induced memory dysfunction in mice. Gut 60 (3):307–317

Goehler LE, Park SM, Opitz N, Lyte M, Gaykema RP (2008) Campylobacter jejuni infection increases anxiety-like behavior in the holeboard: possible anatomical substrates for viscerosensory modulation of exploratory behavior. Brain Behav Immun 22(3):354–366

Greiner TU, Bäckhed F (2016) Microbial regulation of GLP-1 and L-cell biology. Mol Metab 5 (9):753–758

Grenham S, Clarke G, Cryan JF, Dinan TG (2011) Brain–gut–microbe communication in health and disease. Front Physiol 2:94

Han SH, Hong KB, Suh HJ (2017) Biotransformation of monosodium glutamate to gamma-aminobutyric acid by isolated strain Lactobacillus brevis L-32 for potentiation of pentobarbital-induced sleep in mice. Food Biotechnol 31(2):80–93

Hao Z, Wang W, Guo R, Liu H (2019) Faecalibacterium prausnitzii (ATCC 27766) has preventive and therapeutic effects on chronic unpredictable mild stress-induced depression-like and anxiety-like behavior in rats. Psychoneuroendocrinology 104:132–142

Horn T, Klein J (2013) Neuroprotective effects of lactate in brain ischemia: dependence on anesthetic drugs. Neurochem Int 62(3):251–257

Hosoi T, Okuma Y, Nomura Y (2002) The mechanisms of immune-to-brain communication in inflammation as a drug target. Curr Drug Targets Inflamm Allergy 1(3):257–262

Huang SY, Chen LH, Wang MF, Hsu CC, Chan CH, Li JX, Huang HY (2018) Lactobacillus paracasei PS23 delays progression of age-related cognitive decline in senescence accelerated mouse prone 8 (SAMP8) mice. Nutrients 10(7):894

Hughes DT, Sperandio V (2008) Inter-kingdom signalling: communication between bacteria and their hosts. Nat Rev Microbiol 6(2):111–120

Jang HM, Lee KE, Kim DH (2019) The preventive and curative effects of lactobacillus reuteri NK33 and Bifidobacterium adolescentis NK98 on immobilization stress-induced anxiety/ depression and colitis in mice. Nutrients 11(4):819

Janik R, Thomason LA, Stanisz AM, Forsythe P, Bienenstock J, Stanisz GJ (2016) Magnetic resonance spectroscopy reveals oral Lactobacillus promotion of increases in brain GABA, N-acetyl aspartate and glutamate. NeuroImage 125:988–995

Kali A (2016) Psychobiotics: an emerging probiotic in psychiatric practice. Biomed J 39 (3):223–224

Kato-Kataoka A, Nishida K, Takada M, Kawai M, Kikuchi-Hayakawa H, Suda K, Ishikawa H, Gondo Y, Shimizu K, Matsuki T, Kushiro A (2016) Fermented milk containing Lactobacillus casei strain Shirota preserves the diversity of the gut microbiota and relieves abdominal dysfunction in healthy medical students exposed to academic stress. Appl Environ Microbiol 82(12):3649–3658

Kazemi A, Noorbala AA, Azam K, Djafarian K (2019) Effect of prebiotic and probiotic supplementation on circulating pro-inflammatory cytokines and urinary cortisol levels in patients with major depressive disorder: a double-blind, placebo-controlled randomized clinical trial. J Funct Foods 52:596–602

Kelly JR, Kennedy PJ, Cryan JF, Dinan TG, Clarke G, Hyland NP (2015) Breaking down the barriers: the gut microbiome, intestinal permeability and stress-related psychiatric disorders. Front Cell Neurosci 9:392

Kim N, Yun M, Oh YJ, Choi HJ (2018) Mind-altering with the gut: modulation of the gut-brain axis with probiotics. J Microbiol 56(3):172–182

Kimura I, Inoue D, Maeda T, Hara T, Ichimura A, Miyauchi S, Kobayashi M, Hirasawa A, Tsujimoto G (2011) Short-chain fatty acids and ketones directly regulate sympathetic nervous system via G protein-coupled receptor 41 (GPR41). Proc Natl Acad Sci 108(19):8030–8035

Kimura I, Ozawa K, Inoue D, Imamura T, Kimura K, Maeda T, Terasawa K, Kashihara D, Hirano K, Tani T, Takahashi T (2013) The gut microbiota suppresses insulin-mediated fat accumulation via the short-chain fatty acid receptor GPR43. Nat Commun 4(1):1–12

Ko CY, Lin HTV, Tsai GJ (2013) Gamma-aminobutyric acid production in black soybean milk by Lactobacillus brevis FPA 3709 and the antidepressant effect of the fermented product on a forced swimming rat model. Process Biochem 48(4):559–568

Kobayashi Y, Kinoshita T, Matsumoto A, Yoshino K, Saito I, Xiao JZ (2019) Bifidobacterium breve A1 supplementation improved cognitive decline in older adults with mild cognitive impairment: an open-label, single-arm study. J Prev Alzheimers Dis 6(1):70–75

Kunze WA, Mao YK, Wang B, Huizinga JD, Ma X, Forsythe P, Bienenstock J (2009) Lactobacillus reuteri enhances excitability of colonic AH neurons by inhibiting calcium-dependent potassium channel opening. J Cell Mol Med 13(8b):2261–2270

Langkamp-Henken B, Rowe CC, Ford AL, Christman MC, Nieves C, Khouri L, Specht GJ, Girard SA, Spaiser SJ, Dahl WJ (2015) Bifidobacterium bifidum R0071 results in a greater proportion of healthy days and a lower percentage of academically stressed students reporting a day of cold/flu: a randomised, double-blind, placebo-controlled study. Br J Nutr 113(3):426–434

Laval L, Martin R, Natividad JN, Chain F, Miquel S, De Maredsous CD, Capronnier S, Sokol H, Verdu EF, van Hylckama Vlieg JET, Bermudez-Humaran LG (2015) Lactobacillus rhamnosus CNCM I-3690 and the commensal bacterium Faecalibacterium prausnitzii A2-165 exhibit similar protective effects to induced barrier hyper-permeability in mice. Gut Microbes 6(1):1–9

Levenson JM, O'Riordan KJ, Brown KD, Trinh MA, Molfese DL, Sweatt JD (2004) Regulation of histone acetylation during memory formation in the hippocampus. J Biol Chem 279 (39):40545–40559

Lew LC, Hor YY, Yusoff NAA, Choi SB, Yusoff MS, Roslan NS, Ahmad A, Mohammad JA, Abdullah MFI, Zakaria N, Wahid N (2019) Probiotic Lactobacillus plantarum P8 alleviated stress and anxiety while enhancing memory and cognition in stressed adults: a randomised, double-blind, placebo-controlled study. Clin Nutr 38(5):2053–2064

Liang S, Wang T, Hu X, Luo J, Li W, Wu X, Duan Y, Jin F (2015) Administration of Lactobacillus helveticus NS8 improves behavioral, cognitive, and biochemical aberrations caused by chronic restraint stress. Neuroscience 310:561–577

Lim SK, Kwon MS, Lee J, Oh YJ, Jang JY, Lee JH, Park HW, Nam YD, Seo MJ, Roh SW, Choi HJ (2017) Weissella cibaria WIKIM28 ameliorates atopic dermatitis-like skin lesions by inducing tolerogenic dendritic cells and regulatory T cells in BALB/c mice. Sci Rep 7(1):1–9

Liu YW, Liu WH, Wu CC, Juan YC, Wu YC, Tsai HP, Wang S, Tsai YC (2016) Psychotropic effects of Lactobacillus plantarum PS128 in early life-stressed and naïve adult mice. Brain Res 1631:1–12

Luan H, Wang X, Cai Z (2019) Mass spectrometry-based metabolomics: targeting the crosstalk between gut microbiota and brain in neurodegenerative disorders. Mass Spectrom Rev 38 (1):22–33

Lyte M (2011) Probiotics function mechanistically as delivery vehicles for neuroactive compounds: microbial endocrinology in the design and use of probiotics. BioEssays 33(8):574–581

Macpherson AJ, Uhr T (2004) Compartmentalization of the mucosal immune responses to commensal intestinal bacteria. Ann N Y Acad Sci 1029(1):36–43

Maehata H, Kobayashi Y, Mitsuyama E, Kawase T, Kuhara T, Xiao JZ, Tsukahara T, Toyoda A (2019) Heat-killed Lactobacillus helveticus strain MCC1848 confers resilience to anxiety or depression-like symptoms caused by subchronic social defeat stress in mice. Biosci Biotechnol Biochem 83(7):1239–1247

Marin IA, Goertz JE, Ren T, Rich SS, Onengut-Gumuscu S, Farber E, Wu M, Overall CC, Kipnis J, Gaultier A (2017) Microbiota alteration is associated with the development of stress-induced despair behavior. Sci Rep 7:43859

McCusker RH, Kelley KW (2013) Immune–neural connections: how the immune system's response to infectious agents influences behavior. J Exp Biol 216(1):84–98

McLean PG, Borman RA, Lee K (2007) 5-HT in the enteric nervous system: gut function and neuropharmacology. Trends Neurosci 30(1):9–13

McVey Neufeld KA, Kay S, Bienenstock J (2018) Mouse strain affects behavioral and neuroendocrine stress responses following administration of probiotic Lactobacillus rhamnosus JB-1 or traditional antidepressant fluoxetine. Front Neurosci 12:294

Messaoudi M, Lalonde R, Violle N, Javelot H, Desor D, Nejdi A, Bisson JF, Rougeot C, Pichelin M, Cazaubiel M, Cazaubiel JM (2011) Assessment of psychotropic-like properties of a probiotic formulation (Lactobacillus helveticus R0052 and Bifidobacterium longum R0175) in rats and human subjects. Br J Nutr 105(5):755–764

Miller AH, Haroon E, Raison CL, Felger JC (2013) Cytokine targets in the brain: impact on neurotransmitters and neurocircuits. Depress Anxiety 30(4):297–306

Miyaoka T, Kanayama M, Wake R, Hashioka S, Hayashida M, Nagahama M, Okazaki S, Yamashita S, Miura S, Miki H, Matsuda H (2018) Clostridium butyricum MIYAIRI 588 as adjunctive therapy for treatment-resistant major depressive disorder: a prospective open-label trial. Clin Neuropharmacol 41(5):151–155

Mohajeri MH, Brummer RJ, Rastall RA, Weersma RK, Harmsen HJ, Faas M, Eggersdorfer M (2018) The role of the microbiome for human health: from basic science to clinical applications. Eur J Nutr 57(1):1–14

Mohammadi AA, Jazayeri S, Khosravi-Darani K, Solati Z, Mohammadpour N, Asemi Z, Adab Z, Djalali M, Tehrani-Doost M, Hosseini M, Eghtesadi S (2016) The effects of probiotics on mental health and hypothalamic–pituitary–adrenal axis: a randomized, double-blind, placebo-controlled trial in petrochemical workers. Nutr Neurosci 19(9):387–395

Mountzouris KC, Tsirtsikos P, Kalamara E, Nitsch S, Schatzmayr G, Fegeros K (2007) Evaluation of the efficacy of a probiotic containing Lactobacillus, Bifidobacterium, Enterococcus, and Pediococcus strains in promoting broiler performance and modulating cecal microflora composition and metabolic activities. Poult Sci 86(2):309–317

Murata M, Kondo J, Iwabuchi N, Takahashi S, Yamauchi K, Abe F, Miura K (2018) Effects of paraprobiotic Lactobacillus paracasei MCC1849 supplementation on symptoms of the common cold and mood states in healthy adults. Benefic Microbes 9(6):855–864

Obermeier B, Daneman R, Ransohoff RM (2013) Development, maintenance and disruption of the blood-brain barrier. Nat Med 19(12):1584

Ohland CL, Kish L, Bell H, Thiesen A, Hotte N, Pankiv E, Madsen KL (2013) Effects of Lactobacillus helveticus on murine behavior are dependent on diet and genotype and correlate with alterations in the gut microbiome. Psychoneuroendocrinology 38(9):1738–1747

Ohsawa K, Nakamura F, Uchida N, Mizuno S, Yokogoshi H (2018) Lactobacillus helveticus-fermented milk containing lactononadecapeptide (NIPPLTQTPVVVPPFLQPE) improves cognitive function in healthy middle-aged adults: a randomised, double-blind, placebo-controlled trial. Int J Food Sci Nutr 69(3):369–376

Okubo R, Koga M, Katsumata N, Odamaki T, Matsuyama S, Oka M, Narita H, Hashimoto N, Kusumi I, Xiao J, Matsuoka YJ (2019) Effect of Bifidobacterium breve A-1 on anxiety and depressive symptoms in schizophrenia: a proof-of-concept study. J Affect Disord 245:377–385

Oleskin AV, Zhilenkova OG, Shenderov BA, Amerhanova AM, Kudrin VS, Klodt PM (2014) Lactic-acid bacteria supplement fermented dairy products with human behavior-modifying neuroactive compounds. J Pharm Nutr Sci 4:199–206

Overduin J, Schoterman MH, Calame W, Schonewille AJ, Ten Bruggencate SJ (2013) Dietary galacto-oligosaccharides and calcium: effects on energy intake, fat-pad weight and satiety-related, gastrointestinal hormones in rats. Br J Nutr 109(7):1338–1348

Palomar MM, Galdeano CM, Perdigón G (2014) Influence of a probiotic lactobacillus strain on the intestinal ecosystem in a stress model mouse. Brain Behav Immun 35:77–85

Perez-Burgos A, Wang B, Mao YK, Mistry B, Neufeld KAM, Bienenstock J, Kunze W (2013) Psychoactive bacteria Lactobacillus rhamnosus (JB-1) elicits rapid frequency facilitation in vagal afferents. Am J Physiol Gastrointest Liver Physiol 304(2):G211–G220

Perez-Burgos A, Mao YK, Bienenstock J, Kunze WA (2014) The gut-brain axis rewired: adding a functional vagal nicotinic "sensory synapse". FASEB J 28(7):3064–3074

Qureshi IA, Mehler MF (2013) Towards a 'systems'-level understanding of the nervous system and its disorders. Trends Neurosci 36(11):674–684

Réus GZ, Jansen K, Titus S, Carvalho AF, Gabbay V, Quevedo J (2015) Kynurenine pathway dysfunction in the pathophysiology and treatment of depression: evidences from animal and human studies. J Psychiatr Res 68:316–328

Rhee SH, Pothoulakis C, Mayer EA (2009) Principles and clinical implications of the brain–gut–enteric microbiota axis. Nat Rev Gastroenterol Hepatol 6(5):306

Rios AC, Maurya PK, Pedrini M, Zeni-Graiff M, Asevedo E, Mansur RB, Wieck A, Grassi-Oliveira R, McIntyre RS, Hayashi MA, Brietzke E (2017) Microbiota abnormalities and the therapeutic potential of probiotics in the treatment of mood disorders. Rev Neurosci 28 (7):739–749

Ríos-Covián D, Ruas-Madiedo P, Margolles A, Gueimonde M, de los Reyes-Gavilán CG, Salazar N (2016) Intestinal short chain fatty acids and their link with diet and human health. Front Microbiol 7:185

Rudzki L, Ostrowska L, Pawlak D, Małus A, Pawlak K, Waszkiewicz N, Szulc A (2019) Probiotic Lactobacillus Plantarum 299v decreases kynurenine concentration and improves cognitive functions in patients with major depression: a double-blind, randomized, placebo controlled study. Psychoneuroendocrinology 100:213–222

Sanchez M, Darimont C, Panahi S, Drapeau V, Marette A, Taylor VH, Doré J, Tremblay A (2017) Effects of a diet-based weight-reducing program with probiotic supplementation on satiety efficiency, eating behaviour traits, and psychosocial behaviours in obese individuals. Nutrients 9(3):284

Sarkar A, Lehto SM, Harty S, Dinan TG, Cryan JF, Burnet PW (2016) Psychobiotics and the manipulation of bacteria–gut–brain signals. Trends Neurosci 39(11):763–781

Saulnier DM, Ringel Y, Heyman MB, Foster JA, Bercik P, Shulman RJ, Versalovic J, Verdu EF, Dinan TG, Hecht G, Guarner F (2013) The intestinal microbiome, probiotics and prebiotics in neurogastroenterology. Gut Microbes 4(1):17–27

Schnorr SL, Bachner HA (2016) Focus: microbiome: integrative therapies in anxiety treatment with special emphasis on the gut microbiome. Yale J Biol Med 89(3):397

Selhub EM, Logan AC, Bested AC (2014) Fermented foods, microbiota, and mental health: ancient practice meets nutritional psychiatry. J Physiol Anthropol 33(1):2

Sherwin E, Dinan TG, Cryan JF (2018) Recent developments in understanding the role of the gut microbiota in brain health and disease. Ann N Y Acad Sci 1420(1):5–25

Silva YP, Bernardi A, Frozza RL (2020) The role of short-chain fatty acids from gut microbiota in gut-brain communication. Front Endocrinol 11:25

Slykerman RF, Hood F, Wickens K, Thompson JMD, Barthow C, Murphy R, Kang J, Rowden J, Stone P, Crane J, Stanley T (2017) Effect of Lactobacillus rhamnosus HN001 in pregnancy on postpartum symptoms of depression and anxiety: a randomised double-blind placebo-controlled trial. EBioMedicine 24:159–165

Smith CJ, Emge JR, Berzins K, Lung L, Khamishon R, Shah P, Rodrigues DM, Sousa AJ, Reardon C, Sherman PM, Barrett KE (2014) Probiotics normalize the gut-brain-microbiota axis in immunodeficient mice. Am J Physiol Gastrointest Liver Physiol 307(8):G793–G802

Sobol CV, Belostotskaya GB (2016) Product fermented by Lactobacilli induces changes in intracellular calcium dynamics in rat brain neurons. Biochem (Moscow) Suppl Ser A Membr Cell Biol 10(1):37–45

Stilling RM, Dinan TG, Cryan JF (2014) Microbial genes, brain & behaviour–epigenetic regulation of the gut–brain axis. Genes Brain Behav 13(1):69–86

Sudo N, Chida Y, Aiba Y, Sonoda J, Oyama N, Yu XN, Kubo C, Koga Y (2004) Postnatal microbial colonization programs the hypothalamic–pituitary–adrenal system for stress response in mice. J Physiol 558(1):263–275

Sun Y, Geng W, Pan Y, Wang J, Xiao P, Wang Y (2019) Supplementation with Lactobacillus kefiranofaciens ZW3 from Tibetan Kefir improves depression-like behavior in stressed mice by modulating the gut microbiota. Food Funct 10(2):925–937

Takada M, Nishida K, Kataoka-Kato A, Gondo Y, Ishikawa H, Suda K, Kawai M, Hoshi R, Watanabe O, Igarashi T, Kuwano Y (2016) Probiotic Lactobacillus casei strain Shirota relieves stress-associated symptoms by modulating the gut–brain interaction in human and animal models. Neurogastroenterol Motil 28(7):1027–1036

Tanida M, Imanishi K, Akashi H, Kurata Y, Chonan O, Naito E, Kunihiro S, Kawai M, Kato-kataoka A, Shibamoto T (2014) Injection of Lactobacillus casei strain Shirota affects autonomic nerve activities in a tissue-specific manner, and regulates glucose and lipid metabolism in rats. J Diabetes Investig 5(2):153–161

Thayer JF, Sternberg EM (2009) Neural concomitants of immunity—focus on the vagus nerve. Neuroimage 47(3):908

Tian P, Wang G, Zhao J, Zhang H, Chen W (2019) Bifidobacterium with the role of 5-hydroxytryptophan synthesis regulation alleviates the symptom of depression and related microbiota dysbiosis. J Nutr Biochem 66:43–51

Vighi G, Marcucci F, Sensi L, Di Cara G, Frati F (2008) Allergy and the gastrointestinal system. Clin Exp Immunol 153:3–6

Wei CL, Wang S, Yen JT, Cheng YF, Liao CL, Hsu CC, Wu CC, Tsai YC (2019) Antidepressant-like activities of live and heat-killed Lactobacillus paracasei PS23 in chronic corticosterone-treated mice and possible mechanisms. Brain Res 1711:202–213

Yano JM, Yu K, Donaldson GP, Shastri GG, Ann P, Ma L, Nagler CR, Ismagilov RF, Mazmanian SK, Hsiao EY (2015) Indigenous bacteria from the gut microbiota regulate host serotonin biosynthesis. Cell 161(2):264–276

Gut-Brain Axis: Probiotic Interactions and Implications for Human Mental Health

11

Devaraja Gayathri, Mahanthesh Vasudha, and Chakra Siddappa Prashantkumar

Abstract

Microbial colonization commences during birth, and establishment of stable gut microbiota takes place in the first 3–5 years of life. A diverse group of microbiome including virome, facultative anaerobic bacteria (Proteobacteria), microaerophilic bacteria (*Lactobacillus*) and anaerobic bacteria (*Bifidobacterium* and *Bacteroides*) colonize the intestine. The gut comes across various different types of components like diet, allergens, microbial toxins and infectious agents and its interactions with endocrine, circulatory, neural and immune systems resulting in host physiological responses. The autonomic nervous system (NS) and the enteric NS play an important role in the neural control of gastrointestinal function. Probiotics are live microbes, which while administered in adequate amounts provide a benefit to their host. The usage of probiotics along with prebiotics improves intestinal health. In this chapter, we discuss probiotic intervention for the management/control of behaviour disorders of the microbiota-gut-brain axis.

Keywords

Alzheimer's disease · Autism · Gut-brain axis · Multiple sclerosis · Parkinson's disease · Prebiotics · Probiotics

D. Gayathri (✉) · M. Vasudha · C. S. Prashantkumar
Department of Studies in Microbiology, Davangere University, Davangere, Karnataka, India

R. Z. Sayyed, M. Khan (eds.), *Microbiome-Gut-Brain Axis*,
https://doi.org/10.1007/978-981-16-1626-6_11

11.1 Introduction

The gut is an organ system, consisting of a hollow tube stretching from mouth to anus which digests the food and the majority part of the gut of the animal body consists of a diversified group of microorganisms that are living together known as gut microbiota, they exert influence on host physiology, and therefore one's disease susceptibility is altered (Lozupone et al. 2012). In animals, even there is an influence on their emotional behaviour due to the gut microbiota (Forsythe et al. 2010; Cryan and Dinan 2012; Collins et al. 2012; Forsythe and Kunze 2013; Dinan et al. 2013). Changes in the gut microbiome or intestinal specific bacteria often modulate the central and peripheral nervous systems (NS) in the host that result in altered brain function, and this suggests the existence of a microbiota-gut-brain axis (Gayathri and Rashmi 2017, b). There is a bidirectional or two-way communication between the brain and the gut which is called the gut-brain axis. Microorganisms reside as a commensal organism on the skin and other internal surfaces, as normal microflora and a co-evolutionary association with the mammals can be seen with huge quantities around 10^{14} microbial load reside in an adult gastrointestinal (GI) tract, which will be exceeding about 10 times more than the number of human cells in the body. However, variability, and diversity among individual human beings throughout the stages of development and majority of the gut microbiota belong to diverse bacterial species varying from 500 to 1000 different species.

Initial microbial colonization normally occurs during birth and within the first 3–5 years of life, and adult-like composition continuously evolved to fairly stable gut microbiota. Life begins with the colonization of gut microbiota, like facultative anaerobic bacteria (Proteobacteria), microaerophilic bacteria (*Lactobacillus*), anaerobic bacteria such as *Bifidobacterium* species and, later, diverse bacterial community among *Bacteroides* (Rodriguez et al. 2015). During the neonatal stage, one can get severe infections due to pathogenic organism's invasion as the neonate immature immune system to fight against the pathogens, development of the immune system in the neonates occurs through maternal interactions like breastfeeding which gives passive immunity to the neonate (Walker 2017). Milk sugar lactose, a disaccharide, promotes *Lactobacillus* growth and shapes up the gut microbiome in young ones (Francavilla et al. 2012).

Gut environmental factors play a crucial role in the determination of the composition of gut microbiota by subsequent introduction of solid food (Rodriguez et al. 2015). The entire physiological and immunological protection restricts the invasion of disease-causing agents and toxins into the circulation, and many other functions are characterized by the gut-blood barrier. Even though environmental factors and diet alters the gut microbiome, however still microbial functional metabolic pathways remain quite stable (Schmidt et al. 2018).

In the elderly population alteration in the gut microbiota has been noticed, where the reduction in the diversity of saccharolytic bacteria and Bifidobacterium is noticed, in contrast, there is an increase in proteolytic bacteria and certain types of Proteobacteria (Claesson et al. 2011). Due to the disruption of intestinal barrier function, plasma markers have shown to increase intestinal permeability in old-age population (Qi et al. 2017). However, probiotics are functional foods to have healthy lifespan-enhancing effects, which may include suppression of chronic low-grade inflammations reported in a mouse model (Matsumoto et al. 2011), signifying the importance of the gut microbiota in the maintenance of overall health. The gut virome shows more inter-individual variation and is less affected by environmental changes than the gut microbiome (Minot et al. 2011). The human gut environment virome has a group of hypervariable sequences which has been thought to be a repository for viral evolution or adaption to a fresh environment (Minot et al. 2012, 2013). Replacement of the gut virome in diabetes mellitus seems to precede the expansion of autoimmunity (Zhao et al. 2017), reflecting a virome role in the disease. A fungal group in the gut does not appear to initiate illness evidently but would exhibit dysbiosis leading to systemic inflammation (Iliev and Leonardi 2017).

11.2 Gut Physiology

The small intestine is divided into three parts, viz. the duodenum, jejunum and ileum, and the large intestine includes only the colon; both the small intestine and the large intestine are distinct in structure as well as composition; in the proximal colon, a large number of goblet cells exist, while Peyer's patches are primarily sited in the small intestine (Nguyen et al. 2015; Atuma et al. 2001); and the mucin layers are thinner and the microvilli are numerous in the small intestine when compared to the colon. About 70% of the immunological cells are localized in the gut; these cells help in balancing immune activation and tolerance to the gut microbiome (McDermott and Huffnagle 2014). The gut is the second most galvanizing organ, facilitating communication with the brain (Furness et al. 2014). The main function of the complex vascular layers of the gut is efficient absorption of nutrients and water and also maintaining a gradient of oxygen along with the GIT (gastrointestinal tract) (Zheng et al. 2015). The impact on host physiological response depends on the gut interacting with environmental factors (diet, toxins and pathogens) and its interactions with endocrine, circulatory, neural and immune systems.

11.2.1 Neural Control of the Gut

Neurodevelopment is a complex process dependent on both intrinsic and extrinsic signals. The ANS (autonomic nervous system) and the ENS (enteric nervous system) play an important role in the neural control of gastrointestinal function. The physiological conditions of the gut like acidity, levels of nutrients, osmolarity and pain are conveyed by the ANS to the brain (Berthoud et al. 2004). Submucosal plexus also called as Meissner's Plexus and Myenteric Plexus are the enteric nervous system, which contributes to in situ neural communication in the intestine and the ANS (Furness et al. 2014). The development of the brain depends on key pre- and post-natal events that assimilate environmental cues, such as molecular signals from the gut. These cues are mainly originated from the gut microbiome, as the gut is our largest portal to the molecular universe, numerous dilatory components directly interact for the neurodevelopment and induce functional alteration in the mature brain with long-term implication to health (Chang et al. 2009; Zeisel 2004). Research in animal models and humans has inextricably linked gut bacteria to the development and function of the immune system. Indeed, germ-free (GF) mice, devoid of all associated microorganisms, exhibit increased risk-taking behaviours and hyperactivity, while also displaying learning and memory deficits compared to conventional (specific-pathogen-free (SPF)) mice (Clarke et al. 2013; Gareau et al. 2011; Heijtz et al. 2011; Neufeld et al. 2011). Further, GF mice show changes in the expression of the 5-hydroxytryptamine receptor (5-HT1A), neurotrophic factors (e.g. BDNF) and NMDA receptor subunits in the hippocampus (Bercik et al. 2011a; Heijtz et al. 2011; Sudo et al. 2004), while also displaying impaired blood-brain barrier function, as well as increased myelination in the prefrontal cortex (Braniste et al. 2014; Hoban et al. 2016). Treatment with the probiotic bacteria *Bifidobacterium longum* reduced anxiety and decreased the excitability of the *ileal* myenteric plexus neurons in mice with infectious colitis (Bercik et al. 2011b), indicating communication of probiotics with the CNS via the ENS and the vagal nerve. Further investigations are required to identify the neurons that are affected by probiotics and the signals that are involved in this communication and to identify other alterations in gut microbiota that may also affect the ENS. Also, the ENS sends sensory signals from the gut to the nucleus tractus solitarii (NTS) in the CNS, and it communicates bidirectionally with the brain through the vagus nerve. Changes in the gut microbiota induced by an energy-dense diet have been associated with alterations in brain-gut vagal communication in a rat model of obesity (Vaughn et al. 2017), which may alter vagal satiety signalling and stimulate energy intake (de Lartigue et al. 2011). There are many beneficial effects of treatment with probiotics (*Lactobacillus rhamnosus* and *B. longum*) on stress and anxiety which have been demonstrated to be vagus nerve dependent (Bercik et al. 2011a; Bravo et al. 2011). There are many pieces of evidence studied in animal models for a potential role of the microbiome in neuropsychiatric conditions, including depression and anxiety, autism spectrum disorder (ASD), schizophrenia and even Parkinson's disease (PD) and Alzheimer's disease (AD). Mental disorders of the gut-brain axis and bidirectional communication between the gut and the brain are depicted in Figs. 11.1 and 11.2, respectively (Burokas et al. 2015; Mayer et al. 2015).

11.3 Role of the Gut in the Immune System

In human body, the immune system protects us from pathogens and other foreign particles; our immune system comprises lymphocytes and different types of innate immune cells, like macrophages and dendritic cells. Whereas the epithelial layer provides mucosal immunity which is called as gut-associated lymphoid tissues (GALTs) which forms a boundary between the gut and the blood, so the gut is called as the largest immune organ having a complex mucosal immune system

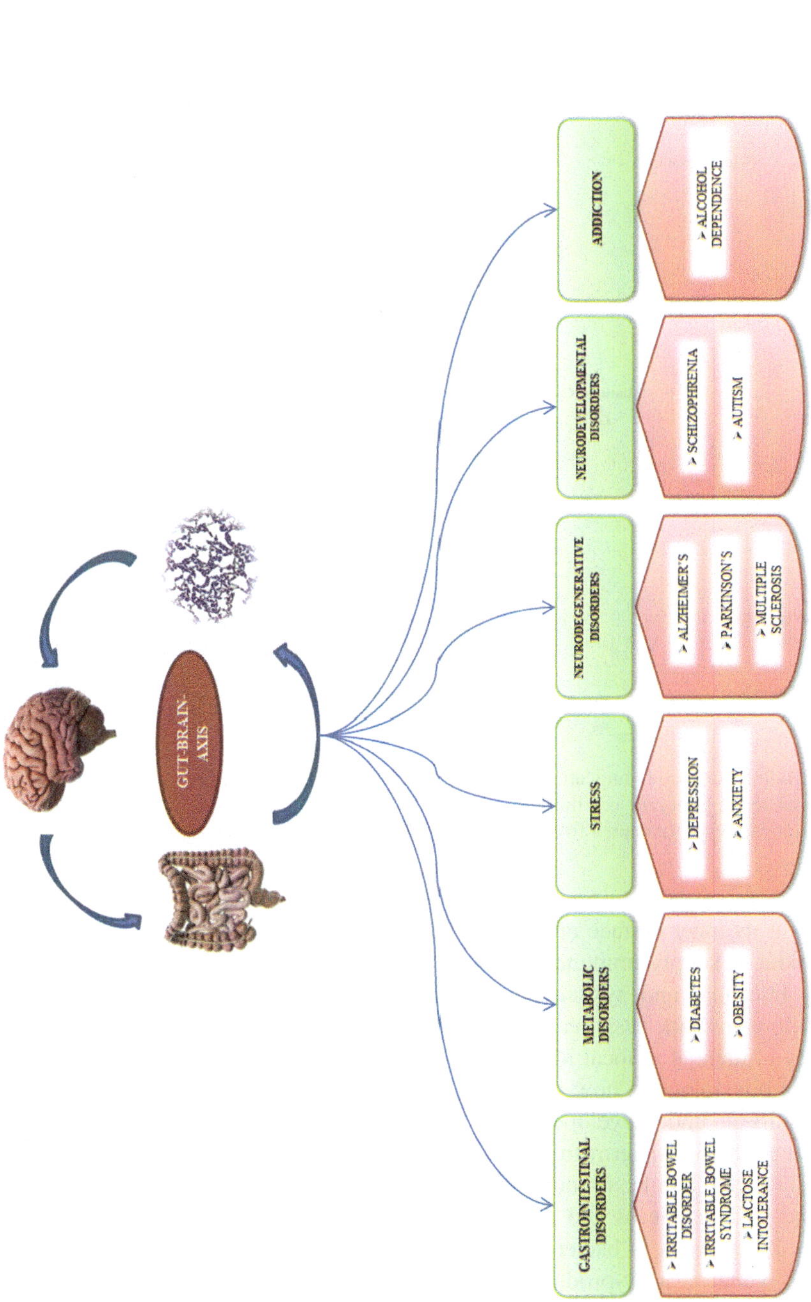

Fig. 1 Mental disorders of gut-brain axis. The gut microbiota/probiotic helps in maintaining homeostasis and its dysfunction has been related to various psychiatric disorders. (Adapted from Burokas et al. (2015))

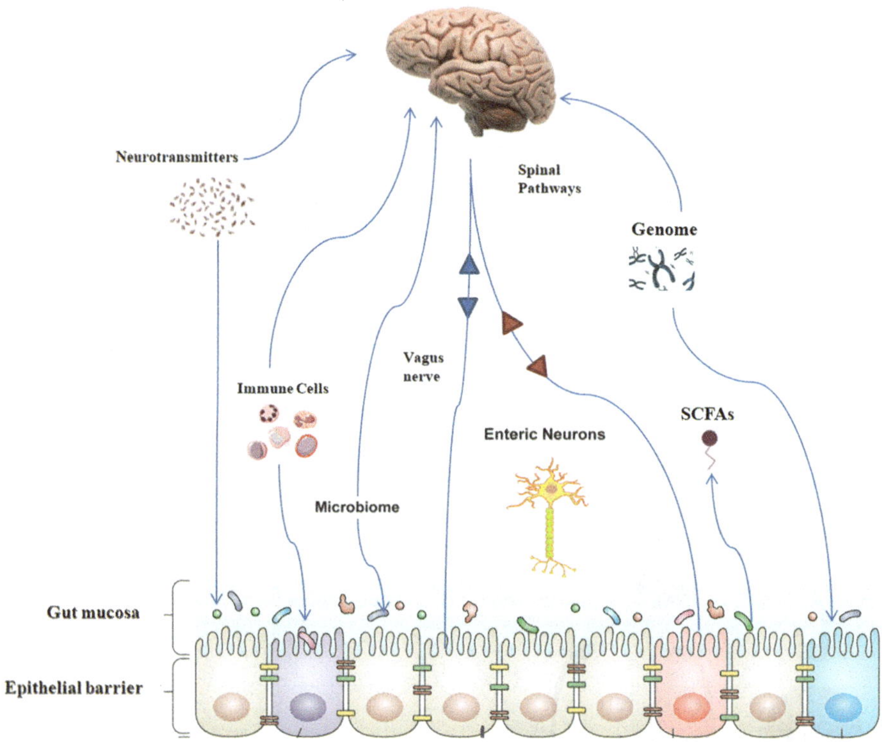

Fig. 2 Bidirectional communication between the gut microbiota and the brain. The gut microbiota can modulate the gut-brain axis through many direct and indirect pathways. They include immune and neural pathways. (Adapted from Mayer et al. (2015))

located at its inner surface and exposed to the lumen which interacts with the gut microbiota to supply immune responses and tolerance (McDermott and Huffnagle 2014). Intestinal and systemic homeostasis ensures the physiological range of harmonious immune responses. Therefore, the critical role of the gut microbiota is not only influential in local immune outcomes but along with it is also maintaining systemic physiology (Chow et al. 2010). Altered or deficiency of normal gut flora will results in underdeveloped GALT, abnormal systemic and central immunity (Erny et al. 2015). In germ-free animals there is a reduction in the levels of T helper 17 (T_H17) cells, B cells, immunoglobulin A (IgA), plasma cells and an imbalance of T_H1 and T_H2 responses and impaired T_{reg} cell function is reported. T_{reg} cells are induced by a variety of bacterial groups (Atarashi et al. 2011) by the SCFA such as butyrate which is a by-product of bacterial fermentation (Furusawa et al. 2013). The germ-free mice when colonized with Lactobacillus (L.) rhamnosus LOCK0900, L. rhamnosus LOCK0908 and L. casei LOCK0919 shows significant alterations in enterocytes, wherein restoring of microfilaments, building up of apical junction improves loose intestinal barriers (Kozakova et al. 2016). Different alterations in physiological parameters in germ-free animals include impaired blood-brain barrier integrity (Braniste et al. 2014), an inflated hypothalamic-pituitary-adrenal response to stress and adjusted neurotransmitter levels (Sudo et al. 2004; Mayer et al. 2014; Yano et al. 2015).

11.4 Probiotics

Probiotics are live microbes, which while administered in adequate amounts provide a benefit to their host (WHO). There are many research works that have been studied to establish and check the importance and role of probiotics in behaviours (Bravo et al. 2011; Desbonnet et al. 2010; Dinan et al. 2013). In the animal model, a variety of probiotic bacteria are assessed for effectiveness in behaviour modulation. Two genera of probiotics (*Bifidobacterium* and *Lactobacillus*) are investigated for useful effects on health. *L. helveticus* R0052, when administered in control and Western diet-fed mice, has shown to ease anxiety-like behaviour and improve reminiscence dysfunction in the Barnes maze (Ohland et al. 2013). The dextran sodium sulphate-induced colitis model, when treated with *Bifidobacterium longum*, showed normalized anxiety-like behaviour (Bercik et al. 2011a). C57BL/6 mice when infected with Citrobacter rodentium showed exaggerated acute stress resulting in memory dysfunction. This was prevented and treated by the daily treatment of infected mice with probiotics *L. rhamnosus* (R0011) and *L. helveticus* (R0052) (Gareau et al. 2011). The visceral pain is alleviated by using probiotic treatment (Rousseaux et al. 2007; Verdu et al. 2006). To study and understand the consequences of probiotics on brain function in healthy humans, researchers used magnetic resonance imaging to measure the function and response to an emotional task, particularly in sensory and interoceptive regions which were seen to be reduced in female patients who have been administered with used fermented milk product which was fermented by four different probiotic strains like *Bifidobacterium animalis* subsp. *lactis, Streptococcus thermophilus, Lactobacillus bulgaricus* and *Lactococcus lactis* subsp. *lactis* in a consortium when compared to the control female patient (Tillisch et al. 2013). The Hospital Anxiety and Depression Scale (HADS) was used to conduct a study on two different groups: one group was administered with *Lactobacillus*- and *Bifidobacterium*-containing probiotics, while another group was administered with placebo as the control; it was found that global psychological distress and anxiety symptoms were reduced significantly in the treated group (Messaoudi et al. 2011). Importantly, an infant's microbiota has been altered by supplementation of probiotics to the mother during and after pregnancy (Lahtinen et al. 2009; Mueller et al. 2015). In future studies, we need to there is a need for further trials focused on testing the efficacy of prebiotics and probiotics, timing and amount of probiotics to be used, combinations of probiotics for the synergistic effect, understanding the mechanism of action of probiotics on different organs of the human body by using more specialized techniques, compiling the previous data and drawing a suitable way of treatment for different diseases (Mueller et al. 2015).

11.5 Prebiotics

Prebiotics are non-digestible dietary fibre food ingredients which induce growth activity in probiotic bacteria (Saulnier et al. 2013). The usage of prebiotics stimulate the colonization of probiotic bacteria like, Lactobacillus and Bifidobacteria in the gut of the animal body. Prebiotics like galacto-oligosaccharides and fructo-oligosaccharides with probiotic bacteria have numerous advantageous effects on the immune system of the gut and on brain function, specifically, in mental health and also help in the treatment of psychiatric disorders wherein increased brain-derived neurotrophic factor (BDNF) expression and *N*-methyl-D-aspartate (NMDA) receptor signalling, providing support in treating these disorders (Drakoularakou et al. 2010; Savignac et al. 2013; van Vlies et al. 2012). A recent study has confirmed that supplementing prebiotic galacto-oligosaccharides showed an early anxiolytic-like profile which suppresses the neuroendocrine stress response and a boost in the processing of positive versus negative attention towards vigilance in the patients (Schmidt et al. 2015). Additionally, insulin-type fructans and lactulose prebiotics are administered to patients which modulate gut transit, decrease putrefactive activity within the gut lumen, which in turn prevent GI infections, and inflammatory response diminish subsequently (Casellas et al. 2007; Lewis et al. 2005; dePreter et al. 2008).

11.6 Disorders of the Microbiota-Gut-Brain Axis

11.6.1 Stress, Anxiety and Depression

Despite well scientific advancement and a well-established association between stress and psychiatric disorders, we lack an understanding on how the complex processes by which stress mediates pathological changes that increase susceptibility to disease are ongoing (Hornig 2013). Recently, due to better understanding and intensive research on gut microbiota and GI disorders, great attention has been towards microbiota-gut-brain axis dysregulation in stress-associated CNS disorders (Bested et al. 2013; Bravo et al. 2012; Cryan and O'Mahony 2011; Foster and McVey Neufeld 2013; Sherman et al. 2014). In the prefrontal cortex and other higher cortical regions produce modulatory signals to other regions of brain-like amygdala, the hippocampus and the paraventricular together through complex integration they generate the stress response (Moloney et al. 2012, 2014). However, researchers conducted in vivo studies on germ-free mice and rodents to ascertain the role played through microbiota in the programming of the stress response, a germ-free mouse that was colonized with *Bifidobacterium infantis* which reversed the exaggerated hypothalamic-pituitary-adrenal (HPA) stress response in contrast to enteropathogenic *Escherichia coli*; this confirmed that more suitable HPA axis activity in germ-free mice following acute psychological stress, providing first convincing evidence of the critical role of probiotic bacteria in early development for the HPA system to become fully susceptible to inhibitory neural regulation (Sudo et al. 2004). Ait-Belgnaoui et al. (2012) found that acute stress in rats when treated with a probiotic strain, *Lactobacillus farciminis*, attenuates intestinal permeability and the HPA axis.

Due to the fast and demanding lifestyle, most of the individuals experience chronic stress, depression and anxiety in early life (Burokas et al. 2014; Caspi et al. 2003; Kendler et al. 2000). The major role in the regulation of mood, anxiety and stress is played by microbiota which was proved by animal studies (Fond et al. 2015). Grigoleit et al. (2011) conducted an experiment using endotoxin lipopolysaccharide, which when administered into healthy individuals results in increased levels of pro-inflammatory cytokines, salivary cortisol and plasma norepinephrine and exaggerated anxiety and depression. Another study proves that in maternally separated rat offspring, the levels of corticosterone can be normalized using a probiotic treatment (Gareau et al. 2007). Expression levels of GABA receptor gene in the brain and the stress-induced corticosterone were significantly reduced when the mice were administered with *L. rhamnosus* (Bravo et al. 2011). *Bifidobacterium* can confer promising protection in individuals before stress exposure by altering tryptophan, peripheral cytokine levels and concentrations of the serotonin precursor (Desbonnet et al. 2008). A person who is suffering from major depression and anxiety will have major alterations in the gut which is called as the leaky gut phenomenon wherein intestinal permeability leads to translocation of enteropathogenic bacteria as they cross the gut mucosal barrier and interact along with the enteric nervous system and the locally residing immune cells, and many reports state that activation of immune cells leads to inflammation which is mediated by elevated levels of IgM and IgA in the serum lipopolysaccharide of enterobacteria of highly depressed individuals when compared with healthy individuals (Gareau et al. 2008; Maes et al. 2008). When considering to present antidepressant treatment options, emerging and promising treatment option is psychobiotics which consist of live organisms which provide a good healthy gut by producing neuroactive compounds and reduce the HPA activity; along with this, they reduce the population of harmful pathogenic microbiota which indirectly reduce the inflammatory response and impact positively on behavioural, neurochemical and immunological measures relevant to the brain-gut axis disorders (Dinan et al. 2013; Gayathri and Rashmi 2017, b).

11.6.2 Alzheimer's Disease

Alzheimer's disease (AD) can be explained as dementia seen in elderly persons; it is a sort of short-term memory loss or difficulty in remembering recent events (Querfurth and LaFerla 2010). Many symptoms include behavioural issues, problem with language, disorientation, mood swings, loss of self-care management and loss of motivation (Burns and Iliffe 2009). AD patients suffer from slow degeneration of neurons due to abnormality in inflammatory signals within the brain and deposition of amyloid protein leading to dysfunction in the brain; this is strongly proved by epidemiological and clinical evidence (Huang and Mucke 2012). Evidence from previous studies has led to the belief that the mediators of neurodegeneration behind the cognitive decline and memory loss (Perry and Holmes 2014). Interestingly, when in vivo experiments were conducted on mice, they were induced with AD and it was found that AD pathogenesis was associated with an elevated inflammatory response in the peripheral system, i.e. in the brain and blood of mice (Aso et al. 2015; Jiang et al. 2009). Dysregulation of serotonergic and kynurenine routes of tryptophan metabolism influences the CNS pathological conditions of dementia, Huntington's disease and AD (Ruddick et al. 2006). In AD-affected cells, there is a well-established scientific data about phosphorylation and expression of tau protein which are intracellular tangles containing tau protein which is hyperphosphorylated and are regulated by insulin and insulin growth factor signalling cascades and therefore if patients have impaired insulin signalling which will be also the crucial aspects for AD (de la Monte and Wands 2008). The in vivo experiment conducted on AD-infected C57BL/6 wild-type were compared to AppNL-G-F mice, wild type was administered with vehicle (same solvent of th stock solution which is used for test group) whereas test group with a commercial consortium of probiotics VSL#3 for 8 weeks, and they conducted faecal microbiome analysis and UPLC-MS/MS for quantifying the amount SCFA in the serum and brain, and they found that there was an increased level of lactate, acetate and c-Fos gene expression in AppNL-G-F mice when compared to C57BL/6 wild type; this proves that there was increased neuronal activity in AD mice (Kaur et al. 2020). Bonfili et al. (2020) found that AD-infected mice (3xTg-AD) when treated with probiotics showed a sign of increased glucose transporters and modulation of the pAkt and pAMPK pathways that lead to decreased hyperphosphorylation of tau protein which in turn delayed disease progression and helped in maintaining glucose homeostasis.

11.6.3 Parkinson's Disease

Parkinson's disease (PD) is a progressive nervous system disorder that affects the movement or motor system. Symptoms start gradually, starting with a barely noticeable tremor in only one hand. Tremors are common, but the disorder also commonly causes stiffness or slowing of movement. The death of dopamine-generating cells within the substantia nigra (basal ganglia structure) plays a crucial role in movements (Dickson et al. 2009). The important feature of PD may be a broad range of non-motor symptoms as recognized by the olfactory (loss of smell), gastrointestinal (GI), cardiovascular and urogenital systems (Mulak and Bonaz 2015). In another study, dysregulation of the brain-gut-microbiota axis in PD results in GI dysfunction, which is observed in 80% of PD subjects (Cersosimo and Benarroch 2012). The bidirectional brain-gut-microbiota axis interactions modulate pro- and anti-inflammatory responses (Hollister et al. 2014). It has been suggested that the gut microbiota changes associated with intestinal inflammation may contribute to the initiation of α-syn misfolding (Devos et al. 2013; Olanow et al. 2014). The interesting concept of molecular mimicry involving the microbiota in neurodegeneration has been proposed. Further, Friedland (2015) suggested that bacterial proteins may elicit cross-seeded misfolding, inflammation and oxidative stress, and cellular toxicity in neurodegeneration, influencing the event of PD. In a germ-free animal model, the gut microbiota influences the blood-brain barrier permeability associated with reduced expression of the tight junction proteins in a homological way as it affects the intestinal epithelial barrier (Braniste et al. 2014). Sui et al. (2014) proved that a bidirectional transport of α-syn into and out of the brain by the blood-brain barrier is feasible and suggested that LPS-induced inflammation could increase α-syn uptake by the brain by disrupting the blood-brain barrier. In an animal model of PD, inducing the microglial complement pathway to damage dopaminergic neurons results in inflammation (Bodea et al. 2014). More research into a new therapeutic approach for Parkinson's disease based on changing the gut microbiota with probiotics, prebiotics, or maybe faecal microbiota transplantation is needed.

11.6.4 Multiple Sclerosis

Multiple sclerosis (MS) is a destructive autoimmune disorder that is characterized by the progressive deterioration of neurological function. This damage leads to symptoms like disrupting the power of nervous system parts to communicate, including physical, mental and sometimes psychiatric problems (Compston and Coles 2008). Recent studies suggested that the gut microbiota may have a role in MS (Berer et al. 2011); in germ-free mice, it has been manifested that the induction of experimental autoimmune encephalomyelitis (EAE), by myelin oligodendrocyte glycoprotein (MOG) peptide, was greatly attenuated. In germ-free animals, it shows the reduced immune responses to MOG due to this relative resistance (Lee et al. 2011); further, the gut microbiota shows the effect on CNS function via the immune system. In another study, a similar effect is shown in which mice that were genetically predisposed to spontaneously develop EAE were housed under germ-free or specific-pathogen-free conditions and, as a result, remained fully protected from EAE throughout their life, and this protection degenerates upon colonization with conventional microbiota in adulthood. There are some pieces of evidence of information demonstrating a key role of the gut microbiota in immunomodulatory mechanisms underlying MS, and further research studies should also investigate whether other aspects of MS pathophysiology, especially at the spinal cord level, and the beneficial role of the gut microbiota towards MS.

11.6.5 Autism

Autism is a neurodevelopmental disorder characterized by impaired social communication (verbal and nonverbal communication) and constrained and recurring behaviour. Microbiota dysbiosis and GI abnormalities have been identified in children with autism symptoms (Finegold et al. 2010; Williams et al. 2011). The gut profile of the children suffering from autism reveals that, increased levels Clostridium, Bacteroides and Desulfovibrio species and incontrast decreased levels of Firmicutes and Bifidobacterium species (Song et al. 2004). Some studies showing high intestinal permeability in autistic subjects may be involved in the pathogenesis of the disease rather than in the consequences of autistic behaviours (Finegold et al. 2010; Yap et al. 2010). However, a recent study demonstrated that microbiota of the animals play a crucial role in social behaviours they used two different types of mice group among them one was germ-free mice characterized by social avoidance and deficits in social cognition additionally to increases in repetitive rooming behaviours, on the contrary, another group of germ-free mice was colonized with specific bacteria this group showed improved social behaviours which prove that gut microbiota has an indirect role in reducing autism (Desbonnet et al. 2014). These studies provide promising evidence indicating a more direct role of the microbiota-gut-brain axis in the pathogenesis of autism.

11.6.6 Schizophrenia

Schizophrenia is a serious neuropsychiatric ailment that has been identified for over a century, yet the disease's mechanism remains unknown. Its symptoms consist of false viewpoint, uncertain or perplexed thinking and auditory, visual, olfactory and gustatory hallucinations and usually respond well to medication (Picchioni and Murray 2007). Recent clinical studies demonstrated an unregulated immune and inflammatory status in patients with schizophrenia and a correlation between the extents of inflammatory markers (Hope et al. 2013). It occurs due to the action of pro-inflammatory cytokines that results in uncontrolled neuroinflammation involved in the pathogenesis of schizophrenia (Dennison et al. 2012; Nemani et al. 2014). Recently, a human virus called *Chlorovirus* (family Phycodnaviridae) has been identified that affects cognitive function relevant to schizophrenia in animal models (Yolken et al. 2014). On the other hand, animal models used for the clinical study of schizophrenia demonstrated that the gut microbiota profile is linked to memory performance, signifying an influence of the microbiota on cognition in the model, which was supported by the restoration of cognition through oral ampicillin administration (Jorgensen et al. 2015). In some results, the alternation of microbiota in schizophrenia includes structural damage to the GI tract, a heightened immune reaction to infectious pathogens and food antigens and other neuropsychiatric disorders (Nemani et al. 2014).

11.6.7 Alcohol Dependence

Alcohol dependence is a kind of disorder wherein the individual shows psychological dependency on alcohol. In preclinical trials, alcohol was administered to mice and rats, which caused microbial dysbiosis and gut profile revealed a decrease in Firmicutes while increasing Bacteroidetes levels (Mutlu et al. 2009; Yan et al. 2011). Stool cultures of alcoholic individuals when compared with healthy individuals showed a significant decline in the *Bifidobacterium* and *Lactobacillus* (Kirpich et al. 2008). Individuals with the habit of chronic alcoholism suffer from dysbiosis and overgrowth of pathogenic bacteria which leads to gut mucosal damage and intestinal permeability (Keshavarzian et al. 2009; Leclercq et al. 2012; Yan et al. 2011). Alcoholic individuals with high alcohol-craving scores have upregulated pro-inflammatory pathways, especially IL-8 and IL-1β, which possibly will initiate inflammation in the gut (Leclercq et al. 2014a, b). There is a connection between the gut microbiota, depression and anxiety, and that connection leads to negative reinforcement of drinking tendency in actively drinking alcohol-dependent subjects (Koob and Le Moal 2005). These factors and disorders are strongly related to the urge of drinking; without doubt, the gut microbiota seems to be a previously unidentified target in the management of alcohol dependence, but recent development with regard to gut microbiota concerning stress and anxiety management leads to a new possibility towards the treatment of alcohol dependence (deTimary et al. 2013; Leclercq et al. 2014a, b).

11.6.8 Cognition/Behaviour

Cognition is "the mental action or process of acquiring knowledge and understanding through thought, experience, and the senses". These processes include thinking, knowing, remembering and judging. These are higher-level functions of the brain and encompass language, imagination, perception and planning. The gut microbiota is essential for normal cognitive development in germ-free mice (Gareau et al. 2011). An in vivo study conducted using *L. rhamnosus* (JB-1) showed marked increase and alteration of expression of $GABA_{B1b}$ in the different regions of the brain and found increased expression in prelimbic and cingulate regions present in the cortical region of the brain; alternatively, there was marked decrease of expression in the locus coeruleus, hippocampus and amygdala; besides, another gene ($GABA_{A\alpha2}$) responsible for regulating many physiological and psychological processes was also assessed, which showed increased expression in the hippocampus but decreased expression in the amygdala and prefrontal cortex when compared with control-fed mice; it also found that there was a significant reduction in corticosterone which is produced during stress, anxiety and depression-related behaviour (Bravo et al. 2011). In humans, administration of a probiotic consortium (composed of *L. helveticus* and *B. longum*) impacted normal behaviour to healthy human volunteers significantly. This suggests that administration of probiotics may potentially provide benefits on overall mood and cognition in a healthy control population.

References

Ait-Belgnaoui A, Durand H, Cartier C, Chaumaz G, Eutamene H, Ferrier L, Houdeau E, Fioramonti J, Bueno L, Theodorou V (2012) Prevention of gut leakiness by a probiotic treatment leads to attenuated HPA response to an acute psychological stress in rats. Psychoneuroendocrinology 37(11):1885–1895

Aso E, Sanchez-Pla A, Vegas-Lozano E, Maldonado R, Ferrer I (2015) Cannabis based medicine reduces multiple pathological processes in AbetaPP/PS1 mice. J Alzheimers Dis 43(3):77–991

Atarashi K, Tanoue T, Shima T, Imaoka A, Kuwahara T, Momose Y, Cheng G, Yamasaki S, Saito T, Ohba Y, Taniguchi T (2011) Induction of colonic regulatory T cells by indigenous *Clostridium* species. Science 331:337–341

Atuma C, Strugala V, Allen A, Holm L (2001) The adherent gastrointestinal mucus gel layer: thickness and physical state in vivo. Am J Physiol Gastrointest Liver Physiol 280:G922–G929

Bercik P, Denou E, Collins J, Jackson W, Lu J, Jury J, Deng Y, Blennerhassett P, Macri J, McCoy KD et al (2011a) The intestinal microbiota affect central levels of brain-derived neurotropic factor and behavior in mice. Gastroenterology 141:599–609

Bercik P, Park AJ, Sinclair D, Khoshdel A, Lu J, Huang X, Deng Y, Blennerhassett PA, Fahnestock M, Moine D et al (2011b) The anxiolytic effect of Bifidobacterium longum NCC3001 involves vagal pathways for gut–brain communication. Neurogastroenterol Motil 23:1132–1139

Berer K, Mues M, Koutrolos M, Rasbi ZA, Boziki M, Johner C, Wekerle H, Krishnamoorthy G (2011) Commensal microbiota and myelin autoantigen cooperate to trigger autoimmune demyelination. Nature 479(7374):538–541

Berthoud HR, Blackshaw LA, Brookes SJ, Grundy D (2004) Neuroanatomy of extrinsic afferents supplying the gastrointestinal tract. Neurogastroenterol Motil 16:28–33

Bested AC, Logan AC, Selhub EM (2013) Intestinal microbiota, probiotics and mental health: from Metchnikoff to modern advances: part I—autointoxication revisited. Gut Pathog 5(1):5

Bodea LG, Wang Y, Linnartz-Gerlach B, Kopatz J, Sinkkonen L, Musgrove R, Kaoma T, Muller A, Vallar L, Di Monte DA, Balling R, Neumann H (2014) Neurodegeneration by activation of the microglial complement-phagosome pathway. J Neurosci 34:8546–8556

Bonfili L, Cecarini V, Gogoi O, Berardi S, Scarpona S, Angeletti M, Rossi G, Eleuteri AM (2020) Gut microbiota manipulation through probiotics oral administration restores glucose homeostasis in a mouse model of Alzheimer's disease. Neurobiol Aging 87:35–43

Braniste V, Al-Asmakh M, Kowal C, Anuar F, Abbaspour A, Toth M, Korecka A, Bakocevic N, Ng LG, Kundu P, Gulyás B, Halldin C, Hultenby K, Nilsson H, Hebert H, Volpe BT, Diamond B, Pettersson S (2014) The gut microbiota influences blood-brain barrier permeability in mice. Sci Transl Med 6:263–158

Bravo JA, Forsythe P, Chew MV, Escaravage E, Savignac HM, Dinan TG (2011) Ingestion of Lactobacillus strain regulates emotional behavior and central GABA receptor expression in a mouse via the vagus nerve. Proc Natl Acad Sci U S A 108(38):16050–16055

Bravo JA, Julio-Pieper M, Forsythe P, Kunze W, Dinan TG, Bienenstock J, Cryan JF (2012) Communication between gastrointestinal bacteria and the nervous system. Curr Opin Pharmacol 12(6):667–672

Burns A, Iliffe S (2009) Alzheimer's disease. BMJ 338:b158

Burokas A, Martin-Garcia E, Gutierrez-Cuesta J, Rojas S, Herance JR, Gispert JD, Serra MÁ, Maldonado R (2014) Relationships between serotonergic and cannabinoid system in depressive like behavior: a PET study with [11C]-DASB. J Neurochem 130(1):126–135

Burokas A, Moloney RD, Dinan TG, Cryan JF (2015) Microbiota regulation of the mammalian gut–brain axis. Adv Appl Microbiol 91:1–62

Casellas F, Borruel N, Torrejon A, Varela E, Antolin M, Guarner F, Malagelada JR (2007) Oral oligofructose-enriched inulin supplementation in acute ulcerative colitis is well tolerated and associated with lowered faecal calprotectin. Aliment Pharm Ther 25(9):1061–1067

Caspi A, Sugden K, Moffitt TE, Taylor A, Craig IW, Harrington H, McClay J, Mill J, Martin J, Braithwaite A, Poulton R (2003) Influence of life stress on depression: moderation by a polymorphism in the 5-HTT gene. Science 301(5631):386–389

Cersosimo MG, Benarroch EE (2012) Pathological correlates of gastrointestinal dysfunction in Parkinson's disease. Neurobiol Dis 46(3):559–564

Chang CY, Ke DS, Chen JY (2009) Essential fatty acids and human brain. Acta Neurol Taiwanica 18:231–241

Chow J, Lee SM, Shen Y, Khosravi A, Mazmanian SK (2010) Host-bacterial symbiosis in health and disease. Adv Immunol 107:243–274

Claesson MJ, Cusack S, O'Sullivan O, Greene-Diniz R, de Weerd H, Flannery E, Marchesi JR, Falush D, Dinan T, Fitzgerald G, Stanton C (2011) Composition, variability, and temporal stability of the intestinal microbiota of the elderly. Proc Natl Acad Sci U S A 108:4586–4591

Clarke G, Grenham S, Scully P, Fitzgerald P, Moloney RD, Shanahan F, Dinan TG, Cryan JF (2013) The microbiome-gut-brain axis during early life regulates the hippocampal serotonergic system in a sex-dependent manner. Mol Psychiatry 18:666–673

Collins SM, Surette M, Bercik P (2012) The interplay between the intestinal microbiota and the brain. Nat Rev Microbiol 10:735–742

Compston A, Coles A (2008) Multiple sclerosis. Lancet 372(9648):1502–1517

Cryan JF, Dinan TG (2012) Mind-altering microorganisms: the impact of the gut microbiota on brain and behaviour. Nat Rev Neurosci 13:701–712

Cryan JF, O'Mahony SM (2011) The microbiome-gut-brain axis: from bowel to behavior. Neurogastroenterol Motil 23(3):187–192

de la Monte SM, Wands JR (2008) Alzheimer's disease is type 3 diabetes-evidence reviewed. J Diabetes Sci Technol 2(6):1101–1113

de Lartigue G, Barbier de la Serre C, Espero E, Lee J, Raybould HE (2011) Diet-induced obesity leads to the development of leptin resistance in vagal afferent neurons. Am J Physiol Endocrinol Metab 301:E187–E195

Dennison U, McKernan D, Cryan J, Dinan T (2012) Schizophrenia patients with a history of childhood trauma have a pro-inflammatory phenotype. Psychol Med 42(9):1865–1871

dePreter V, Vanhoutte T, Huys G, Swings J, Rutgeerts P, Verbeke K (2008) Baseline microbiota activity and initial bifidobacteria counts influence responses to prebiotic dosing in healthy subjects. Aliment Pharmacol Ther 27(6):504–513

Desbonnet L, Garrett L, Clarke G, Bienenstock J, Dinan TG (2008) The probiotic Bifidobacteria infantis: an assessment of potential antidepressant properties in the rat. J Psychiatr Res 43 (2):164–174

Desbonnet L, Garrett L, Clarke G, Kiely B, Cryan JF, Dinan TG (2010) Effects of the probiotic Bifidobacterium infantis in the maternal separation model of depression. Neuroscience 170 (4):1179–1188

Desbonnet L, Clarke G, Shanahan F, Dinan TG, Cryan JF (2014) Microbiota is essential for social development in the mouse. Mol Psychiatry 19(2):146–148

deTimary P, Cordovil de Sousa Uva M, Denoel C, Hebborn L, Derely M, Desseilles M, Luminet O (2013) The associations between self-consciousness, depressive state and craving to drink among alcohol dependent patients undergoing protracted withdrawal. PLoS One 8(8):e71560

Devos D, Lebouvier T, Lardeux B, Biraud M, Rouaud T, Pouclet H, Coron E, Bruley des Varannes S, Naveilhan P, Nguyen JM, Neunlist M, Derkinderen P (2013) Colonic inflammation in Parkinson's disease. Neurobiol Dis 50:42–48

Dickson DW, Fujishiro H, Orr C, DelleDonne A, Josephs KA, Frigerio R, Burnett M, Parisi JE, Klos KJ, Ahlskog JE (2009) Neuropathology of non-motor features of Parkinson disease. Parkinsonism Relat Disord 15:S1–S5

Dinan TG, Stanton C, Cryan JF (2013) Psychobiotics: a novel class of psychotropic. Biol Psychiatry 74:720–726

Drakoularakou A, Tzortzis G, Rastall RA, Gibson GR (2010) A double-blind, placebo-controlled, randomized human study assessing the capacity of a novel galactooligosaccharide mixture in reducing traveller's diarrhoea. Eur J Clin Nutr 64(2):146–152

Erny D, de Angelis ALH, Jaitin D, Wieghofer P, Staszewski O, David E, Keren-Shaul H, Mahlakoiv T, Jakobshagen K, Buch T, Schwierzeck V (2015) Host microbiota constantly control maturation and function of microglia in the CNS. Nat Neurosci 18(7):965–977

Finegold SM, Dowd SE, Gontcharova V, Liu C, Henley KE, Wolcott RD, Youn E, Summanen PH, Granpeesheh D, Dixon D, Liu M (2010) Pyrosequencing study of fecal microflora of autistic and control children. Anaerobe 16(4):444–453

Fond G, Boukouaci W, Chevalier G, Regnault A, Eberl G, Hamdani N, Dickerson F, Macgregor A, Boyer L, Dargel A, Oliveira J (2015) The "psychomicrobiotic": targeting microbiota in major psychiatric disorders: a systematic review. Pathol Biol 63:35–42

Forsythe P, Kunze W (2013) Voices from within: gut microbes and the CNS. Cell Mol Life Sci 70:55–69

Forsythe P, Sudo N, Dinan T, Taylor VH, Bienenstock J (2010) Mood and gut feelings. Brain Behav Immun 24:9–16

Foster JA, McVey Neufeld KA (2013) Gut-brain axis: how the microbiome influences anxiety and depression. Trends Neurosci 36(5):305–312

Francavilla R, Calasso M, Calace L, Siragusa S, Ndagijimana M, Vernocchi P, Brunetti L, Mancino G, Tedeschi G, Guerzoni E, Indrio F (2012) Effect of lactose on gut microbiota and metabolome of infants with cow's milk allergy. Pediatr Allergy Immunol 23(5):420–427

Friedland RP (2015) Mechanisms of molecular mimicry involving the microbiota in neurodegeneration. J Alzheimers Dis 45:349–362

Furness JB, Callaghan BP, Rivera LR, Cho HJ (2014) The enteric nervous system and gastrointestinal innervation: integrated local and central control. Adv Exp Med Biol 817:39–71

Furusawa Y, Obata Y, Fukuda S, Endo TA, Nakato G, Takahashi D, Nakanishi Y, Uetake C, Kato K, Kato T, Takahashi M (2013) Commensal microbe-derived butyrate induces the differentiation of colonic regulatory T cells. Nature 504(7480):446–450

Gareau MG, Jury J, MacQueen G, Sherman PM, Perdue MH (2007) Probiotic treatment of rat pups normalises corticosterone release and ameliorates colonic dysfunction induced by maternal separation. Gut 56(11):1522–1528

Gareau MG, Silva MA, Perdue MH (2008) Pathophysiological mechanisms of stress-induced intestinal damage. Curr Mol Med 8(4):274–281

Gareau MG, Wine E, Rodrigues DM, Cho JH, Whary MT, Philpott DJ, MacQueen G, Sherman PM (2011) Bacterial infection causes stress-induced memory dysfunction in mice. Gut 60 (3):307–317

Gayathri D, Rashmi BS (2017) Mechanism of development of depression and probiotics as adjuvant therapy for its prevention and management. Ment Health Prev 5:40–51

Grigoleit JS, Kullmann JS, Wolf OT, Hammes F, Wegner A, Jablonowski S, Engler H, Gizewski E, Oberbeck R, Schedlowski M (2011) Dose-dependent effects of endotoxin on neurobehavioral functions in humans. PLoS One 6(12):28330

Heijtz RD, Wang S, Anuar F, Qian Y, Björkholm B, Samuelsson A, Hibberd ML, Forssberg H, Pettersson S (2011) Normal gut microbiota modulates brain development and behavior. Proc Natl Acad Sci 108:3047–3052

Hoban AE, Stilling RM, Ryan FJ, Shanahan F, Dinan TG, Claesson MJ, Clarke G, Cryan JF (2016) Regulation of prefrontal cortex myelination by the microbiota. Transl Psychiatry 6:e774

Hollister EB, Gao C, Versalovic J (2014) Compositional and functional features of the gastrointestinal microbiome and their effects on human health. Gastroenterology 146:1449–1458

Hope S, Ueland T, Steen NE, Dieset I, Lorentzen S, Berg AO, Agartz I, Aukrust P, Andreassen OA (2013) Interleukin 1 receptor antagonist and soluble tumor necrosis factor receptor 1 are associated with general severity and psychotic symptoms in schizophrenia and bipolar disorder. Schizophr Res 145(1–3):36–42

Hornig M (2013) The role of microbes and autoimmunity in the pathogenesis of neuropsychiatric illness. Curr Opin Rheumatol 25(4):488–495

Huang Y, Mucke L (2012) Alzheimer mechanisms and therapeutic strategies. Cell 148 (6):1204–1222

Iliev ID, Leonardi I (2017) Fungal dysbiosis: immunity and interactions at mucosal barriers. Nat Rev Immunol 17:635–646

Jiang X, Zhang D, Shi J, Chen Y, Zhang P, Mei B (2009) Increased inflammatory response both in brain and in periphery in presenilin 1 and presenilin 2 conditional double knock-out mice. J Alzheimers Dis 18(3):515–523

Jorgensen B, Krych L, Pedersen TB, Plath N, Redrobe JP, Hansen AK, Nielsen DS, Pedersen CS, Larsen C, Sørensen DB (2015) Investigating the long-term effect of subchronic phencyclidine-treatment on novel object recognition and the association between the gut microbiota and behavior in this animal model of schizophrenia. Physiol Behav 141:32–39

Kaur H, Golovko S, Golovko MY, Singh S, Darland DC, Combs CK (2020) Effects of probiotic supplementation on short chain fatty acids in the App NL-GF mouse model of Alzheimer's disease. J Alzheimers Dis 76(3):1083–1102

Kendler KS, Thornton LM, Gardner CO (2000) Stressful life events and previous episodes in the etiology of major depression in women: an evaluation of the "kindling" hypothesis. Am J Psychiatry 157(8):1243–1251

Keshavarzian A, Farhadi A, Forsyth CB, Rangan J, Jakate S, Shaikh M, Banan A, Fields JZ (2009) Evidence that chronic alcohol exposure promotes intestinal oxidative stress, intestinal hyperpermeability and endotoxemia prior to development of alcoholic steatohepatitis in rats. J Hepatol 50(3):538–547

Kirpich IA, Solovieva NV, Leikhter SN, Shidakova NA, Lebedeva OV, Sidorov PI, Bazhukova TA, Soloviev AG, Barve SS, McClain CJ, Cave M (2008) Probiotics restore bowel flora and improve liver enzymes in human alcohol-induced liver injury: a pilot study. Alcohol 42 (8):675–682

Koob GF, Le Moal M (2005) Plasticity of reward neurocircuitry and the 'dark side' of drug addiction. Nat Neurosci 8(11):1442–1444

Kozakova H, Schwarzer M, Tuckova L, Srutkova D, Czarnowska E, Rosiak I, Hudcovic T, Schabussova I, Hermanova P, Zakostelska Z, Aleksandrzak-Piekarczyk T (2016) Colonization of germ-free mice with a mixture of three lactobacillus strains enhances the integrity of gut mucosa and ameliorates allergic sensitization. Cell Mol Immunol 13:251–262

Lahtinen SJ, Boyle RJ, Kivivuori S, Oppedisano F, Smith KR, Robins-Browne R, Salminen SJ, Tang ML (2009) Prenatal probiotic administration can influence Bifidobacterium microbiota development in infants at high risk of allergy. J Allergy Clin Immunol 123(2):499–501

Leclercq S, Cani PD, Neyrinck AM, Starkel P, Jamar F, Mikolajczak M, Delzenne NM, de Timary P (2012) Role of intestinal permeability and inflammation in the biological and behavioral control of alcohol-dependent subjects. Brain Behav Immunol 26(6):911–918

Leclercq S, De Saeger C, Delzenne N, de Timary P, Starkel P (2014a) Role of inflammatory pathways, blood mononuclear cells, and gut-derived bacterial products in alcohol dependence. Biol Psychiatry 76(9):725–733

Leclercq S, Matamoros S, Cani PD, Neyrinck AM, Jamar F, Starkel P, Windey K, Tremaroli V, Bäckhed F, Verbeke K, de Timary P (2014b) Intestinal permeability, gut-bacterial dysbiosis, and behavioral markers of alcohol-dependence severity. Proc Natl Acad Sci U S A 111(42): E4485–E4493

Lee YK, Menezes JS, Umesaki Y, Mazmanian SK (2011) Proinflammatory T-cell responses to gut microbiota promote experimental autoimmune encephalomyelitis. Proc Natl Acad Sci U S A 108:4615–4622

Lewis S, Burmeister S, Brazier J (2005) Effect of the prebiotic oligofructose on relapse of Clostridium difficile-associated diarrhea: a randomized, controlled study. Clin Gastroenterol Hepatol 3(5):442–448

Lozupone CA, Stombaugh JI, Gordon JI, Jansson JK, Knight R (2012) Diversity, stability and resilience of the human gut microbiota. Nature 489:220–230

Maes M, Kubera M, Leunis JC (2008) The gut-brain barrier in major depression intestinal mucosal dysfunction with an increased translocation of LPS from gram negative enterobacteria (leaky gut) plays a role in the inflammatory pathophysiology of depression. Neuro Endocrinol Lett 29 (1):117–124

Matsumoto M, Kurihara S, Kibe R, Ashida H, Benno Y (2011) Longevity in mice is promoted by probiotic-induced suppression of colonic senescence dependent on upregulation of gut bacterial polyamine production. PLoS One 6:e23652

Mayer EA, Knight R, Mazmanian SK, Cryan JF, Tillisch K (2014) Gut microbes and the brain: paradigm shift in neuroscience. J Neurosci 34:15490–15496

Mayer EA, Tillisch K, Gupta A (2015) Gut/brain axis and the microbiota. J Clin Invest 125 (3):926–938

McDermott AJ, Huffnagle GB (2014) The microbiome and regulation of mucosal immunity. Immunology 142:24–31

Messaoudi M, Lalonde R, Violle N, Javelot H, Desor D, Nejdi A, Bisson JF, Rougeot C, Pichelin M, Cazaubiel M, Cazaubiel JM (2011) Assessment of psychotropic-like properties of a probiotic formulation (Lactobacillus helveticus R0052 and Bifidobacterium longum R0175) in rats and human subjects. Br J Nutr 105(5):755–764

Minot S, Sinha R, Chen J, Li H, Keilbaugh SA, Wu GD, Lewis JD, Bushman FD (2011) The human gut virome: inter-individual variation and dynamic response to diet. Genome Res 21 (10):1616–1625

Minot S, Grunberg S, Wu GD, Lewis JD, Bushman FD (2012) Hypervariable loci in the human gut virome. Proc Natl Acad Sci 109(10):3962–3966

Minot S, Bryson A, Chehoud C, Wu GD, Lewis JD, Bushman FD (2013) Rapid evolution of the human gut virome. Proc Natl Acad Sci 110(30):12450–12455

Moloney RD, O'Leary OF, Felice D, Bettler B, Dinan TG, Cryan JF (2012) Early-life stress induces visceral hypersensitivity in mice. Neurosci Lett 512(2):99–102

Moloney RD, Desbonnet L, Clarke G, Dinan TG, Cryan JF (2014) The microbiome: stress, health and disease. Mamm Genome 25(1–2):49–74

Mueller NT, Bakacs E, Combellick J, Grigoryan Z, Dominguez-Bello MG (2015) The infant microbiome development: mom matters. Trends Mol Med 21(2):109–117

Mulak A, Bonaz B (2015) Brain-gut-microbiota axis in Parkinson's disease. World J Gastroenterol 21(37):10609–10620

Mutlu E, Keshavarzian A, Engen P, Forsyth CB, Sikaroodi M, Gillevet P (2009) Intestinal dysbiosis: a possible mechanism of alcohol-induced endotoxemia and alcoholic steatohepatitis in rats. Alcohol Clin Exp Res 33(10):1836–1846

Nemani K, Hosseini Ghomi R, McCormick B, Fan X (2014) Schizophrenia and the gut-brain axis. Prog Neuropsychopharmacol Biol Psychiatry 56C:155–160

Neufeld KM, Kang N, Bienenstock J, Foster JA (2011) Reduced anxiety-like behavior and central neurochemical change in germ-free mice. Neurogastroenterol Motil 23:255–264

Nguyen TL, Vieira-Silva S, Liston A, Raes J (2015) How informative is the mouse for human gut microbiota research? Dis Model Mech 8:1–16

Ohland CL, Kish L, Bell H, Thiesen A, Hotte N, Pankiv E, Madsen KL (2013) Effects of Lactobacillus helveticus on murine behavior are dependent on diet and genotype and correlate with alterations in the gut microbiome. Psychoneuroendocrinology 38(9):1738–1747

Olanow CW, Wakeman DR, Kordower JH (2014) Peripheral alpha synuclein and Parkinson's disease. Mov Disord 29:963–966

Perry VH, Holmes C (2014) Microglial priming in neurodegenerative disease. Nat Rev Neurol 10 (4):217–224

Picchioni MM, Murray RM (2007) Schizophrenia. BMJ 335(7610):91–95

Qi Y, Goel R, Kim S, Richards EM, Carter CS, Pepine CJ, Raizada MK, Buford TW (2017) Intestinal permeability biomarker zonulin is elevated in healthy aging. J Am Med Dir Assoc 18:810

Querfurth HW, LaFerla FM (2010) Alzheimer's disease. N Engl J Med 362(4):329–344

Rodriguez JM, Murphy K, Stanton C, Ross RP, Kober OI, Juge N, Avershina E, Rudi K, Narbad A, Jenmalm MC, Marchesi JR (2015) The composition of the gut microbiota throughout life, with an emphasis on early life. Microb Ecol Health Dis 26(1):26050

Rousseaux C, Thuru X, Gelot A, Barnich N, Neut C, Dubuquoy L, Dubuquoy C, Merour E, Geboes K, Chamaillard M, Ouwehand A (2007) Lactobacillus acidophilus modulates intestinal pain and induces opioid and cannabinoid receptors. Nat Med 13(1):35–37

Ruddick JP, Evans AK, Nutt DJ, Lightman SL, Rook GA, Lowry CA (2006) Tryptophan metabolism in the central nervous system: medical implications. Expert Rev Mol Med 8(20):1–27
Saulnier DM, Ringel Y, Heyman MB, Foster JA, Bercik P, Shulman RJ, Versalovic J, Verdu EF, Dinan TG, Hecht G, Guarner F (2013) The intestinal microbiome, probiotics and prebiotics in neurogastroenterology. Gut Microbes 4(1):17–27
Savignac HM, Corona G, Mills H, Chen L, Spencer JP, Tzortzis G, Burnet PW (2013) Prebiotic feeding elevates central brain derived neurotrophic factor, N-methyl-D-aspartate receptor subunits and D-serine. Neurochem Int 63(8):756–764
Schmidt K, Cowen PJ, Harmer CJ, Tzortzis G, Errington S, Burnet PW (2015) Prebiotic intake reduces the waking cortisol response and alters emotional bias in healthy volunteers. Psychopharmacology 232(10):1793–1801
Schmidt TS, Raes J, Bork P (2018) The human gut microbiome: from association to modulation. Cell 172(6):1198–1215
Sherman MP, Zaghouani H, Niklas V (2014) Gut microbiota, the immune system, and diet influence the neonatal gut-brain axis. Pediatr Res 77(1–2):127–135
Song Y, Liu C, Finegold SM (2004) Real-time PCR quantitation of clostridia in feces of autistic children. Appl Environ Microbiol 70(11):6459–6465
Sudo N, Chida Y, Aiba Y, Sonoda J, Oyama N, Yu XN, Kubo C, Koga Y (2004) Postnatal microbial colonization programs the hypothalamic-pituitary-adrenal system for stress response in mice. J Physiol 558:263–275
Sui YT, Bullock KM, Erickson MA, Zhang J, Banks WA (2014) Alpha synuclein is transported into and out of the brain by the blood brain barrier. Peptides 62:197–202
Tillisch K, Labus J, Kilpatrick L, Jiang Z, Stains J, Ebrat B, Guyonnet D, Legrain-Raspaud S, Trotin B, Naliboff B, Mayer EA (2013) Consumption of fermented milk product with probiotic modulates brain activity. Gastroenterology 144(7):1394–1401
van Vlies N, Hogenkamp A, Thijssen S, Dingjan GM, Knipping K, Garssen J, Knippels LMJ (2012) Effects of short-chain galacto- and long-chain fructo-oligosaccharides on systemic and local immune status during pregnancy. J Reprod Immunol 94(2):161–168
Vaughn AC, Cooper EM, DiLorenzo PM, O'Loughlin LJ, Konkel ME, Peters JH, Hajnal A, Sen T, Lee SH, de La Serre CB, Czaja K (2017) Energy-dense diet triggers changes in gut microbiota, reorganization of gut-brain vagal communication and increases body fat accumulation. Acta Neurobiol Exp 77(1):18–30
Verdu EF, Bercik P, Verma-Gandhu M, Huang XX, Blennerhassett P, Jackson W, Mao Y, Wang L, Rochat F, Collins SM (2006) Specific probiotic therapy attenuates antibiotic induced visceral hypersensitivity in mice. Gut 55(2):182–190
Walker WA (2017) The importance of appropriate initial bacterial colonization of the intestine in newborn, child, and adult health. Pediatr Res 82(3):387–395
Williams BL, Hornig M, Buie T, Bauman ML, Cho Paik M, Wick I, Bennett A, Jabado O, Hirschberg DL, Lipkin WI (2011) Impaired carbohydrate digestion and transport and mucosal dysbiosis in the intestines of children with autism and gastrointestinal disturbances. PLoS One 6 (9):e24585
Yan AW, Fouts DE, Brandl J, Starkel P, Torralba M, Schott E, Tsukamoto H, Nelson KE, Brenner DA, Schnabl B (2011) Enteric dysbiosis associated with a mouse model of alcoholic liver disease. Hepatology 53(1):96–105
Yano JM, Yu K, Donaldson GP, Shastri GG, Ann P, Ma L, Nagler CR, Ismagilov RF, Mazmanian SK, Hsiao EY (2015) Indigenous bacteria from the gut microbiota regulate host serotonin biosynthesis. Cell 161(2):264–276
Yap IK, Angley M, Veselkov KA, Holmes E, Lindon JC, Nicholson JK (2010) Urinary metabolic phenotyping differentiates children with autism from their unaffected siblings and age-matched controls. J Prot Res 9(6):2996–3004

Yolken RH, Jones-Brando L, Dunigan DD, Kannan G, Dickerson F, Severance E, Sabunciyan S, Talbot CC, Prandovszky E, Gurnon JR, Agarkova IV (2014) Chlorovirus ATCV-1 is part of the human oropharyngeal virome and is associated with changes in cognitive functions in humans and mice. Proc Natl Acad Sci U S A 111(45):16106–16111

Zeisel SH (2004) Nutritional importance of choline for brain development. J Am Coll Nutr 23:621S–626S

Zhao G, Vatanen T, Droit L, Park A, Kostic AD, Poon TW, Vlamakis H, Siljander H, Härkönen T, Hämäläinen AM, Peet A (2017) Intestinal virome changes precede autoimmunity in type I diabetes-susceptible children. Proc Natl Acad Sci U S A 114:E6166–E6175

Zheng L, Kelly CJ, Colgan SP (2015) Physiologic hypoxia and oxygen homeostasis in the healthy intestine. A review in the theme: cellular responses to hypoxia. Am J Physiol Cell Physiol 309: C350–C360

Probiotic: A Sustainable Approach Towards Healthy Food 12

Salma Khan, Asia Nosheen, Paul A. Correa, and Pierre A. Mendy

Abstract

Probiotic microorganisms play a very important role in food and medicines. These microorganisms improve the food quality, enhance bioavailability of the nutrients, produce antimicrobial and antioxidant compounds, degrade toxic compounds such as phytic acid and mycotoxins, and improve the digestion process. Probiotic microorganisms are extensively used in food and food products such as in yogurt, cheese, kefir, kimchi, formula milk, fermented food, medicines, and many other applications. There are two most important genera of the probiotic bacteria such as *Lactobacillus* and *Bifidobacterium* which are widely used in food products. Probiotic microorganisms help to cure many diseases such as irritable bowel syndrome and inflammatory bowel disease, reduce allergy and diarrhea, and reduce the saturated fatty acid level from the bloodstream. They reduce the chances of occurrence of breast cancer in females. These bacteria compete for the necessary nutrients and leave a very small amount of nutrients for the pathogenic bacteria and bind to the intestinal epithelium through adhesion sites and colonize and prevent the pathogenic bacteria to bind by reducing the surface area. Probiotic bacteria boost the immune system through signaling mechanism by releasing the cytokines for the destruction of pathogenic bacteria. Although there have been remarkable uses and application of probiotics from the last three decades, it still needs a lot of research to ensure the safety and stability of these food products.

S. Khan · A. Nosheen (✉) · P. A. Correa · P. A. Mendy
Department of Biosciences, COMSATS University, Islamabad, Pakistan
e-mail: asia.nosheen@comsats.edu.pk

R. Z. Sayyed, M. Khan (eds.), *Microbiome-Gut-Brain Axis*,
https://doi.org/10.1007/978-981-16-1626-6_12

Keywords

Probiotics · Food security · Sustainable food · Digestion · Medicines · *Lactobacillus* · *Bifidobacteria*

12.1 Introduction

As population is increasing, the demand for the food is also increasing day by day. Thus, the food industry is the biggest one which generates largest profits in the world above $500 billion per year (Granato et al. 2011). The importance of the food industry is to produce a diversity of foods that contains those constituents which are beneficial for human health (Granato et al. 2011). The use of beneficial bacteria/bacterial products in foods is called as probiotics or useful foods and is one of the rapidly developing domains of the food manufacturing industry. A lot of research has been carried out on probiotics that helped researchers to identify, isolate, and characterize probiotic bacteria and their benefits on human health (Denkova et al. 2013; Hill et al. 2014; Chakraborty and Bhowal 2015; Rahman 2015; Wang et al. 2017; Raghuwanshi et al. 2018). Probiotic bacteria are considered as a sustainale source for food industry and can play an important role in the food security of a country.

The word probiotic means "for life" and was first coined by two scientists: Kollath and Vergin (Rijkers et al. 2011; Hill et al. 2014; Wedajo 2015). Elie Metchnikoff at Pasteur Institute in France was the first Russian scientist who documented the suitable properties of *Lactobacillus* in fermented dairy products (Tripathi and Giri 2014). The present definition of probiotics is that these are live microorganisms such as *Lactobacillus* and *Bifidobacterium* species which when added to the food in sufficient quantity are beneficial for human health and ameliorate the intestinal stability (FAO/WHO 2002). Elie Metchnikoff stated that the *Lactobacillus* bacteria in acrid milk have the ability to prevent the progression and noxiousness of anaerobic, spore-forming bacteria in the alimentary canal (Tripathi and Giri 2014). The two most important genera of probiotic bacteria include *Lactobacillus* and *Bifidobacterium* that can be used in fermented food products and help in the digestion of food (Swain et al. 2014; Ashraf and Smith 2015; Begum et al. 2017).

Lactobacillus is a Gram-positive, non-motile, and rod-shaped fermentative bacterium that survives well in anaerobic conditions and is moderately acid resistant and is able to effectively survive passage through the stomach. Moreover, the lack of lipopolysaccharides in their cell wall virtually eliminates the risk of endotoxic shock (Khalil and Anwar 2016).

On the basis of fermentative capability, *Lactobacillus* can be divided into two groups: homofermentative and heterofermentative species (Halasz 2011). The second most important genus of probiotic bacteria includes *Bifidobacterium* which is a Gram-positive and anaerobic bacterium and has the capability to grow well at the pH ranging from 4.5 to 8.5. *Bifidobacterium breve* plays an important role in curing constipation in toddler stage (Afzaal et al. 2013). Other species of probiotic bacteria include *Lactococcus*, *Leuconostoc mesenteroides*, *Lactobacillus acidophilus*, *Saccharomyces boulardii*, and *Streptococcus thermophilus* (Fijian 2014; Adeniyi et al. 2015; Sornplang and Piyadeatsoontorn 2016; Begum et al. 2017). The main sources of probiotic bacteria include fermented dairy products such as fresh yogurt, milk, and cheese. Yogurt and cheese are the main transporters of probiotic bacteria in humans which are beneficial for human health and balance the intestinal microflora (Pyar and Peh 2014; Yadav et al. 2015; Ashraf and Smith 2015).

Other fermented dairy-based products that serve as transporters of probiotic bacteria include chocolates, mousse, and ice cream (Begum et al. 2017). Supplements in the form of capsules and tablets also contain probiotic bacteria (Granato et al. 2011; Raghuwanshi et al. 2018). According to Granato et al. (2011), probiotic foods include fresh fruit and vegetable juices such as pineapple juice, orange juice, carrot juice, cabbage juice, and ginger juice as well as pickles, probiotic beverages, and meat products. Kitchen waste can also be used as a main source of probiotics (Yin et al. 2013).

Probiotics have beneficial effects in curing diseases like inflammatory bowel disease (IBD) and irritable bowel syndrome (IBS) and minimize the levels of saturated fatty acids in the blood (Fox et al. 2015; Raghuwanshi et al. 2018). Some strains of *Lactobacillus* such as *Lactobacillus acidophilus* have the ability to take up lipids from the blood. Probiotics also play an important role in preventing diseases like diarrhea and reducing allergy symptoms in children (Kaur et al. 2014). Probiotic bacteria termed as "friendly or useful bacteria" are also used in food industries in order to improve the taste of fermented food products in a diversity of means (Kaur et al. 2014). These beneficial bacteria are used in preservation of different fruits and vegetables and as appetizers of different fermented dairy products by food manufacturing industries to improve the taste and shelf life of these products which when consumed by humans are beneficial for their health and reduce the occurrence of different stomach diseases (Chen et al. 2013).

12.2 Characteristics of Probiotic Bacteria

Probiotics are anti-oxidative in nature. When oxidative stress arises, abnormal amount of reactive oxygen species such as superoxide anion radicals, hydroxyl radicals, and hydrogen peroxide are produced, which results in cell and DNA damage. These free oxygen radicals accumulate in the abdominal tract of humans and damage the lining of the abdominal wall and thus cause the state of chronic infection/disease (Kushugulova et al. 2014). According to Wang et al. (2017), *Lactobacillus* and *Bifidobacterium* strains have been observed to limit the progression of oxygen free radicals in the abdominal tract of humans by producing the short-chain fatty acids such as acetate and butyrate. These molecules have antioxidant properties that limit the growth of reactive oxygen species (ROS) and aid to lessen the chronic inflammation and repair the lining of the abdominal wall that has been damaged by ROS. According to Park et al. (2011) and Abubakr et al. (2012),

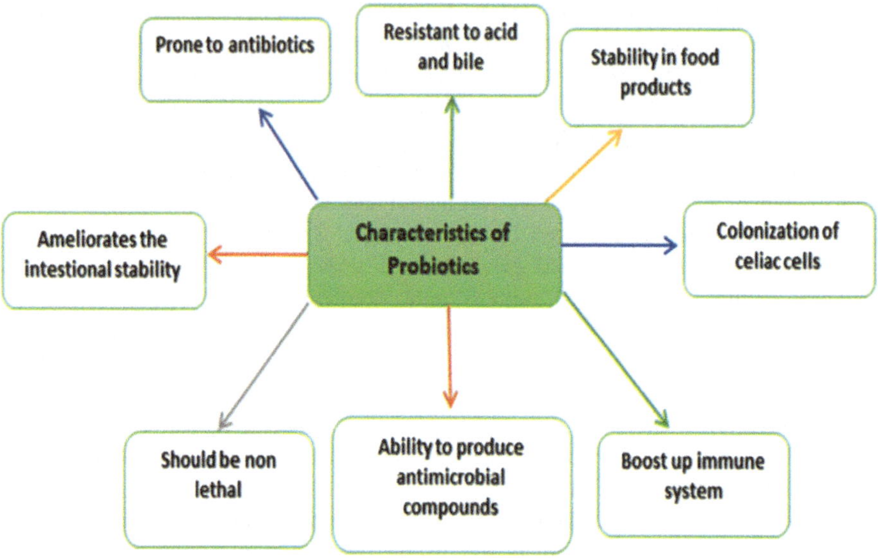

Fig. 12.1 Characteristics of probiotic bacteria. (Modified from Afzaal et al. 2013)

fermented food products such as kimchi and yogurt also have antioxidant activities like DPPH (2,2-diphenyl-1-picrylhydrazyl) and ABTS (azino-bis-3-ethylbenzothiazoline-6-sulfonic acid) radical scavenging activities that limit the accumulation of active free radicals in the abdominal tract of humans and prevent the progression of different diseases like diarrhea, food poisoning, and systemic and enteric infections.

Probiotics are also antibacterial in nature. The antibacterial nature of probiotics depends upon the production of active metabolites which are accountable for preventing and killing the human pathogenic bacteria (Tonekabon 2013). These active antibacterial metabolites include end products of sugar fermentation such as acetic acid and lactic acid that in the presence of low pH have antibacterial activity. The low molecular weight organic compounds such as acetaldehyde, ethanol, acetoin, reuterin, and carbon dioxide produced also prevent the growth of bacteria by disrupting their membrane. Production of peptides such as bacteriocin also has direct antimicrobial activity as it interferes during the cell wall synthesis of target pathogenic bacteria and causes the pore formation in its cell wall and thus inhibits the expression of its virulence gene (Raghuwanshi et al. 2018). According to Pundir et al. (2013), the activity of probiotic bacteria leads towards the development of fermented food products such as yogurt that inhibit or kill the pathogenic bacteria and aid in the safety and improvement of intestinal microbiota and play an important role in the health of consuming community (Fig. 12.1).

Antibiotic susceptibility test is one of the major selection criteria for probiotics. The bacterial strains that should be used as probiotic bacteria in fermented food products should be precisely checked for antibiotic resistance which could be useful for restoring the gut micrbiota after the treatment of antibiotic (Gueimonde et al. 2013). Beneficial bacteria such as *Lactobacillus* and *Bifidobacterium* have natural as well as acquired antibiotic resistance which seems to be safe for use in functional foods for the health of consuming community (Gueimonde et al. 2013) (Fig. 12.1).

The important characteristic feature of probiotic bacteria is that it should be able to tolerate the harsh conditions of the human gut such as high osmotic salt (NaCl) concentration. This feature gives an indication of the osmotolerance level of the probiotic bacteria such as *Lactobacillus* strains. The bacterial cells which are cultured in high salt concentration could lose their turgor pressure which in turn affects their physiology, enzymatic activity, and metabolism (Menconi et al. 2014). The high osmotic tolerance of NaCl is one of the most important requirements for probiotic bacteria such as *Lactobacillus* to be used as commercial strains or in probiotic foods. The reason behind is that when lactic acid is produced by these bacteria, the alkali would be pushed into the broth to avoid an excessive decline in pH and the free acid would be transformed into salt, thus increasing the osmotic pressure on bacterial cells (Menconi et al. 2014) (Fig. 12.1).

Another important feature of probiotic bacteria is that it should be able to survive in an acidic environment of the human gut. Probiotic bacteria such as *Lactobacillus* are able to survive in an acidic environment due to its capability to produce lactic acid and ferment lactose (Mishra and Sharma 2014). The survival of probiotic bacteria in an acidic environment is a prerequisite for the colonization of the intestinal epithelium and to perform its metabolic activity efficiently in the intestine of humans. Moreover, by the consumption of probiotic foods such as yogurt, cheese, kimchi, and kefir, the probiotic bacteria reach the abdominal tract of humans where they survive in an acidic environment and play an important role in balancing the normal microbiota, thus preventing from different stomach diseases as well as from systemic and enteric infections (Yepez and Tenea 2015) (Fig. 12.1).

12.2.1 Species of Probiotic Bacteria

Some of the most important probiotic bacteria that are beneficial for human health and are most commonly used in food products are given below.

12.2.1.1 *Lactobacillus acidophilus*

Lactobacillus is a Gram-positive, non-motile, and rod-shaped fermentative bacterium that survives well in anaerobic conditions and is moderately acid resistant. They are able to survive effectively in the stomach; moreover, the lack of lipopolysaccharides in their cell wall virtually eliminates the risk of endotoxic shock (Khalil and Anwar 2016).

Lactobacillus acidophilus is one of the most commonly used beneficial bacteria in the fermentation of yogurt (Sornplang and Piyadeatsoontorn 2016). These bacteria can also be obtained by eating yogurt in an adequate amount that contains the live cultures of these bacteria which plays an important role in the assimilation process in the gut (Ashraf and Smith 2015). During the assimilation process, it plays an important role by creating the unfavorable circumstances for the development of pathogenic bacteria by producing lactic acid and hydrogen peroxide (Sornplang and Piyadeatsoontorn 2016). *Lactobacillus acidophilus* also lowers the blood pressure in people and relieves the signs of IBS (irritable bowel syndrome). In the case of children, it also minimizes the symptoms of diarrhea triggered by different antibiotics (Prasanna et al. 2014; Fox et al. 2015).

12.2.1.2 *Lactobacillus rhamnosus*

Lactobacillus rhamnosus is a Gram-positive, anaerobic, non-motile, and heterofermentative bacterium that is a portion of ordinary gut microbiota in human beings. It is a probiotic bacterium that can be obtained by eating fermented food products in an adequate amount such as simple yogurt (Segers and Lebeer 2014; Yadav et al. 2015). It is generally regarded as safe (GRAS) and is effective in decreasing the virus-related respiratory damage caused by the pulmonary viruses (Fijian 2014). It reduces obesity in women and is useful for the treatment of bacterial vaginosis in females. It is also useful for the treatment of stomach infections and minimizes the risk of diarrhea triggered by different antibiotics in patients (Toiviainen et al. 2015). *Lactobacillus rhamnosus* also reduces the feelings of nervousness and depression in people (Bravo et al. 2011).

12.2.1.3 *Leuconostoc mesenteroides*

Leuconostoc mesenteroides is a lactic acid bacterium that is present in naturally fermented food products. It is a Gram-positive, non-spore-forming, facultative anaerobic, catalase-negative, and rod-shaped bacterium (Shukla et al. 2014).

It is usually present on the peels of fruits and vegetables and is accountable for starting the fermentation of different food products such as sauerkraut, cheese, sausage, yogurt, buttermilk, and pickles. It is also used in making bread dough (Shukla et al. 2014). It has the ability to survive at low pH and in the presence of bile salts and pepsin (Benmechernene et al. 2014). *Leuconostoc mesenteroides* also produces an environmental-friendly polymer of glucose (dextran) that has numerous applications in food and cosmetic industries (Aman et al. 2012). It is beneficial for human health and has the ability to endure passage through the digestive track of humans and significantly raises the quantity of probiotic bacterial cells in the gut (Milani et al. 2015).

12.2.1.4 *Weissella confusa*

Weissella confusa (*W. confusa*) is a member of lactic acid bacteria. It is a Gram-positive, catalase-negative, non-spore-forming, heterofermentative, and rod-shaped bacterium. It is present in numerous habitats such as on the skin and the gastrointestinal tract of humans, milk, peels of fruits and vegetables, and in fermented food products such as in European sourdough (Fusco et al. 2015).

W. confusa also produces biodegradable polymers that have numerous applications in food, clinical, and cosmetic industries (Abriouel et al. 2015; Kamboj et al. 2015). Antimicrobial activity of *W. confusa* and its useful role in fermentation of food make it as a probiotic. It is resistant to vancomycin. It can also be used as a probiotic for oral health, preventing the glucan biofilm creation of *Streptococcus mutans* (Fusco et al. 2015).

12.2.1.5 *Bifidobacterium bifidum and breve*

Bifidobacterium is the second most important probiotic bacterium that is a Gram-positive and anaerobic bacterium and has the capability to grow well at the pH ranging from 4.5 to 8.5 (Afzaal et al. 2013). *Bifidobacteria* have the ability to survive in the gastrointestinal tract of humans and considerably increase the number of probiotic bacterial cells in the stomach (Milani et al. 2015). The two most important species of *Bifidobacteria* that act as probiotic bacteria include *Bifidobacterium bifidum* and *Bifidobacterium breve*. *Bifidobacterium bifidum* helps in overhauling the stomach ulcers caused by pathogenic bacteria (*Helicobacter pylori*), whereas *Bifidobacterium breve* plays an important role in curing constipation in toddler stage and reduces the menace of kidney stones (Afzaal et al. 2013).

12.2.1.6 *Streptococcus thermophilus and salivarius*

During the fermentation process of dairy products, *Streptococcus thermophilus* is used as a starter culture to initiate the process of fermentation (Ashraf and Smith 2015). *Streptococcus thermophilus* prevents the stomach ulcers and reduces the risk of kidney stones. It is also found to relieve the pain of abdominal cramps, nausea, and diarrhea triggered by different antibiotics (Prasanna et al. 2014; Fox et al. 2015). One of the most important probiotic bacteria of the *Streptococcus* genus is *Streptococcus salivarius* which is found to ameliorate the symptoms of halitosis, a bad sniff caused by the bad microorganisms that reside in the mouth (Sornplang and Piyadeatsoontorn 2016).

12.2.1.7 *Bacillus coagulans*

Bacillus coagulans is also the most important probiotic bacterium that is a Gram-positive bacterium and is given to patients having severe immune disorders. It is available in the market in the form of different dietary supplements (Ashraf and Shah 2014). This probiotic bacterium is used to treat stomach disorders such as diarrhea triggered by different antibiotics and cures constipation in toddler stage. Moreover, it is also used to treat irritable bowel syndrome (Sornplang and Piyadeatsoontorn 2016). *Bacillus coagulans* in the form of dietary supplement is also useful in preventing lung infections and boosts up the immune system of patients (Ashraf and Shah 2014).

12.2.1.8 *Bacillus subtilis*

Bacillus subtilis is a Gram-positive, facultative anaerobic, spore-forming, and rod-shaped bacterium that is found in the gut of human beings (Kubo et al. 2011). It is a useful bacterium and is added in the formula of probiotic supplements (Kubo et al. 2011). It is generally regarded as safe and useful for the consumption of humans both by the Food and Drug Administration of the United States and by the EFSA (European Food Safety Authority). It reduces stress and improves gut microbiota in humans (McKenney et al. 2013). It aids in digestion and stimulates the immune system (Khatri et al. 2016). It is also used to treat the symptoms of nausea, fatigue, irritable bowel syndrome (IBS), and urinary tract infections in people (McKenney et al. 2013). *Bacillus subtilis* can also be added with other beneficial bacteria in probiotic supplements which create a competitive environment for pathogenic bacteria in the intestine (Bermudez-Brito et al. 2012; McKenney et al. 2013).

12.3 Mode of Action of Probiotics

Probiotics compete against the pathogenic bacteria for the same necessary nutrients and in turn leave a very small amount of nutrients for pathogenic bacteria to utilize. Probiotic bacteria bind to the adhesion sites of the intestinal epithelium and thus prevent the pathogenic bacteria to bind to these sites and colonize by reducing the surface area as shown in Fig. 12.2.

Probiotic bacteria then activate the immune system cells by signaling mechanism which releases various cytokines for the destruction of pathogenic bacteria. Finally, probiotic bacteria attack the pathogenic bacteria by releasing the bacteriocins that are toxins and thus kill them directly (Bermudez-Brito et al. 2012).

12.3.1 Mechanism of Action of Probiotics in the Human Gut

Probiotic bacteria compete with microbial pathogens in the human gut for a limited number of receptors present on the surface of the intestinal epithelium. Probiotics affect the bacterial groups of the abdominal tract and suppress their growth by inducing the production of β-defensin and IgA (Hemaiswarya et al. 2013). Probiotic

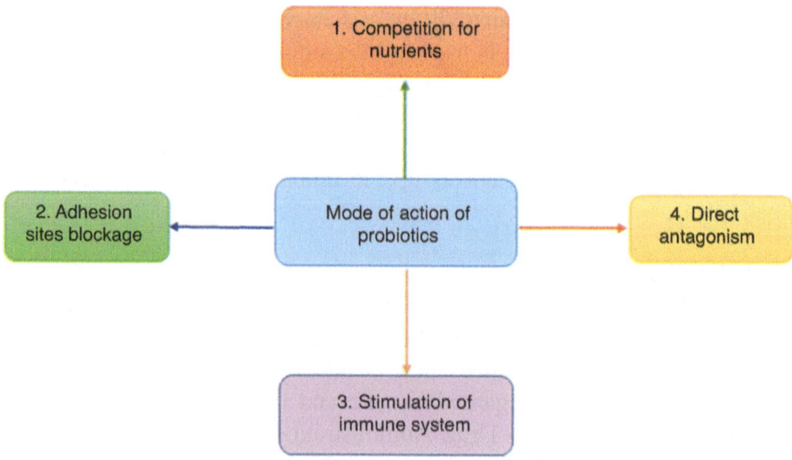

Fig. 12.2 Mode of action of probiotic bacteria. (Modified from Bermudez-Brito et al. 2012)

bacteria then strengthen the gut barrier by maintaining the tight junctions and induce the production of mucin. Probiotic-mediated immunomodulation then occurs through arbitration of cytokine secretion through signaling pathways such as NF-κβ (nuclear factor-kappa β) and MAPK (mitogen-activated protein kinase) pathways (Hemaiswarya et al. 2013).

Modulation of the immune system takes place through immune modulatory cells such as dendritic cells and the induction of protective cytokines such as IL-10 (interleukin-10) and TGF-β (tumor growth factor-β) and suppresses pro-inflammatory cytokines such as TNF (tumor necrosis factor). In the last step, the induction of T regulatory cells and T cell apoptosis takes place in the mucosal immune section to prevent inflammation (Hemaiswarya et al. 2013).

12.3.2 Stability and Sustainability of Probiotic Strains

The stability and sustainability of probiotic strains depend upon the following aspects:

- The probiotic strains must have the capacity to endure and sustain themselves in storage without the forfeiture of viability (Shewale et al. 2014).
- The probiotic strains must have the capacity to develop and propagate to utmost amount in an inexpensive fermentation medium (Shewale et al. 2014).
- The probiotic strains must have the capacity to develop and propagate in microaerophilic and oxygen-consuming conditions (Shewale et al. 2014).
- The probiotic strains must have the stability in food products and should be able to produce antimicrobial compounds (Ashraf and Smith 2015).
- The probiotic strains should be nonlethal and should have the ability to resist physical treatment without substantial forfeiture of viability (Ashraf and Smith 2015).

12.3.3 Probiotic Foods

Some of the probiotic foods that are consumed worldwide are given below (Fig. 12.3).

12.3.3.1 Yogurt

Yogurt is one of the natural sources of the probiotic bacteria that is made by the fermentation of milk through beneficial bacteria such as *Lactobacillus* and *Bifidobacterium* (Damunupola et al. 2014). Yogurt is also beneficial for human health as it lowers the blood pressure in people and relieves the signs of IBS (irritable bowel syndrome). Moreover, in children, it also minimizes the symptoms of diarrhea triggered by different antibiotics (Fox et al. 2015; Prasanna et al. 2014).

Fig. 12.3 Probiotic foods that are consumed worldwide. (Source: https://www.eatthis.com/best-probiotic-foods)

12.3.3.2 Cheese

Yogurt and cheese are the main transporters of probiotic bacteria in humans which are beneficial for human health and balance the intestinal microflora (Pyar and Peh 2014; Yadav et al. 2015; Ashraf and Smith 2015). Different kinds of cheese such as mozzarella, cheddar, and Gouda contain probiotic bacteria such as *Lactobacillus* and *Bifidobacterium* that have the ability to survive in the gastrointestinal tract of humans and considerably increase the number of probiotic bacterial cells in the stomach (Milani et al. 2015). Cheese is a highly rich source of vitamins, minerals, and proteins and lowers the risk of osteoporosis if taken in reasonable amounts (Bonjour et al. 2013).

12.3.3.3 Kefir

Kefir is a Turkish term meaning "feeling good" after ingestion (Leite et al. 2013). It is a fermented dairy drink that contains *Lactobacillus* and *Bifidus* bacteria and is produced by mixing the fermented kefir grains in milk (Leite et al. 2013). As compared to yogurt, it is a healthier source of probiotic bacteria and reduces a lot of stomach problems (Ritchie and Romanuk 2012).

12.3.3.4 Kimchi

Kimchi is a fermented peppery cabbage food that is commonly consumed by Korean people (Park et al. 2014). It is the best source of probiotic bacteria such as *Lactobacillus kimchii* which is a new strain of probiotic bacteria isolated from this food. Kimchi is also composed of some other strains of lactic acid bacteria that are beneficial for human health and ameliorate the intestinal stability (Park et al. 2014).

12.3.3.5 Miso Soup

Miso soup is prepared by mixing soybeans, rye, barley, and rice in order to make a paste, and then this paste is added to a bowl containing hot water to make a soup that is used as a medicinal soup by the Japanese people. It is a rich source of probiotics such as *Lactobacillus* and *Bifidus* bacteria (Tripathi and Giri 2014).

12.3.3.6 Green Olives

Fermented green olives are a rich source of lactic acid bacteria which give them a unique taste (Bautista-Gallego et al. 2013). The two most important probiotic bacteria that have been used in the fermentation process are *Lactobacillus plantarum* and *Lactobacillus pentosus* which provide aroma and flavor to the food.

12.3.4 Storage of Probiotic Bacterial Supplement

The probiotic bacterial supplement should be stored at 4–5 °C in refrigerator to retain the viability of bacteria and should be used before the expiration date of supplement (Shewale et al. 2014).

There are some other probiotic supplements or products available in the market that are shelf stable according to their industrialists, and their packing and delivery requirements must be encountered (Shewale et al. 2014).

12.3.5 Health Benefits of Probiotics

Probiotics have a lot of health benefits, comprising the capability to adjust the gut, improve assimilation, and diminish the painful side effects of severe antibiotics

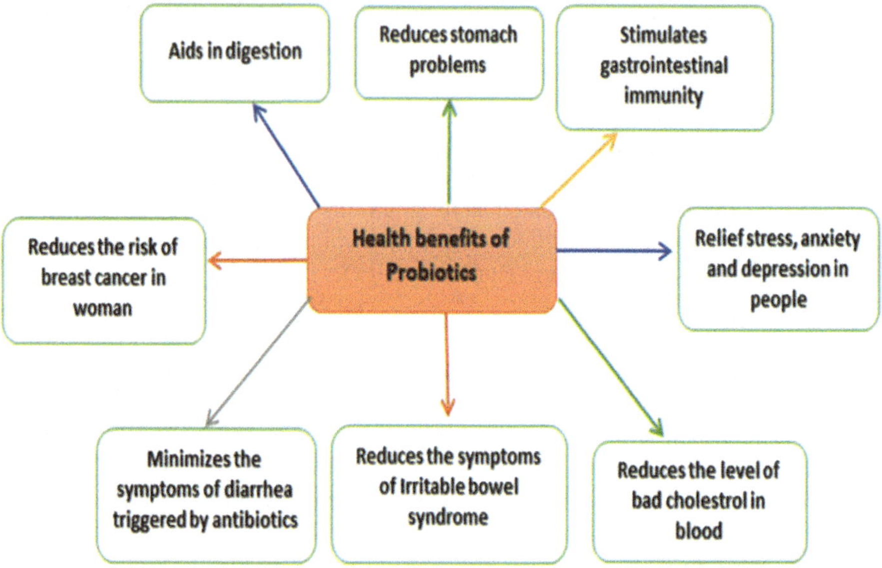

Fig. 12.4 Health benefits of probiotics. (Modified from Kitamoto 2015, Fox et al. 2015, Akkasheh et al. 2016, Corgneau et al. 2017)

(Syngai et al. 2016). Among the numerous advantages of consuming probiotics, some are listed as follows (Fig. 12.4).

12.3.5.1 Cure of Stomach Problems

Probiotics reduce the symptoms of irritable bowel syndrome in people through the coordination of immune responses. Probiotics help in the digestion of food, increase the lactose tolerance, and reduce a lot of stomach problems (Ritchie and Romanuk 2012; Corgneau et al. 2017).

Probiotic foods containing bacteria such as *Lactobacillus rhamnosus* and *Lactobacillus casei* and the yeast *Saccharomyces boulardii* are most frequently associated with minimizing the symptoms of diarrhea triggered by different antibiotics (Fox et al. 2015; Prasanna et al. 2014).

12.3.5.2 Reduce Anxiety and Stress

According to the research that has been conducted by Akkasheh et al. (2016), consuming probiotic supplements containing the strains of *Lactobacillus acidophilus*, *Lactobacillus casei*, and *Bifidobacterium bifidum* regularly for 8–10 weeks reduces the symptoms of anxiety, stress, and depression in patients with major depressive disorder (MDD).

12.3.5.3 Reduce the Risk of Breast Cancer

Probiotic foods/supplements also reduce the risk of breast cancer in women (Kitamoto 2015). This potential has been confirmed by animal experiments and human breast cancer cell trials. Probiotics have property to enhance the systemic immune system, have anticancer property, and can be used to control the progression of breast cancer (Mendoza 2019).

12.3.5.4 Reduce Bad Cholesterol

Probiotics protect the heart from different heart diseases by reducing the levels of low-density lipoprotein called as "bad" cholesterol in the blood and modestly reduce the blood pressure in people (Fox et al. 2015). The supplements containing probiotics proved to be potent novel nonpharmacological alternative to reduce the risk of cardiovascular diseases. A meta-analysis was conducted to explore the effects of different probiotics on serum total cholesterol, which showed a potential role of probiotic bacteria in reducing cardiovascular diseases (Wang et al. 2018).

12.3.5.5 Reduce Allergy

The quality of life has been affected by the prevalence of allergic disorders worldwide and has created interest to explore the role of probiotic bacteria to treat such allergies. Probiotics also provide defense against diseases such as common cold and flu and reduce the menace of certain allergic inflammations such as eczema in teenagers (Kang et al. 2013; Cuello et al. 2015). Literature showed that probiotics are being studied to cure diseases like asthma, food allergy, atopic dermatitis, and allergic rhinitis (Wang et al. 2019).

Since the last two decades, a remarkable increase has been seen in the use of probiotics and their applications in food manufacturing industries. Probiotics still need a lot of research as simply adding the beneficial bacteria to the different food products cannot be predicted that either they will be helpful in transferring the health benefits to humans or not. New research must be carried out to discover more ways in order to ensure safety and stability of probiotic bacteria in different food products (Wedajo 2015, 2015).

References

Abriouel H, Lerma LL, Casado Munoz MDC, Montoro BP, Kabisch J, Pichner R, Galvez A (2015) The controversial nature of the Weissella genus: technological and functional aspects versus whole genome analysis-based pathogenic potential for their application in food and health. Front Microbiol 6(1):1197–1105

Abubakr MA, Hassan Z, Imdakim MMA (2012) Antioxidant activity of lactic acid bacteria (LAB) fermented skim milk as determined by 1,1-diphenyl-2-picrylhydrazyl (DPPH) and ferrous chelating activity (FCA). Afr J Microbiol Res 6(34):6358–6364

Adeniyi BA, Adetoye A, Ayeni FA (2015) Antibacterial activities of lactic acid bacteria isolated from cow faeces against potential enteric pathogens. Afr Health Sci 15(3):888–895

Afzaal M, Zahoor T, Arshad MU, Maan AA, Shahbaz M, Zafarullah M, Abid T (2013) Probiotics: health claims, potential and realities. Pak J Food Sci 23(3):139–143

Akkasheh G, Kashani-Poor Z, Tajabadi-Ebrahimi M, Jafari P, Akbari H, Taghizadeh M, Esmaillzadeh A (2016) Clinical and metabolic response to probiotic administration in patients with major depressive disorder: a randomized, double-blind, placebo-controlled trial. Nutrition 32(3):315–320

Aman A, Siddiqui NN, Qader SAU (2012) Characterization and potential applications of high molecular weight dextran produced by *Leuconostoc mesenteroides* AA1. Carbohydr Polym 87 (1):910–915

Ashraf R, Shah NP (2014) Immune system stimulation by probiotic microorganisms. Crit Rev Food Sci Nutr 54(7):938–956

Ashraf R, Smith SC (2015) Selective enumeration of dairy based strains of probiotic and lactic acid bacteria. Int Food Res J 22(6):2576–2586

Bautista-Gallego J, Arroyo-Lopez FN, Rantsiou K, Jimenez-Diaz R, Garrido-Fernandez A, Cocolin L (2013) Screening of lactic acid bacteria isolated from fermented table olives with probiotic potential. Food Res Int 50(1):135–142

Begum PS, Madhavi G, Rajagopal S, Viswanath B, Razak MA, Venkataratnamma V (2017) Probiotics as functional foods: potential effects on human health and its impact on neurological diseases. Int J Nutr Pharmacol Neurol Dis 7(2):23–31

Benmechernene Z, Fernandez-No I, Quintela-Baluja M, Böhme K, Kihal M, Calo-Mata P, Barros-Velazquez J (2014) Genomic and proteomic characterization of bacteriocin-producing *Leuconostoc mesenteroides* strains isolated from raw camel milk in two southwest Algerian arid zones. Biomed Res Int 2014(4):1–10

Bermudez-Brito M, Plaza-Díaz J, Munoz-Quezada S, Gómez-Liorente C, Gil A (2012) Probiotic mechanisms of action. Ann Nutr Metab 61(2):160–174

Bonjour JP, Benoit V, Payen F et al (2013) Consumption of yogurts fortified in vitamin D and calcium reduces serum parathyroid hormone and markers of bone resorption: a double-blind randomized controlled trial in institutionalized elderly women. J Clin Endocrinol Metab 98:2915–2921

Bravo JA, Forsythe P, Chew MV, Escaravage E, Savignac HM, Dinan TG, Cryan JF (2011) Ingestion of *Lactobacillus* strain regulates emotional behavior and central GABA receptor expression in a mouse via the vagus nerve. Proc Natl Acad Sci 108(38):16050–16055

Chakraborty A, Bhowal J (2015) Isolation, identification and analysis of probiotic properties of *Lactobacillus spp.* from selected regional dairy product. Int J Curr Microbiol Appl Sci 4 (6):621–628

Chen YS, Wu HC, Wang CM, Lin CC, Chen YT, Jhong YJ, Yanagida F (2013) Isolation and characterization of lactic acid bacteria from pobuzihi (fermented cummingcordia), a traditional fermented food in Taiwan. Folia Microbiol 58(2):103–109

Corgneau M, Scher J, Ritie-Pertusa L, Le DT, Petit J, Nikolova Y, Gaiani C (2017) Recent advances on lactose intolerance: tolerance thresholds and currently available answers. Crit Rev Food Sci Nutr 57(15):3344–3356

Cuello CA, Brozek JL, Fiocchi A, Pawankar R, Yepes-Nunez JJ, Terracciano L, Schunemann HJ (2015) Probiotics for the prevention of allergy: a systematic review and meta-analysis of randomized controlled trials. J Allergy Clin Immunol 136(4):952–961

Damunupola DAPR, Weerathilake WADV, Sumanasekara G (2014) Evaluation of quality characteristics of goat milk yogurt incorporated with beetroot juice. Int J Sci Res Publ 4 (10):2250–3153

Denkova R, Denkova Z, Yanakieva V, Blazheva D (2013) Antimicrobial activity of probiotic *Lactobacilli, Bifidobacteria* and *Propionic acid bacteria*, isolated from different sources. In: Microbial pathogens and strategies for combating them: science, technology and education, vol 2. Formatex Research Center, Badajoz, pp 857–864

FAO/WHO (2002) Guidelines for the evaluation of probiotics in food. Food and Agriculture Organization of the United Nations and World Health Organization Working Group Report

Fijian S (2014) Microorganisms with claimed probiotic properties: an overview of recent literature. Int J Environ Res Public Health 11(5):4745–4767

Fox MJ, Ahuja KD, Robertson IK, Ball MJ, Eri RD (2015) Can probiotic yogurt prevent diarrhea in children on antibiotics? A double-blind, randomized, placebo-controlled study. Br Med J Open 5(1):64–74

Fusco V, Quero GM, Cho GS, Kabisch J, Meske D, Neve H, Franz CM (2015) The genus Weissella: taxonomy, ecology and biotechnological potential. Front Microbiol 6(155):259–262

Granato D, Branco GF, Cruz AG, Faria JDAF, Shah NP (2011) Probiotic dairy products as functional foods. Compr Rev Food Sci Food Saf 9(5):455–470

Gueimonde M, Sánchez B, de los Reyes-Gavilan CG, Margolles A (2013) Antibiotic resistance in probiotic bacteria. Front Microbiol 4(1):202–210

Halasz A (2011) Lactic acid bacteria. Food Qual Std 3(1):70–82

Hemaiswarya S, Raja R, Ravikumar R, Carvalho IS (2013) Mechanism of action of probiotics. Braz Arch Biol Technol 56(1):113–119

Hill C, Guarner F, Reid G, Gibson GR, Merenstein DJ, Pot B, Calder PC (2014) Expert consensus document: the international scientific association for probiotics and prebiotics consensus statement on the scope and appropriate use of the term probiotic. Nat Rev Gastroenterol Hepatol 11 (8):506–510

Kamboj K, Vasquez A, Balada-Llasat JM (2015) Identification and significance of Weissella species infections. Front Microbiol 6(254):1204–1210

Kang EJ, Kim SY, Hwang IH, Ji YJ (2013) The effect of probiotics on prevention of common cold: a meta-analysis of randomized controlled trial studies. Korean J Fam Med 34(1):2–10

Kaur A, Arora M, Pandove G (2014) Probiotics and its health benefits. J Glob Biosci 3(3):686–693

Khalil MI, Anwar MN (2016) Isolation, identification and characterization of lactic acid bacteria from milk and yoghurts. Res Rev J Food Dairy Technol 4(3):17–26

Khatri I, Sharma S, Ramya TNC, Subramanian S (2016) Complete genomes of *Bacillus coagulans* and *Bacillus subtilis*, two phylogenetically distinct probiotics. Public Libr Sci One 11 (6):156745

Kitamoto K (2015) Cell biology of the Koji mold *Aspergillus oryzae*. Biosci Biotechnol Biochem 79(6):863–869

Kubo Y, Rooney AP, Tsukakoshi Y, Nakagawa R, Hasegawa H, Kimura K (2011) Phylogenetic analysis of *Bacillus subtilis* strains applicable to natto (fermented soybean) production. Appl Environ Microbiol 77(18):6463–6469

Kushugulova A, Saduakhasova S, Kozhakhmetov S, Shakhabayeva G, Tynybayeva I, Nurgozhin T, Zhumadilov Z (2014) Antioxidant activity of the probiotic consortium in vitro. Int J Probiot Prebiot 9(2):55–60

Leite AMDO, Miguel MAL, Peixoto RS, Rosado AS, Silva JT, Paschoalin VMF (2013) Microbiological, technological and therapeutic properties of kefir: a natural probiotic beverage. Braz J Microbiol 44(2):341–349

McKenney PT, Driks A, Eichenberger P (2013) The *Bacillus subtilis* endospore: assembly and functions of the multilayered coat. Nat Rev Microbiol 11(1):33–44

Menconi A, Kallapura G, Latorre JD, Morgan MJ, Pumford NR, Hargis BM, Tellez G (2014) Identification and characterization of lactic acid bacteria in a commercial probiotic culture. Biosci Microbiota Food Health 33(1):25–30

Mendoza L (2019) Potential effect of probiotics in the treatment of breast cancer. Oncol Rev 13 (2):422. https://doi.org/10.4081/oncol.2019.422

Milani C, Turroni F, Duranti S, Lugli GA, Mancabelli L, Ferrario C, Ventura M (2015) Genomics of the genus *Bifidobacterium* reveals species-specific adaptation to the glycan-rich gut environment. Appl Environ Microbiol 82(15):3500–3515

Mishra A, Sharma KP (2014) Isolation and characterization of probiotic microorganism from fermented dairy products. Bull Biosci 5(1):10–14

Park JM, Shin JH, Gu JG, Yoon SJ, Song JC, Jeon WM et al (2011) Effect of antioxidant activity in kimchi during a short-term and over-ripening fermentation period. J Biosci Bioeng 112:356–359. https://doi.org/10.1016/j.jbiosc.2011.06.003

Park KY, Jeong JK, Lee YE, Daily JW III (2014) Health benefits of kimchi (Korean fermented vegetable) as a probiotic food. J Med Food 17(1):6–20

Prasanna PHP, Grandison AS, Charalampopoulos D (2014) *Bifidobacteria* in milk products: an overview of physiological and biochemical properties, exopolysaccharide production, selection criteria of milk products and health benefits. Food Res Int 55(1):247–262

Pundir RK, Rana S, Kashyap N, Kaur A (2013) Probiotic potential of lactic acid bacteria isolated from food samples: an in vitro study. J Appl Pharm Sci 3(3):85–93

Pyar HASSAN, Peh K (2014) Characterization and identification of *Lactobacillus acidophilus* using biolog rapid identification system. Int J Pharm Pharm Sci 6(1):975–1491

Raghuwanshi S, Misra S, Sharma R, Bisen PS (2018) Probiotics: nutritional therapeutic tool. J Probiot Health 6(194):2–8

Rahman SMK (2015) Probiotic properties analysis of isolated lactic acid bacteria from buffalo milk. Arch Clin Microbiol 7(1):5–10

Rijkers GT, De Vos WM, Brummer RJ, Morelli L, Corthier G, Marteau P (2011) Health benefits and health claims of probiotics: bridging science and marketing. Br J Nutr 106(9):1291–1296

Ritchie ML, Romanuk TN (2012) A meta-analysis of probiotic efficacy for gastrointestinal diseases. Public Libr Sci One 7(4):34938

Segers ME, Lebeer S (2014) Towards a better understanding of *Lactobacillus rhamnosus* GG-host interactions. Microb Cell Factories 13(1):1–7

Shewale RN, Sawale PD, Khedkar CD, Singh A (2014) Selection criteria for probiotics: a review. Int J Probiot Prebiot 9(1):17–22

Shukla R, Iliev I, Goyal A (2014) *Leuconostoc mesenteroides* NRRL B-1149 as probiotic and its dextran with anticancer properties. J Biosci Biotechnol 3(1):79–87

Sornplang P, Piyadeatsoontorn S (2016) Probiotic isolates from unconventional sources: a review. J Anim Sci Technol 58(1):26

Swain MR, Anandharaj M, Ray RC, Parveen Rani R (2014) Fermented fruits and vegetables of Asia: a potential source of probiotics. Biotechnol Res Int 2014(1):19

Syngai GG, Gopi R, Bharali R, Dey S, Lakshmanan GA, Ahmed G (2016) Probiotics—the versatile functional food ingredients. J Food Sci Technol 53(2):921–933

Toiviainen A, Jalasvuori H, Lahti E, Gursoy U, Salminen S, Fontana M, Soderling E (2015) Impact of orally administered lozenges with *Lactobacillus rhamnosus* GG and *Bifidobacterium animalis subsp*. Lactis BB-12 on the number of salivary mutans *Streptococci*, amount of plaque, gingival inflammation and the oral microbiome in healthy adults. Clin Oral Investig 19 (1):77–83

Tonekabon I (2013) Evaluation of antimicrobial activity of three *Lactobacillus spp*. against antibiotic resistance *Salmonella typhimurium*. Adv Stud Biol 5(2):61–70

Tripathi MK, Giri SK (2014) Probiotic functional foods: survival of probiotics during processing and storage. J Funct Foods 9(1):225–241

Wang Y, Wu Y, Wang Y, Xu H, Mei X, Yu D, Li W (2017) Antioxidant properties of probiotic bacteria. Nutrients 9(5):521

Wang L, Guo MJ, Gao Q, Yang JF, Yang L, Pang XL, Jiang XJ (2018) The effects of probiotics on total cholesterol: a meta-analysis of randomized controlled trials. Medicine 97(5):e9679. https://doi.org/10.1097/MD.0000000000009679

Wang HT, Anvari S, Anagnostou K (2019) The role of probiotics in preventing allergic disease. Children (Basel, Switzerland) 6(2):24. https://doi.org/10.3390/children6020024

Wedajo B (2015) Lactic acid bacteria: benefits, selection criteria and probiotic potential in fermented food. J Probiot Health 3(2):129

Yadav A, Jaiswal P, Jaiswal M, Kumar N, Sharma R, Raghuwanshi S, Bisen PS (2015) Concise review: importance of probiotics yogurt for human health improvement. J Environ Sci Toxicol Food Technol 9(1):25–30

Yepez L, Tenea GN (2015) Genetic diversity of lactic acid bacteria strains towards their potential probiotic application. Rom Biotechnol Lett 20(2):10191–10199

Yin CH, Dong X, Lv L, Wang ZG, Xu QQ, Liu XL, Yan H (2013) Economic production of probiotics from kitchen waste. Food Sci Biotechnol 22(1):59–63

Production and Biofunctionality of Milk-Derived Bioactive Peptides

13

Hiral Chaudhari and Subrota Hati

Abstract

Milk is known as a source of macro- and micronutrients. Milk proteins comprise a wide range of bioactive peptides. The bioactive peptides have specific amino acid sequences in the form of hydrolysates. These types of bioactive peptides are vital for proper bodily function. Bioactivities of milk proteins depend on the release of various fragments of peptides with specific amino acid sequences. In the gastrointestinal tract, peptides are digested by the proteolytic enzyme, or during fermentation and food processing, to liberate and activate encoded bioactive peptides from the native protein. Milk proteins encrypted with bioactive peptides exhibit various biofunctionalities, such as antibacterial, antioxidative, opioid-like, antihypertensive, immunomodulatory, antithrombotic and cytomodulatory activities. Diarrhea, thrombosis, dental carries, oxidative stress, hypertension, mineral malabsorption, and immunodeficiency diseases can be treated by these types of bioactive peptides. These bioactive peptides are used in the formulation of functional foods, natural drugs, and nutraceuticals because of their beneficial health effects.

The food industry is particularly interested in bioactive peptides extracted from milk proteins because of the functional and physiological roles they play. This current chapter summarizes the production of milk-derived bioactive peptides along with their general characteristics, physiological functions, and potential applications in functional health food developments.

Keywords

Biofunctional properties · Bioactive peptides · Milk protein

H. Chaudhari · S. Hati (✉)
Department of Dairy Microbiology, SMC College of Dairy Science, Anand Agricultural University, Anand, Gujarat, India

R. Z. Sayyed, M. Khan (eds.), *Microbiome-Gut-Brain Axis*,
https://doi.org/10.1007/978-981-16-1626-6_13

13.1 Introduction

The fragments of protein known as bioactive peptides are gaining worldwide recognition for their physiological health benefits. It is acknowledged that milk proteins contain all fundamental amino acids; however, in specific occasions, it has been affirmed that milk proteins perform various functionalities in vivo through the aid of bioactive peptides (Ricci-Cabello et al. 2012).

Milk proteins are known as precursors of bioactive peptides. Proteolytic enzymes are responsible for the production of hydrolysates that contain unique amino acid sequences known as bioactive peptides and that can provide potential benefits to human health. Bioactive compounds have numerous health benefits that can help prevent disease. There is a rising interest in the helpful capabilities of bioactive peptides (Ricci-Cabello et al. 2012). Currently, many researchers in the scientific community are exploring new studies on bioactive peptides released from milk and their derivatives.

According to Fitzgerald and Murray (2006), bioactive peptides are characterized as peptides with chemical or useful action that affects physiological function. They interact with explicit receptors on track cells prompting the occurrence of physiological reactions. Bioactive peptides are classified according to their functional properties, such as opioid, immunomodulatory, antimicrobial, antithrombotic, antihypertensive, mineral binding, and antioxidative. These peptides perform an important role in human health.

13.2 Bioactive Peptides

The word "bioactive" is composed by two words: **bio-** and **–active**. In origin: bio-comes from the Greek (βío-) "bios," meaning life, and –active from the Latin "activus," meaning dynamic (Guaadaoui et al. 2014). Bioactive peptides are protein portions created through specific alterations or breaks from parent proteins. These bioactive peptides are known as dynamic peptides or physiological dynamic peptides and are equipped to perform distinctive body functions. The size of bioactive peptides ranges from 2 to 20 amino corrosive buildups, and it relies on the kind, creation, and nature of bioactive peptides (Wijesekara and Kim 2010). There are several factors on which activity of bioactive peptides depend, such as amino acid sequences of protein, pre-treatment, enzyme specificity on action, condition of hydrolysis (e.g., pH, temperature, degree of hydrolysis), enzyme inactivation treatment, separation, and purification (Korhonen and Pihlanto 2006).

Milk peptides are obtained from milk proteins through an enzymatic breakdown by stomach-related enzymes or by *Lactobacilli* that produces proteinases catalysts during the fermentation of milk (Jauhiainen and Korpela 2007). Bioactive peptides obtained from milk proteins are dynamic when these are delivered from the antecedent proteins. Bioactive peptides are created by processing or proteolysis both in vivo and in vitro.

Bioactive peptides may act as regulatory compounds with hormone-like activity when they are liberated from the protein chain, as reported since 1979, and numerous peptides exhibit various activities such as opioid-like, antithrombotic or anti-hypertension activity, immunomodulation, or mineral utilization properties. Milk, plant, and animal proteins are the main wellspring of bioactive peptides. Antimicrobial peptides are the main naturally dynamic peptide found in milk followed by immunomodulatory peptides (Sharma et al. 2011).

13.2.1 Production of Bioactive Peptides

Bioactive peptides produced from milk are present in both casein (α-, β-, and γ-casein) and whey proteins (β-lactoglobulin, α-lactalbumin, serum albumin, immunoglobulins, lactoferrin, and protease-peptone fractions) and can be released from their parent proteins by four ways:

(1) Enzymatic hydrolysis using digestive enzymes, such as alcalase, trypsin and pepsin, (2) fermentation by starter cultures by release of proteolytic enzyme, (3) hydrolysis using enzymes obtained from proteolytic microorganisms, and (4) fusion of fermentation and hydrolysis by adding specific enzymes (Phelan et al. 2009) (Fig. 13.1).

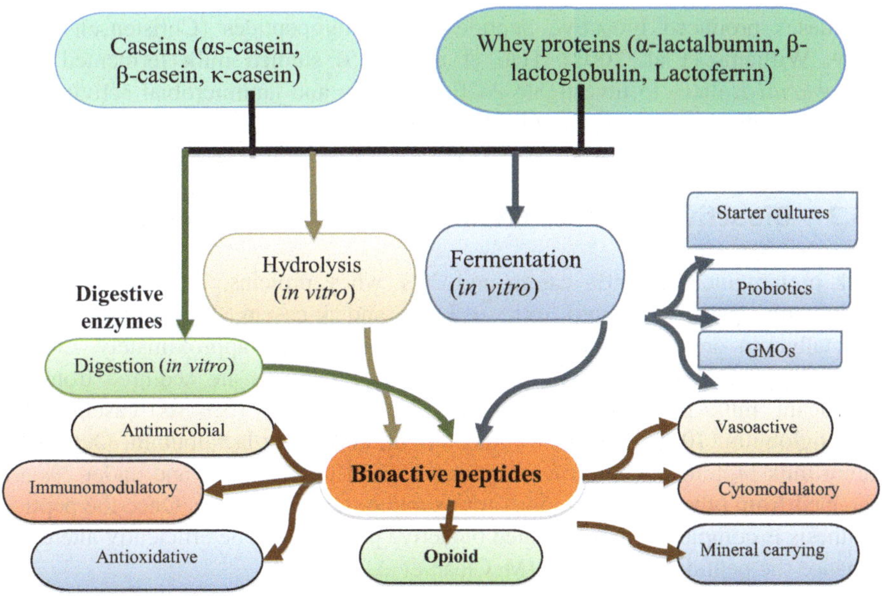

Fig. 13.1 The generation of bioactive peptides is depicted in a schematic diagram

13.2.1.1 Enzymatic Hydrolysis

In gastrointestinal digestion, enzymes such as pepsin, trypsin, or chymotrypsin are significant and responsible for protein breakdown (Korhonen and Pihlanto 2003). Numerous bioactive peptides are delivered from milk proteins through the activity of pepsin, trypsin, and chymotrypsin enzymes (Gobbetti et al. 2007), and other proteolytic catalyst such as alcalase, thermolysin, and subtilisin are also used to produce bioactive peptides (Korhonen 2009). Preheating is utilized to improve the hydrolysis of whey protein (Reddy et al. 1988; Guo et al. 1995). When whey protein is heated at a temperature of 65 °C, a partially unfolded conformation occurs, known as a molten globule state, and is characterized by the presentation of hydrophobic clusters. These alterations permit more prominent access of the enzymes to specific sites, which are already out of reach to the enzyme action (Hirose 1993).

Alcalase is utilized for the mechanical creation of whey protein hydrolysate (WPH) using the enzymatic hydrolysis strategy because of its endoprotease with wide particularity and minimal effort requirement. Alcalase is a serine alkaline protease created by the bacterial strain of *Bacillus licheniformis*. It has an ideal pH between 8 and 9 for action and wide pH dependability. Alcalase has high specificity for aromatic (Phe, Trp, and Tyr), acidic (Glu), sulfur-containing (Met), aliphatic (Leu and Ala), hydroxyl (Ser), and basic (Lys) residues (Doucet et al. 2003).

13.2.1.2 Microbial Fermentation

Bioactive peptides are created by the proteolytic activities of starter and non-starter bacteria used for the formation of fermented dairy items. Lactic acid bacteria generate bioactive peptides through their proteinases and peptidases. Extracellular proteinases generate oligopeptides from longer chains of protein. Intracellular peptidases produced bioactive peptides from oligopeptides (Christensen et al. 1999; Williams et al. 2002). Hati et al. (2015) studied milk fermented with *Lactobacilli* isolates to investigate ACE-inhibitory and antimicrobial activity produced by milk-derived peptides (Table 13.1).

13.3 Bioactive Peptides Obtained from Milk Proteins

Milk protein includes 80% casein and 20% whey proteins. The milk casein is additionally partitioned into αs1, αs2, β, and κ-casein. β-lactoglobulin and α-lactalbumin are significant whey proteins, comprising approximately 75% of complete outright whey proteins. Most bioactive peptides are acquired from the significant milk proteins, for example, 36% β-casein, 13% αs1-casein, 11% β-lactoglobulin, 10% κ-casein, 8% αs2-casein, and 5% α-lactalbumin. Out of the minor milk proteins, lactoferrin comprises 15% of the database, and under 2% are obtained from other minor proteins, for example, serum albumin. Once the peptide synthesis is completed, distinguished bioactive peptides can be efficiently altered to enhance the peptides bioactivity (McClean et al. 2014).

A few bioactive peptides are encoded in the essential groupings of milk proteins (Fig. 13.1). A segment of these bioactive peptides nominally affects gastrointestinal

Table 13.1 Proteolytic enzymes from various organism species that release bioactive peptides

Microorganism	Peptide sequence	Protein precursor	Bioactivity
Lactobacillus helveticus and *Saccharomyces cerevisiae*	Val-Pro-Pro, Ile-Pro-Pro	β-casein, α-casein	ACE inhibitor, antihypertensive
Lactobacillus GG and enzymes + pepsin and trypsin	Tyr-Pro-Phe-Pro, Ala-Val-Pro-Tyr-Pro-Gln-Arg, Thr-Thr-Met-Pro-Leu-Trp	β-casein, αs1 casein	Opioid, ACE inhibitor, immunostimulating
Lb. helveticus CP90 proteinase	Lys-Val-Leu-Pro-Val-Pro-(Glu)	β-casein	ACE inhibitor
Lb. helveticus CPN 4	Tyr-Pro	Whey proteins	ACE inhibitor
Lb. delbrueckii ssp. *bulgaricus* IFO13953	Ala-Arg-His-Pro-His-Pro-His-Leu-Ser-Phe-Met	α-casein	Antioxidative
Lb. rhamnosus + hydrolysis with pepsin and Colorase PP	Asp-Lys-Ile-His-Pro-Phe, Tyr-Gln-Glu-Pro-Val-Leu, Val-Lys-Glu-Ala-Met-Ala-Pro-Lys	β-casein	ACE inhibitor, antioxidative
Lb. delbrueckii ssp. *bulgaricus* + pepsin and trypsin	Ser-Lys-Val-Tyr-Pro-Phe-Pro-Gly-Pro-Ile	β-casein	ACE inhibitor

Dziuba and Dziuba (2014)

capacities. For instance, phosphopeptides improve gastrointestinal exploitation of calcium and *p*-casomorphins to limit gastrointestinal compression and liquid emission. Immuno-stimulating peptides and antihypertensive peptides (ACE inhibitors) effectively affect general well-being (Meisel 1997; Schanbacher et al. 1998; Clare and Swaisgood 2000; Park et al. 2010; Mohanty et al. 2016a, b). These bioactive peptides encoded in milk protein may be delivered during the regular hydrolytic measure in the gastrointestinal lumen or during food handling, and may likewise have significant organic and well-being impacts (Fig. 13.2 and Table 13.2).

13.4 Biofunctional Properties of Bioactive Peptides

13.4.1 Effect on Cardiovascular System

13.4.1.1 Antihypertensive Peptides

Hypertension (blood pressure greater than 140/90) is linked to coronary artery disease and stroke, and it is responsible for causes of mortality in developing countries (Lin et al. 2012). Protein hydrolysates are likewise a wellspring of bioactive peptides that are not active in the original native protein but become active when hydrolysis occurs. These peptides have numerous beneficial activities in human health, for example, antihypertensive properties (Hartmann and Meisel 2007). Angiotensin converting enzymes (ACE) perform a vital function in the balancing

1	10	20
R---P---K---H---P---I---K---H---Q---G---L---P---Q---E---V---L---N---E---N---L---L---R---F---F---V---		
30	40	50
A---P---F---P---Q---V---F---G---K---E---K---V---N---E---L---S---K---D---I---G---S---E---S---T---E---		
	60	70
D---Q---A---M---E---D---I---K---Q---M---E---A---E---S---I---S---S---S---E---E---I---V---P---N---S---		
80	90	100
V---E---Q---K---H---I---Q---K---E---D---V---P---S---E---R---Y---L---G---Y---L---E---Q---L---L---R--		
	110	120
L---K---K---Y---K---V---P---Q---L---E---I---V---P---N---S---A---E---E---R---L---H---S---M---K---E---		
130	140	150
G---I---H---A---Q---Q---K---E---P---M---I---G---V---N---Q---E---L---A---Y---F---Y---P---E---L---F---		
	160	170
R---Q---F---Y---Q---L---N---A---Y---P---S---G---A---W---Y---Y---V---P---L---G---T---Q---Y---T---N---		
180	190	199
A---P---S---F---S---N---I---P---N---P---I---G---S---E---N---S---E---K---T---T---M---P---L---W---		

Antibacterial peptide: 1-23.

ACE inhibitors: 23-34; 194-199.

α- Casomorphin: 90-96.

Immunostimulating peptide: 92-94.

Bitter peptides: 23-34; 91-100.

Fig. 13.2 The primary structure of bovine α_{s1}-casein and the position of bioactive peptides derived from α_{s1}-casein (Schlimme and Meisel 1995)

of blood pressure by converting angiotensin I into angiotensin II, a powerful vasoconstrictor, and by concurrently inactivating bradykinin, a vasodilator (Guyton et al. 2006). Peptides obtained from milk proteins may show this activity (IA) against ACE. In this way, these proteins could be brought into the eating regimen as another non-pharmacological way to prevent and treat blood vessel hypertension (Costa et al. 2007; Miguel et al. 2009; Otte et al. 2007).

Milk proteins are a wellspring of bioactive peptides, and ACE inhibitory peptides form through the enzymatic hydrolysis and fermentation with lactic acid bacteria from milk protein. A few milk peptides deterred ACE in vitro (Hati et al. 2018). The antihypertensive impact of hydrolysates is normally investigated in vitro by their ability to impede ACE. This enzyme is a key part of the blood pressure regulation cycle, and its hindrance helps control hypertension (Espejo-Carpio et al. 2013). The renin–angiotensin system (RAS) and the kinin–nitric oxide system (KNOS),

Table 13.2 The roles of the major biologically active milk components

Milk precursors or components	Bioactive compounds	Bioactivities observed
α-, β-caseins	Casomorphins	Opioid agonist (decrease gut mobility, gastric emptying rate; increase amino acids and electrolytes uptake)
α-, β-caseins	Casokinins	ACE inhibitory (increase blood flow to intestinal epithelium)
α-, β-caseins	Phosphopeptides	Mineral binding (Ca binding; increase mineral absorption, i.e., Ca, P, Zn)
α-, β-caseins	Immunopeptides	Immunomodulatory (increase immune response and phagocytic activity)
	Casomorphins	
	Casokinins	
α_{s1}-casein	Isracidin	Antimicrobial
α_{s2}-casein	Casocidin	Antimicrobial
κ-casein	Casoxins	Opioid antagonist
κ-casein	Casoplatelins	Antithrombotic
α-lactalbumin (α-La), β-lactoglobulin (β-La)	Lactorphins	Opioid agonist
Serum albumin	Serorphin	Opioid agonist
α-La, β-La and Serum albumin	Lactokinins	ACE inhibitory
Immunoglobulins	IgG, IgA	Immunomodulatory (passive immunity)
Lactoferrin	Lactoferrin	Immunomodulator, antimicrobial
Lactoferrin	Lactoferroxins	Opioid antagonist
Oligosaccharides	Oligosaccharides	Probiotic (increase growth of Bifidobacteria in GI tract)
Glycolipids	Glycolipids	Antimicrobial
Oligosaccharides	Oligosaccharides	
Prolactin	Prolactin	Immunomodulatory
Growth factors	IGF-1, TGF-α, EGF, TGF-β	Organ development and functions
Parathromone-P	PTHrP	Increase Ca^{2+} metabolism and uptake

Schanbacher et al. (1998), Clare and Swaisgood (2000), and Park et al. (2010)

involving different metabolic pathways, are the two main systems responsible for high blood pressure (Martínez-Maqueda et al. 2012) (Fig. 13.3).

In the RAS, the conversion of angiotensin I (Ang I) to angiotensin II (Ang II) (Vasocostrictory peptide) is catalyzed by renin and ACE during intermediate steps. Accordingly, inhibition of ACE considerably prompts blood pressure reduction (FitzGerald and Meisel 2000). In the KNOS, ACE deactivates the vasodilatory peptides, i.e., bradykinin and kallidin. Bradykinin binds to β-receptors and stimulation of nitric oxide synthase (NOS) produces vasodilation (FitzGerald and Meisel 2000). Consequently, renin inhibitors prevent the formation of Ang I and Ang II. Angiotensinogen is the main known substrate of renin. Renin inhibitory peptides

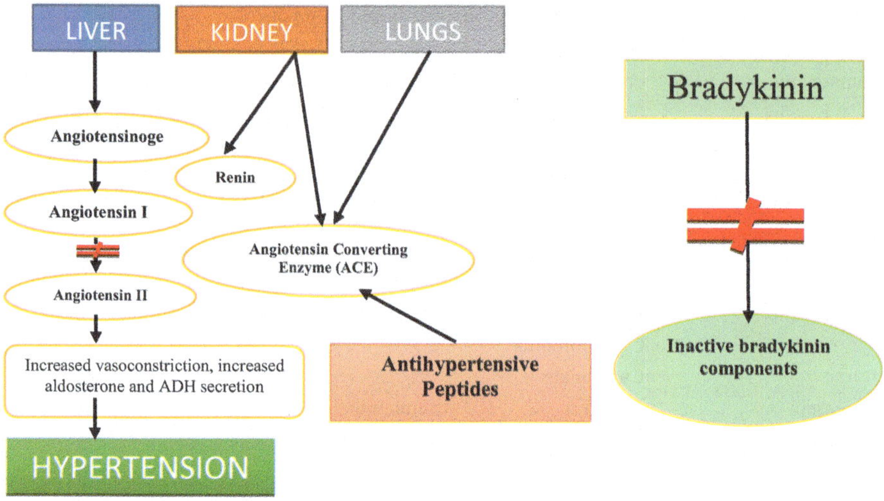

Fig. 13.3 The kinin–nitric oxide (NO) system and the renin–angiotensin system

are more specific (Staessen et al. 2006; Udenigwe et al. 2012). Calcium channel blockers are associated with voltage-gated calcium channels (VGCCs) in cardiovascular muscle and vein dividers, decreasing intracellular calcium and therefore bringing down vasoconstriction. Several studies have indicated that peptides can work as calcium channel blockers. His-Arg-Trp peptides were shown to have a vasorelaxation effect on the phenylephrine-contracted thoracic aorta (Tanaka et al. 2009).

To date, peptides derived from milk proteins are considered as the best characterized ACE inhibition elements and have shown antihypertensive activity in vitro (Escudero et al. 2014) and in vivo (Fitzgerald and Murray 2006). The first ever identified and characterized ACE inhibition peptides were Val-Pro-Pro (VPP) and Ile-Pro-Pro (IPP) that were isolated from *L. helveticus* fermented milk products (Nakamura et al. 1995), followed by the identification of other milk-derived antihypertensive peptides such as Ser-Lys-Val-Tyr-Pro (SLVTP) from *Streptococcus salivarius* ssp. *thermophilus* and *Lactococcus lactis* biovar *diacetylactis* fermented milk (Ashar and Chand 2004).

NK9 (*L. casei*) used for goat milk fermentation (10 kDa permeates) showed peptide sequence AFPEHK, which has been indicated to promote ACE inhibitory activity (Parmar et al. 2019). Solanki and Hati (2018) studied the potential of *Lactobacillus rhamnosus* for producing angiotensin I converting enzyme (ACE) inhibitory peptides in fermented camel milk (Indian breed) and identified a QTDIMIFTIGPA peptide sequence, which has been confirmed to have ACE inhibitory activity.

Goat milk fermented with M5 *(L. fermentum)* (3 and 10 kDa permeates) showed peptide sequences LARPKHPINHRGLSPE and TEEEKNRLNFLKKISQY, respectively. When using M16 *(L. paracasei)* to ferment goat milk (3 and 10 kDa

Table 13.3 Antihypertensive peptides and their functions

ACE inhibitory peptides	Functions	References
DVWY, FQ, VVG. DVWY, VAE, WTFR, DPYKLRP, PYKLRP, YKLRP, GILRP	Inhibits ACE in thoracic aorta tissue and suppresses angiotensin II-mediated vasoconstriction.	Koyama et al. (2014)
VPP, IPP, GAAGGAF LIVTQ, LIVT, LLKPY, AHLL, FISNHAY, AAATP, LGL, SFVTT, IIT	Competitively bind and inhibit ACE and results in blood pressure reduction	Xu and Gao (2015), Li et al. (2012)
ADVFNPR, VVLYK, LPILR, VIGPR	Lower endothelia-1 levels significantly	Zheng et al. (2017)

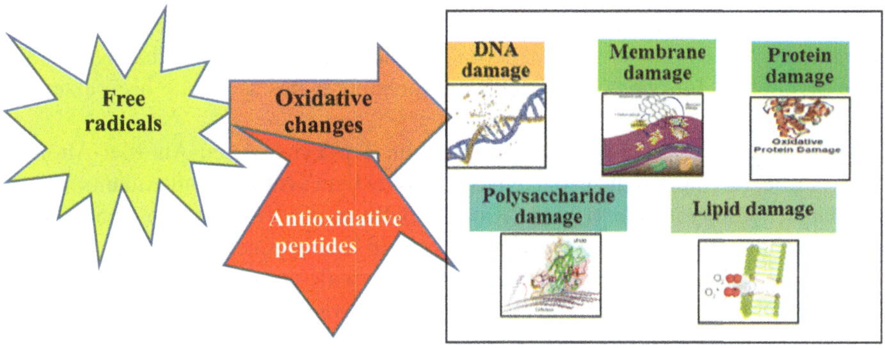

Fig. 13.4 Schematic demonstration oxidative changes and action of antioxidative peptides

permeates), the peptide sequences of ENSGKTTMPLW and PEEIKITVDDKHYQKALNEI were found, respectively, which had ACE inhibitory activity confirmed by Protein Information Resource (PIR) and Antihypertensive Inhibiting Peptide Database (AHTPDB) (Parmar et al. 2019).

The type of enzyme, enzyme:substrate (E:S) proportion, and the utilization of ultrafiltration (UF) influenced the peptide profiles and the ACE inhibitory activity of the whey protein concentrate (WPC) hydrolysates. The best peptide profile was acquired by utilizing papain with an E:S ratio of 2:100 and UF, though pancreatin delivered the best ACE inhibitory activity with an E:S ratio of 0.5:100 (Silvestre et al. 2012) (Table 13.3).

13.4.1.2 Antioxidative Peptides

Oxidation metabolism is fundamental for the endurance of cells; however, it produces free radicals and other reactive oxygen species (ROS) as a result, which can cause oxidative damage (Fig. 13.4). The body has its assurance system against ROS; oxidative pressure occurs when ROS overburden the body's cell reinforcement safeguard component, which may be a primary causative factor of several lifestyle-caused disorders (Hernandez-Ledesma et al. 2005). Dairy products and their fragment are antioxidative, e.g., milk, skim milk, whey, casein, and lactoferrin (Steijins and Van 2000; Cervato et al. 1999; Colbert and Decker 1991; Taylor and

Richardson 1980). Antioxidative peptides from milk proteins play a fundamental role in the support of cell reinforcement safeguard frameworks by preventing the development of free radicals or by scavenging free radicals and dynamic oxygen species, which initiate oxidative harm to biomolecules and cause aging, cancer, heart disease, stroke, and arteriosclerosis. So far, many antioxidant peptides obtained from both casein and whey proteins have been portrayed (Power et al. 2013).

Whey protein hydrolysates (WPHs) have shown a wide scope of cancer prevention agent movement in an iron-catalyzed liposome oxidation framework or a copper-catalyzed liposome emulsion (Colbert and Decker 1991), dependent upon the proteases used. Adriena et al. (2010) uncovered that on hydrolysis with microbial proteases (alcalase, flavourzyme, protamex, and neutrase), the cell reinforcement action of whey protein extended from 7–19.8 to 40–54.2%. There are questions about the potential prosperity effects of manufactured cancer prevention agents, and thus the interest for characteristic cell reinforcements has as of late been extended (Park et al. 2001).

A peptide derived from the β-lactoglobulin Trp-Tyr-Ser-Leu-Ala-Met-Ala-Ala-Ser-Asp-Ile, has more radical-scavenging activity than synthetic antioxidants such as butylated hydroxyanisole (BHA) (Ricci-Cabello et al. 2012). Mann et al. (2015) reported the antioxidant activity of whey protein hydrolysates in the milk beverage system showed maximum antioxidant activity with corolase enzyme (1.42 μM Trolox/mg of protein) as compared to enzyme flavourzyme and alcalase (0.81 and 1.16 μM Trolox/mg of protein respectively).

Padghan et al. (2018) presented a process for purification and characterization of antioxidative peptides derived from fermented milk (lassi) by lactic cultures. To achieve this, they prepared lassi by utilizing standard *dahi* culture NCDC-167(BD4) and the other one was made with the equivalent *dahi* culture joined with *Lactobacillus acidophilus* NCDC-15 as an adjunct culture. They observed that lassi manufactured by utilizing the *dahi* culture in addition to adjunct culture (0.66 μM Trolox/mg of protein) showed a greater antioxidant activity than *dahi* culture alone (0.20 μM Trolox/mg of protein). Panchal et al. (2019) also reported the production and characterization of novel antioxidative peptides obtained from fermented goat milk by using *L. fermentum* (M4), and they discovered maximum antioxidant activity (ABTS assay) (52.27%), hydroxyl free radical scavenging activity (55.73%), and superoxide free radical scavenging activity (43.03%) at 37 °C after 48 h. They also found the sequences of SPAQTLQWQVLPNTVPAK (2D-PAGE), YIPIQYVLSR (2D-PAGE), and IAKYIPIQYVLSR (10 kDa permeate) were somewhat similar with YQEPVLGPVRGFPIL (query sequence), and sequences VPLFVQVGEVIK (2D-PAGE), SCQDQPTTLAR (2D-PAGE), HPHPHLSFMAIPPK (2D-PAGE), MASFISLSSK (2D-PAGE), YIPIQYVLSR (2D-PAGE), IAKYIPIQYVLSR (10 kDa permeate), and SPAQTLQWQVLPNTVPAK (2DPAGE) were somewhat matched with VQSWMHQPPQPLSPT (query sequence), which serves as an antioxidant.

13.4.1.3 Hypocholesterolemic Peptides

It is necessary to maintain an appropriate ratio of blood lipids, as it is one of the most significant danger factors leading to cardiovascular diseases (CVD) (Claas and Arnett 2016). Milk proteins, particularly whey proteins hydrolysates or peptides, have been shown to apply hypocholesterolemic impacts in various animal models. Whey protein fragment f (71–75) with sequence IIAEK, known as lactostatin, was the main factor responsible for the observed effect (Nagaoka et al. 2001). Another peptide, β-lactotensin, obtained from chymotrypsin β-lactoglobulin hydrolysate, diminished complete cholesterol, LDL, and VLDL cholesterol content in mice fed a cholesterol enhanced eating routine (Shin et al. 1998). An epic peptide (Ile-Ile-Ala-Glu-Lys) from tryptic hydrolysate of β-lactoglobulin showed hypocholesterolemic impact (Nagaoka et al. 2001).

13.4.1.4 Antithrombotic Peptides

Thrombosis is local coagulation of blood in the circulatory system, leading to CVD (Grundy et al. 1998). Thrombosis is caused by fibrinogen and thrombin; they both react with blood platelets and give bold platelet coagulation. Antithrombotic peptides are characterized by their ability to inhibit platelet aggregation and fibrinogen binding (Hanjaya-Putra et al. 2018).

Fibrinogen (factor I) is a glycoprotein complex that occurs in the blood of vertebrates. During tissue and vascular injury, it is converted enzymatically by thrombin to "fibrin" (fibrin is a stringy, non-globular protein associated with blood coagulation) and afterward to a fibrin-based blood cluster. Fibrin clusters work fundamentally to block veins with blood clots to prevent blood loss.

The known antithrombotic peptides obtained from milk proteins are casoplatelin (MAIPPKKNQDDK) from κ-casein, peptide YQEPVLGPVRGPFPIIV from β-casein, and KRDS peptide from lactoferrin (Rutherfurd and Gill 2000; Erdmann et al. 2008; Rojas-Ronquillo et al. 2012) (Fig. 13.5). Fractions corresponding to k-CN f (152–160) and f (155–160) isolated as ACE inhibitors may also possess antithrombotic activity (Gobbetti et al. 2007). Guzmán-Rodríguez et al. (2019) found iron restricting and antithrombotic peptides delivered during the fermentation of milk by *Lactobacillus casei shirota* that produced antithrombotic action by hindering fibrin cross-connecting, which is required for the coagulating movement, utilizing the technique reported by Zhang et al. (2008) and observed that fermentation with pH 6.0 and incubation temperature 39.5 °C for 12 h presented the highest antithrombotic activity, with a value of 79.1%, as compared to other pH and temperatures.

13.4.2 Effect on the Gastrointestinal System

13.4.2.1 Mineral Binding Peptides

Mineral deficiencies are crucial nutritional problems worldwide. In this specific circumstance, mineral fortification is truly an outstanding and basic procedure to forestall this deficiency (Zimmermann and Hurrell 2007). It has been suggested that

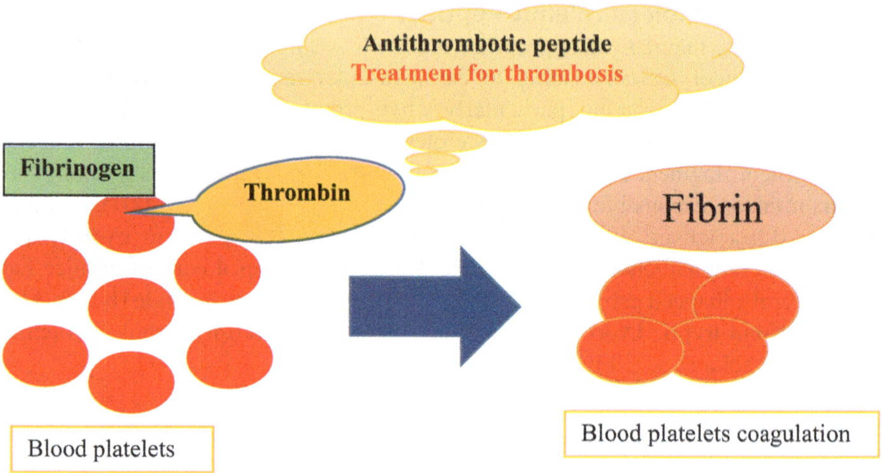

Fig. 13.5 The antithrombotic peptide's mechanism of action

the phosphorylated areas are passed from casein during digestion. CPPs refer to casein-derived phosphorylated peptides, which have single and multiple phosphoryl residues, and these phosphopeptides are delivered by enzymatic hydrolysis of α-, β- and κ-caseins both in vitro and in vivo (Clare and Swaisgood 2000). Because of the high amount of negative charges, these peptides productively bond divalent cations and go about as biocarriers for minor components, for example, Fe, Mn, Cu, and Se. CPPs generally refer to peptides created after enzymatic treatment with trypsin and which upgrade the ingestion of calcium over the distal small digestive tract (Pfeuffer and Schrezenmeir 2000). Casein phosphopeptides (CPPs), which may function as carriers for different minerals, especially calcium, are utilized in the food industry as ingredients or fortifiers in some low mineral containing foods and beverages. It has been proposed that CPPs, which form soluble complexes with calcium phosphate in vitro, may lead to enhanced calcium absorption by limiting the precipitation of calcium in the distal ileum (Meisel and FitzGerald 2003).

13.4.2.2 Antidiabetic Peptides

Diabetes is a chronic disease that happens either when the pancreas does not deliver enough insulin or when the body cannot successfully utilize the insulin it produces. Insulin is a hormone that manages blood sugar. Hyperglycemia, or high blood glucose (blood sugar), is a typical impact of uncontrolled diabetes and after some time induces serious damage to a considerable portion of the body's frameworks, particularly the nerves and veins. There are two significant kinds of diabetes: type 1, which results from a failure to convey insulin; and type 2, which results from insulin resistance (Ensor et al. 2014). Results from one study demonstrated that peptides with inhibitory property against α-glucosidase activity can be generated from the peptic digestion of whey proteins (Lacroix and Li-Chan 2013).

Different peptides were identified for the dipeptidyl peptidase-IV (DPP-IV) inhibitory activity from the Gouda-type cheese of the water-soluble fraction. The β-casein peptide residue 70–77 (β-CN f70–77; LPQNIPPL) demonstrated the maximum DPP-IV inhibitor activity (Uenishi et al. 2012). Trypsin-derived camel milk protein hydrolysates have displayed potent DPP-IV inhibitory properties in vitro (Nongonierma et al. 2018). To check the potentiality of *Lactobacillus* cultures on the production of milk-derived bioactive peptides with antidiabetic activity, ten *Lactobacillus* isolates were evaluated using three methods, including (1) α-amylase inhibitory activity, (2) α-glucosidase inhibitory activity, and (3) pancreatic lipase inhibitory activity, and it was reported that the highest α-amylase inhibition activity, α-glucosidase inhibitory activity, and pancreatic lipase inhibitory activity were observed in *Lb. rhamnosus* (M9) culture, *Lb. fermentum* (M7), and *Lb. casei* (NK9), respectively, as compared to other cultures. Likewise, bovine milk fermented with M2 (3 kDa permeate) indicated peptide sequence LFVPALLSLGALGLCLAA, which is obtained from lactotrasferrin, and milk fermented with M2 (10 kDa permeate) indicated peptide sequence NAGPFTPTV, which is obtained from αs_2 casein encoded in the antidiabetic peptides (ALG) (Kinariwala et al. 2019).

13.4.2.3 Antiobesity Peptides

In the regulation of food intake, satiety plays an important role and has significance in the control of obesity. It is widely acknowledged that protein is the most satiating part of food. The satiating impact of whey protein is fundamental because of a high grouping of branch chain amino acids, especially L-leucine. As to the casein portion of milk, it was recommended that peptides from casein hydrolysates activate the peripheral opioid and cholecystokinin receptors and block the antagonist receptors, which diminishes their impact on food entry (Hernández-Ledesma et al. 2014). Several studies showed that the whey protein-derived peptide, i.e., glycomacropeptide (GMP), stimulates the release of cholecystokinin (CCK), which may advance satiety in rodents (Pedersen et al. 2000). Mudgil et al. (2018) studied characterization and identification of novel antidiabetic and antiobesity peptides from camel milk protein hydrolysates. Antiobesity activities of camel whey protein hydrolysates (CWPHs) were investigated according to (1) pancreatic lipase inhibition activity and (2) cholesteryl esterase inhibition activity, and it was found that pepsin generated hydrolysates showed the highest pancreatic lipase inhibitory activity at each hydrolysis time in contrast with the unhydrolyzed camel whey (CW) (4.3%); 6 h of pepsin hydrolysis produced CWPH (P6), which was the most elevated (21.3% ± 0.2) cholesteryl esterase inhibition, 6 h of trypsin hydrolysis created CWPH (T6) (17.9 ± 0.5), and 3 h of chymotrypsin hydrolysis generated CWPH (C3).

13.4.2.4 Antimicrobial Peptides

The antimicrobial activity of milk is due to the synergistic action of naturally occurring peptides and defense proteins other than immunoglobulins, for example, lactoferrin, lactoperoxidase, and lysozyme. The first antimicrobial peptide isolated from milk through rennet, named lactenin, was recognized by Simmes and Jones in

1930. This peptide displayed antimicrobial activity against pathogenic strains of streptococci. Antimicrobial peptides (AMPs) are increasingly recognized as a critical first line of defense against many pathogens (Mulero et al. 2008). Most antimicrobial bioactive peptides act either by penetrating and disrupting microbial membrane integrity or by translocating across the membrane and acting on internal targets (Steinstraesser et al. 2011).

The peptide called casecidins from chymosin-treated casein hydrolysates showed antimicrobial activity against pathogenic *Staphylococcus aureus* and a few lactobacilli (Mohanty et al. 2016a, b). The antimicrobial action of the peptides may be because of interruption of microbial membranes, prompting particle and metabolite spillage, depolarization, disturbance of membrane coupled respiration, and eventually cell demise (Phadke et al. 2005). Hati et al. (2018) studied the influence of whey protein concentrate on the production of antibacterial peptides derived from fermented milk by lactic acid bacteria and found skim milk supplemented with 1.5% WPC fermented by *Lactobacillus rhamnosus* MTCC 5945 (NS4) produced ETVPYMFEN peptide sequence, identified as lactoferrin, which is a multifunctional, iron-binding glycoprotein that contains various antimicrobial peptides.

13.4.3 Effect on the Immune System

13.4.3.1 Immunomodulatory Peptides

Milk protein hydrolysates and peptides enhance immune cell functions (Horiguchi et al. 2005). Caseinomacropeptide (CMP) promotes the growth of bifidobacteria or lactobacilli that help to prevent enteric disease (Bruck et al. 2003). Several peptides, including f63–68 and f191–193 from bovine β-casein and f194–199 from bovine αs_1-casein have been identified as stimulating phagocytosis in mice and humans in vitro and protecting mice in vivo against *Klebsiella pneumonia* infection (Tidona et al. 2009) (Tables 13.4 and 13.5).

13.5 Conclusions and Future Directions

Scientific studies imply that milk has a plethora of bioactive peptides that can emphatically affect human well-being. The enthusiasm for bioactive milk peptides is expanding because milk proteins are accessible in incredible amounts with a high level of purity at a low cost. These bioactive peptides can be created and made bioavailable by proteolysis, which is caused by digestive enzymes, or through fermentation, which is caused by bacteria.

Further work is required to make a mixture of modified peptide sequences to discover pharmacologically active peptides with more strength and longer activity duration. In addition, there is a need to create advanced technologies for the production of peptide-improved functional foods with explicit health claims.

Table 13.4 Biological activities of bioactive peptides obtained from fermented milk

Biological activity	Name of product	Name of starter	Bioactive compound	References
ACE inhibitory	Cheese	Lactic acid	FFVAP	Korhonen (2009)
ACE inhibitory	Fermented camel milk	*Lb. rhamnosus (NS4)*	QTDIMIFTIGPA	Solanki and Hati (2018)
Lipid lowering	Fermented milk	Lc. Lactis NRRL B-50571	YPSYGL, SLPQNIPPL, TVQVTSTAV	Galland et al. (2017)
Immunomodulation	Fermented milk	*Lb. helveticus* R389	Peptide not characterized	LeBlanc et al. (2002)
Antimicrobial	Sodium caseinate fermented	*Lactobacillus acidophilus* DPC6026	VLNENLLR, IKHQGLPQE	Hayes et al. (2006)

Table 13.5 Bioactive peptides are used in commercial dairy products and additives to make health foods

Brand name	Country	Bioactive peptide	Health claims
Evolus	Finland	VPP, IPP from β-CN and k-CN in calcium-enriched fermented milk	Blood pressure reduction
Calpis	Japan	VPP, IPP from β-CN and k-CN in sour milk	
BioZate	USA	B-lg from hydrolyzed whey protein	
BioPUREGMP	USA	k-CN f(106–169)	Prevention of dental caries, influences the clotting of blood, protection against viruses and bacteria
PRODIET F200/lactium	France	αs1-CN (91–100) in flavored milk and confectionery	Reduction of stress effects
Cysteine peptide	Netherlands	Milk protein-derived peptides	Aids to raise the energy level and sleep
C12 peptides	Netherlands	Casein-derived peptides	Reduction of blood pressure
Vivinal Alpha	Netherlands	Whey-derived peptides	Aids for relaxation and sleep
PeptoPro	Netherlands	Casein-derived peptides	Improve muscle recovery and athletic performance
Capolac	Sweden	Casein-derived peptides	Improve mineral absorption

Korhonen and Pihlanto (2006)

References

Adriena DK, Anne P, Pertti M, Ladislav C, Hannu JTK (2010) Antioxidant properties of whey protein hydrolysates as measured by three methods. Eur Food Res Technol 230:865–874. https://doi.org/10.1007/s00217-010-1231-9

Ashar MN, Chand R (2004) Fermented milk containing ACE-inhibitory peptides reduces blood pressure in middle aged hypertensive subjects. Milchwissenschaft 59(7–8):363–366

Bruck WM, Graverholt G, Gibson GR (2003) A two-stage continuous culture system to study the effect of supplemental α-lactalbumin and glycomacropeptide on mixed cultures of human gut bacteria challenged with enteropathogenic Escherichia coli and Salmonella serotype Typhimurium. J Appl Microbiol 95(1):44–53. https://doi.org/10.1046/j.1365-2672.2003.01959

Cervato G, Cazzola R, Cestaro B (1999) Studies on the antioxidant activity of milk caseins. Int J Food Sci Nutr 50(4):291–296. https://doi.org/10.1080/096374899101175

Christensen JE, Dudley EG, Pederson JA, Steele JL (1999) Peptidases and amino acid catabolism in lactic acid bacteria. Antonie Van Leeuwenhoek 76(1–4):217–246. https://doi.org/10.1023/A:1002001919720

Claas SA, Arnett DK (2016) The role of healthy lifestyle in the primordial prevention of cardiovascular disease. Curr Cardiol Rep 18(6):56. https://doi.org/10.1007/s11886-016-0728-7

Clare DA, Swaisgood HE (2000) Bioactive milk peptides: a prospectus. J Dairy Sci 83 (6):1187–1195. https://doi.org/10.3168/jds.S0022-0302(00)74983-6

Colbert LB, Decker EA (1991) Antioxidant activity of an ultrafiltration permeate from acid whey. J Food Sci 56(5):1248–1250. https://doi.org/10.1111/j.1365-2621.1991.tb04744

Costa EL, da Rocha Gontijo JA, Netto FM (2007) Effect of heat and enzymatic treatment on the antihypertensive activity of whey protein hydrolysates. Int Dairy J 17(6):632–640. https://doi.org/10.1016/j.idairyj.2006.09.003

Doucet D, Otter DE, Gauthier SF, Foegeding EA (2003) Enzyme-induced gelation of extensively hydrolyzed whey proteins by Alcalase: peptide identification and determination of enzyme specificity. J Agric Food Chem 51(21):6300–6308. https://doi.org/10.1021/jf026242v

Dziuba B, Dziuba M (2014) Milk proteins-derived bioactive peptides in dairy products: molecular, biological and methodological aspects. Acta Sci Pol Technol Aliment 13(1):5–26

Ensor J, Boyd E, Juhola S, Castan Broto, V. (2014) Building adaptive capacity in the informal settlements of Maputo: lessons for development from a resilience perspective. In: Inderberg, T. H., Eriksen, S., O'Brien, K. and Synga, L. (eds.) Climate change adaptation and development. Routledge, pp. 19–38. ISBN: 9781138025967

Erdmann K, Cheung BW, Schröder H (2008) The possible roles of food-derived bioactive peptides in reducing the risk of cardiovascular disease. J Nutr Biochem 19(10):643–654. https://doi.org/10.1016/j.jnutbio.2007.11.010

Escudero E, Mora L, Toldrá F (2014) Stability of ACE inhibitory ham peptides against heat treatment and in vitro digestion. Food Chem 161:305–311. https://doi.org/10.1016/j.foodchem.2014.03.117

Espejo-Carpio FJ, De Gobba C, Guadix A, Guadix EM, Otte J (2013) Angiotensin I-converting enzyme inhibitory activity of enzymatic hydrolysates of goat milk protein fractions. Int Dairy J 32(2):175–183. https://doi.org/10.1016/j.idairyj.2013.04.002

FitzGerald RJ, Meisel H (2000) Milk protein-derived peptide inhibitors of angiotensin-I-converting enzyme. Br J Nutr 84(S1):33–37. https://doi.org/10.1017/S0007114500002221

Fitzgerald RJ, Murray BA (2006) Bioactive peptides and lactic fermentations. Int J Dairy Technol 59(2):118–125. https://doi.org/10.1111/j.1471-0307.2006.00250

Galland BV, Córdova AFG, Figueroa JCR (2017) U.S. patent no. 9,533,015. U.S. Patent and Trademark Office, Washington, DC

Gobbetti M, Minervini F, Rizzello CG (2007) Bioactive peptides in dairy products. In: Hui YH (ed) Handbook of food products manufacturing. John Wiley & Sons, Inc., New Jersey, pp 489–517

Grundy SM, Balady GJ, Criqui MH, Fletcher G, Greenland P, Hiratzka LF, Ockene IS (1998) Primary prevention of coronary heart disease: guidance from Framingham: a statement for healthcare professionals from the AHA Task Force on Risk Reduction. Circulation 97 (18):1876–1887

Guaadaoui A, Benaicha S, Elmajdoub N, Bellaoui M, Hamal A (2014) What is a bioactive compound, a combined definition for a preliminary consensus. Int J Nutr Food Sci 3 (3):174–179. https://doi.org/10.11648/j.ijnfs.20140303.16

Guo MR, Fox PF, Flynn A, Kindstedt PS (1995) Susceptibility of β-lactoglobulin and sodium caseinate to proteolysis by pepsin and trypsin. J Dairy Sci 78(11):2336–2344. https://doi.org/10.3168/jds.S0022-0302(95)76860-6

Guyton AC, Hall JE, Guyton AC (2006) Tratado de Fisiologia Medica. Elsevier Brasil, Rio de Janeiro

Guzmán-Rodríguez F, Gómez-Ruiz L, Rodríguez-Serrano G, Alatorre-Santamaría S, García-Garibay M, Cruz-Guerrero A (2019) Iron binding and antithrombotic peptides released during the fermentation of milk by *Lactobacillus casei shirota*. Revista Mexicana de Ingeniería Química 18(3):1161–1165. https://doi.org/10.24275/uam/izt/dcbi/revmexingquim/2019v18n3/Guzman

Hanjaya-Putra D, Haller C, Wang X, Dai E, Lim B, Liu L, Hagemeyer CE (2018) Platelet-targeted dual pathway antithrombotic inhibits thrombosis with preserved hemostasis. JCI Insight 3 (15):1–18. https://doi.org/10.1172/jci.insight.99329

Hartmann R, Meisel H (2007) Food-derived peptides with biological activity: from research to food applications. Curr Opin Biotechnol 18(2):163–169. https://doi.org/10.1016/j.copbio.2007.01.013

Hati S, Sreeja V, Solanki J, Prajapati JB (2015) Significance of proteolytic microorganisms on ACE-inhibitory activity and release of bioactive peptides during fermentation of milk. Indian J Dairy Sci 68:584–591

Hati S, Patel N, Sakure A, Mandal S (2018) Influence of whey protein concentrate on the production of antibacterial peptides derived from fermented milk by lactic acid bacteria. Int J Pept Res Ther 24(1):87–98. https://doi.org/10.1007/s10989-017-9596-2

Hayes M, Ross RP, Fitzgerald GF, Hill C, Stanton C (2006) Casein-derived antimicrobial peptides generated by *Lactobacillus acidophilus* DPC6026. Appl Environ Microbiol 72(3):2260–2264. https://doi.org/10.1128/AEM.72.3.2260-2264.2006

Hernandez-Ledesma B, Davalos A, Bartolome B, Amigo L (2005) Preparation of antioxidant enzymatic hydrolysates from alphalactalbumin and beta-actoglobulin. Identification of active peptides by HPLC–MS/MS. J Agric Food Chem 53:588–593. https://doi.org/10.1021/jf048626m

Hernández-Ledesma B, García-Nebot MJ, Fernández-Tomé S, Amigo L, Recio I (2014) Dairy protein hydrolysates: peptides for health benefits. Int Dairy J 38(2):82–100. https://doi.org/10.1016/j.idairyj.2013.11.004

Hirose M (1993) Molten globule state of food proteins. Trends Food Sci Technol 4(2):48–51

Horiguchi N, Horiguchi H, Suzuki Y (2005) Effect of wheat gluten hydrolysate on the immune system in healthy human subjects. Biosci Biotechnol Biochem 69(12):2445–2449. https://doi.org/10.1271/bbb.69.2445

Jauhiainen T, Korpela R (2007) Milk peptides and blood pressure. J Nutr 137(3):825S–829S. https://doi.org/10.1093/jn/137.3.825S

Kinariwala D, Panchal G, Sakure A, Hati S (2019) Exploring the potentiality of lactobacillus cultures on the production of milk-derived bioactive peptides with antidiabetic activity. Int J Pept Res Ther 26:1613–1627. https://doi.org/10.1007/s10989-019-09958-5

Korhonen H (2009) Milk-derived bioactive peptides: from science to applications. J Funct Foods 1 (2):177–187. https://doi.org/10.1016/j.jff.2009.01.007

Korhonen H, Pihlanto A (2003) Bioactive peptides: new challenges and opportunities for the dairy industry. Aust J Dairy Technol 58(2):129

Korhonen H, Pihlanto A (2006) Bioactive peptides: production and functionality. Int Dairy J 16 (9):945–960. https://doi.org/10.1016/j.idairyj.2005.10.012

Koyama M, Hattori S, Amano Y, Watanabe M, Nakamura K (2014) Blood pressure-lowering peptides from neo-fermented buckwheat sprouts: a new approach to estimating ACE-inhibitory activity. Peer Rev J 9(9):e105802. https://doi.org/10.1371/journal.pone.0105802

Lacroix IM, Li-Chan EC (2013) Inhibition of dipeptidyl peptidase (DPP)-IV and α-glucosidase activities by pepsin-treated whey proteins. J Agric Food Chem 61(31):7500–7506. https://doi.org/10.1021/jf401000s

LeBlanc JG, Matar C, Valdez JC, LeBlanc J, Perdigon G (2002) Immunomodulating effects of peptidic fractions issued from milk fermented with *Lactobacillus helveticus*. J Dairy Sci 85 (11):2733–2742. https://doi.org/10.3168/jds.S0022-0302(02)74360-9

Li Y, Zhou J, Huang K, Sun Y, Zeng X (2012) Purification of a novel angiotensin I-converting enzyme (ACE) inhibitory peptide with an antihypertensive effect from loach (Misgurnus anguillicaudatus). J Agric Food Chem 60(5):1320–1325. https://doi.org/10.1021/jf204118n

Lin L, Lv S, Li B (2012) Angiotensin-I-converting enzyme (ACE)-inhibitory and antihypertensive properties of squid skin gelatin hydrolysates. Food Chem 131(1):225–230. https://doi.org/10.1016/j.foodchem.2011.08.064

Mann B, Kumari A, Kumar R, Sharma R, Prajapati K, Mahboob S, Athira S (2015) Antioxidant activity of whey protein hydrolysates in milk beverage system. J Food Sci Technol 52 (6):3235–3241. https://doi.org/10.1007/s13197-014-1361-3

Martínez-Maqueda D, Miralles B, Recio I, Hernández-Ledesma B (2012) Antihypertensive peptides from food proteins: a review. Food Funct 3(4):350–361. https://doi.org/10.1039/c2fo10192k

McClean S, Beggs LB, Welch RW (2014) Antimicrobial activity of antihypertensive food-derived peptides and selected alanine analogues. Food Chem 146:443–447. https://doi.org/10.1016/j.foodchem.2013.09.094

Meisel H (1997) Biochemical properties of regulatory peptides derived from mil proteins. Pept Sci 43(2):119–128

Meisel H, FitzGerald RJ (2003) Biofunctional peptides from milk proteins: mineral binding and cytomodulatory effects. Curr Pharm Des 9(16):1289–1296. https://doi.org/10.2174/1381612033454847

Miguel M, Contreras MM, Recio I, Aleixandre A (2009) ACE-inhibitory and antihypertensive properties of a bovine casein hydrolysate. Food Chem 112(1):211–214. https://doi.org/10.1016/j.foodchem.2008.05.041

Mohanty DP, Mohapatra S, Misra S, Sahu PS (2016a) Milk derived bioactive peptides and their impact on human health—a review. Saudi J Biol Sci 23(5):577–583. https://doi.org/10.1016/j.sjbs.2015.06.005

Mohanty D, Jena R, Choudhury PK, Pattnaik R, Mohapatra S, Saini MR (2016b) Milk derived antimicrobial bioactive peptides: a review. Int J Food Prop 19(4):837–846. https://doi.org/10.1080/10942912.2015.1048356

Mudgil P, Kamal H, Yuen GC, Maqsood S (2018) Characterization and identification of novel antidiabetic and anti-obesity peptides from camel milk protein hydrolysates. Food Chem 259:46–54. https://doi.org/10.1016/j.foodchem.2018.03.082

Mulero I, Noga EJ, Meseguer J, García-Ayala A, Mulero V (2008) The antimicrobial peptides piscidins are stored in the granules of professional phagocytic granulocytes of fish and are delivered to the bacteria-containing phagosome upon phagocytosis. Dev Comp Immunol 32 (12):1531–1538. https://doi.org/10.1016/j.dci.2008.05.015

Nagaoka S, Futamura Y, Miwa K, Awano T, Yamauchi K, Kanamaru Y, Kuwata T (2001) Identification of novel hypocholesterolemic peptides derived from bovine milk β-lactoglobulin. Biochem Biophys Res Commun 281(1):11–17. https://doi.org/10.1006/bbrc.2001.4298

Nakamura Y, Yamamoto N, Sakai K, Takano T (1995) Antihypertensive effect of sour milk and peptides isolated from it that are inhibitors to angiotensin Iconverting enzyme. J Dairy Sci 78 (6):1253–1257

Nongonierma AB, Paolella S, Mudgil P, Maqsood S, FitzGerald RJ (2018) Identification of novel dipeptidyl peptidase IV (DPP-IV) inhibitory peptides in camel milk protein hydrolysates. Food Chem 244:340–348. https://doi.org/10.1016/j.foodchem.2017.10.033

Otte J, Shalaby SM, Zakora M, Pripp AH, El-Shabrawy SA (2007) Angiotensin-converting enzyme inhibitory activity of milk protein hydrolysates: effect of substrate, enzyme and time of hydrolysis. Int Dairy J 17(5):488–503. https://doi.org/10.1016/j.idairyj.2006.05.011

Padghan PV, Mann B, Hati S (2018) Purification and characterization of antioxidative peptides derived from fermented milk (Lassi) by lactic cultures. Int J Pept Res Ther 24(2):235–249. https://doi.org/10.1007/s10989-017-9608-2

Panchal G, Hati S, Sakure A (2019) Characterization and production of novel antioxidative peptides derived from fermented goat milk by L. fermentum. Lebensmittel-Wissenschaft & Technologie (LWT) 119:108887. https://doi.org/10.1016/j.lwt.2019.108887

Park PJ, Jung WK, Nam KS, Shahidi F, Kim SK (2001) Purification and characterization of antioxidative peptides from lecithin-free egg yolk protein. J Am Oil Chem Soc 78:651–656. https://doi.org/10.1007/s11746-001-0321-0

Park SY, Lee JS, Baek HH, Lee HG (2010) Purification and characterization of antioxidant peptides from soy protein hydrolysate. J Food Biochem 34(S1):120–132. https://doi.org/10.1111/j.1745-4514.2009.00313

Parmar H, Hati S, Panchal G, Sakure AA (2019) Purification and production of novel angiotensin I-converting enzyme (ACE) inhibitory bioactive peptides derived from fermented goat milk. Int J Pept Res Ther 26:997–1011. https://doi.org/10.1007/s10989-019-09902-7

Pedersen NLR, Nagain-Domaine C, Mahé S, Chariot J, Rozé C, Tomé D (2000) Caseinomacropeptide specifically stimulates exocrine pancreatic secretion in the anesthetized rat. Peptides 21(10):1527–1535. https://doi.org/10.1016/S0196-9781(00)00307-7

Pfeuffer M, Schrezenmeir J (2000) Bioactive substances in milk with properties decreasing risk of cardiovascular diseases. Br J Nutr 84(S1):155–159. https://doi.org/10.1017/S0007114500002385

Phadke SM, Deslouches B, Hileman SE, Montelaro RC, Wiesenfeld HC, Mietzner TA (2005) Antimicrobial peptides in mucosal secretions: the importance of local secretions in mitigating infection. J Nutr 135(5):1289–1293. https://doi.org/10.1093/jn/135.5.1289

Phelan M, Aherne A, FitzGerald RJ, O'Brien NM (2009) Casein-derived bioactive peptides: biological effects, industrial uses, safety aspects and regulatory status. Int Dairy J 19 (11):643–654. https://doi.org/10.1016/j.idairyj.2009.06.001

Power O, Jakeman P, FitzGerald RJ (2013) Antioxidative peptides: enzymatic production, in vitro and in vivo antioxidant activity and potential applications of milk-derived antioxidative peptides. Amino Acids 44(3):797–820. https://doi.org/10.1007/s00726-012-1393-9

Reddy IM, Kella NK, Kinsella JE (1988) Structural and conformational basis of the resistance of beta-lactoglobulin to peptic and chymotryptic digestion. J Agric Food Chem 36(4):737–741. https://doi.org/10.1021/jf00082a015

Ricci-Cabello I, Olalla Herrera M, Artacho R (2012) Possible role of milk-derived bioactive peptides in the treatment and prevention of metabolic syndrome. Nutr Rev 70(4):241–255. https://doi.org/10.1111/j.1753-4887.2011.00448

Rojas-Ronquillo R, Cruz-Guerrero A, Flores-Nájera A, Rodríguez-Serrano G, Gómez-Ruiz L, Reyes-Grajeda JP, García-Garibay M (2012) Antithrombotic and angiotensin-converting enzyme inhibitory properties of peptides released from bovine casein by *Lactobacillus casei Shirota*. Int Dairy J 26(2):147–154. https://doi.org/10.1016/j.idairyj.2012.05.002

Rutherfurd KJ, Gill HS (2000) Peptides affecting coagulation. Br J Nutr 84(S1):99–102. https://doi.org/10.1017/S0007114500002312

Schanbacher FL, Talhouk RS, Murray FA, Gherman LI, Willett LB (1998) Milk-borne bioactive peptides. Int Dairy J 8(5–6):393–403. https://doi.org/10.1016/S0958-6946(98)00062-4

Schlimme E, Meisel H (1995) Bioactive peptides derived from milk proteins. Structural, physiological and analytical aspects. Food Nahrung 39(1):1–20. https://doi.org/10.1002/food.19950390102

Sharma S, Singh R, Rana S (2011) Bioactive peptides: a review. Int J Bioautom 15(4):223–250

Shin K, Yamauchi K, Teraguchi S, Hayasawa H, Tomita M, Otsuka Y, Yamazaki S (1998) Antibacterial activity of bovine lactoferrin and its peptides against enterohaemorrhagic Escherichia coli O157: H7. Lett Appl Microbiol 26(6):407–411

Silvestre MPC, Silva MR, Silva VDM, Souza MWSD, Junior L, de Oliveira C, Afonso WDO (2012) Analysis of whey protein hydrolysates: peptide profile and ACE inhibitory activity. Braz J Pharm Sci 48(4):747–757. https://doi.org/10.1590/S1984-82502012000400019

Solanki D, Hati S (2018) Considering the potential of *Lactobacillus rhamnosus* for producing Angiotensin I-Converting Enzyme (ACE) inhibitory peptides in fermented camel milk (Indian breed). Food Biosci 23:16–22. https://doi.org/10.1016/j.fbio.2018.03.004

Staessen JA, Li Y, Richart T (2006) Oral renin inhibitors. Lancet 368(9545):1449–1456. https://doi.org/10.1016/S0140-6736(06)69442-7

Steijins JM, Van HAC (2000) Occurrence, structure, biochemical properties and technological characteristics of lactoferrin. Br J Nutr 84:S11–S17. https://doi.org/10.1017/S0007114500002191

Steinstraesser L, Kraneburg U, Jacobsen F, Al-Benna S (2011) Host defense peptides and their antimicrobial-immunomodulatory duality. Immunobiology 216(3):322–333. https://doi.org/10.1016/j.imbio.2010.07.003

Tanaka M, Watanabe S, Wang Z, Matsumoto K, Matsui T (2009) His-Arg-Trp potently attenuates contracted tension of thoracic aorta of Sprague-Dawley rats through the suppression of extracellular Ca^{2+} influx. Peptides 30(8):1502–1507. https://doi.org/10.1016/j.peptides.2009.05.012

Taylor MJ, Richardson T (1980) Antioxidant activity of skim milk: effect of heat and resultant sulfhydryl groups. J Dairy Sci 63(11):1783–1795. https://doi.org/10.3168/jds.S0022-0302(80)83140-7

Tidona F, Criscione A, Guastella AM, Zuccaro A, Bordonaro S, Marletta D (2009) Bioactive peptides in dairy products. Ital J Anim Sci 8(3):315–340. https://doi.org/10.4081/ijas.2009.315

Udenigwe CC, Li H, Aluko RE (2012) Quantitative structure–activity relationship modeling of renin-inhibiting dipeptides. Amino Acids 42(4):1379–1386. https://doi.org/10.1007/s00726-011-0833-2

Uenishi H, Kabuki T, Seto Y, Serizawa A, Nakajima H (2012) Isolation and identification of casein-derived dipeptidyl-peptidase 4 (DPP-4)-inhibitory peptide LPQNIPPL from gouda-type cheese and its effect on plasma glucose in rats. Int Dairy J 22(1):24–30. https://doi.org/10.1016/j.idairyj.2011.08.002

Wijesekara I, Kim SK (2010) Angiotensin-I-converting enzyme (ACE) inhibitors from marine resources: prospects in the pharmaceutical industry. Mar Drugs 8(4):1080–1093. https://doi.org/10.3390/md8041080

Williams AG, Noble J, Tammam J, Lloyd D, Banks JM (2002) Factors affecting the activity of enzymes involved in peptide and amino acid catabolism in non-starter lactic acid bacteria isolated from Cheddar cheese. Int Dairy J 12(10):841–852. https://doi.org/10.1016/S0958-6946(02)00063-8

Xu X, Gao Y (2015) Purification and identification of angiotensin I-converting enzyme-inhibitory peptides from alphalbumin 2 during simulated gastrointestinal digestion. J Sci Food Agric 95 (5):906–914. https://doi.org/10.1016/j.foodchem.2014.11.013

Zhang SB, Wang Z, Xu SY (2008) Antioxidant and antithrombotic activities of rapeseed peptides. J Am Oil Chem Soc 85:521–527. https://doi.org/10.1007/s11746-008-1217

Zheng Y, Li Y, Zhang Y, Ruan X, Zhang R (2017) Purification, characterization, synthesis, in vitro ACE inhibition and in vivo antihypertensive activity of bioactive peptides derived from oil palm kernel glutelin-2 hydrolysates. J Funct Foods 28:48–58. https://doi.org/10.1016/j.jff.2016.11.021

Zimmermann MB, Hurrell RF (2007) Nutritional iron deficiency. Lancet 370(9586):511–520. https://doi.org/10.1016/S0140-6736(07)61235-5

Role of Probiotics in Allergic Diseases 14

Swati T. Gadhe and Vipul P. Patel

Abstract

The prevalence of allergic disorders has been increasing worldwide and significantly impacts the quality of life of the atopic individual. There has been an increased interest in the role of probiotics for the prevention and treatment of allergic disorders, given the recent evidence that atopy risk may be associated with a dysbiosis of the gut microbiome. Research in this area is ongoing with some studies showing possible benefits of probiotics, with seemingly little to no risk. While these studies suggest that there may be a promise in probiotic use for the prevention or treatment of allergy, further evidence is needed to determine its efficacy, optimal dosing, and strains needed for treatment. In this review, we discuss recently published studies examining the benefits, risks, and role of probiotics in preventing atopic dermatitis, asthma, allergic rhinitis, and food allergy.

Keywords

Allergies · Clinical applications · Food allergy · Mechanisms · Probiotics

S. T. Gadhe · V. P. Patel (✉)
SRES' Sanjivani College of Pharmaceutical Education and Research, Kopargaon, Maharashtra, India

R. Z. Sayyed, M. Khan (eds.), *Microbiome-Gut-Brain Axis*,
https://doi.org/10.1007/978-981-16-1626-6_14

14.1 Introduction

Allergic diseases are increasing in prevalence globally and cause a significant burden on society both economically and physiologically (Pawankar et al. 2013). Despite their universality, the etiology of allergic disorders remains unknown. In 1989, it was hypothesized that reducing exposure to microbes resulted in an immune system variation, favoring a shift toward an allergic response. This hypothesis was based on an observation of reduced incidence of hay fever and eczema in children living with older siblings within greater families resulting in increased microbial exposure. In more recent times, it has been demonstrated that dysbiosis of the gut microbiome can be associated with an increased risk of atopy (Lambrecht and Nhammad 2017). Increasingly, probiotics (the "good bacteria") have been used in an attempt to correct this. Probiotics are defined as "live microorganisms which confer a useful effect on the host" according to the World Health Organization (WHO) (Pawankar et al. 2013). Probiotics in allergic diseases affect phagocytosis and synthesis of pro-inflammatory cytokines and thus have been proposed as modulators of the allergic response and advocated as therapeutic and preventive medications for allergic diseases. It has been suggested that probiotics may prevent the allergic response due to their anti-inflammatory effects, although this area remains controversial. The intestine is the body's largest immune organ; most of the antibody-producing cells reside in the intestine (Brandtzaeg 2002). The intestinal microbiota represents the body's greatest microbial exposure by a substantial extent and in part works to provide stimulation of the immune system. The specific composition of the intestinal microbiota may affect the risk of developing allergic diseases (Wang and Anvari 2019; Penders et al. 2007).

Allergic diseases are complex multifactorial disorders, with interactions of genetic, environmental, and socioeconomic factors determining disease expression and leading to different phenotypes.

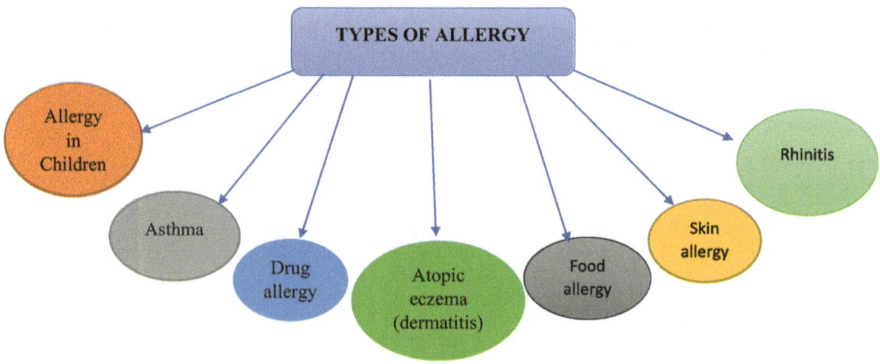

14.2 Probiotics

Probiotics are live microorganisms aided with a demand that they give health benefits when consumed, generally by enhancing or replacing the gut flora (Lvory et al. 2008). Probiotics are considered generally harmless to consume but may cause bacterial-host interactions and unwanted side effects in rare cases. A microorganism introduced into the body for its useful qualities is complete sentence.

14.2.1 Characteristics of Probiotics

There are several generally accepted characteristics that define probiotic bacteria:

- Are microbial organisms.
- Remain inconstant and reliable after culture, manipulation, and storage before use.
- Survive gastric, biliary, and pancreatic digestion.
- Are able to produce a host response once they enter the intestinal microbial ecosystem.
- Yield a functional or clinical use to the host when administered.

14.2.2 Mechanisms of Action of Probiotics in Allergic Disease

The most important probiotic mechanisms of action include enhancement of the epithelial barrier, increased adhesion to intestinal mucosa, concomitant inhibition of pathogen cohesion, competitive exclusion of pathogenic microorganisms, production of antimicrobial substances, and attenuation of the immune system. Allergic disorders are related with a move of the Th1/Th2 cytokine balance leading to activation of Th2 cytokines and the release of interleukin-4 (IL-4), IL-5, and IL-13 as well as IgE manufacturing (Tang 2005; Coskun and Klinikleri 2007). Probiotics can potentially modulate the toll-like receptors and the proteoglycan recognition proteins of enterocytes, leading to activation of dendritic cells and a Th1 response (Flinterman 2007). The resulting stimulation of Th1 cytokines can suppress Th2 responses. Pediatric studies suggest that probiotic use in children with atopic conditions such as atopic dermatitis results in the increase of IFN production and decrease of IgE- and antigen-induced TNF-α, IL-5, and IL-10 secretion (Prescott 2005; Kalliomäki et al. 2001) (Fig. 14.1; Table 14.1).

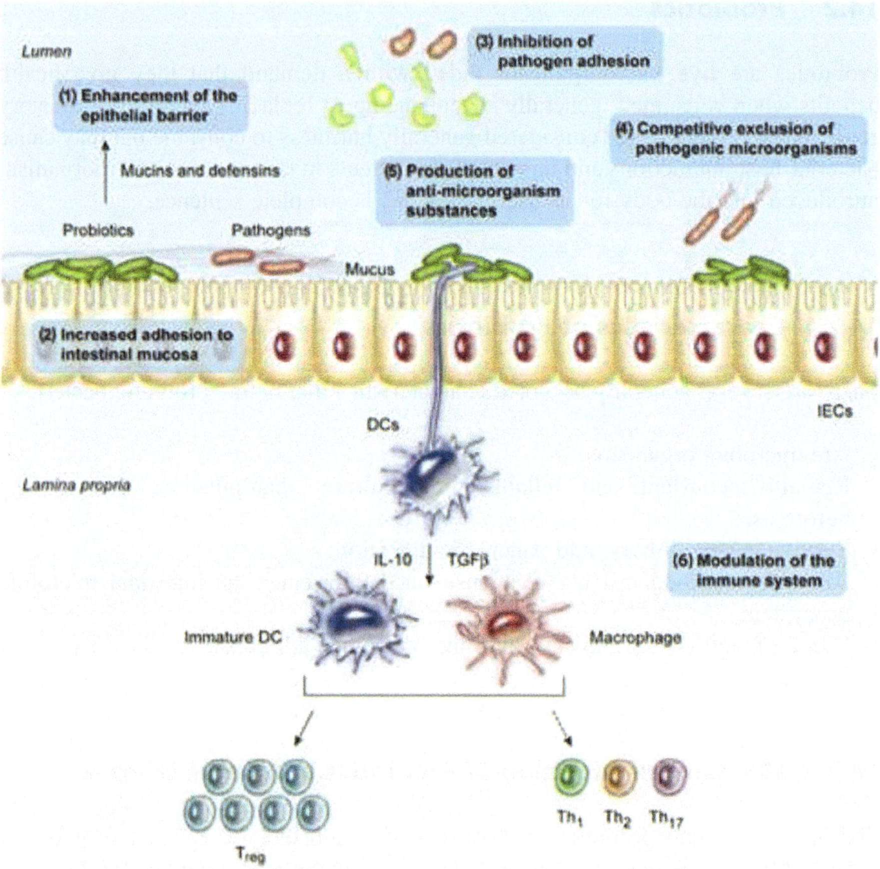

Fig. 14.1 Mechanism of action of probiotic

Table 14.1 Application of probiotics in various allergies

Disease	Probiotics	Timing	Dose	Result	References
Atopic dermatitis	LGG LGG	Prenatal + postnatal Childhood	≥109-10 ≥109-10	Decrease with effect Decrease with effect	Kalliomäki et al. (2003), Kopp et al. (2008), Kalliomäki et al. (2007), Kim et al. (2014)
Nutritional allergy	Various LGG	Prenatal + postnatal Infancy	≥109-10 ≥109-10	Ineffectiveness Decrease with effect	Osborn (2007), Majamaa and Isolauri (1997), Kim et al. (2014), Tang (2005), Majamaa Hand Isolauri (1997)
Allergic rhinitis	Various *Bifidobacterium*	Prenatal + postnatal Early childhood	≥109-10 ≥109-10	Ineffectiveness Decrease with effect	Kalliomäki et al. (2003), Kopp et al. (2008), Kalliomäki et al. (2007), Lvory et al. (2008), Wang et al. (2004)
Asthma	Various *Lactobacillus*	Prenatal + postnatal Game child	≥109-10 ≥109-10	Ineffectiveness Ineffectiveness	Kalliomäki et al. (2003), Kopp et al. (2008), Kalliomäki et al. (2007), Miraglia Del Giudice et al. (2012), Van De Pol et al. (2011)

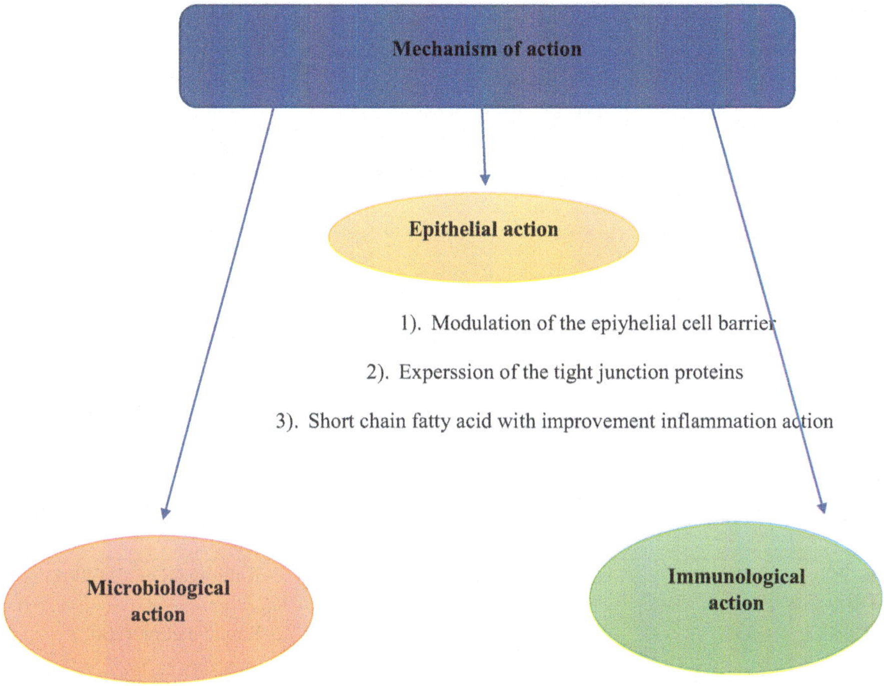

1). Modulation of the distribution of microbiota rate	1). Innate immunition modulation the
2). Composition adhesion to the receptorswith the stopage of the pathogens invasion	2). Modulation of Th1 /Th2rate
	3). Increase of number & activity of T regulatory cell

14.2.3 Role of Probiotic in Atopic Dermatitis

Atopic dermatitis is the most common chronic inflammatory skin disease and is often the primary step in the atopic march. The two primary theories on the origin of atopic dermatitis are the "inside-out" hypothesis, which speculates that instability in the enteric microbiota results in inflammatory processes, and the "outside-in" hypothesis, which suggests that the disrupted skin microbiome is the primary triggering event for atopic dermatitis (Giovannini et al. 2007). Numerous animal and in vitro studies, as well as several human trials, suggest a useful effect of probiotics in allergic diseases. Various randomized studies demonstrated that when *Lactobacillus* GG or placebo was given to pregnant mothers with a strong family history of eczema, allergic rhinitis, or asthma and to their infants for the first 6 months after delivery, the frequency of developing atopic dermatitis in the offspring was decreased in 2 years, 4 years, and 7 years by 50%, 44%, and 36%, respectively

Table 14.2 Prevention of sensitization and allergic diseases

Study type	Probiotic type	Conclusion	References
R, C, DB	*L. reuteri*	Decreased IgE-associated eczema	Abrahamsson (2007)
R, PC, DB	LGG	Decreased atopic dermatitis	Kalliomäki et al. (2007)
R, PC, DB	LGG, LC705 *L. rhamnosus*	Lower IgE-associated disease	Kukkonen (2007)
R, PC, DB	*L. acidophilus* LAVRI-A1	No change in atopic dermatitis	Taylor (2006)
R, PC, DB	*L. acidophilus* LAVRI-A1	No change in atopic dermatitis	Taylor (2006)
C	*E. coli*	Decreased long-term influence of allergy	Lodinova et al. (2004)
R, PC, DB	LGG	Decreased atopic dermatitis	Kalliomäki et al. (2003)
R, PC, DB	*E. coli*	Decreased allergic development	Lodinova-Zadnikova et al. (2003)
R, PC, DB	LGG	Decreased atopic dermatitis	Kalliomäki et al. (2001)
R, PC, DB	LGG	Decreased atopic dermatitis	Rautava et al. (2002)

(Taylor 2006; Kalliomäki et al. 2001). The idea of probiotic supplementation to restore balance in the microbiome of humans is the foundational argument for the use of probiotics in primary prevention of atopic dermatitis. Three systematic reviews have so far investigated probiotic supplementation regarding eczema prevention (Slattery et al. 2016) (Table 14.2).

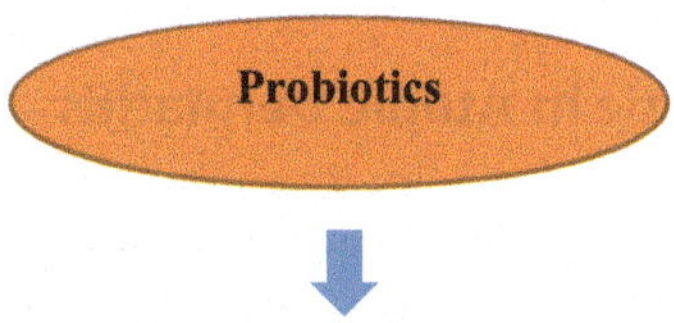

- Reverse increased intestinal permeability.
- Enhance gut-specific IgA responses.
- Promote gut barrier function.
- Modulation of immune response.
- Enhance production of IL-10 and cytokines that promote production of IgE antibodies.

14.2.4 Role of Probiotic in Asthma

A small number of studies happen that try to address the efficacy of probiotic supplementation in the treatment or prevention of asthma. A study using fermented milk containing *L. casei* and studying its effect on the number of parts of asthma and allergic rhinitis found no statistical difference between intervention and control groups of asthmatic children. However, the number of rhinitis parts was lower in the probiotic group, leading the authors to conclude that *L. casei* may be useful for children with allergic rhinitis but not asthmatic children (Coskun and Klinikleri 2007; Food and Agriculture Organization 2001). Atopy is a common precursor for asthma, a chronic inflammatory condition of the airways that, when uncontrolled, can result in a poor quality of life or even death. The indication for use of probiotics as a preventive or therapeutic agent for respiratory allergies shows less. The efficacy of probiotics in asthma as a preventive measure has not been evaluated and may be worthwhile studying. However, to date, there is no evidence to justify the use of probiotics for the treatment or prevention of asthma.

- Preventive vital for asthma and allergic disorders have been introduced in 2014:
  - General health education: escape of tobacco smoke exposure during pregnancy and after birth.
  - First prevention for infants at higher risk: several longitudinal birth-associated studies have clearly demonstrated an increased harm of allergic expression if one or two parents are or have been affected themselves.
  - Secondary prevention strategies for children who have already developed allergic sensitization or the first indication of allergic diseases: these strategies aim to reduce the incidence of clinical indication, such as rhinitis, food allergy, or asthma.

14.2.5 Role of Probiotic in Allergic Rhinitis

The efficacy of probiotics in treating allergic rhinitis is incompatible (Tang and Chen 2001). According to Wang and colleagues, where *Lactobacillus paracasei* 33 was given for 30 days to 80 children with perennial rhinoconjunctivitis, the quality of life questionnaire score significantly improved relative to placebo (Coskun and Klinikleri 2007). In seasonal and perennial allergic rhinitis in addition to those who state that probiotics are ineffective in the treatment of periodic and persistent allergic rhinitis, there are also researchers who state that probiotics are effective. In a study of probiotic infuse bed quilt/pillow cases, an improvement was found in the symptoms of the patients with allergic rhinitis, and the quality of life increased (Wang et al. 2004).

14.2.6 Role of Probiotic in Food Allergy

Recent studies propose that probiotics may have a role in the treatment of food allergy by maintaining the intestinal epithelial barrier integrity, overcoming intestinal inflammatory responses, and inducing mucosal IgA production and tolerogenic immune response (Lvory et al. 2008; Bering et al. 2007). Hol et al. found no effect of *L. casei* CRI 431 and *B. lactis* Bb12 supplements for 12 months on the procurenment of tolerance in 119 infants with cow's milk allergy. In another study on children with egg, peanut, or cow's milk allergy who were treated with a probiotic mix for 3 months showed that the treatment did not impact sensitization or ex vivo immune responses. This would further desire that the confirmation as to whether probiotics can induce tolerance in allergy is currently lacking (Giovannini et al. 2007).

The most common types of food allergens include eggs, milk, peanuts, tree nuts, fish, shellfish, wheat, and soy.

References

Abrahamsson TR (2007) Probiotics in prevention of IgE associated eczema. A double blind, randomized, placebo controlled trail. J Allergy Clin Immunol 119(5):1174–11780. https://doi.org/10.1016/000071488

Bering M, Jult A, Vermeulen H, Derycke L, Ucer H (2007) Probiotics impregnated bedding cover for house dust mite allergic rhinitis: a pilot randomized clinical trial. Clin Exp Allergy 47:1092–1096

Brandtzaeg P (2002) Current understanding of Gastrointestinal immunoregulation and its relation to food allergy. Ann N Y Acad Sci 964:13–45

Coskun T, Klinikleri T (2007) Pro-pre and symbiotic. J Pediatr Sci 3:82–98

Flinterman AE (2007) Probiotics have a different immunomodulatory potential in vitro versus ex vivo upon oral administration in children with food allergy. Int Arch Allergy Immunol 143 (3):237–244. https://doi.org/10.1159/000099467

Food and Agriculture Organization (2001) World Health Organization report of joint FAO/WHO expert consultation on evaluation of health and nutraceutical properties of probiotics in food including powder milk with live lactic acid bacteria. FAD/WHO Report no. 10-1-2001. WHO Int, Cordoba

Giovannini M, Agostoni C, Riva E, Salvini F, Ruscitto A (2007) A randomized prospective double blind controlled trial on effects of long term consumption of fermented milk containing lactacillus cases in pre school children with allergic asthma and/or rhinitis. Pediatrras 62:215–220

Kalliomäki M, Salminen S, Arvilommi H, Karo P, Koskinen P, Isolauri E (2001) Probiotics in primary prevention of atopic disease: a randomized placebo-controlled trial. Lancet 357:1076–1079

Kalliomäki M, Salminen S, Poussa T, Arvilommi H, Isolauri E (2003) Probiotics and prevention of atopic disease: 4-year follow-up of a randomized placebo-controlled trial. Lancet 361:1869–1871

Kalliomäki M, Salminen S, Poussa T, Isolauri E (2007) Probiotics during the first 7 years of life: a cumulative risk reduction of eczema in a randomized, placebo-controlled trial. J Allergy Clin Immunol 119:1019–1021

Kim SO, Ah YM, Yu YM, Choi KH, Shin WG, Lee JY (2014) Effects of probiotics for the treatment of atopic dermatitis: a meta-analysis of randomized controlled trials. Ann Allergy Asthma Immunol 113:217–226

Kopp MV, Hennemuth I, Heinzmann A, Urbanek R (2008) Randomized, double-blind, placebo-controlled trial of probiotics for primary prevention: no clinical effects of Lactobacillus GG supplementation. Pediatrics 121:850–856

Kukkonen K (2007) Probiotics & prebiotics galacto-oligosaccharide in the prevention of allergic disease a randomized, double blind, placebo controlled trials. J Allergy Clin Immunol 119 (1):192–198

Lambrecht B, Nhammad H (2017) The immunology of the allergy epidemic and the hygiene hypothesis. Nat Immunol 18:1076–1083

Lodinova ZR (2004) Influence of oral colonization with probiotics E.coli strain after birth on frequency of recurrent infection, allergy and development of some immunologic parameter. Ceskagynekol 69(Suppl 1):91–97

Lodinova-Zadnikova R, Cukrowska B, Tlaskalova-Hogenova H (2003) Oral administration of probiotic Escherichia coli after birth reduces frequency of allergies and repeated infections later in life (after 10 and 20 years). Int Arch Allergy Immunol 131:209–211

Lvory K, Chamber SJ, Pin C, Prieto E, Arques JL, Nicoletti C (2008) Oral delivery of lactobacillus casei Shirota modifies allergen – induced immune response in allergic rhinitis. Clin Exp Allergic 38:1282–1289

Majamaa H, Isolauri E (1997) Probiotics: a novel approach in the management of food allergy. J Allergy Clin Immunol 99:179–185

Miraglia Del Giudice M, Maiello N, Decimo F, Fusco N, Sullo N et al (2012) Airways allergic inflammation and L. reuterii treatment in asthmatic children. J Biol Regul Homeost Agents 26:35–40

Osborn DA (2007) Probiotics in infants for prevention of allergic disease and food hypersensitivity. Cochrane Database Syst Rev 4:CD006475

Pawankar R, Canonica G, Holgate S, Lackey R, Blaiss M (2013) White book on allergy: world allergy organization. World Allergy Organization (WAO), Milwaukee

Penders J, Stabberingh EE, Van den Brandt PA, Thijs C (2007) The role of the intestinal microbiota in the development of atopic disorder. Allergy 62:1223–1236

Prescott SL (2005) Clinical effects of probiotics are associated with increased interferon-gamma responses in very young children with atopic dermatitis. Clin Exp Allergy 35(12):1557–1564

Rautava S, Kalliomäki M, Isolauri E (2002) Probiotics during pregnancy and breast-feeding might confer immunomodulatory protection against atopic disease in the infant. J Allergy Clin Immunol 109(1):119–121. https://doi.org/10.1067/mai.2002.120273

Slattery J, MacFabe, Frye RE (2016) The Significance of the enteric microbiome on the development of childhood disease: A review of prebiotics and probiotics therapies in disorder in childhood. Clin Med Insights Pediatr 10:91–107. https://doi.org/10.4137/cmped.s38338

Tang RB (2005) Risk factors associated with the development of asthma. J Chin Med Assoc 68:199–201

Tang RB, Chen SJ (2001) Soluble interleukin 2 receptor and interleukin 4 in sera of asthmatic children before and after a prednisolone course. Ann Allergy Asthma Immunol 86:314–317

Taylor AL (2006) Effects of probiotic supplementation for the first 6 months of life on allergen-and vaccine-specific immune responses. Clin Exp Allergy 36(10):1227–1235. https://doi.org/10.1111/j.1365-2222.2006.02553

Van De Pol MA, Lutter R, Smids BS, Weersink EJ, Van der Zee JS (2011) Symbiotic reduce allergen-induced T-helper 2 response and improve peak expiratory flow in allergic asthmatics. Allergy 66:39–47

Wang TH, Anvari S (2019) The probiotics in preventing in allergic disease. Allergy, Asthma Clin Immunol 6(2):24. https://doi.org/10.3390/children6020024

Wang MF et al (2004) Treatment of perennial allergic rhinitis with lactic acid bacteria. Pediat Allergic Immunol 15(2):152–158

Probiotics Suppress the Depression: A Look at the Possible Mechanisms of Action

15

Leila Khalili and R. Z. Sayyed

Abstract

Manipulating the intestinal microbiota for the benefit of the mental health is a concept that has become widely acknowledged. Emerging evidence suggests that modifying the composition of the gut microbiota via probiotic supplementation may be a viable adjuvant treatment option for individuals with depression. The aim of this chapter is to illustrate the possible pathways through which gut microbiota may influence depression. PubMed, Scopus, and Web of Science databases were searched by using "probiotics", "depression", and "mechanism" key words for searching the studies aiming the application of probiotics and the beneficial effects of them in depression control and/or treatment. Findings of relevant studies suggest that probiotics could be considered as a promising adjuvant treatment to improve depression. The results of previous investigations suggest that modulation of inflammation, affecting the hypothalamic-pituitary-adrenal (HPA) axis, and interference with neurotransmitter signaling are the potential pathways through which probiotics may influence depression. Probiotics can alleviate depressive symptoms through several mechanisms; however, additional studies are necessary.

Keywords

Depression · Mechanisms · Probiotics

L. Khalili (✉)
Department of Community Nutrition, Faculty of Nutrition and Food Sciences, Tabriz University of Medical Sciences, Tabriz, Iran
e-mail: khalilil@tbzmed.ac.ir

R. Z. Sayyed
Department of Microbiology, PSGVP Mandal's Arts, Science and Commerce College, Shahada, Maharashtra, India

R. Z. Sayyed, M. Khan (eds.), *Microbiome-Gut-Brain Axis*,
https://doi.org/10.1007/978-981-16-1626-6_15

15.1 Introduction

Depression is a common mental disorder, which can be long-lasting or recurrent, substantially impairing an individual's ability to function in their daily life (Vilagut et al. 2016). In recent years, there has been major interest in exploring the link between the health of the gut and mental health (Schmidt 2015a, b). Several pathways have been identified that describe how gut microbiota may influence depression (Uher and McGuffin 2010). Altered microbiota has been linked to neuropsychological disorders such as depression. Traditionally, depression has been treated with a range of therapies including antidepressants and talking therapies; however, research has started to emerge which suggests that probiotics, live microorganisms that exert health benefit on the host when ingested in adequate amounts, may significantly reduce the symptoms of depression. Fortunately, studies have indicated that gut microbiota may be modulated with the use of probiotics, antibiotics, and fecal microbiota transplants as a prospect for therapy in microbiota-associated diseases. Probiotics are regulated as dietary supplement foods and now are available in capsules, tablets, packets, or powders and are contained in various fermented foods, most commonly yoghurt or dairy fermented drinks. The primary rationale for using probiotics involves restoring microbial balance. The administration of probiotics which contains beneficial bacteria may restore the microbial balance in the gastrointestinal tract (Li et al. 2020). It has been argued that gut microbiota may play a role in bidirectional communication between the gut and the central nervous system (Arneth 2018). The aim of this chapter is to illustrate the possible pathways through which gut microbiota may influence depression.

15.2 Neurological Disorders and Gut-Brain Axis

The gut is closely connected to the brain via 200–600 million neurons (Furness 2006). Bidirectional communication between the gut and the brain has long been recognized; that is, signals from the brain can influence the motor, sensory, and secretory modalities of the gastrointestinal (GI) tract and, in turn, visceral messages from the gut can influence brain function (Grenham et al. 2011; Tabrizi et al. 2019) (Fig. 15.1). Recently, there is expanding evidence for the view rethinking the gut-brain axis as the concept of a gut microbiota-brain axis due to the crucial role of gut microbiota in the bidirectional gut-brain axis (Cryan and Dinan 2012). It is now well recognized that the organisms of the gastrointestinal tract make important contributions to health and disease, including mood and cognition, and psychopathology. Nevertheless, we are still a long way from understanding the potential mechanisms underlying this connection complexity.

Although it has long been recognized that major disturbances in gut flora can affect central nervous system function, it is only now emerging that "normal" gut microbiota might have a role in mood and psychopathology (Forsythe et al. 2010). Both endocrine and neural pathways are involved in signaling gut immune responses to the brain. The neural pathways involved in the microbiome-gut-brain axis include

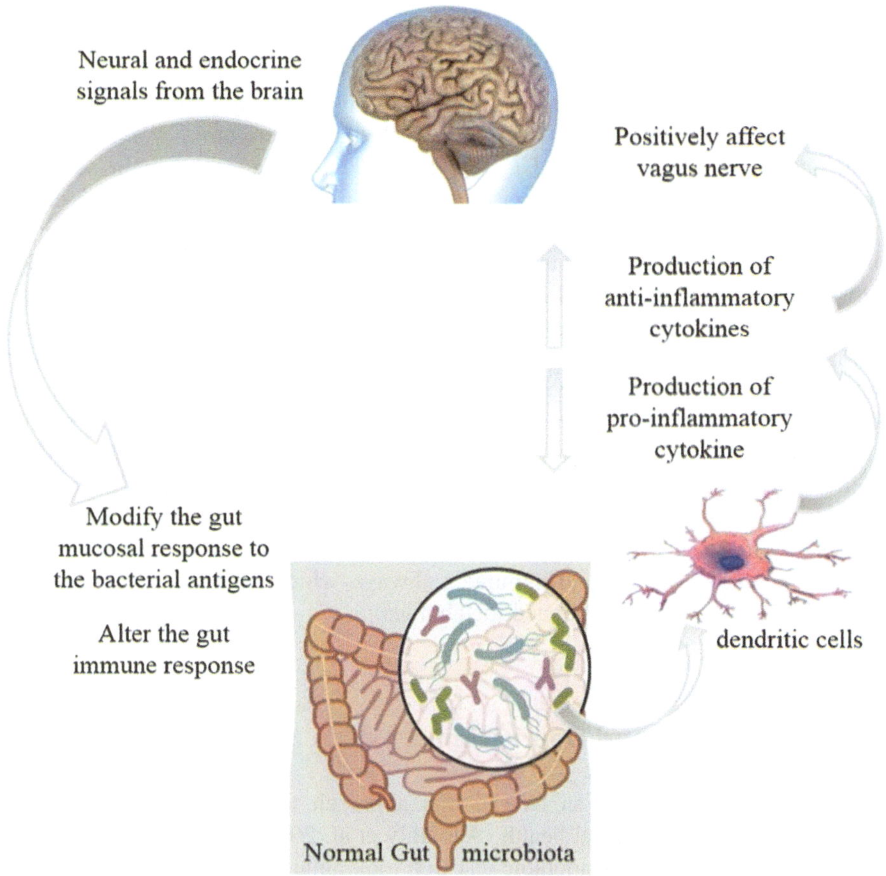

Fig. 15.1 The bidirectional communication between the gut and the nervous system (Grenham et al. 2011)

the sympathetic and parasympathetic autonomic nervous system and the local enteric nervous system.

Modulation of gut microbiota through consumption of probiotics might improve health by replacing harmful microbes with useful ones (Hemarajata and Versalovic 2013). It is believed that the primary mechanisms of action of probiotics are to contribute to modify the composition and function of the gut microbiota. The effect of gut microbiota on the brain can be established through several mechanisms. One of these involves the inhibition of histone deacetylase activity induced by the short-chain fatty acids (SCFAs) which are the end products of prebiotic fermentation by intestinal microorganisms. This may be responsible for the imbalance in histone acetylation levels and transcriptional dysregulations observed in neurodegenerative disorders (Dinan and Cryan 2017). Another proposed mechanism of interconnection of the gut and the brain is related to a direct effect of SCFAs on GI cells. This induces the production of hormones, such as leptin, which have a beneficial impact on the central nervous system and, consequently, on memory and cognition (Rea et al. 2016). Another mechanism to be considered when linking gut microbiota and brain activity involves the interference of gut microbiota in the levels of different neurotransmitters and neuromodulators, particularly serotonin, γ-aminobutyric acid, and dopamine (Dinan and Cryan 2017). Dysregulation of brain activities promoted by dysbiosis may have a tremendous impact on a number of diseases, notably in mood disorders (Umbrello and Esposito 2016). The hypothalamic-pituitary-adrenal (HPA) axis is another interesting mechanism that makes the bridge between the gut and the brain (Berding and Donovan 2016). The HPA axis regulates the adaptive responses to stressors, such as environmental stress or systemic pro-inflammatory cytokines, in vertebrates. Activation of the HPA axis leads to the secretion and release of the corticotropin-releasing factor (CRF) from the hypothalamus and of the adrenocorticotropic hormone from the pituitary gland, resulting in the production of cortisol from the adrenal glands (Carabotti et al. 2015). It has been reported that gut microbiota may modulate the HPA axis, which, in turn, may regulate gut microbiota (Carabotti et al. 2015). However, the routes of communication between the gut microbiota and the brain are not fully elucidated, possibly through neural, endocrine, and immune pathways, which could be affected by gut microbiota or microbiota-generated metabolites.

15.3 Probiotics and Depression

Several studies have used an overall diet approach to evaluate the association between nutrition and mental health (Akbaraly et al. 2009; Sánchez-Villegas et al. 2009), but there is also considerable research looking at isolated nutrients and their impact on mental health. Central to this research are probiotics (Dinan and Quigley 2011). Probiotics are transient entities that colonize the GI tract and influence various pathways. It has been well established that probiotics have therapeutic effects on many GI disorders (Elangovan et al. 2019); however, with the emergence of the gut-brain axis, it has been discovered that their therapeutic effects extend beyond the gut and into the central nervous system (Mörkl et al. 2020). In recent years, there has been major interest in exploring the link between the health of the gut and mental health (Schmidt 2015a, b). Modulation of inflammation, affecting the hypothalamic-pituitary-adrenal (HPA) axis, and interference with neurotransmitter signaling are the potential pathways through which gut microbiota may influence depression (Uher and McGuffin 2008). It has been found that treatment with probiotics may improve symptoms associated with MDD (major depressive disorder) by increasing neurotransmitters' availability and/or decreasing levels of inflammatory markers. The potential of probiotics to be used as a novel treatment for MDD could have a major impact on those seeking antidepressant treatment by reducing the stigma, latency, and side effects associated with typical antidepressants (Wallace and Milev 2017). Despite extensive preclinical data, the clinical effects of probiotics on mental health have yet to be studied comprehensively in a sample of depressed patients.

15.4 Mechanisms of Probiotics' Antidepressant Effect

15.4.1 Modulation of Inflammation

During the past decade, there has been renewed interest in the relationship between the brain, gut microbiota, and immune system, as well as in the study of microbiota changes as a possible source of inflammatory activity in mood disorders. In other chronic conditions, such as irritable bowel syndrome (IBS), inflammatory bowel disease (IBD), diabetes, and obesity, the association between gut microbiota composition, inflammation, and depressive symptoms has been attributed to a process known as intestinal dysbiosis (Pflughoeft and Versalovic 2012). Intestinal dysbiosis is conceptualized as a state in which there is an alteration of normal intestinal microbiota and has been highly associated with chronic low-grade inflammation in humans (Cani and Delzenne 2009). Consequently, it has been hypothesized to be involved in the pathophysiology of MDD (Rogers et al. 2016).

The results of animal studies, particularly those involving manipulation of the microbiota, support the association between microbiota abnormalities and depressive-like behaviors. Such studies have opened up new avenues of investigation for the pathophysiology of MDD as well as for the development of novel treatment interventions. The cumulative evidence suggests that modifying the composition of the gut microbiota, for example, using a probiotic, might be a viable treatment option for individuals with MDD (Park et al. 2018).

One possible pathway through which probiotics initiate their psychotropic effects is the link between gut bacteria and immunity. Immunoglobulin A and immunoglobulin M mediate inflammation and responses to lipopolysaccharide which have been shown to be elevated in depressed patients (Maes 2011). Moreover, a link has also been made that implicates higher inflammatory interleukin-6 (IL-6) and tumor necrosis factor-alpha (TNF-α) in depressed patients (Dowlati et al. 2010). Research within an animal model has shown that gastrointestinal inflammation appears to induce anxious behavior and cause alterations to the central nervous system biochemistry (Bercik et al. 2011).

Previously, the effects of probiotic supplementation on the biomarkers of inflammation have been reported (Badehnoosh et al. 2018). As the circulating levels of inflammatory biomarkers might be reduced by modulating gut bacteria composition, the therapeutic application of probiotics in mood disorders would seem a reasonable proposition. It has been demonstrated that fecal microbiota transplantation (FMT) from depressed humans to germ-free (GF) mice increased depressive-like behavior in recipient mice (Zheng et al. 2016). The findings from this study provided the rationale for an ongoing clinical trial with the primary aim of evaluating the effect of FMT capsules versus placebo on depressive symptoms in patients with MDD. One clinical trial has been conducted that examined the effects of probiotic supplementation on symptoms of depression in patients with MDD (Akkasheh et al. 2016). The study was an 8-week randomized, double-blind, placebo-controlled trial that included 40 patients with DSM-IV-defined MDD. The probiotic capsule contained three viable, freeze-dried strains of *Lactobacillus casei*, *Lactobacillus acidophilus*, and *Bifidobacterium bifidum*. The results indicated that patients receiving the probiotic intervention had significantly lower Beck Depression Inventory scores compared to placebo. Notably, the researchers also found significant reductions in inflammatory marker, serum hs-CRP, in the probiotic intervention group compared to placebo.

An anti-inflammatory mechanism is underlying the antidepressant effects of probiotics (Park et al. 2018). There is adequate evidence supporting that (1) inflammation is implicated in the pathophysiology of depression and (2) probiotic consumption reduces inflammation. Considering the potential link between peripheral and brain inflammatory activation, a corollary of the finding that probiotics reduce peripheral inflammation is that probiotics also reduce brain inflammation. As such, it could be conjectured that probiotics have therapeutic efficacy in other disorders characterized by brain inflammatory activation. However, brain inflammation is a complicated notion with disparate etiological roots and therefore overlapping etiology may be a prerequisite in this regard. The anti-inflammatory mechanism was evaluated in a recent study conducted by Abildgaard et al. (2017). In the study, rats treated with a probiotic mixture containing eight different *Bifidobacterium* and *Lactobacillus* species displayed significantly reduced depressive-like behaviors compared to rats treated with a vehicle control. Importantly, this reduction in depressive-like behavior was correlated with a reduction in the level of circulating pro-inflammatory cytokines (i.e., TNF-α, IL-6) (Abildgaard et al. 2017). Considering the anti-inflammatory properties of probiotics, it is possible that probiotic treatment may be effective in a subgroup of depressed individuals with elevated inflammation. However, continued research in this domain is warranted.

15.4.2 Hypothalamic-Pituitary-Adrenal (HPA) Axis and Neurotransmitter Signaling

The hypothalamic-pituitary-adrenal (HPA) axis principal purpose is to maintain homoeostasis to physical and psychological stress. Disruption of the HPA axis has been implicated in the pathogenesis of mood disorders (Cleare 2004). Research using rats has found that probiotics are able to interfere with the HPA response to acute physiological stress, and according to Naseribafrouei et al. (2014), this would indicate a mechanistic connection linking the gut microbiota, HPA, and mood disorders.

Increased expression of pro-inflammatory cytokines (IL-1β, IL-6, TNF-α, interferon gamma (IFN-γ), and C-reactive protein (CRP)) is repeatedly observed in patients suffering from depression and has been associated with specific symptoms of depression (Wallace and Milev 2017). This overall increase in inflammation contributes to depressive symptoms by activating the HPA axis, as well as reducing the availability of neurotransmitter precursors and altering neurotransmitter metabolism. This inflammation can be caused by increased intestinal permeability. When the tight junctions of the gastrointestinal lining become compromised and permeability increases, it allows toxins and other forms of waste to leak into the bloodstream. Namely, gut-derived endotoxins called lipopolysaccharide (LPS) molecules are found in the outer membrane of gram-negative bacteria. These endotoxins trigger immune activation through Toll-like receptor 4 (TLR4) (Kawai et al. 2001), causing the body to mount a global immune response. It is hypothesized that probiotics may exert their therapeutic effects on the central nervous system by improving the integrity of the gastrointestinal lining, reducing the ability of endotoxins to leak into the bloodstream, and, in turn, decreasing global inflammation. The reduction of this inflammation may result in improved regulation of the HPA axis and neurotransmitter activity.

Direct interference with transmitter signaling may also be linked to depressive states. Gamma-aminobutyric acid (GABA), an inhibitory neurotransmitter, can be produced by intestinal bacteria, and probiotics can modify depressive behavior from GABA signaling, at least in the rat model (Bendtsen et al. 2012). These findings appear to indicate the potential benefits of the normalization of intestinal microbiota in the regulation of mood and suggest that probiotic bacteria may serve as a therapeutic treatment for depression. Moreover, there is robust evidence that demonstrates probiotics' ability to change behavior and improve the mood, anxiety, and cognition of rodents by altering neurotransmitter activity. Findings suggest that probiotics have a positive impact on the central nervous system by regulating critical neurotransmitters implicated in depression.

Serotonin, a monoamine neurotransmitter, is biosynthesized from the essential amino acid tryptophan, both in the central nervous system and the gastrointestinal tract. In the central nervous system, it is involved primarily in regulating stress and emotions, appetite, and sleep. In the gastrointestinal tract, it is responsible for key functions such as gastrointestinal motility and intestinal secretions. Alterations in the microbiome have been shown to profoundly influence neurotransmission of serotonin in both the peripheral and central nervous systems. It is hypothesized that probiotics in the GI tract improve central nervous system symptoms associated with MDD by increasing production of free tryptophan and, in turn, increasing serotonin availability (Wallace and Milev 2017). This increase in serotonin may facilitate regulation of the HPA axis and reduce depressive symptoms caused by a depletion of the neurotransmitter.

15.5 Conclusion

Probiotics are proposed to have a range of health benefits. There is an increasing body of research which has reported that the microbiota of the intestines may function beyond the gut. And it is clear from research that probiotics might have favorable effects on mood and psychological problems. Through normalizing basal intestinal microbiota, applications of particular probiotics appear to improve immune response and reverse the behavioral effects of depression.

Conflict of Interest There is no conflict of interest.

References

Abildgaard A, Elfving B, Hokland M, Wegener G, Lund S (2017) Probiotic treatment reduces depressive-like behaviour in rats independently of diet. Psychoneuroendocrinology 79:40–48

Akbaraly TN, Brunner EJ, Ferrie JE, Marmot MG, Kivimaki M, Singh-Manoux A (2009) Dietary pattern and depressive symptoms in middle age. Br J Psychiatry 195(5):408–413

Akkasheh G, Kashani-Poor Z, Tajabadi-Ebrahimi M, Jafari P, Akbari H, Taghizadeh M, Memarzadeh MR, Asemi Z, Esmaillzadeh A (2016) Clinical and metabolic response to probiotic administration in patients with major depressive disorder: a randomized, double-blind, placebo-controlled trial. Nutrition 32(3):315–320

Arneth BM (2018) Gut–brain axis biochemical signalling from the gastrointestinal tract to the central nervous system: gut dysbiosis and altered brain function. Postgrad Med J 94 (1114):446–452

Badehnoosh B, Karamali M, Zarrati M, Jamilian M, Bahmani F, Tajabadi-Ebrahimi M, Jafari P, Rahmani E, Asemi Z (2018) The effects of probiotic supplementation on biomarkers of inflammation, oxidative stress and pregnancy outcomes in gestational diabetes. J Maternal-Fetal Neonatal Med 31(9):1128–1136

Bendtsen KMB, Krych L, Sørensen DB, Pang W, Nielsen DS, Josefsen K, Hansen LH, Sørensen SJ, Hansen AK (2012) Gut microbiota composition is correlated to grid floor induced stress and behavior in the BALB/c mouse. PLoS One 7(10):e46231

Bercik P, Park A, Sinclair D, Khoshdel A, Lu J, Huang X, Deng Y, Blennerhassett P, Fahnestock M, Moine D (2011) The anxiolytic effect of Bifidobacterium longum NCC3001 involves vagal pathways for gut–brain communication. Neurogastroenterol Motil 23 (12):1132–1139

Berding K, Donovan SM (2016) Microbiome and nutrition in autism spectrum disorder: current knowledge and research needs. Nutr Rev 74(12):723–736

Cani PD, Delzenne NM (2009) The role of the gut microbiota in energy metabolism and metabolic disease. Curr Pharm Des 15(13):1546–1558

Carabotti M, Scirocco A, Maselli MA, Severi C (2015) The gut-brain axis: interactions between enteric microbiota, central and enteric nervous systems. Ann Gastroenterol 28(2):203

Cleare AJ (2004) The HPA axis and the genesis of chronic fatigue syndrome. Trends Endocrinol Metab 15(2):55–59

Cryan JF, Dinan TG (2012) Mind-altering microorganisms: the impact of the gut microbiota on brain and behaviour. Nat Rev Neurosci 13(10):701

Dinan TG, Cryan JF (2017) The microbiome-gut-brain axis in health and disease. Gastroenterol Clin 46(1):77–89

Dinan TG, Quigley EM (2011) Probiotics in the treatment of depression: science or science fiction? Austr N Z J Psychiatry 45(12):1023–1025

Dowlati Y, Herrmann N, Swardfager W, Liu H, Sham L, Reim EK, Lanctôt KL (2010) A meta-analysis of cytokines in major depression. Biol Psychiatry 67(5):446–457

Elangovan A, Allegretti JR, Fischer M (2019) Microbiota modulation-based therapy for luminal GI disorders: current applications of probiotics and fecal microbiota transplantation. Expert Opin Biol Ther 19(12):1343–1355

Forsythe P, Sudo N, Dinan T, Taylor VH, Bienenstock J (2010) Mood and gut feelings. Brain Behav Immun 24(1):9–16

Furness JB (2006) Novel gut afferents: Intrinsic afferent neurons and intestinofugal neurons. Auton Neurosci 125(1-2):81–85
Grenham S, Clarke G, Cryan JF, Dinan TG (2011) Brain–gut–microbe communication in health and disease. Front Physiol 2:94
Hemarajata P, Versalovic J (2013) Effects of probiotics on gut microbiota: mechanisms of intestinal immunomodulation and neuromodulation. Ther Adv Gastroenterol 6(1):39–51
Kawai T, Takeuchi O, Fujita T, Inoue J-I, Mühlradt PF, Sato S, Hoshino K, Akira S (2001) Lipopolysaccharide stimulates the MyD88-independent pathway and results in activation of IFN-regulatory factor 3 and the expression of a subset of lipopolysaccharide-inducible genes. J Immunol 167(10):5887–5894
Li C, Niu Z, Zou M, Liu S, Wang M, Gu X, Lu H, Tian H, Jha R (2020) Probiotics, prebiotics, and synbiotics regulate the intestinal microbiota differentially and restore the relative abundance of specific gut microorganisms. J Dairy Sci 103(7):5816–5829
Maes M (2011) An intriguing and hitherto unexplained co-occurrence: depression and chronic fatigue syndrome are manifestations of shared inflammatory, oxidative and nitrosative (IO&NS) pathways. Prog Neuro-Psychopharmacol Biol Psychiatry 35(3):784–794
Mörkl S, Butler MI, Holl A, Cryan JF, Dinan TG (2020) Probiotics and the microbiota-gut-brain axis: focus on psychiatry. Curr Nutr Rep 9:171–182
Naseribafrouei A, Hestad K, Avershina E, Sekelja M, Linløkken A, Wilson R, Rudi K (2014) Correlation between the human fecal microbiota and depression. Neurogastroenterol Motil 26 (8):1155–1162
Park C, Brietzke E, Rosenblat JD, Musial N, Zuckerman H, Ragguett R-M, Pan Z, Rong C, Fus D, McIntyre RS (2018) Probiotics for the treatment of depressive symptoms: an anti-inflammatory mechanism? Brain Behav Immun 73:115–124
Pflughoeft KJ, Versalovic J (2012) Human microbiome in health and disease. Annu Rev Pathol 7:99–122
Rea K, Dinan TG, Cryan JF (2016) The microbiome: a key regulator of stress and neuroinflammation. Neurobiol Stress 4:23–33
Rogers G, Keating D, Young R, Wong M, Licinio J, Wesselingh S (2016) From gut dysbiosis to altered brain function and mental illness: mechanisms and pathways. Mol Psychiatry 21 (6):738–748
Sánchez-Villegas A, Delgado-Rodríguez M, Alonso A, Schlatter J, Lahortiga F, Majem LS, Martínez-González MA (2009) Association of the Mediterranean dietary pattern with the incidence of depression: the Seguimiento Universidad de Navarra/University of Navarra follow-up (SUN) cohort. Arch Gen Psychiatry 66(10):1090–1098
Schmidt C (2015a) Mental health: thinking from the gut. Nature 518(7540):S12–S15
Schmidt C (2015b) Thinking from the gut. Nature 518(7540):S12–S14
Tabrizi A, Khalili L, Homayouni-Rad A, Pourjafar H, Dehghan P, Ansari F (2019) Prebiotics, as promising functional food to patients with psychological disorders: a review on mood disorders, sleep, and cognition. NeuroQuantology 17:6
Uher R, McGuffin P (2008) The moderation by the serotonin transporter gene of environmental adversity in the aetiology of mental illness: review and methodological analysis. Mol Psychiatry 13(2):131–146
Uher R, McGuffin P (2010) The moderation by the serotonin transporter gene of environmental adversity in the etiology of depression: 2009 update. Mol Psychiatry 15(1):18
Umbrello G, Esposito S (2016) Microbiota and neurologic diseases: potential effects of probiotics. J Transl Med 14(1):298
Vilagut G, Forero CG, Barbaglia G, Alonso J (2016) Screening for depression in the general population with the center for epidemiologic studies depression (CES-D): a systematic review with meta-analysis. PLoS One 11(5):e0155431
Wallace CJ, Milev R (2017) The effects of probiotics on depressive symptoms in humans: a systematic review. Ann General Psychiatry 16(1):1–10
Zheng P, Zeng B, Zhou C, Liu M, Fang Z, Xu X, Zeng L, Chen J, Fan S, Du X (2016) Gut microbiome remodeling induces depressive-like behaviors through a pathway mediated by the host's metabolism. Mol Psychiatry 21(6):786–796

Role of Probiotics in Diabetes Control

16

Leila Khalili, Khadijeh Eslamnezhad, Ali Barzegar, Azadeh Dehghani, Nazanin Zakeri, and Esmat Mehrabi

Abstract

Diabetes mellitus, a metabolic disorder recognized by high blood glucose, is caused by insufficient production of insulin and/or insulin resistance. A natural and safe solution is needed for controlling the vast increase in the prevalence of this disorder. Intestinal microbiota can affect the host pro-inflammatory status, insulin resistance, and body weight. Moderating gut microbiota by the use of prebiotics, probiotics, and antibiotics can provide positive effects on insulin resistance and glucose metabolism. The live microorganisms present in probiotics provide beneficial effects on the host health. This chapter highlights the current evidences in probiotic effectiveness and future prospects for exploring probiotic therapy in the prevention and control of diabetes. Probiotics can improve insulin sensitivity and reduce autoimmune responses through modulating intestinal microbiota and then decreasing oxidative stress and inflammatory reactions. Probiotics affect the host by modulating intestinal permeability and mucosal immune response, managing eating behaviors by appetite-regulating hormones,

L. Khalili · A. Barzegar · A. Dehghani
Department of Community Nutrition, Faculty of Nutrition and Food Sciences, Tabriz University of Medical Sciences, Tabriz, Iran

K. Eslamnezhad
Obstetrics and Gynecology, Faculty of Medicine, Ardabil University of Medical Sciences, Ardabil, Iran

N. Zakeri
Department of Nutrition, Faculty of Nutrition and Food Sciences, Tabriz University of Medical Sciences, Tabriz, Iran

E. Mehrabi (✉)
Faculty of Nursing and Midwifery, Department of Midwifery, Tabriz University of Medical Sciences, Tabriz, Iran
e-mail: mehrabie@tbzmed.ac.ir

R. Z. Sayyed, M. Khan (eds.), *Microbiome-Gut-Brain Axis*,
https://doi.org/10.1007/978-981-16-1626-6_16

and regulating inflammatory-associated disorders. Moreover, probiotics control host metabolism by affecting food intake through biochemically converting molecules derived from the host or from gut microbes themselves. Previous evidences support the hypothesis that the modulation of the gut microbiota by probiotics could be effective in the management of several types of diabetes.

Keywords

Chronic diseases · Diabetes · Gut microbes · Intestinal microbiota · Probiotics

16.1 Introduction

Diabetes mellitus (DM), a group of metabolic disorders, is recognized by increased blood glucose level. Patients with DM have a higher risk of morbidity and mortality than the general population. The worldwide prevalence of DM has been increasing over recent decades. In 1964, it was estimated that 30 million people had DM. Less than 40 years later, the WHO estimated that there were 171 million people living with DM. The international diabetes federation (IDF) estimated the global prevalence to be 151 million in 2000, 194 million in 2003, 246 million in 2006, 285 million in 2009, 366 million in 2011, and 382 million in 2013. The dramatic increase in DM has occurred in all countries, and in rural as well as urban areas. The prevalence of DM in adults aged 20–79 years was estimated to be 8.8% in 2015 and has been predicted to rise to 10.4% in 2040 worldwide (Ogurtsova et al. 2017).

Controlling this serious global health-related problem by natural food without side effects is a challenge for medical nutrition therapy (MNT) of DM. Intestinal microbiota can affect the host by influencing pro-inflammatory status, insulin resistance, bile acid metabolism, body weight, and modulating the gut hormones. Modulating gut microbiota by consumption of prebiotics, probiotics, and antibiotics can provide positive effects on improvement of glucose metabolism and insulin resistance. According to the findings of animal studies, probiotics can improve insulin-binding potential, inhibit damages of Langerhans islets' β-cells, and increase insulin sensitivity by improving the glucose transporter 4 (GLUT-4) transcription (Lin et al. 2014). Most of the studies showed improvements in at least one of the blood glucose markers (Razmpoosh et al. 2016). More studies are needed to improve our knowledge about the complex relationship between intestinal microbiota and hosts with DM. This chapter is conducted on the mechanisms of probiotics' action in preventing or managing DM.

16.2 Gut Microbiota

The human distal gut harbors nearly 1.5 kg microorganisms that provides important functions for the human hosts (Allin et al. 2015). Gut microbiota is known as an important factor linking genes, the environment, and the immune system (Musso

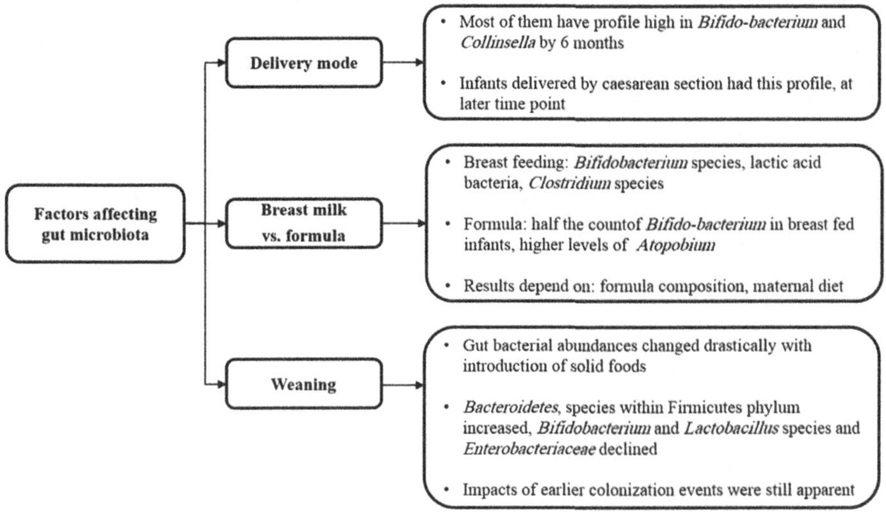

Fig. 16.1 Factors affecting gut microbiota

et al. 2010). The genome size of human gut microbial organ, named microbiome, is twice the size of the human nuclear genome and carries vital biological and metabolic functions (Musso et al. 2010). The human gut microbiota is mostly dominated by Bacteroidetes and Firmicutes (Turnbaugh et al. 2006). The variety of bacteria in the infant gut is primarily very low (Adlerberth and Wold 2009). By way of growing and changing from infancy to old age, the microbiota will be changed (Fig. 16.1) (Clemente et al. 2012). Babies are exposed to a vast of microbes from various environments from birth and are rapidly colonized by the microbes they first met by chance, either from the microbes of skin or vagina of their mothers, depending on delivery mode (Dominguez-Bello et al. 2010). Vaginally born babies have microbe communities similar to their mothers' vaginal microbiota. However, infants delivered by cesarean section have a microbiota characteristic of skin that is mainly consists of *Propionibacterium* and *Staphylococcus* spp. (Dominguez-Bello et al. 2010). The neonates with cesarean delivery have delayed microbial colonization by *Lactobacillus*, *Bifidobacterium*, and *Bacteroides*; therefore, the incidence of type 1 diabetes (T1DM) has been shown to happen more frequently in this condition (Vehik and Dabelea 2012). Other factors including the antibiotic therapy in the newborn and non-breast-milk diets are also connected with changes in the gut microbiota (Dominguez-Bello et al. 2010). Some factors affect the composition of gut microbiome species in adult life such as dietary pattern, use of antibiotic in early life, luminal pH and osmolarity, and environmental factors (Blaser 2011). Despite the vast inter-individual variability in the gut microbial composition, most individuals harbor microbiota that can be categorized into one of these three main genera: *Bacteroides*, *Ruminococcus*, or *Prevotella* (Arumugam et al. 2011). Ingestion of probiotics—"live microorganisms which, when administered in sufficient amounts, present health benefits to the host"—is a way to modify gut microbiota composition (Homayouni 2009; Jafar-Abadi et al. 2020). Dairy products such as fermented foods, yogurts, and some cheeses contain various amounts of probiotics. However, it is not clear to what degree these food sources can alter the gut microbiota and provide beneficial biological effects outside research settings (Allen et al. 2010).

16.3 Diabetes and Probiotics

Diabetes, a chronic disease that affects several systems, contains important complications and involves several molecular mechanisms connected with the intestinal microbiota (Gomes et al. 2014). According to recent studies, the use of probiotics can be connected with stimulation of the immune system, reduction of inflammation, protection against intestinal and respiratory disorders, reduction of blood cholesterol levels, and anti-tumorigenic effects (Jafar-Abadi et al. 2020). These beneficial effects originate from the probiotics' ability in producing antimicrobial substances, improving the intestinal barrier function, combating against the other pathogens, and modulating the immune system (Markowiak and Śliżewska 2017). It has been revealed that the therapeutic effects of probiotics on blood glucose could be associated with alteration of the intestinal microbiota composition in patients with diabetes (Larsen et al. 2010). There are growing interests in the prevention and control of diabetes by probiotic interventions. Current studies supposed probiotic bacteria as antidiabetic agents, since consumption of probiotics has been found to normalize glucose homeostasis in diabetic animal models (Bonfili et al. 2020). The probable mechanisms are summarized in Fig. 16.2. As probiotics are found in foods such as dairy products and consumed in vast amounts worldwide, such beneficial effects may have a huge influence in clinical practice, particularly with regard to dietary recommendations to both healthy subjects at risk of diabetes and patients with diabetes (Yadav et al. 2008). Probiotics can modulate gut microbiota and are known as effective treatment for insulin resistance (Moroti et al. 2012). The most frequently used strains of probiotics in functional foods and dietary supplements are *Lactobacilli* and *Bifidobacteria*. It has been shown that the consumption of *L. casei*, *L. lactis*, and *L. acidophilus* can improve the glycemic response and HbA1c (Yadav et al. 2007). *L. plantarum* DSM15313 is also proposed to decrease glycemia, reduce insulin resistance, and promote glucose tolerance (Andersson et al. 2010). Here, we are going to highlight the effects of probiotics on different types of diabetes separately.

16.3.1 Gestational Diabetes Mellitus (GDM)

Gestational diabetes mellitus (GDM) is known as a complication of pregnancy and is characterized by glucose intolerance. Throughout pregnancy, the gut microbiota undergoes significant changes. From the first (T1) to the third trimester (T3), the

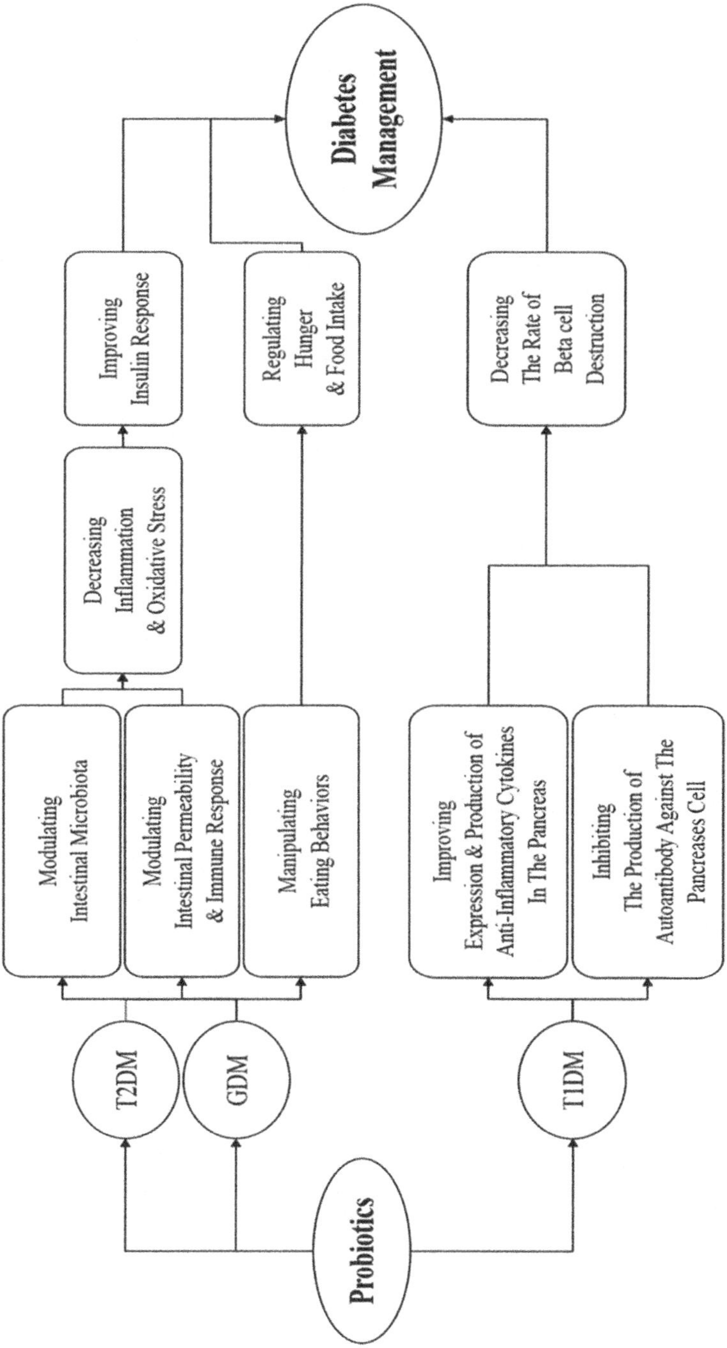

Fig. 16.2 Diabetes management by probiotics

species richness of the gut microbiome decreases (Koren et al. 2012), although this has not been observed in all studies (DiGiulio et al. 2015). There are an increase in Proteobacteria and Actinobacteria phyla and a reduction in the beneficial bacterial species *Roseburia intestinalis* and *Faecalibacterium prausnitzii* (Koren et al. 2012; Tilg and Moschen 2015). These changes in gut microbial composition cause inflammation and correlate with increases in fat mass, blood glucose, insulin resistance, and circulating pro-inflammatory cytokines in the expectant mother (Gohir et al. 2015). This "diabetic-like" state observed during the later stages of all healthy pregnancies is thought to maximize nutrient provision to the developing fetus (Wang et al. 2016). However, increased insulin resistance combined with an inability to secrete the additional insulin required to maintain glucose homeostasis can result in the development of gestational diabetes mellitus (GDM) in the mother and macrosomia in the baby. GDM is a potential risk factor for short-term and long-term morbidity of mothers and babies (Rad et al. 2017). In the short term, women with GDM may experience preeclampsia and delivery by cesarean section (Dempsey et al. 2005). Moreover, GDM raises the risk of adiposity, macrosomia (a birth weight of >4000 g), shoulder dystocia, admission to the neonatal intensive care unit (NICU), and neonatal hypoglycemia (Wendland et al. 2012). In the long term, GDM is connected with high risk of obesity and metabolic and cardiovascular disorders in both mother and baby. About 7% of all pregnancies in the United States are complicated by GDM, and its prevalence in Iran is approximately 6% of pregnancies (Harlev and Wiznitzer 2010; Almasi and Salehiniya 2014).

Safe and inexpensive interventions for the prevention and treatment of GDM are needed. Considering that certain microorganisms in the gastrointestinal tract can produce a positive effect on host metabolism, probiotic supplements can help maintain bacterial diversity and homeostasis in people with metabolic disorders (Sekirov et al. 2010; Gregor and Hotamisligil 2011). Probiotics can control GDM through changing metabolism (Barrett et al. 2014). Diet can change the gut microbiota composition and gene expression along with affecting host metabolism. Modifying the gut microbiome has several beneficial effects on the host, such as altering nutrient absorption and influencing inflammatory pathways and lipid and glucose metabolism (Rad et al. 2017). Several studies showed benefits of probiotic use for improving blood glucose control in patients with GDM. The effect of consumption of 300 mg/day probiotic yogurt containing 10^6 CFU *Lactobacillus acidophilus* and *Bifidobacterium lactis* on glycemic response and the outcome of pregnancy in GDM patients was evaluated by Ebrahimi et al. (2019). They found that using probiotic yogurt caused a significant improvement in blood glucose levels and reduced the risk of macrosomia (Ebrahimi et al. 2019). Asemi et al. evaluated the effects of daily consumption of probiotic yogurt on insulin resistance and levels of insulin in the serum of pregnant women in the third trimester of gestation (Asemi et al. 2013). The probiotic yogurt used in their study was enriched with a probiotic culture of *L. acidophilus* LA5 and *B. animalis* BB12 with at least 10^7 CFU. Daily consumption of probiotic yogurt for 9 weeks was effective in maintaining normal serum insulin levels in pregnant women and thus contributing to prevent the development of insulin resistance, which usually develops during the last trimester in pregnant women. The study demonstrated an improvement in glycemic control during the last trimester of pregnancy, extending in the postpartum period for 12 months. In the study conducted by Badehnoosh et al. on 60 subjects with GDM, they found that consumption of probiotic capsule containing *L. acidophilus*, *L. casei*, and *B. bifidum* (2×10^9 CFU/g each) for 6 weeks had beneficial effects on glycemic response and serum inflammatory and oxidative stress biomarkers (Badehnoosh et al. 2018). Luoto et al. found that a combined dietary/probiotic supplementation decreased the rate of GDM from 34% to 13% in pregnant women with normal weight (Luoto et al. 2010). Pregnant women receiving dietary counseling and probiotic supplement containing *L. rhamnosus* GG and *B. lactis* BB12 showed better glucose response and HbA1c levels compared with the control group receiving only healthy diet (Luoto et al. 2010). Similarly, another study showed that dietary recommendations together with probiotic supplementation, during pregnancy and up to 12 months after delivery, improved insulin sensitivity, decreased blood glucose, and reduced insulin concentrations and was more effective than dietary recommendations alone (Laitinen et al. 2008). The safety of probiotic use for pregnant women was examined in a trial in which 1×10^{10} CFU probiotic supplement containing two strains of *Lactobacilli* (*Lactobacillus paracasei* CUL08 and *Lactobacillus salivarius* CUL61) and *Bifidobacteria* (*Bifidobacterium bifidum* CUL20 and *Bifidobacterium animalis* subsp. *lactis* CUL34) was consumed daily by women during the last month of pregnancy and by infants aged 0–6 months (Allen et al. 2010). The findings support the safety of probiotic consumption during pregnancy and early infancy. So, probiotics can be used as a safe method for GDM prevention and/or control in the high-risk overweight and obese pregnant women (Nitert et al. 2013).

16.3.2 Type 2 Diabetes Mellitus (T2DM)

Type 2 diabetes mellitus (T2DM), a metabolic disorder known by high blood glucose, is caused by the combination of insufficient secretion of insulin and insulin resistance (Hassanalilou et al. 2017; Khalili et al. 2019a, b). This metabolic illness along with three other diseases (cardiovascular diseases, cancer, and respiratory diseases) accounts for over 80% of all premature noncommunicable diseases' deaths. Around 425 million people worldwide (approximately 9% of adults aged 20–79 years) were estimated to have diabetes in 2017 (IDF estimates) (Forouhi and Wareham 2019). Obesity and T2DM are linked with the gut microbiome alteration (Larsen et al. 2010). It has been shown that the gut microbiome diversity reduced in obese animal models and humans, with a rise in Firmicutes and a drop in Bacteroidetes (Turnbaugh et al. 2008). Plasma glucose levels in patients with T2DM are positively correlated with the ratio of Bacteroidetes/Firmicutes (Larsen et al. 2010). Probiotics may effectively and safely modify the composition and function of human gut microbiota to lessen the adverse metabolic effects associated with pathogenic microbial communities (Sun and Chang 2014). The outcomes of a randomized controlled trial showed that 8 weeks of 10^8 CFU *L. casei* supplementation improved glycemic response in patients with T2DM (Khalili et al. 2019a, b). Razmpoosh et al. revealed that consumption of probiotic supplement consisted of *Lactobacillus*, *Bifidobacterium*, and *Streptococcus strains* caused a significant decrease in FPG concentration in patients with T2DM (Razmpoosh et al. 2019). Raygan et al. showed that after a 12-week intervention, probiotic supplementation containing *Bifidobacterium* and *Lactobacillus* strains had beneficial effects on glycemic control (Raygan et al. 2019). In another trial, improvement of insulin sensitivity was observed through a 4-week supplementation with the probiotic strain *Lactobacillus acidophilus* NCFM™ in men with T2DM (Andreasen et al. 2010). Moroti et al. found a significant reduction in FPG and rise in HDL cholesterol in T2DM patients who received a daily dose of 200 mL of a symbiotic drink containing 10^8 CFU/mL *Lactobacillus acidophilus*, 10^8 CFU/mL *Bifidobacterium bifidum*, and 2 g oligofructose over 30 days (Moroti et al. 2012). It has also been revealed that consumption of probiotic yogurt containing *B. lactis* BB12 and *L. acidophilus* LA5 for 6 weeks reduced FPG and HbA1c levels and improved the activity of superoxide dismutase and glutathione peroxidase in individuals with T2DM (Ejtahed et al. 2012). Considering the results of several studies in this field, probiotics, mainly *Bifidobacteria* and *Lactobacilli*, arose as the potential bio-therapeutics with confirmed efficiency in the prevention and treatment of T2DM (Markle et al. 2013).

16.3.3 Type 1 Diabetes Mellitus (T1DM)

The gut microbiota modulates the autoimmune pathogenesis of type 1 diabetes mellitus (T1DM) via mechanisms that remain largely unknown. T1DM is caused by autoimmune destruction of pancreatic β-cells (Burrack et al. 2017). Environmental factors cause the disease through numerous mechanisms that may either trigger the initial autoimmune response in genetically susceptible individuals or modify the destructive processes at several points throughout the history of the disease (Lin et al. 2014). The inflammatory components are innate immune sensors that are highly influenced by the gut environment and play pivotal roles in maintaining intestinal immune homeostasis. In humans, T1DM susceptibility has been linked to compositional changes in the gut microbiota and, specifically, with a significant increased representation of bacteria of the Bacteroidetes phylum and a decrease in the number of *Bifidobacterium*, *Lactobacillus*, and *Clostridium* strains (Murri et al. 2013). Studies in preclinical models of T1DM have indicated that the gut microbiota plays a key role in controlling disease onset and severity. For example, the absence of MyD88 (myeloid differentiation factor 88), an adaptor molecule involved in TLR signaling, protects NOD (nonobese diabetic) mice from autoimmune T1DM by inducing a protective microbiota profile characterized by a low Firmicutes/Bacteroidetes ratio and *Lactobacillus* strain enrichment (Wen et al. 2008). The gut microbiota, a complex microenvironment, is connected with the immune system and can regulate the immune responses (Yue et al. 2019). Microbiome intervention in young T1DM-prone rodents protected the islet autoimmunity and disease (Markle et al. 2013). According to evidences, microbial therapy can provide protection of individuals with high genetic risk of T1DM. Calcinaro et al. showed that diabetes development in NOD mice was prevented by oral consumption of VSL#3 (Calcinaro et al. 2005). Moreover, the rate of β-cells destruction and insulitis was decreased in protected mice. The prevention was due to the enhanced expression and production of IL-10 in the pancreas, where IL-10-positive islet-infiltrating mononuclear cells were detected. Dolpady et al. demonstrated that administration of the *Lactobacillus*-enriched VSL#3 probiotic prevents T1DM in NOD mice by enriching the local microbiota with *Lactobacillus* strains and by inducing substantial modifications in the microbiota composition (Dolpady et al. 2016). Ljungberg et al. evaluated the effect of probiotic consumption during the first 6 months of life on the emergence of T1DM-associated autoantibodies in children with genetic risk for T1DM. Moreover, they did a pilot study on 200 subjects to show the safety and feasibility of the use of probiotics during the first 6 months of life. They showed that the concentration of autoantibodies at 6, 12, and 24 months of age was at expected levels. No subject was detected positive for autoantibody at 12 months of age. Although one subject was recognized positive for autoantibody at 6 and 24 months of age, no sample was detected positive for more than one autoantibody (Ljungberg et al. 2006). Therefore, we can conclude that probiotics can inhibit the production of autoantibody against the pancreases cell. More studies are needed to examine the effectiveness of probiotic combination on pathogenesis and improvement of T1DM in animal models or potential clinical trials. Most of the studies in this field are done on animal models; so, the need for clinical trials evaluating the efficacy of probiotic use in preventing or controlling autoimmune responses against β-cells is remaining.

16.4 Mechanisms of Probiotics' Action

16.4.1 Modulation of Inflammation and Oxidative Stress

Probiotics have the ability to improve insulin resistance and decrease the blood glucose by improving inflammation (Zhang et al. 2016). Probiotics can improve function of intestinal barrier and reduce the microorganisms and their derivatives (e.g. lipopolysaccharide (LPS)) transmission (Cui et al. 2017) to the systemic circulation, so decrease the release of pro-inflammatory cytokines via Toll-like receptor-4 (TLR-4) signaling (Yang et al. 2017). The TLRs, a vast group of cell membrane proteins existing in various types of cells, can detect microbe-associated molecular patterns (MAMPs) during inflammatory responses (Gomes et al. 2014). TLR-4 can be found in insulin target tissues. These actions could be settled by stimulation of TLR-4, through activation of cytokine signaling cascades together with increased reactive oxygen species (ROS) concentration (Cristofaro and Opal 2006). This chronic low-grade inflammation with high levels of pro-inflammatory cytokines is known as the main pathogenic factor of insulin resistance and diabetes (Shoelson et al. 2006). Therefore, the antidiabetic capability of probiotics may be due to their immunomodulatory effects. Furthermore, probiotics have antioxidant efficacy through mechanisms that could be related to enzyme inhibition, reactive oxygen species hunting, metal ion chelation, and inhibition of ascorbate autoxidation (Milind and Jyoti 2014). Previously, the effects of probiotic supplementation on the biomarkers of oxidative stress and inflammation have also been reported (Vehik and Dabelea 2012). The study conducted by Badehnoosh et al. demonstrated that the 6-week intervention of probiotic supplements among women with GDM had beneficial effects on FPG, serum hs-CRP, plasma TAC, MDA, and oxidative stress index (Badehnoosh et al. 2018). Mohamadshahi et al. showed that consumption of probiotic yogurt containing *B. Animalis* and *L. acidophilus* caused a significant decrease in HbA1c and TNF-α levels in the intervention group (Mohamadshahi et al. 2014). It has been revealed that VSL#3 modulated the reduction of hepatic natural killer cells and minimized the activation of NFκB in high-fat-diet-fed male C57BL6 mice (Ma et al. 2008). Other similar studies showed improved activity of the antioxidant enzymes including superoxide dismutase, glutathione peroxidase, and catalase (Kleniewska et al. 2016; Mirmiranpour et al. 2019). So, regulation of inflammation and oxidative stress can be considered as one of the potential mechanisms of probiotics' action in metabolic management.

16.4.2 Probiotics and Endocannabinoid (eCB) System

Inflammation and diabetes are proposed to be connected with the endocannabinoid (eCB) system (Cani et al. 2014). It has been shown that intestinal microbiota regulates gut eCB expression that can modify gut permeability and plasma LPS levels through CB1 receptor (Muccioli et al. 2010). Gut microbiota modification could decrease gut permeability in obese mice. Blocking CB1 receptor in obese mice progressed gut barrier function by increasing distribution and localization of tight junction proteins (ZO-1 and occludin). So, the eCB system can modulate the gut permeability via the mentioned mechanism (Cani et al. 2012). Activation of CB2 receptor moderates glucose tolerance in rats, and CB1 receptor blockade mimics the effects of CB2 receptor agonists (Bermudez-Silva et al. 2007). Modulation of glucose homeostasis by the eCB system is due to the interaction of CB1 and CB2 receptors. The alterations in CB2 receptor expression are positively associated with intestinal quantity of *Lactobacillus* and negatively with counts of *Clostridium* (Aguilera et al. 2013). It has been shown that modulation of the gut microbiota by probiotics, especially strains belonging to *Lactobacillus*, can upregulate CB2 receptor expression in rodents (Rousseaux et al. 2007). Considering the available evidence, the host biological systems can be regulated by specific gut microbes that can lead to the control of energy homoeostasis, glucose metabolism, and inflammation in obesity and T2DM (Cani et al. 2014). So, probiotics may be used as an alternative preventive approach and a complementary treatment method to control diabetic complications (Mohamadshahi et al. 2014).

16.4.3 Appetite-Regulating Hormones and Eating Behavior

The challenge to resist desires for high-sucrose and high-fat foods is part of many people's daily life. Unhealthy diet is one of the main reasons of metabolic problems such as obesity (Muccioli et al. 2010), heart disease, cancer, diabetes, and sleep apnea (Anderson et al. 2003; Calle and Kaaks 2004). Although unhealthy diet has harmful effects on health, it is often difficult to be changed. There are several cognitive modules for control over eating behavior (Kurzban and Athena Aktipis 2007). Evolutionary encounter between host and gut microbes leads them to divergent desires over host eating behavior. Gut microbes can affect eating behavior in a way that improves their fitness at the expense of host fitness. Some other hypotheses suggest that microbes may influence eating behaviors, though not in the context of evolutionary conflict and competing fitness interests (Lyte 2011). Conflict over the achievement of resource can happen as a result of conflict between different genetic interests within an organism. Meta-genomic conflict between microbiome and host can be known as an extension of this genetic conflict context. The microbial control is managed by the vagus nerve; so, microbial signals may interfere with the physiological regulation organized by the vagus nerve. Food preferences will be modified by blocking the vagus then reducing microbial signaling via the vagus nerve. It was shown that weight loss could be achieved by blocking the vagus nerve (Sarr et al. 2012). Microbial communities with low alpha (intra-sample) diversity overgrow by one or more species with improved capability to produce behavior-altering neurochemicals and hormones. However, any single microbial species in microbial communities with high alpha diversity will have a tendency to occur at lower abundance. Highly diverse gut microbiota will try to be more resistant to pathogenic invasions in comparison with the less diverse microbiota (Ursell et al. 2013). Probiotics that can increase microbiota diversity in humans are expected to decrease craving compared to control interventions. The higher microbial diversity is the fewer cravings will be. Another potential way for management of the mammalian eating behavior is through appetite-regulating hormones. Supplementation with VSL#3, containing *Lactobacillus* strains, in mice decreased appetite-inducing hormones neuropeptide Y and AgRP (agouti-related protein) in the hypothalamus (Yadav et al. 2013). Moreover, the levels of leptin, cholecystokinin, and other satiety peptides, which regulate hunger and food intake by affecting vagus nerve signaling, were reduced. Commensal and pathogenic bacteria produce peptides similar to mammalian hormones such as ghrelin, neuropeptide Y, peptide YY, and leptin that control satiety and hunger (Duca et al. 2012). As shown by Batra et al., insulin sensitivity was improved by *Bifidobacterium adolescentis* by increased production of glucagon-like peptide 1 (GLP-1) (Batra et al. 2013). GLP-1 improves glucose tolerance by several mechanisms involving modulation of food intake, insulin secretion, and pancreatic cell mass (Kawai et al. 2018). Moreover, antibodies produced by humans and other mammals may be essential in maintaining the fidelity of host signaling systems. These antibodies also act as autoantibodies against mammalian hormones; so, microbes can manage human eating behavior (1) directly with peptides that are similar to satiety-regulating hormones or (2) indirectly by motivating production of autoantibodies interfering with appetite regulation (Duca et al. 2012). The antibody response confirms the hypothesis that the regulation of eating behavior would be affected by conflict between host and microbiota. A randomized controlled trial on patients with T2DM showed that *L. casei* supplementation could affect dietary intake and body weight in a way that improved glycemic response (Khalili et al. 2019a, b). So, affecting dietary intake is one of the important mechanisms of probiotics' action in diabetes management.

16.5 Conclusion

The dynamic interactions between diet and the gut microbiota and their metabolic consequences play a key role in the pathogenesis of diabetes. Modulating the gut microbiota by probiotics is expected to improve glycemic response and insulin resistance. Probiotics are believed to have beneficial effects on improving several metabolic disorders such as nonalcoholic fatty liver disease, hyperlipidemia, and diabetes as well as some other complications including cancer and immune-related diseases. The collective evidence revealed that probiotics may act as an important mediator of environmental factors triggering diabetes. Future studies will probably unravel the underlying mechanisms by which probiotics can prevent or improve diabetes. More researches are needed to study the effects of different probiotic strains on the prevention or improvement of several types of diabetes. In addition, the host-gut microbiota dynamics and their metabolic consequences are important issues to be addressed for the design of intervention studies aimed at preventing or improving diabetes.

Conflict of Interest There is no conflict of interest.

References

Adlerberth I, Wold A (2009) Establishment of the gut microbiota in Western infants. Acta Paediatr 98(2):229–238

Aguilera M, Vergara P, Martinez V (2013) Stress and antibiotics alter luminal and wall-adhered microbiota and enhance the local expression of visceral sensory-related systems in mice. Neurogastroenterol Motil 25(8):515–529

Allen SJ, Jordan S, Storey M, Thornton CA, Gravenor M, Garaiova I, Plummer SF, Wang D, Morgan G (2010) Dietary supplementation with lactobacilli and bifidobacteria is well tolerated and not associated with adverse events during late pregnancy and early infancy. J Nutr 140 (3):483–488

Allin KH, Nielsen T, Pedersen O (2015) Mechanisms in endocrinology: gut microbiota in patients with type 2 diabetes mellitus. Eur J Endocrinol 172(4):167–177

Almasi S, Salehiniya H (2014) The prevalence of gestational diabetes mellitus in Iran (1993–2013): a systematic review. J Isfahan Med Sch 32(299):1396–1412

Anderson JW, Kendall CW, Jenkins DJ (2003) Importance of weight management in type 2 diabetes: review with meta-analysis of clinical studies. J Am Coll Nutr 22(5):331–339

Andersson U, Bränning C, Ahrné S, Molin G, Alenfall J, Önning G, Nyman M, Holm C (2010) Probiotics lower plasma glucose in the high-fat fed C57BL/6J mouse. Benefic Microbes 1 (2):189–196

Andreasen AS, Larsen N, Pedersen-Skovsgaard T, Berg RM, Møller K, Svendsen KD, Jakobsen M, Pedersen BK (2010) Effects of Lactobacillus acidophilus NCFM on insulin sensitivity and the systemic inflammatory response in human subjects. Br J Nutr 104(12):1831–1838

Arumugam M, Raes J, Pelletier E, Le Paslier D, Yamada T, Mende DR, Fernandes GR, Tap J, Bruls T, Batto J-M (2011) Enterotypes of the human gut microbiome. Nature 473 (7346):174–180

Asemi Z, Samimi M, Tabassi Z, Rad MN, Foroushani AR, Khorammian H, Esmaillzadeh A (2013) Effect of daily consumption of probiotic yoghurt on insulin resistance in pregnant women: a randomized controlled trial. Eur J Clin Nutr 67(1):71–74

Badehnoosh B, Karamali M, Zarrati M, Jamilian M, Bahmani F, Tajabadi-Ebrahimi M, Jafari P, Rahmani E, Asemi Z (2018) The effects of probiotic supplementation on biomarkers of inflammation, oxidative stress and pregnancy outcomes in gestational diabetes. J Matern Fetal Neonatal Med 31(9):1128–1136

Barrett HL, Nitert MD, Conwell LS, Callaway LK (2014) Probiotics for preventing gestational diabetes. Cochrane Database Syst Rev 2:CD009951

Batra P, Das SK, Salinardi T, Robinson L, Saltzman E, Scott T, Pittas AG, Roberts SB (2013) Relationship of cravings with weight loss and hunger. Results from a 6 month worksite weight loss intervention. Appetite 69:1–7

Bermudez-Silva FJ, Sanchez-Vera I, Suárez J, Serrano A, Fuentes E, Juan-Pico P, Nadal A, de Fonseca FR (2007) Role of cannabinoid CB2 receptors in glucose homeostasis in rats. Eur J Pharmacol 565(1-3):207–211

Blaser M (2011) Stop the killing of beneficial bacteria. Nature 476(7361):393–394

Bonfili L, Cecarini V, Gogoi O, Berardi S, Scarpona S, Angeletti M, Rossi G, Eleuteri AM (2020) Gut microbiota manipulation through probiotics oral administration restores glucose homeostasis in a mouse model of Alzheimer's disease. Neurobiol Aging 87:35–43

Burrack AL, Martinov T, Fife BT (2017) T cell-mediated beta cell destruction: autoimmunity and alloimmunity in the context of type 1 diabetes. Front Endocrinol 8:343

Calcinaro F, Dionisi S, Marinaro M, Candeloro P, Bonato V, Marzotti S, Corneli R, Ferretti E, Gulino A, Grasso F (2005) Oral probiotic administration induces interleukin-10 production and prevents spontaneous autoimmune diabetes in the non-obese diabetic mouse. Diabetologia 48 (8):1565–1575

Calle EE, Kaaks R (2004) Overweight, obesity and cancer: epidemiological evidence and proposed mechanisms. Nat Rev Cancer 4(8):579–591

Cani PD, Osto M, Geurts L, Everard A (2012) Involvement of gut microbiota in the development of low-grade inflammation and type 2 diabetes associated with obesity. Gut Microbes 3 (4):279–288

Cani PD, Geurts L, Matamoros S, Plovier H, Duparc T (2014) Glucose metabolism: focus on gut microbiota, the endocannabinoid system and beyond. Diabetes Metab 40(4):246–257

Clemente JC, Ursell LK, Parfrey LW, Knight R (2012) The impact of the gut microbiota on human health: an integrative view. Cell 148(6):1258–1270

Cristofaro P, Opal SM (2006) Role of toll-like receptors in infection and immunity. Drugs 66 (1):15–29

Cui Y, Liu L, Dou X, Wang C, Zhang W, Gao K, Liu J, Wang H (2017) Lactobacillus reuteri ZJ617 maintains intestinal integrity via regulating tight junction, autophagy and apoptosis in mice challenged with lipopolysaccharide. Oncotarget 8(44):77489

Dempsey JC, Ashiny Z, Qiu C-F, Miller RS, Sorensen TK, Williams MA (2005) Maternal pre-pregnancy overweight status and obesity as risk factors for cesarean delivery. J Maternal-Fetal Neonat Med 17(3):179–185

DiGiulio DB, Callahan BJ, McMurdie PJ, Costello EK, Lyell DJ, Robaczewska A, Sun CL, Goltsman DS, Wong RJ, Shaw G (2015) Temporal and spatial variation of the human microbiota during pregnancy. Proc Natl Acad Sci 112(35):11060–11065

Dolpady J, Sorini C, Di Pietro C, Cosorich I, Ferrarese R, Saita D, Clementi M, Canducci F, Falcone M (2016) Oral probiotic VSL#3 prevents autoimmune diabetes by modulating microbiota and promoting indoleamine 2,3-dioxygenase-enriched tolerogenic intestinal environment. J Diabetes Res 2016:7569431

Dominguez-Bello MG, Costello EK, Contreras M, Magris M, Hidalgo G, Fierer N, Knight R (2010) Delivery mode shapes the acquisition and structure of the initial microbiota across multiple body habitats in newborns. Proc Natl Acad Sci 107(26):11971–11975

Duca FA, Swartz TD, Sakar Y, Covasa M (2012) Increased oral detection, but decreased intestinal signaling for fats in mice lacking gut microbiota. PLoS One 7(6):e39748

Ebrahimi FS, Rad AH, Mosen M, Abbasalizadeh F, Tabrizi A, Khalili L (2019) Effect of L. acidophilus and B. lactis on blood glucose in women with gestational diabetes mellitus: a randomized placebo-controlled trial. Diabetol Metab Syndr 11(1):75

Ejtahed HS, Mohtadi-Nia J, Homayouni-Rad A, Niafar M, Asghari-Jafarabadi M, Mofid V (2012) Probiotic yogurt improves antioxidant status in type 2 diabetic patients. Nutrition 28 (5):539–543

Forouhi NG, Wareham NJ (2019) Epidemiology of diabetes. Medicine 47(1):22–27

Gohir W, Whelan FJ, Surette MG, Moore C, Schertzer JD, Sloboda DM (2015) Pregnancy-related changes in the maternal gut microbiota are dependent upon the mother's periconceptional diet. Gut Microbes 6(5):310–320

Gomes AC, Bueno AA, de Souza RGM, Mota JF (2014) Gut microbiota, probiotics and diabetes. Nutr J 13(1):60

Gregor MF, Hotamisligil GS (2011) Inflammatory mechanisms in obesity. Annu Rev Immunol 29:415–445

Harlev A, Wiznitzer A (2010) New insights on glucose pathophysiology in gestational diabetes and insulin resistance. Curr Diab Rep 10(3):242–247

Hassanalilou T, Payahoo L, Shahabi P, Abbasi MM, Jafar-abadi MA, Bishak YK, Khordadmehr M, Esnaashari S, Barzegar A (2017) The protective effects of Morus nigra L. leaves on the kidney function tests and histological structures in streptozotocin-induced diabetic rats. Biomed Res 28:14

Homayouni A (2009) Letter to the editor. Food Chem 114:1073

Jafar-Abadi MA, Dehghani A, Khalili L, Barzegar A, Mesrizad M, Hassanalilou T (2020) A meta-analysis of randomized controlled trials of the effect of probiotic food or supplement on glycemic response and body mass index in patients with type 2 diabetes, updating the evidence. Curr Diabetes Rev 2020:16

Kawai T, Tanino F, Fukazawa M, Ogawa K, Nagao S, Yoshino H, Komatsu S-I, Suzuki Y, Kawabe Y (2018) OWL833, an orally active nonpeptide GLP-1 receptor agonist, improves glucose tolerance by increasing insulin secretion and reduces food intake of cynomolgus monkeys. Am Diabetes Assoc 67:90439718

Khalili L, Alipour B, Jafar-Abadi MA, Faraji I, Hassanalilou T, Abbasi MM, Vaghef-Mehrabany E, Sani MA (2019a) The effects of lactobacillus casei on glycemic response, serum sirtuin1 and fetuin-a levels in patients with type 2 diabetes mellitus: a randomized controlled trial. Iran Biomed J 23(1):68

Khalili L, Alipour B, Jafarabadi MA, Hassanalilou T, Abbasi MM, Faraji I (2019b) Probiotic assisted weight management as a main factor for glycemic control in patients with type 2 diabetes: a randomized controlled trial. Diabetol Metab Syndr 11(1):5

Kleniewska P, Hoffmann A, Pniewska E, Pawliczak R (2016) The influence of probiotic Lactobacillus casei in combination with prebiotic inulin on the antioxidant capacity of human plasma. Oxidative Med Cell Longev 2016:1340903

Koren O, Goodrich JK, Cullender TC, Spor A, Laitinen K, Bäckhed HK, Gonzalez A, Werner JJ, Angenent LT, Knight R (2012) Host remodeling of the gut microbiome and metabolic changes during pregnancy. Cell 150(3):470–480
Kurzban R, Athena Aktipis C (2007) Modularity and the social mind: are psychologists too self-ish? Personal Soc Psychol Rev 11(2):131–149
Laitinen K, Poussa T, Isolauri E (2008) Probiotics and dietary counselling contribute to glucose regulation during and after pregnancy: a randomised controlled trial. Br J Nutr 101 (11):1679–1687
Larsen N, Vogensen FK, Van Den Berg FW, Nielsen DS, Andreasen AS, Pedersen BK, Al-Soud WA, Sørensen SJ, Hansen LH, Jakobsen M (2010) Gut microbiota in human adults with type 2 diabetes differs from non-diabetic adults. PLoS One 5(2):e9085
Lin C-H, Lin C-C, Shibu MA, Liu C-S, Kuo C-H, Tsai F-J, Tsai C-H, Hsieh C-H, Chen Y-H, Huang C-Y (2014) Oral lactobacillus reuteri GMN-32 treatment reduces blood glucose concentrations and promotes cardiac function in rats with streptozotocin-induced diabetes mellitus. Br J Nutr 111(4):598–605
Ljungberg M, Korpela R, Ilonen J, Ludvigsson J, Vaarala O (2006) Probiotics for the prevention of beta cell autoimmunity in children at genetic risk of type 1 diabetes—the PRODIA study. Ann N Y Acad Sci 1079(1):360–364
Luoto R, Laitinen K, Nermes M, Isolauri E (2010) Impact of maternal probiotic-supplemented dietary counselling on pregnancy outcome and prenatal and postnatal growth: a double-blind, placebo-controlled study. Br J Nutr 103(12):1792–1799
Lyte M (2011) Probiotics function mechanistically as delivery vehicles for neuroactive compounds: microbial endocrinology in the design and use of probiotics. Bioessays 33(8):574–581
Ma X, Hua J, Li Z (2008) Probiotics improve high fat diet-induced hepatic steatosis and insulin resistance by increasing hepatic NKT cells. J Hepatol 49(5):821–830
Markle JG, Frank DN, Mortin-Toth S, Robertson CE, Feazel LM, Rolle-Kampczyk U, Von Bergen M, McCoy KD, Macpherson AJ, Danska JS (2013) Sex differences in the gut microbiome drive hormone-dependent regulation of autoimmunity. Science 339 (6123):1084–1088
Markowiak P, Śliżewska K (2017) Effects of probiotics, prebiotics, and synbiotics on human health. Nutrients 9(9):1021
Milind P, Jyoti M (2014) Curd: a sedative with a bonus bowl of useful side effect. Int Res J Pharm 5:131–135
Mirmiranpour H, Huseini HF, Derakhshanian H, Khodaii Z, Tavakoli-Far B (2019) Effects of probiotic, cinnamon, and synbiotic supplementation on glycemic control and antioxidant status in people with type 2 diabetes; a randomized, double-blind, placebo-controlled study. J Diabetes Metab Disord 2019:1–8
Mohamadshahi M, Veissi M, Haidari F, Shahbazian H, Kaydani G-A, Mohammadi F (2014) Effects of probiotic yogurt consumption on inflammatory biomarkers in patients with type 2 diabetes. Bioimpacts 4(2):83
Moroti C, Magri LFS, de Rezende Costa M, Cavallini DC, Sivieri K (2012) Effect of the consumption of a new symbiotic shake on glycemia and cholesterol levels in elderly people with type 2 diabetes mellitus. Lipids Health Dis 11(1):29
Muccioli GG, Naslain D, Bäckhed F, Reigstad CS, Lambert DM, Delzenne NM, Cani PD (2010) The endocannabinoid system links gut microbiota to adipogenesis. Mol Syst Biol 6:1
Murri M, Leiva I, Gomez-Zumaquero JM, Tinahones FJ, Cardona F, Soriguer F, Queipo-Ortuño MI (2013) Gut microbiota in children with type 1 diabetes differs from that in healthy children: a case-control study. BMC Med 11(1):46
Musso G, Gambino R, Cassader M (2010) Obesity, diabetes, and gut microbiota: the hygiene hypothesis expanded? Diabetes Care 33(10):2277–2284
Nitert MD, Barrett HL, Foxcroft K, Tremellen A, Wilkinson S, Lingwood B, Tobin JM, McSweeney C, O'Rourke P, McIntyre HD (2013) SPRING: an RCT study of probiotics in

the prevention of gestational diabetes mellitus in overweight and obese women. BMC Pregnancy Childbirth 13(1):50

Ogurtsova K, da Rocha Fernandes J, Huang Y, Linnenkamp U, Guariguata L, Cho NH, Cavan D, Shaw J, Makaroff L (2017) IDF diabetes atlas: global estimates for the prevalence of diabetes for 2015 and 2040. Diabetes Res Clin Pract 128:40–50

Rad H, Abbasalizadeh S, Vazifekhah S, Abbasalizadeh F, Hassanalilou T, Bastani P, Ejtahed H-S, Soroush A-R, Javadi M, Mortazavian AM (2017) The future of diabetes management by healthy probiotic microorganisms. Curr Diabetes Rev 13(6):582–589

Raygan F, Ostadmohammadi V, Asemi Z (2019) The effects of probiotic and selenium co-supplementation on mental health parameters and metabolic profiles in type 2 diabetic patients with coronary heart disease: a randomized, double-blind, placebo-controlled trial. Clin Nutr 38(4):1594–1598

Razmpoosh E, Javadi M, Ejtahed HS, Mirmiran P (2016) Probiotics as beneficial agents in the management of diabetes mellitus: a systematic review. Diabetes Res Rev 32(2):143–168

Razmpoosh E, Javadi A, Ejtahed HS, Mirmiran P, Javadi M, Yousefinejad A (2019) The effect of probiotic supplementation on glycemic control and lipid profile in patients with type 2 diabetes: a randomized placebo controlled trial. Diabetes Metab Synd 13(1):175–182

Rousseaux C, Thuru X, Gelot A, Barnich N, Neut C, Dubuquoy L, Dubuquoy C, Merour E, Geboes K, Chamaillard M (2007) Lactobacillus acidophilus modulates intestinal pain and induces opioid and cannabinoid receptors. Nat Med 13(1):35–37

Sarr MG, Billington CJ, Brancatisano R, Brancatisano A, Toouli J, Kow L, Nguyen NT, Blackstone R, Maher JW, Shikora S (2012) The EMPOWER study: randomized, prospective, double-blind, multicenter trial of vagal blockade to induce weight loss in morbid obesity. Obes Surg 22(11):1771–1782

Sekirov I, Russell SL, Antunes LCM, Finlay BB (2010) Gut microbiota in health and disease. Physiol Rev 90(3):859–904

Shoelson SE, Lee J, Goldfine AB (2006) Inflammation and insulin resistance. J Clin Invest 116 (7):1793–1801

Sun J, Chang EB (2014) Exploring gut microbes in human health and disease: pushing the envelope. Genes Dis 1(2):132–139

Tilg H, Moschen AR (2015) Food, immunity, and the microbiome. Gastroenterology 148 (6):1107–1119

Turnbaugh PJ, Ley RE, Mahowald MA, Magrini V, Mardis ER, Gordon JI (2006) An obesity-associated gut microbiome with increased capacity for energy harvest. Nature 444(7122):1027

Turnbaugh PJ, Bäckhed F, Fulton L, Gordon JI (2008) Diet-induced obesity is linked to marked but reversible alterations in the mouse distal gut microbiome. Cell Host Microbe 3(4):213–223

Ursell LK, Treuren WV, Metcalf JL, Pirrung M, Gewirtz A, Knight R (2013) Replenishing our defensive microbes. Bioessays 35(9):810–817

Vehik K, Dabelea D (2012) Why are C-section deliveries linked to childhood type 1 diabetes? Diabetes 61(1):36–37

Wang Q, Würtz P, Auro K, Mäkinen V-P, Kangas AJ, Soininen P, Tiainen M, Tynkkynen T, Jokelainen J, Santalahti K (2016) Metabolic profiling of pregnancy: cross-sectional and longitudinal evidence. BMC Med 14(1):205

Wen L, Ley RE, Volchkov PY, Stranges PB, Avanesyan L, Stonebraker AC, Hu C, Wong FS, Szot GL, Bluestone JA (2008) Innate immunity and intestinal microbiota in the development of Type 1 diabetes. Nature 455(7216):1109–1113

Wendland EM, Torloni MR, Falavigna M, Trujillo J, Dode MA, Campos MA, Duncan BB, Schmidt MI (2012) Gestational diabetes and pregnancy outcomes-a systematic review of the World Health Organization (WHO) and the International Association of Diabetes in Pregnancy Study Groups (IADPSG) diagnostic criteria. BMC Pregnancy Childbirth 12(1):23

Yadav H, Jain S, Sinha P (2007) Antidiabetic effect of probiotic dahi containing Lactobacillus acidophilus and Lactobacillus casei in high fructose fed rats. Nutrition 23(1):62–68

Yadav H, Jain S, Sinha P (2008) The effect of probiotic dahi containing Lactobacillus acidophilus and Lactobacillus casei on gastropathic consequences in diabetic rats. J Med Food 11(1):62–68

Yadav H, Lee J-H, Lloyd J, Walter P, Rane SG (2013) Beneficial metabolic effects of a probiotic via butyrate-induced GLP-1 hormone secretion. J Biol Chem 288(35):25088–25097

Yang X, Gao X-C, Liu J, Ren H-Y (2017) Effect of EPEC endotoxin and bifidobacteria on intestinal barrier function through modulation of toll-like receptor 2 and toll-like receptor 4 expression in intestinal epithelial cell-18. World J Gastroenterol 23(26):4744

Yue B, Luo X, Yu Z, Mani S, Wang Z, Dou W (2019) Inflammatory bowel disease: a potential result from the collusion between gut microbiota and mucosal immune system. Microorganisms 7(10):440

Zhang Q, Wu Y, Fei X (2016) Effect of probiotics on glucose metabolism in patients with type 2 diabetes mellitus: a meta-analysis of randomized controlled trials. Medicina 52(1):28–34

Role of Probiotics in Autism Spectrum Disorders

17

Aravind Sundararaman, Kashika Sethi, and Prakash Halami

Abstract

Autism is a spectrum of conditions leading to challenges in speech, non-verbal communication, repetitive behaviour and social skills. Autism has many subtypes predominantly influenced by environmental and genetic factors. The development of autism is associated with sensitive sensory, gastrointestinal disorders, depression and sleep disorders. Autism predominantly affects children, of age 2 or 3, early diagnosis leads to potential outcomes. Children with autism are affected by gastrointestinal problems such as abdominal pain, constipation and diarrhoea. Microbial communities of the gut can influence many aspects of human physiology and gut-linked disorders. In addition, fermented foods consisting of probiotics have shown to reduce the impacts on neurology in humans. In this chapter, we will discuss the role of probiotics in combating autism. Further, the mode of action of gut microbes and their interaction with the disease to reverse the physiology, neurology and immunological disorders is discussed.

Keywords

Gut brain axis · Probiotics · Autism · Neurodegenerative disease

A. Sundararaman · P. Halami (✉)
Department of Microbiology and Fermentation Technology, CSIR—Central Food Technological Research Institute, Mysuru, India
e-mail: Prakashalami@cftri.res.in

K. Sethi
Department of Nutrition Biology, Central University of Haryana, Mahendragarh, Haryana, India

R. Z. Sayyed, M. Khan (eds.), *Microbiome-Gut-Brain Axis*,
https://doi.org/10.1007/978-981-16-1626-6_17

17.1 Introduction

Human body constitutes many microbial communities, bacterial cells present in the system, nearly over number the human cells. It is broadly understood that highly diverse and stable microbiota encourage overall human health. In particular, gut microbes emerge as a major factor influencing the host's health status (Ghaisas et al. 2016). The diverse consortia of microorganisms inhabiting the gastrointestinal tract (GIT), collectively known as gut microbiota, profoundly influence host physiology through resistance to infection, nutrient metabolism and immune system development (Fung et al. 2017).

The gut microflora is necessary for maintaining health, and many studies have reported to treat several chronic disorders such as diabetes, obesity, metabolic syndrome and neurodegenerative-related disorders via bacterial cell-based biotherapies. Probiotic ingestion is believed to form a healthy gut environment by keeping a balance between bacterial populations and encouraging their favourable metabolic actions. The gut microbiome interacts with the host by a number of biochemical and functional associations and hence by and large influencing host homeostasis and health. The composition of gut microbiota is established early during the development of host and experience a myriad of changes throughout the lifetime.

Gut flora inhabited with the harmful pathogenic bacteria leads to a condition called dysbiosis. Gut dysbiosis is a major factor responsible for several GI disorders that may enhance T-helper cells, inflammatory cytokines and monocytes inducing elevated intestinal and blood-brain barrier (BBB) permeability through the microbiota-gut-brain axis. Microbial dysregulation has been observed in various neurological conditions including Parkinson's disease, Alzheimer's disease and autism spectrum disorder (ASD). In addition, depletion of gut microbiota can have an influence on the criticality of the fundamental pathology or behavioural issues seen in various brain disorders. However, the mechanisms involved for such effects are very slowly unfolding. This reveals an indirect or direct interaction among gut microbes and the central nervous system.

The complex interactions of host and gut microbiota lead to prevention of various gut-linked disorders including ulcerative colitis (UC) and other central nervous system (CNS) diseases. Furthermore, many recent clinical and preclinical studies proposed that focusing the gut microbiota via probiotic, prebiotic or nutritional interventions may be an effective "psychobiotic" approach for dealing with the symptoms in NDD (Sherwin et al. 2018). Thus, strategies including maintenance and regulation of healthy intestinal microbiota could potentially lower the prevalence and individual risk of neurodegenerative diseases.

17.2 Gut-Brain Axis

The gut-brain axis is gaining popularity in recent times, and the microbiome plays a key role in physiological and biological basis for age-related neurodevelopment and neurodegenerative disorders. The interaction between the gut and the brain initiates on birth and acts as a considerable factor in profiling how the brain is structured. The gut is also known as the second brain due to its involvement in the brain activities, and the similarities within the two systems are in terms of their structures, mechanisms and biochemical pathways: the gut immune barrier (GIB) and the BBB (Vojdani et al. 2016). Although there is structural separation, recent findings have demonstrated the presence of two-way communication between the gut microflora and the brain.

The interactions greatly influence neurodegenerative processes initially in NDD and tumours of the CNS (Ma et al. 2019). The gut-brain axis has demonstrated a critical role of gut microbes in orchestrating behaviour and brain development where the immune system regulates these interactions. Gut microbes modulate the functions and maturation of immune cells residing in the tissues in the CNS. They also influence peripheral immune cell activation, regulating responses to autoimmunity, brain injury, neuroinflammation and neurogenesis. Consequently, the immune system and gut microbiota are implicated in the etiopathogenesis of neurodevelopmental diseases, such as Alzheimer's disease, autism spectrum disorder and depression (Fung et al. 2017). Studying the communication between the CNS and the GIT (brain-gut-microbiome axis) can provide insights into why the GIT disorders are more common in children with ASD (Fig. 17.1) (Luna et al. 2016).

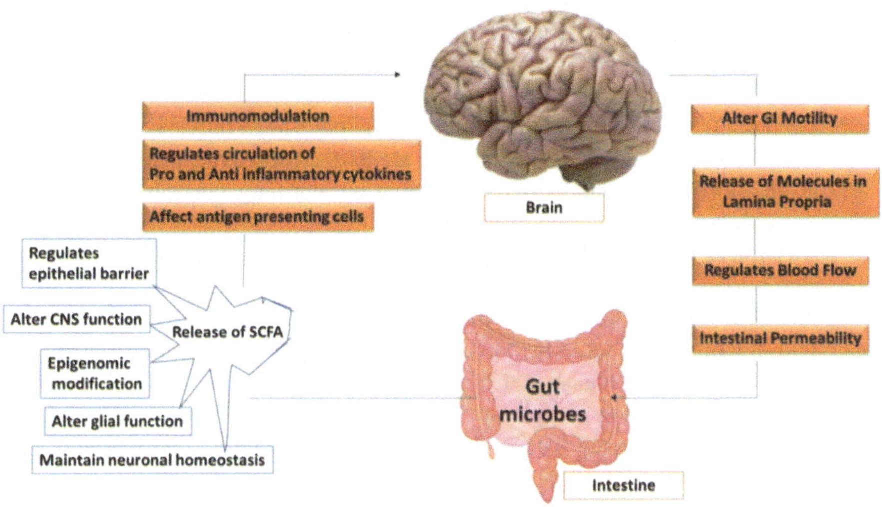

Fig. 17.1 The two-way interactions of the gut and the brain modulating the microbial flora to alleviate neurodegenerative diseases

Over the past decade, gut microbiome as a modulator for brain activities has been a hot topic of research particularly in animal models. Besides, rising preclinical and clinical studies proved that gut microbiome acts as a potent factor for neurological diseases such as AD, PD, ASD, IBS and stroke (Zhu et al. 2020). However, further studies including randomized controlled trials in humans are required to formulate novel therapeutics using specific microbes. Therefore, this chapter discusses the significant relationship between gut microbiome and brain activities that play an important role in neurological diseases with an emphasis on autism. Further, it highlights the role of probiotic supplementation in harnessing neuronal disorders through the interactions of the gut-brain axis.

17.3 History of Autism and Causes

The popular Swiss psychiatrist Eugen Bleuler describes the indifferent and reserved conditions of some people with the word called autism. The term autism was coined in 1908. Later, Leo Kanner in 1943 described 11 children by saying "an autistic distraction of perceptual contact". Similarly, in 1944, Hans Asperger used the term "autistic psychopaths" to described the disorder in his study. In both studies, there was sheer resemblance to schizophrenia. Patients suffering from ASD have a different neural network (Liu et al. 2019). The main cause of this disorder was assumed to be impaired relationships, abnormality in verbal communication and behaviours, but presently they are regarded as signs of this problem. These features should be viewed as an acclimatization rather than a problem, indicating distinct structural nervous systems in the body. In spite of the fact that aloofness is not distinct to the complex developmental disorders of ASD, it has become common to exterminate individuals having impairments in social development, communication, distinguished and rigid selection of interest and behaviours (van der Gaag 2017).

At one point in time, ASD was so rare that other than clinicians, it was not known to the general public. Currently, the cases of ASD have been observed in 1 in every 59 children in the United States (Nazeer et al. 2019). Moreover, the prevalence of ASD in the general population is increasing at an alarming rate which is directly influencing the structure of child psychology, education and family life. The role of the environment in ASD is still highly unknown. Every day, psychologists and educators associated with both general and special education are facing issues to recognize and treat kids having autism. Technologies in the past decade on assessment and treatment are greatly advanced. However, development of new and effective therapeutic strategies is the need of the hour to combat the devastating disease. Models including unicellular organism to complex animals prove to be useful tool to illuminate the mechanisms of neurodegenerative diseases and their promising therapeutics.

17.3.1 Neurodegenerative Disorders (NDDs)

Neurodegenerative disorders are characterized by the depletion of neurons within the brain and spinal cord. The distinct type of neurological disorders and its clinical symptoms depend on the region of the CNS involved. These influence about 10% of the old-age population around the world, making it one of the most crucial medical and social concerns. The common NDDs include Parkinson's disease (PD), Alzheimer's disease (AD) and autism spectrum disorder (ASD) (Westfall et al. 2017). NDDs pose a set of pathological conditions initiating loss of neurons and irreversible dysfunction that determine clinical presentation and course. The neuro-degenerative mechanisms are multifactorial mainly caused by environmental, genetic and endogenous factors of the host (Jellinger 2009). NDDs are currently referred to the genetic mechanism relating to protein deposits, designated as proteinopathies or "protein misfolding" based on critical conformational changes of proteins.

The common pathogenic mechanisms of NDDs include:

1. Abnormal protein dynamics, aggregation, defective degradation and mutations in molecular chaperones
2. Formation of free radicals (ROS) and oxidative stress (OS)
3. Mitochondrial dysfunctions, impaired bioenergetics and DNA damage
4. Dysfunction of neurotrophins
5. Neuroinflammatory processes

The secondary effects of the disorders include disruption of cellular/axonal transport and fragmentation of neuronal Golgi apparatus. However, these mechanisms are interconnected, leading to abnormality and cell death.

Neuroanatomical changes frequently observed in the brains of autistic individuals are the overgrowth of the brain termed macrocephaly and abnormal neuronal connectivity. The defects in synaptic proteins are also known to influence ASD through changes in synaptic function, structure and neural circuits. This suggests "synaptopathy" is an important component of ASD.

Abnormalities in cell organization of autistic brains are observed early during brain development in the regions including the cerebellum, the frontal lobe and subcortical limbic structures. The cerebellar activation is significantly reduced during selective attention tasks, whereas the activity is elevated during motor task (Allen et al. 2004). However, the potential role of the cerebellum in NDD is restricted to motor and sensory dysfunctions. Nevertheless, it is distinct that core symptoms of autism are associated with the cerebellum.

Genes with functions of epigenetic modulations are involved in ASD susceptibility. The analysis of 215 putative genes identified 19.5% of genes are epigenetic regulators and suggests that few pathogenic variants have potential for diverse disease phenotypes (Duffney et al. 2018). The high risk genes with high penetrance are present in the nucleus and involved in harnessing expression, protein-protein interactions important in CNS developmental patterning (Casanova et al. 2016).

In vivo studies in mouse models for genetic implications observed a variety of neuronal proteins in the development of ASD (Won et al. 2013). The selective deletion of *Tsc1* (tuberous sclerosis 1) in cerebellar Purkinje cells sufficiently caused core autism-like behaviours including reduced excitability in Purkinje cells of mice (Tsai et al. 2012). In addition, mice lacking the neuroligin-3 gene (*Nlgn3*$^{-/-}$ mice), identified in autistic patients, showed occluded metabotropic glutamatergic receptor (mGluR)-dependent long-term depression (LTD) at synapses between parallel fibres and Purkinje cells along with motor coordination deficits (Baudouin et al. 2012). Purkinje cell-specific *Nlgn3* in mice can rescue both synaptic and behavioural perturbations, suggesting the potential to rectify the neural circuits even after complete development.

Studies have also identified *NEUROD2* pathogenic mutations associated with ASD. During embryogenesis, cortical projection neurons (CPNs) relocate and hence cause impaired thickness and laminar positioning of layers of cortical. In children, dendritic spine change and internal extractability show an upward trend in L5 CPNs. This elucidates the significance of *Neurod2* in overall cortical development and activities, whose modifications can also be related to ASD and associated signs in the novel *NEUROD2* mutation syndrome (Runge et al. 2020).

Although genetic causes of ASD are significantly determined, the mechanism of pathogens regulating the genetic susceptibility is unclear. Same variants of pathogens exhibit heterogeneous diseases and levels of disability. The heterogeneity may possibly be caused due to the presence of second modulating variants which interact with other susceptibility loci. To date, germline second hits have been primarily found as genetic evidence supporting a multiplex theory of autism.

17.3.2 Role of Microbes Present in Autistic Patients

It has been already noted that ASD leads to modifications in human behaviours including verbal interactions and various hand stereotypes. With various genetic impairments, there is a difference in the composition of microbiome in normal individual and an ASD patient.

The analysis of faecal microbiota of children suffering from ASD identified eminent gut dysbiosis and nomenclatures consisting of a higher relative myriad of families: Bifidobacteriaceae, Lactobacillaceae and Veillonellaceae. On the other hand, the gut microbiome of children without ASD showed higher rates of abundance of the family Prevotellaceae, *Roseburia* belonging to the family Lachnospiraceae and *Faecalibacterium* of the family Clostridiaceae. The ratio of probiotics in healthy and diseased individuals needs to be analysed (Pulikkan et al. 2018).

Comparative analysis among healthy and ASD subjects revealed low abundance of Erysipelotrichaceae, Enterococcaceae and Desulfovibrionaceae in healthy children in comparison to children with ASD. Similarly, children with ASD were found to have abundant *Butyrivibrio* and *Coprococcus* from the family Lachnospiraceae,

Klebsiella from the family Enterococcaceae and *Ruminococcus* from the family Ruminococcaceae (Pulikkan et al. 2018).

In another study, ASD group was determined to have a downward trend in the relative abundance of Bacteroidetes belonging to the genera *Bilophila*, *Alistipes*, *Parabacteroides*, *Dialister* and *Veillonella* and a considerable increase in Firmicutes belonging to the genera *Corynebacterium*, *Collinsella* and *Dorea*. This shows autism is related to alteration in intestinal microbial community structure. Autistic patients are also identified by the signs of constipation. The characteristic relates to the presence of higher-level bacterial taxa belonging to the genera *Escherichia*, *Shigella* and *Clostridium* (Strati et al. 2017).

Interestingly, protein digestion and absorption pathway although not directly related to the disease progression is upregulated in valproic acid-injected mice to determine environmental risk factors causing ASD. This pathway implicates the serotonin production from enterochromaffin cells is stimulated by the production of short-chain fatty acids (SCFAs) (Reigstad et al. 2015). The comorbidity in ASD is observed to be elevated serum serotonin and increased SCFA levels which h is attributed to the presence of *Clostridia* class of bacteria including *Clostridiales*, *Tissierellaceae* and *Sporanaerobacter* (Lim et al. 2017).

The autistic disease severity is associated with the higher occurrence of *Clostridium* spp. in the gut (Iovene et al. 2017). Especially, the production of beta2-toxin by *Clostridium perfringens* is significantly enhanced in autistic children, and the abundance of the gene is related to the occurrence of ASD (Finegold et al. 2017). The study also suggests that *Clostridium* spp. inducing sub-acute tetanus infection might be the cause in some cases of ASD (Srikantha and Mohajeri 2019). The infection with *Clostridium tetani* occurs in dysbiotic GI tract. It produces tetanus neurotoxin, which permeates the intestinal barrier and enters the nucleus solitarius via the vagal nerve and subsequently the CNS entirely. The release of synaptic vesicles consisting of neurotransmitters is inhibited by tetanus neurotoxin through cleaving of the membrane-associated protein synaptobrevin significant for the stability of vesicles. The presence of cleaved synaptobrevin in synapses reduces and degenerates synaptic activity thereby diminished social behaviour is observed in ASD patients.

Similarly, studies have also identified a higher relative number of mycobiota in patients including the genus *Candida*, especially *Candida albicans*, in the faecal samples when compared to neurotypical subjects. Reportedly, *Candida* is associated with some autistic behaviour as it proliferates and produces ammonia and toxins. These also cause malabsorption of carbohydrates and minerals significant in ASD pathophysiology. However, 60% of healthy populations are assessed to be asymptomatic carrier of *Candida* spp. (Kantarcioglu et al. 2016). The reduced levels of gut microbiota and mycobiota are also associated with ASD, paving further research for novel strategies to combat neurological disorders.

17.3.3 Role of Probiotics in Overcoming the Disease

Generally, all the cognitive activities are regulated by the CNS, which is further regulated by the brain. Any type of injury to the neuronal system leads to the severe abnormality to the host as it may further cause NDD.

The overall physical and mental health of humans is directly associated with eating habits and brain health. Many recent studies have reported the critical role of intestinal microbiota in the regulation of the neuroimmune system, the central nervous system and the neuroendocrine system. The consumption of food including probiotics positively amends gut microbiota, confers health benefits and significantly regulates mental health of the host (Fig. 17.2) (Sivamaruthi et al. 2019). The aberrance in the gut-brain axis leads to serious diseases. The significance of probiotic intervention in children with ASD has been considered as a complementary and alternative therapeutics in ASD treatment.

The probiotic supplementation to ASD subjects demonstrated the severity of ASD is positively linked to GI dysfunction and decreased levels of TNFα (Tomova et al. 2015). Parracho et al. (2010) studied the supplementation of *L. plantarum* WCFS1 to ASD children and observed significant alterations in the faecal microbiota. Observed significant alterations in the fecal microbiota through increased lactobacillus and reduced *clostridum* cluster XIVa. Probiotic intake also improved bowel function compared to placebo feeding.

Supplementation of live mixture of *L. helveticus*, *L. acidophilus*, *B. breve*, *B. longum*, *L. paracasei*, *B. infantis*, *L. plantarum* and *S. thermophilus* (VSL#3) for 4 weeks showed significant improvement of GI symptoms in children with

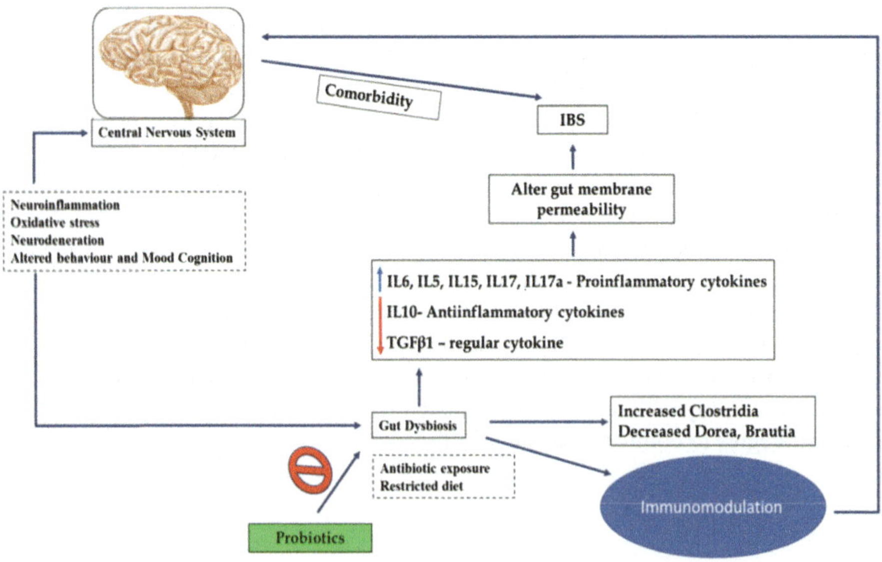

Fig. 17.2 The mechanism of probiotics to combat autism spectrum diseases in the human host

cognitive function disability and ASD (Grossi et al. 2016). Similarly, supplementing 5 g/day consortia of probiotic strains (*L. acidophilus*, *L. rhamnosus* and *B. longum*) for 3 months substantially altered the faecal microbiota in ASD children. The levels of *Lactobacilli* and *Bifidobacteria* were also found to be increased in comparison to the baseline (Shaaban et al. 2018).

Supplementation of *L. acidophilus* Rosell-11 (twice a day) to autistic children for 2 months showed reduction in d-arabinitol levels and the ratio of d-arabinitol/l-arabinitol. The metabolite produced by pathogenic *Candida* spp. and the ratio of d-arabinitol/l-arabinitol in urine is the biomarker of candidiasis. The probiotic intervention reduced the occurrence of candidiasis and improved the hosts' ability to respond to ASD (Kałużna-Czaplińska and Błaszczyk 2012).

The treatment in rats with probiotics identified reduced astrocyte reactivity by decreasing GFAP protein synthesis in posterior brain hemisphere and alleviated motor behaviour of rats after 2 months demonstrating probiotic potential in preventing neurological disease (Ushakova et al. 2009). The beneficial effects on astrocytes disappeared after 6 months treatment suggesting the supplementation of *Lactobacillus* for prolonged period may not be effective because of adaptation to the gastrointestinal and immunological systems.

A randomized controlled trial was performed in humans and animals, supplementing *Bifidobacterium* (*B. breve*, *B. infantis* and *B. longum*) and *Lactobacillus* (*L. rhamnosus* and *L. helveticus*) with the doses between 10^9 and 10^{10} CFU for 2–4 weeks. The results demonstrated the effect of probiotics in improving psychiatric behaviours including anxiety, depression and ASD (Wang et al. 2016).

Children with ASD show considerable reduction in relative abundance of *Bifidobacteriales* and *Bifidobacterium longum*, leading to dysbiosis state in the gut microbiota. Probiotic supplementation with fructoligosaccharide (FOS) in the host increased the levels of *Bifidobacteriales, B. longum* and mitigates *Clostridium* sp. and reduced the occurrence of autism and GI symptoms. Besides, there were high SCFAs, reduced serotonin, elevated homovanillic acid and reduced concentrations of acetic acid, butyric acid and propionic acid in ASD children compared to healthy children (Wang et al. 2020). The severity of ASD assessed by Autism Treatment Evaluation Checklist (ATEC) has revealed that supplementation of probiotics influences the levels of short-chain fatty acids in ASD children (Adams et al. 2011). This suggests the appropriate consumption of probiotics may improve ASD symptoms; however, further research is necessary.

17.4 Conclusions

There are mounting evidences that confirm the aberrations in the gut microbial communities of children suffering from ASD. However, the heterogeneity in patients needs to be assessed, and the unique profile of microbiome has to be fully characterized. Intestinal disorders including GI tract inflammation and bowel dysfunction are frequently observed in most cases of ASD. The amendments of gut microbiota are reported to improve ASD symptoms and reveal beneficial effect of probiotics in the improvement of ASD.

Further studies have recommended dietary regulations may enhance therapeutic advancement in ASD treatment. More research should focus on optimization of probiotic supplementation such as dose and duration for the treatment of ASD, which helps in developing efficient strategies to alleviate ASD symptoms. In addition, the mutual and moral support from parents and society is required for the improvement of the quality of life of ASD children.

Acknowledgements The authors would like to thank the Director, CSIR-Central Food Technological Research Institute, Mysuru, for the facilities provided and Dr. Praveena Mudliar for reviewing the article. AS would like to thank CSIR for providing Nehru Post-Doctoral Fellowship.

References

Adams JB, Johansen LJ, Powell LD, Quig D, Rubin RA (2011) Gastrointestinal flora and gastrointestinal status in children with autism – comparisons to typical children and correlation with autism severity. BMC Gastroenterol 11(1):22. https://doi.org/10.1186/1471-230X-11-22

Allen G, Müller R-A, Courchesne E (2004) Cerebellar function in autism: functional magnetic resonance image activation during a simple motor task. Biol Psychiatry 56(4):269–278. https://doi.org/10.1016/j.biopsych.2004.06.005

Baudouin SJ, Gaudias J, Gerharz S, Hatstatt L, Zhou K, Punnakkal P, Tanaka KF, Spooren W, Hen R, De Zeeuw CI, Vogt K, Scheiffele P (2012) Shared synaptic pathophysiology in syndromic and nonsyndromic rodent models of autism. Science 338(6103):128–132. https://doi.org/10.1126/science.1224159

Casanova EL, Sharp JL, Chakraborty H, Sumi NS, Casanova MF (2016) Genes with high penetrance for syndromic and non-syndromic autism typically function within the nucleus and regulate gene expression. Mol Autism 7(1):18. https://doi.org/10.1186/s13229-016-0082-z

Duffney LJ, Valdez P, Tremblay MW, Cao X, Montgomery S, McConkie-Rosell A, Jiang Y (2018) Epigenetics and autism spectrum disorder: a report of an autism case with mutation in H1 linker histone HIST1H1E and literature review. Am J Med Genet Part B 177(4):426–433. https://doi.org/10.1002/ajmg.b.32631

Finegold SM, Summanen PH, Downes J, Corbett K, Komoriya T (2017) Detection of Clostridium perfringens toxin genes in the gut microbiota of autistic children. Anaerobe 45:133–137. https://doi.org/10.1016/j.anaerobe.2017.02.008

Fung TC, Olson CA, Hsiao EY (2017) Interactions between the microbiota, immune and nervous systems in health and disease. Nat Neurosci 20(2):145–155. https://doi.org/10.1038/nn.4476

Ghaisas S, Maher J, Kanthasamy A (2016) Gut microbiome in health and disease: linking the microbiome–gut–brain axis and environmental factors in the pathogenesis of systemic and neurodegenerative diseases. Pharmacol Ther 158:52–62. https://doi.org/10.1016/j.pharmthera.2015.11.012

Grossi E, Melli S, Dunca D, Terruzzi V (2016) Unexpected improvement in core autism spectrum disorder symptoms after long-term treatment with probiotics. SAGE Open Med Case Rep 4:2050313X16666231. https://doi.org/10.1177/2050313X16666231

Iovene MR, Bombace F, Maresca R, Sapone A, Iardino P, Picardi A, Marotta R, Schiraldi C, Siniscalco D, Serra N, de Magistris L, Bravaccio C (2017) Intestinal dysbiosis and yeast isolation in stool of subjects with autism spectrum disorders. Mycopathologia 182 (3–4):349–363. https://doi.org/10.1007/s11046-016-0068-6

Jellinger KA (2009) Recent advances in our understanding of neurodegeneration. J Neural Transm 116(9):1111–1162. https://doi.org/10.1007/s00702-009-0240-y

Kałużna-Czaplińska J, Błaszczyk S (2012) The level of arabinitol in autistic children after probiotic therapy. Nutrition 28(2):124–126. https://doi.org/10.1016/j.nut.2011.08.002

Kantarcioglu AS, Kiraz N, Aydin A (2016) Microbiota–gut–brain axis: Yeast species isolated from stool samples of children with suspected or diagnosed autism spectrum disorders and in vitro susceptibility against nystatin and fluconazole. Mycopathologia 181(1–2):1–7. https://doi.org/10.1007/s11046-015-9949-3

Lim JS, Lim MY, Choi Y, Ko G (2017) Modeling environmental risk factors of autism in mice induces IBD-related gut microbial dysbiosis and hyperserotonemia. Mol Brain 10(1):14. https://doi.org/10.1186/s13041-017-0292-0

Liu F, Li J, Wu F, Zheng H, Peng Q, Zhou H (2019) Altered composition and function of intestinal microbiota in autism spectrum disorders: A systematic review. Transl Psychiatry 9(1):1–13. https://doi.org/10.1038/s41398-019-0389-6

Luna RA, Savidge TC, Williams KC (2016) The brain-gut-microbiome axis: what role does it play in autism spectrum disorder? Curr Dev Disord Rep 3(1):75–81. https://doi.org/10.1007/s40474-016-0077-7

Ma Q, Xing C, Long W, Wang HY, Liu Q, Wang R-F (2019) Impact of microbiota on central nervous system and neurological diseases: the gut-brain axis. J Neuroinflammation 16(1):53. https://doi.org/10.1186/s12974-019-1434-3

Nazeer A, Hashemi N, Imran N, Azeem MW (2019) Autism spectrum disorder: a concept in evolution. Psychiatr Ann 49(3):103–108. https://doi.org/10.3928/00485713-20190207-01

Parracho HMRT, Gibson GR, Knott F, Bosscher D, Kleerebezem M, McCartney AL (2010) A double-blind, placebo-controlled, crossover-designed probiotic feeding study in children diagnosed with autistic spectrum disorders. Int J Probiot Prebiot 5(2):69–74

Pulikkan J, Maji A, Dhakan DB, Saxena R, Mohan B, Anto MM, Agarwal N, Grace T, Sharma VK (2018) Gut microbial dysbiosis in Indian children with autism spectrum disorders. Microb Ecol 76(4):1102–1114. https://doi.org/10.1007/s00248-018-1176-2

Reigstad CS, Salmonson CE, Iii JFR, Szurszewski JH, Linden DR, Sonnenburg JL, Farrugia G, Kashyap PC (2015) Gut microbes promote colonic serotonin production through an effect of short-chain fatty acids on enterochromaffin cells. FASEB J 29(4):1395–1403. https://doi.org/10.1096/fj.14-259598

Runge K, Mathieu R, Bugeon S, Lafi S, Beurrier C, Sahu S, Schaller F, Loubat A, Herault L, Gaillard S, Cahuc M, Pallesi-Pocachard E, Montheil A, Bosio A, Rosenfeld JA, Hudson E, Lindstrom K, Mercimek-Andrews S, Jeffries L, Chevigny A (2020) Disruption of the transcription factor NEUROD2 causes an autism syndrome via cell-autonomous defects in cortical projection neurons. BioRxiv, 296889. https://doi.org/10.1101/296889

Shaaban SY, El Gendy YG, Mehanna NS, El-Senousy WM, El-Feki HSA, Saad K, El-Asheer OM (2018) The role of probiotics in children with autism spectrum disorder: a prospective, open-label study. Nutr Neurosci 21(9):676–681. https://doi.org/10.1080/1028415X.2017.1347746

Sherwin E, Dinan TG, Cryan JF (2018) Recent developments in understanding the role of the gut microbiota in brain health and disease: the gut microbiota in brain health and disease. Ann N Y Acad Sci 1420(1):5–25. https://doi.org/10.1111/nyas.13416

Sivamaruthi BS, Prasanth MI, Kesika P, Chaiyasut C (2019) Probiotics in human mental health and diseases-a mini-review. Trop J Pharm Res 18:5

Srikantha P, Mohajeri MH (2019) The possible role of the microbiota-gut-brain-axis in autism spectrum disorder. Int J Mol Sci 20(9):2115. https://doi.org/10.3390/ijms20092115

Strati F, Cavalieri D, Albanese D et al (2017) New evidences on the altered gut microbiota in autism spectrum disorders. Microbiome 5:24. https://doi.org/10.1186/s40168-017-0242-1

Tomova A, Husarova V, Lakatosova S, Bakos J, Vlkova B, Babinska K, Ostatnikova D (2015) Gastrointestinal microbiota in children with autism in Slovakia. Physiol Behav 138:179–187

Tsai PT, Hull C, Chu Y, Greene-Colozzi E, Sadowski AR, Leech JM, Steinberg J, Crawley JN, Regehr WG, Sahin M (2012) Autistic-like behaviour and cerebellar dysfunction in Purkinje cell Tsc1 mutant mice. Nature 488(7413):647–651. https://doi.org/10.1038/nature11310

Ushakova G, Fed'kiv O, Prykhod'ko O, Pierzynowski S, Kruszewska D (2009) The effect of long-term lactobacilli (lactic acid bacteria) enteral treatment on the central nervous system of growing rats. J Nutr Biochem 20(9):677–684. https://doi.org/10.1016/j.jnutbio.2008.06.010

van der Gaag RJ (2017) Autism spectrum disorders: developmental history of a concept. In: Barahona CB, van der Gaag RJ (eds) Autism spectrum disorders in adults. Springer, Cham. https://doi.org/10.1007/978-3-319-42713-3_1

Vojdani A, Lambert J, Vojdani E (2016) The gut-brain axis: Autoimmune and neuroimmune disorders. Altern Ther Health Med 22(S3):31–46

Wang H, Lee I-S, Braun C, Enck P (2016) Effect of probiotics on central nervous system functions in animals and humans: a systematic review. J Neurogastroenterol Mot 22(4):589–605. https://doi.org/10.5056/jnm16018

Wang Y, Li N, Yang J-J, Zhao D-M, Chen B, Zhang G-Q, Chen S, Cao R-F, Yu H, Zhao C-Y, Zhao L, Ge Y-S, Liu Y, Zhang L-H, Hu W, Zhang L, Gai Z-T (2020) Probiotics and fructo-oligosaccharide intervention modulate the microbiota-gut brain axis to improve autism spectrum reducing also the hyper-serotonergic state and the dopamine metabolism disorder. Pharmacol Res 157:104784. https://doi.org/10.1016/j.phrs.2020.104784

Westfall S, Lomis N, Kahouli I, Dia SY, Singh SP, Prakash S (2017) Microbiome, probiotics and neurodegenerative diseases: deciphering the gut brain axis. Cell Mol Life Sci 74 (20):3769–3787. https://doi.org/10.1007/s00018-017-2550-9

Won H, Mah W, Kim E (2013) Autism spectrum disorder causes, mechanisms, and treatments: Focus on neuronal synapses. Front Mol Neurosci 6:19. https://doi.org/10.3389/fnmol.2013.00019

Zhu S, Jiang Y, Xu K, Cui M, Ye W, Zhao G, Jin L, Chen X (2020) The progress of gut microbiome research related to brain disorders. J Neuroinflammation 17(1):25. https://doi.org/10.1186/s12974-020-1705-z

From Probiotics to Postbiotics: Key to Microbiome and Health

18

Rajan Walhe, Hina Alim, and Sunita Kumari

Abstract

Humans are consuming probiotic microorganisms through traditionally fermented food since thousands of years, unknowingly. Probiotic microorganisms are usually lactic acid bacteria, are generally recognized as safe, and play a significant role in human health by producing a variety of metabolites. The list of their health benefits is pretty long, but despite of it, probiotics have some limitations to use for therapeutics mainly due to being live. Hence, in the recent past, the focus of health benefits of microorganisms is being shifted from viable live probiotics to nonviable paraprobiotics or probiotic-derived postbiotics. Postbiotics are nonviable metabolic products of probiotic bacteria possessing biological activity in the host. These metabolites are noted to have equivalent health potential as that of probiotics and additionally have advantages over limitations of probiotics that are discussed in this chapter.

Keywords

Human health · Paraprobiotics · Postbiotics · Prebiotics · Probiotic

---

The original version of this chapter was revised: the second author's name has been updated. A correction to this chapter is available at https://doi.org/10.1007/978-981-16-1626-6_23

R. Walhe (✉)
MES Abasaheb Garware College, Pune, Maharashtra, India

H. Alim
Department of Life Sciences, University of Mumbai, Mumbai, Maharashtra, India

S. Kumari
Department of Botany, DDU Gorakhpur University, Gorakhpur, Uttar Pradesh, India

R. Z. Sayyed, M. Khan (eds.), *Microbiome-Gut-Brain Axis*,
https://doi.org/10.1007/978-981-16-1626-6_18

18.1 Introduction

Probiotics have been defined as live microorganisms which when administered in adequate amounts confer health benefit to the host (Markowiak and Slizewska 2017). Though the concept of probiotics has been elaborated in modern science lately, humans have been unknowingly consuming beneficial microorganisms since thousands of years through traditionally fermented foods, which are main sources of probiotic microorganisms (Linares et al. 2017). Probiotic microorganisms are lactic acid bacteria (LAB) and belong to the genera *Lactobacillus*, *Bifidobacterium*, *Pediococcus*, etc. There are many evidences indicating that LAB play a vital role in human health by producing various metabolites (Xu et al. 2019). They are generally recognized as safe and hence important not only in the food sector but also in the pharmaceutical industry.

Thus, LAB are functional components for foods; moreover, their metabolic products like lactic acid and bacteriocins offer property of natural preservative and antimicrobials against contaminating microorganisms (Chuah et al. 2019). Their benefits are reported and include prevention of various infections, immunomodulation, and amelioration of clinical conditions like irritable bowel syndrome (IBS), hypercholesterolemia, various types of cancers, etc. These effects are intervened through mechanisms like alteration of gut microflora, boosting immune response, antiproliferative, anti-oxidative, apoptosis, modulating gut microbiome, etc. The list of health benefits is still incomplete, and despite of it, probiotics have some drawbacks due to their viability status that imparts main limitations for their applications in food and pharmaceutical industries (Nataraj et al. 2020). The limitations include (1) unknown or poor mechanisms at a molecular level, (2) strain-specific effect, (3) threat of antibiotic resistance through horizontal gene transfer, (4) maintaining viability, (5) threat of opportunistic infections, (6) inflammatory responses, and (7) systemic infections (endocarditis, sepsis, etc.). Consequently, though probiotics are nonpathogenic microorganisms offering beneficial effects, carefulness is required when administering in patients with inflammation and severe pancreatitis, and some probiotic strains may result harmful in irritable bowel syndrome (Cicenai et al. 2014).

Fermented foods are important for the gut health, even though the benefit does not characteristically result from the colonizing microbes, but the ferment itself. During fermentation, bacteria form many biomolecules/metabolites (postbiotics) that are beneficial to the gut and immune mechanism (Maguire and Maguire 2019).

Probiotics also maintain and restore the skin microbiota as in the gut, but the use of live bacteria on skin poses some limitations (Majeed et al. 2020). Though since the last four decades or so, the use of probiotics found place in reducing load of common infections in children, scientific community does not hold up probiotic involvement in pediatric diseases due to case findings of probiotic-associated infections like necrotizing enterocolitis, pneumonia, meningitis, bacteremia and their rising trend of adherence, invasion, and cytotoxicity, etc. (Rojas et al. 2020).

In addition to the above-stated disadvantages of probiotics, many physicians are still doubtful in the use of probiotics for pediatric practice because of unusual cases of probiotic unpleasant effects. On the contrary, more reports suggesting that the probiotic strains need not be live to offer benefits to the host, a report indicated inactivated strains can adhere better to the mucosa of the intestine than viable one (Mantziari et al. 2020).

Therefore, recently, the focus of health benefits is progressively displacing from viable probiotic bacteria in the direction of nonviable paraprobiotics or probiotic-derived postbiotics. Postbiotics known to be physiologically rich, with defined chemical structures, safety dosing, and extended shelf life (till 5 years) make them therapeutically appealing. Also, their features include suitability in absorption, synthesis, excretion, and sharp signaling potential with host tissue responses (Puccetti et al. 2020).

18.2 Postbiotics: Definition and Concept

Although, viability is an integral part of definition to label the microbe as probiotics, it is not all the time obligatory to pursue health benefits. The nonviable probiotics that retain their health benefits are generally termed as paraprobiotics, while the term postbiotic is applied for soluble bioactive factors secreted by probiotics or freed after rupture of their cell (Anderson 2019). In recent years, there is emergent attention in probiotic effects shown by these microbial metabolites, also called bioactive postbiotic metabolites (PM), especially in intestinal health and general immunity (Chuah et al. 2019). These metabolites (PM) are reported to be of equivalent potential with probiotics and encompass soluble or secreted factors, metabolites, cell-free supernatant, bacteriocin, etc. The role of these PM in intestinal health and being safer alternative in contrast to live bacteria is well documented (Chuah et al. 2019).

Postbiotics being nonviable bacterial products from probiotic organisms are nontoxic, nonpathogenic, and resistant to hydrolysis by mammalian enzymes (Kerry et al. 2018). The various metabolites of microorganisms claimed for their postbiotic status are shown in Fig. 18.1.

Though several researchers have proposed various terminologies to express postbiotics and paraprobiotics with the relatively similar perception about these terms, as mentioned by Nataraj et al. (2020), the widely accepted definitions are as follows:

- Paraprobiotics (also called as ghost probiotics or inactivated probiotics): nonviable microorganisms (either intact or broken) or crude cell extracts which when administered in adequate amounts confer benefit to the host.
- Postbiotics: Nonviable bacterial metabolic products having biological activity in the host. As mentioned by Gutierrez et al. (2020), postbiotics (also called as metabiotics, pharmacobiotics, or heat-killed probiotics) are bioactive substances produced by probiotic microorganisms, generally LAB. Moreover, many components present in probiotics that are released before death are also recognized as postbiotics. Rojas et al. (2020) mention postbiotics as bioactive

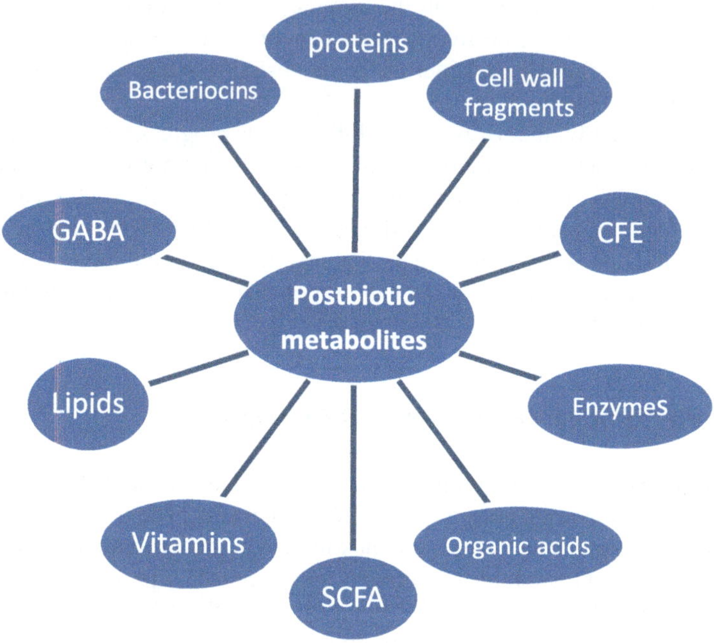

Fig. 18.1 Representative postbiotic metabolites

substances produced during the process of fermentation which maintain health and well-being.

As Rossoni et al. (2020) cited, postbiotics can also be defined as probiotic bacterial products with bioactive property in the host and may include metabolites, cell fractions, fatty acids, proteins, polysaccharides, cell lysates, peptides of peptidoglycan, and adhesion structures like pili.

The recognition of such entities may present an opening to keep away hazards associated with the administration of live microorganisms (Cicenai et al. 2014). Moreover, these things continue offering functional properties to the fermented food those have biogenic benefits resulting from the microbial production of bioactive metabolites during the process of fermentation. Postbiotics comprise extended shelf life safety and hold multiple health benefits (Majeed et al. 2020).

18.3 Sources of Postbiotics

From the discussion above, it is quite clear that postbiotics are products of probiotic microbes, thus the source of postbiotics is probiotics, generally LAB. From these LAB, various cellular components can be the source or origin for postbiotics, as shown in Table 18.1.

Table 18.1 Postbiotics and their location in bacterial cell

Location	Postbiotics	References
Cytoplasm	Enzymes, proteins, vitamins, lipids	Gutierrez et al. (2020)
Cell structure Pili Cytoplasmic membrane	Components of pellicle capsule or cell wall Adhesins Glycerophospholipid	Rojas et al. (2020) Rossoni et al. (2020) Singh et al. (2018)
Secreted products	Organic acids, EPS, SCFA, bacteriocins, polysaccharides, immune modulators, neurotransmitters, etc. Ethanol, diacetyl, acetaldehydes, H_2O_2	Gutierrez et al. (2020) Kerry et al. (2018)

18.4 Forms or Types of Postbiotics

There are varieties of postbiotic biomolecules consisting of secreted metabolic products of probiotic microorganisms like cell-free supernatants (CFS)—vitamins, short-chain fatty acid, organic acid, proteins, bacteriocins, amino acids and their complex or derivatives, etc. Furthermore, the paraprobiotics constituents, i.e., inactivated or dead and nonviable microbial cell preparations containing intact or ruptured cell components like teichoic acids, peptidoglycan, etc., can also be used to evaluate beneficial effects. Usually, paraprobiotics include a broad array of biomolecules like peptidoglycans, surface proteins, and cell wall polysaccharides, whereas postbiotics comprise secreted proteins, peptides, bacteriocins, organic acids, etc. (Teame et al. 2020). Since postbiotics are nonliving entities, their processing and maintenance are easy compared to probiotics. Compiled work out for some representative postbiotic metabolites from source microorganisms with their effects is depicted in Table 18.2.

18.5 Production Methods of Postbiotics

As postbiotics contain inactivated microorganisms or their structures or metabolic products (which are released or secreted during fermentation or poured after rupturing); in many postbiotic preparations used for study, fermented broth is filtered, or heated and the resultant material called cell-free supernatant (CFS) (Mantziari et al. 2020). Alternatively, probiotic cultures are inactivated by heat treatment, filtration (mostly 0.2μm), cell lysis by sonication, followed by centrifugation, and are often exposed to UV. These procedures result in break opening of bacterial cells pouring intracellular biomolecules that can be used as postbiotics. Chuah et al. (2019) mentioned a series of procedures: growth (24 h)—centrifugation (10,000 g, 10 min), CFS—pH adjustment—filtration-storage. Compare et al. (2017) obtained

Table 18.2 Various postbiotic biomolecules and their health benefits

Postbiotic type or form	Example/name	Source organism(s)	Application(s)	References
Amino acids/ derivatives	Selenocysteines and selenomethionines Tryptophan-indoles Gamma-aminobutyric acid (GABA)	*Lactobacillus* spp. Yeasts Mix microbiota *L. plantarum* *L. brevis* *L. paracasei* *L. bulgaricus* *L. bulgaricus* *L. zymae*	Immune modulation, anticancer, oxidative stress, anabolic pathways Intestinal permeability Neurotransmitter inhibitor CVD: hypotensive Epilepsy: gut-brain functionality Anxiety and depressive disorders, diabetes, cancer, asthma, etc.	Singh et al. (2018) Maguire and Maguire (2019) Gutierrez et al. (2020)
Bacteriocins	Ruterin Nisin A Plantaricin PJ4	*Lactobacillus reuteri* *L. lactis* *Lactococcus lactis* *L. helveticus*	Intestinal infections Antibacterial Immune modulator Active against enteric pathogens	Cicenai et al. (2014) Puccetti et al. (2020) Teame et al. (2020)
Cell-free supernatant (CFS)	CFS-PM CFS-PB CFS	*Lactobacillus plantarum* *Lactobacillus casei* DG *Lactobacillus reuteri* *L. acidophilus* *L. casei* *L. rhamnosus*	Anticancer (breast, colon, and cervical) Inflammatory mucosal response LPS-induced liver injury Ant-inflammatory Antioxidant, cancer	Chuah et al. (2019) Compare et al. (2017) Anderson (2019) Jakub et al. (2020)
Cell wall fragments	Lipoteichoic acid (LTA) Peptidoglycan LTA	*Lactobacillus paracasei* D3–5 *L. casei* *Bifidobacterium* sp. *Lactobacillus* sp.	Antiaging Antitumor effect skin mast cell response against bacterial and viral infections	Nataraj et al. (2020) Jakub et al. (2020)

Deoxycholic acid	Bile acids	*Bifidobacterium*, *Lactobacillus*, *Bacteroides*	Inhibition of pro-inflammatory genes, regulates bacterial growth	Puccetti et al. (2020)
Extracellular polymers	EPS	*Lactobacillus plantarum*, *L. gasseri* *L. plantarum*	Antitumor, colon carcinoma Anti-inflammatory Oxidative stress	Chuah et al. (2019) Kwon et al. (2020)
Enzymes	Glutathione peroxidase (GPx), superoxide dismutase (SOD), catalase	*L. fermentum*, *L. plantarum*	Reactive oxygen species (ROS), Crohn's disease, IBS, etc.	Jakub et al. (2020)
Indole/derivatives	5-Hydroxytryptophan, tryptamine, indoleacetic acid, 3-methylindole, indole-3-sulfate	*Lactobacillus*, *Bacteroides*	Antimicrobial peptides, IL-22, epithelial barrier	Puccetti et al. (2020)
Lipid compounds	Glycerophospholipid	*B. animalis* subsp. *lactis*	Down syndrome, antioxidant, Alzheimer disease, etc.	Singh et al. (2018)
Organic acids	Lactic acid, acetic acid	*Lactobacillus plantarum*	Antibiotic replacer in poultry and pig feeds	Chuah et al. (2019)
Polyphenol derivatives or transformed products	Urolithins, equol, and enterolignans	*Bifidobacterium* sp., *Lactobacillus* sp., *Clostridium* sp., *Bacteroides*	Polyphenol bioactivation/ transformations CVD, anticancer Prebiotic effect	Tomova et al. (2019)
Polyphosphates	Polyphosphates	*Lactobacillus brevis*	Colitis, intestinal permeability	Zagato et al. (2014)
Proteins	Protein HM0539 Msp1 and Msp2 (culture supernatant)	*Lactobacillus rhamnosus* LGG (ATCC 53103) *L. rhamnosus* GG	Intestinal barrier injury, colitis, liver injury Intestinal epithelial damage, apoptosis	Gao et al. (2019) Cicenai et al. (2014)
SCFA	Butyrate, acetate, propionate	Bacteroidetes, Firmicutes *Lactobacillus* sp. *Enterococcus* sp.	Mucosal immunity, inhibit pro-inflammatory cytokines, production of IL-18, etc. Diabetes, cancer, obesity	Puccetti et al. (2020) Gutierrez et al. (2020)

(continued)

Table 18.2 (continued)

Postbiotic type or form	Example/name	Source organism(s)	Application(s)	References
Vitamins	Niacin (vitamin B3) Vitamin K B group vitamins Biotin, nicotinic acid, cobalamin, riboflavin, thiamine, pyridoxine, and pantothenic acid	Gut microbiota *Lactococcus lactis*, *L. gasseri*, *L. reuteri* *B. adolescentis* *Propionibacterium freudenreichii*	Intestinal inflammation, colon cancer, colitis, etc. Biofortification Deficiencies	Singh et al. (2015, 2018)

postbiotic preparation for study by growing probiotic culture followed by centrifugation only. Haileselassie et al. (2016) have removed cells from culture broth by centrifugation and the supernatant was filter-sterilized and the preparation stored at pH maintained 7.2–7.5.

18.6 Applications/Benefactions of Postbiotics

One can correlate the benefaction of postbiotics to that of probiotics, since it becomes clear that postbiotics are metabolic products of probiotics. We have enough discussion to convey the limitations of probiotics and how postbiotics can eventually have all those advantages being nonliving. The benefaction of postbiotics has multifaceted coverage as depicted in Fig. 18.2.

Fig. 18.2 Postbiotic effects

18.6.1 Immune Modulation

One of the immune defense responses of the body is inflammation; however, excessive inflammation can harm vital tissues of the organ. Many probiotics are studied to have immunomodulatory effects. Postbiotics are able to fuel the immune system, also involving the intestine, and bowel anti-inflammatory, immunomodulators (Tomasik and Tomasik 2020). Kwon et al. (2020) have shown the immunomodulatory effect of postbiotic EPS isolated from *L. plantarum*. In their study, they found that EPS inhibited pro-inflammatory mediators- NF-8 and MAPK pathways by repressing TLR4 and MyD88 signaling. Diverse postbiotic fractions obtained from *Bacillus coagulans* can provoke anti-inflammatory cytokine production and thus support T helper-dependent immune mechanisms (Jakub et al. 2020). Hence, such properties of postbiotics can bring limitations to TH1-induced immune effects and enhance Th2-arbitrated reactions which are generally seen in persons prone to atopic disease.

18.6.2 Anticancer Properties

Cancer is a global health issue and a chief cause of death affecting many organs by rapidly creating and proliferating transformed cell. GABA (gamma-aminobutyric acid) is one of the postbiotics produced in gut microorganisms, shown to be associated with breast cancer by a prognostic value (Gutierrez et al. 2020). Higher the GABA better is the survival prognosis and GABA was found to restrain colon cancer cells. In the study with PM of *L. plantarum*, Chuah et al. (2019) have noted that apoptosis against cancer cells shows selective toxicity by suppressing proliferation thus indicating potential of anticancer therapeutic value of postbiotics. Induction of apoptosis against cancer cells was observed by SCFA propionate obtained from *P. freudenreichii* (Jakub et al. 2020). The suppression of oncogenes is controlled by these SCFAs.

18.6.3 Luminal and Mucosal Effects

The gut epithelium is considered as the foremost defense line against the huge number of microbes entering the body. A disturbance in this line barrier is called as leaky gut, and the resultant inflammatory reaction poses many clinical conditions of the gut as well as other parts of the body (Anderson 2019). With this concern, a report depicts protein secreted (postbiotic) by *L. rhamnosus* reduced LPS-provoked liver damage through progressing gut uprightness. Gao et al. (2019) observed a promising protective effect on intestinal barrier with a postbiotic protein HM 0539 revealed by stimulating intestinal mucin expression and combating against TNF-α/ LPS-induced gut wound that comprised of disturbed veracity and mucin downward regulation. Haileselassie et al. (2016) have shown a postbiotic CFS of *L. reuteri* influenced retinoic acid forced mucosal like DCs and its effect on regulatory cells with higher IL-10, CD 103, and CD1d expression while suppression of inflammatory genes.

18.6.4 IBS/IBD and Other Conditions

Irritable bowel syndrome (IBS) is one of the commonest GIT problems universally affecting the quality of life of patients. With this concern, postbiotics of *L. casei* have been evaluated and found a promising protective effect in the IBS organ culture model. In the study, TLR4 protein expression and IL-1α, IL-6, and IL-8 mRNA levels were found to be elevated; on the contrary, IL-10 mRNA levels were decreased in both the ileum and the colon (Compare et al. 2017). The postbiotic significantly decreased pro-inflammatory cytokines and provoked the protective effect.

Inflammatory bowel disease (IBD) is a multifaceted chronic inflammatory condition of the GIT, and the intestinal microbiota seems to be the chief etiological factor for its development. Also, studies revealed the role of postbiotics in the reconstruction of impaired interactions of gut microbiota and immune cells. In a mouse model, it was seen that SCFAs and tryptophan postbiotics activated immunomodulatory mechanisms by ordering immune cell generation, trafficking, and functioning (Russo et al. 2019).

18.6.5 Neural Diseases

Experimental reports indicated that metabolites of intestinal microbiota govern the integrity and pathophysiology of the nervous system and thus are associated with neuroimmune clinical conditions. Also, negotiation of integrity of the intestinal tract lining adversely affects the nervous system (Maguire and Maguire 2019). Gutierrez et al. (2020) stated that hypoxic-ischemic actions during fetal development can trigger memory-related shortfall because of neurotransmission disturbance by damage in GABA (postbiotic from *Lactobacillus* spp.) function.

18.6.6 Diabetes

Diabetes has become a great threat to human beings since its occurrence is significantly rising all over. It is due to dysfunctional pancreatic cells that do not produce insulin and result in abnormal glucose levels in the blood. With this connection, it was noted that the therapeutic action of GABA and the found progression of prediabetes and inflammatory response can be inhibited (Gutierrez et al. 2020). This is due to regulatory action of GABA molecule on cells of islets, stressing the depression of insulitis and inflammatory cytokine production. Cavallari et al. (2017) noted that bacterial cell wall-derived muramyl dipeptide (which needs NOD2) is an insulin-sensitizing postbiotic and further showed in obese mice that bacterial cell wall muropeptide can function as postbiotic by improving insulin resistance and metabolic tissue inflammation.

18.6.7 Antimicrobial Potentials

The indiscriminate use of antibiotics for general infections is resulting in antibiotic resistance among pathogens. The surfacing of antimicrobial resistance is alarming and leading to therapeutic failure; hence, there is a need for a new approach to deal with. Many postbiotics from probiotics like bacteriocins, enzymes, organic acids, and small molecules are reported to have antimicrobial activities, for instance, nisin A from *Lactococcus lactis* (Puccetti et al. 2020).

Mantziari et al. (2020) noted important observation: (1) EPS from *Lactobacillus* and *Bifidobacterium* showed protective action against enterotoxigenic *E. coli*, (2) CSF of *B. bifidum* in supplemented medium induced the expression of virulence genes in *E. coli*, (3) postbiotics from *B. bifidum* have shown inhibitory effect against *E. coli*, and (4) CSF from breast milk commensals showed activity to limit HIV infection in vitro. Rossoni et al. (2020) have studied the antifungal activity of postbiotic (crude extract and fraction) of *L. paracasei* and found to inhibit the growth of *Candida auris*.

18.6.8 Miscellaneous Applications

Wegh et al. (2019) reported important findings: (1) a clinical study on postbiotic preparation of *B. breve* revealed that it was well tolerated in infants and decreased the incidence of diarrhea and abdominal distention, (2) postbiotic effect of *L. paracasei* against placebo in adults with atopic dermatitis found better skin severity scores after 12 weeks but not in placebo group, and (3) in old age group study, postbiotics of *L. pentosus* reduce the frequency of common cold compared to placebo on the incidence of common cold. Vandenplas et al. (2020) have also studied in infants (less than 14 days) and found infant formula safe with postbiotics linked prebiotic (ss GOS) from Lactofidus fermentation process.

In a mouse model experiment, infant formula of postbiotics obtained from a specific fermentation process coupled prebiotic study revealed increased functional and morphological intestinal maturation much similar as the mother fed conditions than infant formula lacking postbiotic-prebiotic combination (Salminen et al. 2020).

In addition to the applications and their underlying mechanisms, postbiotics can be exploited vastly with some other areas such as nutrition and health. Postbiotics have potential properties like anti-obesity, antihypertensive, hypocholesterolemic, and many more (Tomasik and Tomasik 2020). Also, in nutrition point of view, food biofortification with postbiotics can be achieved especially for B group vitamins, and this could be an elegant strategy to combat the universal problem of its deficiency-related clinical conditions. Obesity is the chief health concern worldwide that can be dealt with postbiotics through dietary supplements to combat and improve host metabolism (Reynés et al. 2019).

Acne vulgaris is a universal skin disorder generally at peak at puberty and may be now treated with postbiotic formulation LactoSporin as seen in an open-label randomized study (Majeed et al. 2020). It is found that LactoSporin is equally efficient in treating acne lesions in comparison of benzoyl peroxide. The study also revealed the mechanism of amelioration of the condition. The pathology may be associated with the excessive secretion of sebum as the study indicates and LactoSporin could inhibit the 5-alpha reductase enzyme and thus reducing the secretion of sebum.

18.7 Future Prospects as Concluding Remarks

Though the term postbiotics is new compared to the popular terms probiotics and prebiotics, the promising potentials of postbiotics as therapeutics and obvious advantages over probiotics alarm near future pharmaceutical markets to be called for postbiotics. We can find preparations and formulations available easily for probiotics and prebiotics but not for postbiotics. There may be pharmacological, biochemical, and medicolegal aspects to be sorted out for making postbiotics available and useful as therapeutics, but we can certainly expect it in practice to treat a variety of clinical conditions in future.

References

Anderson RC (2019) Are postbiotics the long sought-after solution for a leaky gut? J Nutr 149:1874–1874

Cavallari J, Fullerton MD, Duggan BM, Foley KP, Denou E, Smith BK, Desjardins EM, Henriksbo BD, Kim KJ, Tuinema BR, Stearns JC, Prescott D, Rosensteil P, Coombes BK, Steinberg GR, Schertzer JD (2017) Muramyl dipeptide-based postbiotics mitigate obesity-induced insulin resistance via IRF4. Cell Metab 25:1063–1074

Chuah LO, Foo HL, Loh TC, Alitheen NBM, Yeap SK, Mutalib NEA, Rahim RA, Yusoff K (2019) Postbiotic metabolites produced by Lactobacillus plantarum strains exert selective cytotoxicity effects on cancer cell. BMC Complement Altern Med 19(114):1–12

Cicenai A, Scirocco A, Carabotti M, Pallotta L, Marignani M, Severi C (2014) Postbiotic activities of Lactobacillus-derived factors. Clin Gastroenterol 48:S18–S22

Compare D, Rocco, Coccoli P, Angrisani D, Sgamato C, Lovine B (2017) Lactobacillus casei DG and its postbiotic reduce the inflammatory mucosal response: an ex-vivo organ culture model of post-infectious irritable bowel syndrome. BMC Gastroenterol 7:53. https://doi.org/10.1186/s12876-017-0605-x

Gao J, Li Y, Wan Y, Hu T, Liu L, Yang S, Gong Z, Zeng Q, Wei Y, Yang W, Zeng Z, He X, Huang S-H, Cao H (2019) A novel postbiotic from Lactobacillus rhamnosus GG with a beneficial effect on intestinal barrier function. Front Microbiol 10(477):1–14. https://doi.org/10.3389/fmicb.2019.00477

Gutierrez LD, Vicente LS, Brown LJ, Villaran MC, Chavarria M (2020) Gamma-aminobutyric acid and probiotics: Multiple health benefits and their future in the global functional food and nutraceuticals market. J Funct Foods 64:1–14

Haileselassie Y, Navis M, Vu N, Qazi KR, Rethi B, Sverremark-Ekström E (2016) Postbiotic modulation of retinoic acid imprinted mucosal-like dendritic cells by probiotic Lactobacillus reuteri 17938 in vitro. Front Immunol 7(96):1–11. https://doi.org/10.3389/fimmu.2016.00096

Jakub Z, Marzec A, Ruszczy M, Feleszko W (2020) Postbiotics—a step beyond pre- and probiotics. Nutrients 12(2189):1–17. https://doi.org/10.3390/nu12082189

Kerry R, Patra JK, Gauda S, Park Y, Shin HS, Das G (2018) Benefaction of probiotics for human health: a review. J Food Drug Anal 26:947–939

Kwon M, Lee J, Park S, Kwon O, Seo J, Roh S (2020) Exopolysaccharide Isolated from Lactobacillus plantarum L-14 has anti-inflammatory effects via the toll-like receptor 4 pathway in lps-induced RAW264.7 cells. Int J Mol Sci 21(9283):1–18

Linares DM, Gomez C, Renes E, Fresno J, Tornodijo R, Ross RP, Stanton C (2017) Lactic acid bacteria and Bifidobacteria with potential to design natural biofunctional health-promoting dairy foods. Front Microbiol 8:1–11

Maguire M, Maguire G (2019) Gut dysbiosis, leaky gut, and intestinal epithelial proliferation in neurological disorders: towards the development of a new therapeutic using amino acids, prebiotics, probiotics, and postbiotics. Rev Neurosci 30(2):179–201

Majeed M, Majeed S, Nagabhushanam K, Mundkur L, Rajalakshami H, Shah K, Beede K (2020) Novel topical application of a postbiotic, LactoSporin®, in mild to moderate acne: a randomized, comparative clinical study to evaluate its efficacy, tolerability and safety. Cosmetics 7(70):2–15. https://doi.org/10.3390/cosmetics7030070

Mantziari A, Salminen S, Szajewska H, Nevardo J, Rojas M (2020) Postbiotics against pathogens commonly involved in pediatric infectious diseases. Microorganisms 8(1510):1–22. https://doi.org/10.3390/microorganisms8101510

Markowiak P, Slizewska K (2017) Effects of probiotics, prebiotics, and synbiotics on human health. Nutrients 9(1021):1–30. https://doi.org/10.3390/nu9091021

Nataraj BH, Ali AM, Behare PV, Yadav H (2020) Postbiotics-parabiotics: the new horizons in microbial biotherapy and functional foods. Microb Cell Factories 19(168):1–22

Puccetti M, Xiroudaki S, Ricci M, Giovagnoli S (2020) Postbiotic-enabled targeting of the host-microbiota-pathogen interface: hints of antibiotic decline? Pharmaceutics 12(624):1–31. https://doi.org/10.3390/pharmaceutics12070624

Reynés B, Palou M, Rodríguez AM, Palou A (2019) Regulation of adaptive thermogenesis and browning by prebiotics and postbiotics. Front Physiol 9(1908):1–15. https://doi.org/10.3389/fphys.2018.01908

Rojas JN, Mantziari A, Salminen S, Szajewska H (2020) Postbiotics for preventing and treating common infectious diseases in children: a systematic review. Nutrients 12(389):2–14. https://doi.org/10.3390/nu12020389

Rossoni RD, de Barros PP, Mendonca LC, Menida RP, Silva DH, Fuch BB, Juaueira JC, Mylonakis F (2020) Postbiotic activity of *Lactobacillus paracasei* 28.4 against *Candida auris*. Front Cell Infect Microbiol 10:397. https://doi.org/10.3389/fcimb.2020.00397

Russo E, Giudici F, Fiorindi C, Ficari F, Scaringi S, Amedei A (2019) Immunomodulating activity and therapeutic effects of short chain fatty acids and tryptophan post-biotics in inflammatory bowel disease. Front Immunol 10(2754):1–10. https://doi.org/10.3389/fimmu.2019.02754

Salminen S, Stahl B, Vinderola G, Szajewska H (2020) Infant formula supplemented with biotics: current knowledge and future perspectives. Nutrients 12(1952):1–20. https://doi.org/10.3390/nu12071952

Singh N, Ashish G, Sivaprakasam S, Brady E, Padia R, Shi H, Thangaraju M, Prasad PD, Manicassamy S, Munn D, Lee JR, Offermanns S, Ganapathy V (2015) Activation of the receptor (Gpr109a) for niacin and the commensal metabolite butyrate suppresses colonic inflammation and carcinogenesis. Immunity 40(1):128–139. https://doi.org/10.1016/j.immuni.2013.12.007

Singh A, Vishwakarma V, Singhal B (2018) Metabiotics: the functional metabolic signatures of probiotics: current state-of-art and future research priorities—metabiotics: probiotics effector molecules. Adv Biosci Biotechnol 9:147–189

Teame T, Wang A, Xie M, Zhang Z, Yang Y, Ding Q, Gao C, Olsen RE, Ran C, Zhou Z (2020) Paraprobiotics and postbiotics of probiotic lactobacilli, their positive effects on the host and action mechanisms: a review. Front Nutr 7:570344

Tomasik P, Tomasik P (2020) Probiotics, non-dairy prebiotics and postbiotics in nutrition. Appl Sci 10(1470):1–14

Tomova A, Bukovsky I, Rembert E, Yonas W, Alwarith J, Barnard ND, Kahleova H (2019) The effects of vegetarian and vegan diets on gut microbiota. Front Nutr 6(47):1–10. https://doi.org/10.3389/fnut.2019.00047

Vandenplas Y, Halleux V, Arciszewska M, Lach P, Pokhylko V, Klymenko V, Schoen S, Abrahamse-Berkeveld M, Mulde K, Rubio R (2020) A partly fermented infant formula with postbiotics including 30-GL, specific oligosaccharides, 20-FL, and milk fat supports adequate growth, is safe and well-tolerated in healthy term infants: a double-blind, randomised, controlled, multi-country trial. Nutrients 12(3560):1–17. https://doi.org/10.3390/nu12113560

Wegh C, Geerling SY, Knol J, Roeselers A, Belzer C (2019) Postbiotics and their potential applications in early life nutrition and beyond. Int J Mol Sci 20(4673):1–23. https://doi.org/10.3390/ijms20194673

Xu Y, Cui Y, Yue F, Liu L, Shan Y, Liu B, Zhou Y, Lu X (2019) Exopolysaccharides produced by lactic acid bacteria and *Bifidobacteria*: structures, physiochemical functions and applications in the food industry. Food Hydrocoll 94:475–499

Zagato E, Mileti E, Massimiliano L, Fasano F, Budelli A, Penna G, Rescigno M (2014) Lactobacillus paracasei CBA L74 metabolic products and fermented milk for infant formula have anti-inflammatory activity on dendritic cells in vitro and protective effects against colitis and an enteric pathogen in vivo. PLoS One 9(2):1–14

Effect of Probiotics on Gut Microbiota and Brain Interactions in the Context of Neurodegenerative and Neurodevelopmental Disorders

19

Zakira Naureen, Saima Farooq, Tayyaba Zahoor, and Syed Abdullah Gilani

Abstract

The bidirectional communication between the gut and the brain links emotional and cognitive centers of the brain with peripheral intestinal functions. This interaction between the gut microbiota and the gut-brain axis (GBA) involves signaling from the gut microbiota to the brain and from the brain to the gut microbiota through neural, endocrine, immune, and humoral links as evidenced by germ-free animal models and association of dysbiosis with central nervous system (CNS) disorders (i.e., autism, anxiety-depressive behaviors) and functional gastrointestinal disorders. Probiotics have been reported to influence this interaction by facilitating the colonization of beneficial microorganisms and suppressing the growth of harmful microorganisms, thus improving the gut-brain interactions. Psychobiotics being a novel class of probiotics hold special significance as these affect the central nervous system-related functions and behaviors mediated by the gut-brain axis (GBA) via immune, humoral, neural, and metabolic pathways to improve not only the gastrointestinal (GI) function but also the antidepressant and anxiolytic capacity. In the past few years, some of the psychobiotic strains have been proven scientifically beneficial in suppressing inflammation and reducing cortisol levels, thus improving anxiety and depression. In addition to that, psychobiotics have shown promising results in neurodegenerative and neurodevelopmental disorders, such as Alzheimer's disease (AD), Parkinson's disease (PD), and autism spectrum disorder (ASD). Initial clinical studies have shown that psychobiotics can improve overall GI function, improve symptoms of ASD, and regulate motor

Z. Naureen (✉) · S. Farooq · T. Zahoor · S. A. Gilani (✉)
Department of Biological Sciences and Chemistry, College of Arts and Science, University of Nizwa, Nizwa, Oman
e-mail: zakira@unizwa.edu.om; gilani@unizwa.edu.om

R. Z. Sayyed, M. Khan (eds.), *Microbiome-Gut-Brain Axis*,
https://doi.org/10.1007/978-981-16-1626-6_19

functions of PD patients and cognition behavior in AD patients. This chapter primarily focuses on the effect of psychobiotics on interactions between the gut microbiota and the brain in the context of neurodegenerative and neurodevelopmental disorders.

Keywords

Anxiety · Central nervous system · Depression · Probiotics · Psychobiotics

19.1 Introduction

Human body is home to trillions of bacteria, fungi, parasites, and viruses (collectively named microbiota) that mostly reside on our skin and gut mucosa and do not cause any adverse health issues. Indeed, some of these bacteria are extremely useful to maintain a healthy life. In fact, the human gut contains one of the most complex ecosystems composed of approximately 10^{13}–10^{14} microorganisms belonging to 500–1000 different species. The number of these microbes is one to ten times greater than the number of eukaryotic cells in the body (Qin et al. 2010). The mutual symbiosis between the host and the gut microbiome can be partially attributed to the nutrients present in gut that favor microbiota selection and colonization (Leung and Thuret 2015). The selection and colonization of gut microbiota that begins at birth and establishes within the first 3 years of life is crucial for regulating the development of intestinal physiology, maturation of the nervous and immune system (Palmer et al. 2007), and modulation of the angiogenesis (Andriessen et al. 2016). In addition to that, these microorganisms provide natural biocontrol against the pathogenic microorganisms by their antimicrobial activities which play an important role in maintaining the stability of the gut ecosystem (Bercik et al. 2012). Studies have shown that changes in the microbial colonization of the human gut during early life increase the risk of disease and have a significant impact on the host neurophysiology, behavior, and function of the nervous system (Kamada et al. 2013; Collins and Bercik 2009; Moustafa et al. 2018). Furthermore, these gut microorganisms possess immunomodulatory properties that mediate brain functions and behavior and contribute to etiopathogenesis in various neurodegenerative and behavioral disorders such as anxiety, depression, autism spectrum disorders (ASD), Alzheimer's disease (AD), and Parkinson's disease (PD) (Collins et al. 2012; Fung et al. 2017).

As mentioned above, the gut microbiota starts to colonize the gut during development and continues later on in life. At first, it is mostly composed of *Lactobacillus* sp. and *Bifidobacterium*; however, at later stages in life the microbiome becomes densely populated by the Bacteroidetes and Firmicutes phyla, and in small proportion of Proteobacteria, Verrucomicrobia, Actinobacteria, and Cyanobacteria phyla. In addition, the *Fusobacterium* genus can also be found in gut microbiota (Xu et al. 2019). There are several factors that affect the composition of gut microbiota such as genetic diversity, diet, environment, season, and overall health status, and it is extremely difficult to define a "normal" microbiome for the average human population (Gibson and Roberfroid 1995; Wen and Duffy 2017). In addition to that, the microbial communities in the gut are shaped by the bacteriophages that inhabit this niche (Naureen et al. 2020) and are independent of environmental components such as age, body mass index, gender, and geographic location (Huttenhower et al. 2012; Bajinka et al. 2020).

In addition to these bacteria that colonize our gut by the passage of time, millions of bacteria transit through our gut every day. These bacteria are mostly present in our food and upon entering the gut interact with the gut microbiota in either a healthy or a pathological context. These live microorganisms that provide health benefit to us are termed as probiotics (Hill et al. 2014). Probiotics are the biotic organisms that can alter the gut microbiota composition while having a beneficial effect on the host's health and well-being. These probiotics naturally occur in food items such as olives and dark chocolates and in fermented foods such as sauerkraut, yogurt, cheeses, etc. These microorganisms not only improve the gut microbiota but also help in eliminating the harmful pathogenic microorganisms and improve the overall health (Hemarajata and Versalovic 2013).

19.2 History of Probiotics

The word "probiotic" is derived from the Latin words "pro," meaning to promote, and "biotic," meaning life. In 1907, Élie Metchnikoff observed that the regular consumption of lactic acid bacteria (LAB) in fermented dairy products, such as yogurt, led to enhanced health and relatively longer life in people living in Bulgarian villages (Metchnikoff and Mitchell 1907). However, the evidence on the beneficial effects of microbes was provided by Tissier (1899) when he compared the stool of children suffering from diarrhea with that of the healthy children. Tissier (1899) observed that the concentration of bacteria responsible for diarrhea was low in the stool of healthy children, and this gave him the idea of giving the infected children *Bifidobacteria*, which he had successfully isolated in 1988 from the feces of infants fed on breast milk, in order to rehabilitate the normal gut flora. The distinguishable work done by Henry Tissier led to the modern definition of probiotics by Havenaar and Huis In't Veld (1992) who describe probiotics as "an applicable bacterial culture that grant positive effects to the humans or animals by enhancing the native floral properties, when administered." However, to term a certain bacteria as probiotic, lots of scientific evidence proving its safety for consumption is required. This is extremely important as these live microorganisms have a direct impact on human health. It is worth mentioning here that when probiotics gained popularity, many food and drug companies started to designate bacterial species as probiotics without providing any scientific evidence, thus ending up in banning the word probiotics in the European Union by the European Food Safety Authority (EFSA) (Katan 2012).

However, now there is a huge repertoire of scientific evidence that shows the potential benefits of living microorganisms that are safe for human consumption, e.g., *Lactobacillus* and *Bifidobacterium*, and are beneficial in certain medical conditions, such as irritable bowel syndrome (IBS), dermatitis, high cholesterol levels, eczema, and liver disease. In the past two decades, probiotics have gained much attention with respect to brain health and cognitive function and its effects on the central nervous system (CNS) and mood (Cryan and Dinan 2012). Additionally, probiotics have an important role in improving memory abilities (spatial and non-spatial memory rodents) of rodents and the human beings. Also, probiotics have a dynamic effect on relieving stress, anxiety, and depression. However, to understand this further, we need to first understand the gut-brain microbial axis (Crumeyrolle-Arias et al. 2014).

19.3 Gut-Brain Microbial Axis: Connection of the Gut and the Brain

Probiotics are widely being studied for their role in improving brain health and mental behavior. However to understand that how the microbes living inside the gut influence the brain activity we need to cosider the connection between the gut and the brain. The brain, communicates with the gut through a complex system comprising of the enteric nervous system (division of the peripheral nervous system controlling the gastrointestinal behavior independent of the CNS), the vagus (a large nerve of the central nervous system responsible for sending signals between the brain and the intestine), and the hypothalamus-pituitary axis. All of these components make immunological, neurological, and endocrine bridges allowing the information relay between the brain and the gut (Chandran et al. 2019). The microbes in the gut produce molecules that include neurotransmitters, short-chain fatty acids, and amino acids which travel through this complex system and a communication between the brain and the gut takes place (Fig. 19.1). Moreover, the presence of the gut-brain microbial axis (GBMA) links the emotional and cognitive centers of the brain with peripheral functions of the intestine and also provides communication between the enteric nervous system and the central nervous system. The gut bacteria will influence the brain and the central nervous system by regulating the inflammation and hormone production (Sommer and Bäckhed 2013; Bermúdez-Humarán et al. 2019).

19.4 Role of Microbiota in the Gut-Brain Microbial Axis

There are plenty of clinical and experimental evidences suggesting the importance of gut microbiome interactions with intestinal cells and the enteric nervous system (ENS) and direct involvement with the CNS through neuroendocrine and metabolic pathways (Fig. 19.2). Perhaps the most compelling evidence of GBM interactions comes from the dramatic improvement of patients suffering from hepatic encephalopathy (Bercik et al. 2012). This indicated the involvement of gut microbiota in

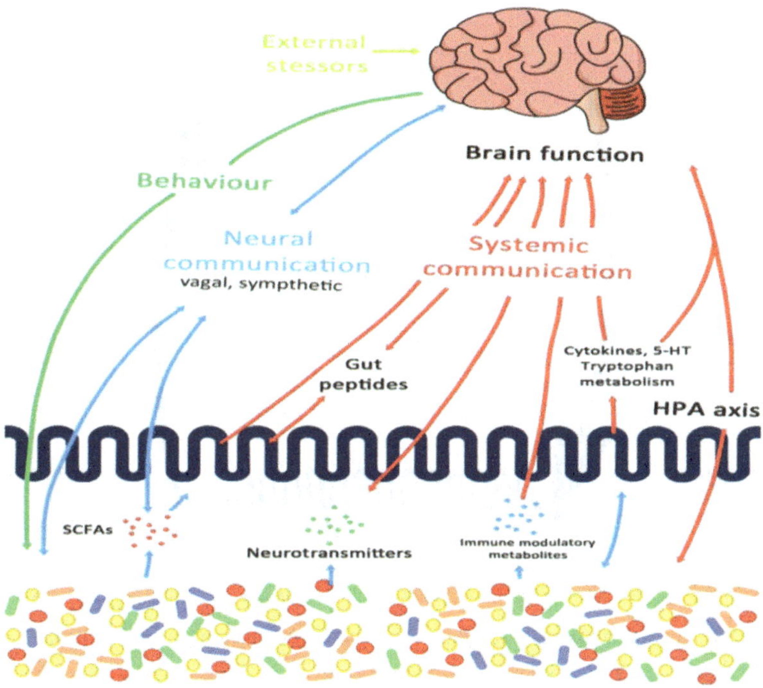

Fig. 19.1 Gut-brain-microbial axis revealing the regulatory interactions established between the gut microbiota, the intestine, and the brain. It includes the communication through vagus and sympathetic nerves; SCFAs, short-chain fatty acids; activation of hypothalamic-pituitary-adrenal (HPA) axis and release of gut peptides (Rogers et al. 2016)

maintaining the health and well-being and highlighted that the gut dysbiosis results in disease. Recent studies have shown that these microorganisms play a role in anxiety and depression (Collins and Bercik 2009; Kamada et al. 2013) and an imbalance in the gut microbiota results in several diseases, specifically autism (Kamada et al. 2013; Collins and Bercik 2009; Moustafa et al. 2018; Fung et al. 2017).

19.5 Gut Dysbiosis and Human Health

Gut dysbiosis refers to a condition in which the physiology of gut microbiome is altered owing to changes in diet, stress, or administration of antibiotics (Clemente et al. 2012). As a result, the intestinal permeability increases and results in leakage of bacteria, bacterial metabolites, and molecules through the mucosa into the systemic circulation, a condition termed as leaky gut syndrome. This in turn has a detrimental impact on the host immune system as demonstrated in diseases such as diabetes, asthma, inflammatory bowel disease (IBD), and psychiatric disorders including depression, anxiety, and autism (Sarkar et al. 2016; Shaaban et al. 2018a, b).

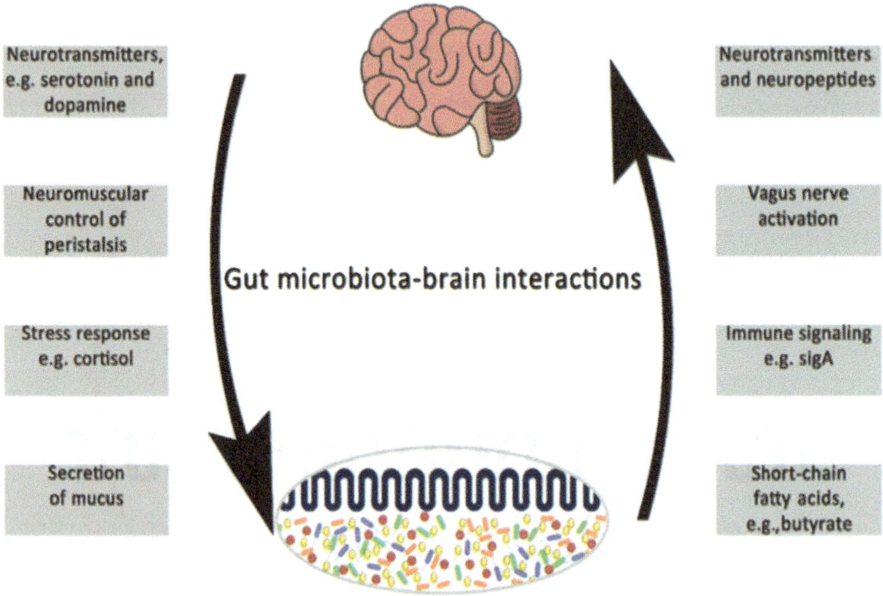

Fig. 19.2 The two-dimensional interactions between the gut microbiota and the brain

Although a huge repertoire of such studies has been focused on bacteria residing in the gut, some studies have emphasized the significance of other microbes such as yeast. For instance, it has been observed that the gut colonization with *Candida* species results in a decrease in carbohydrate and mineral absorption and excessive buildup of toxins that might contribute to development of autism spectrum disorder (Shaaban et al. 2018a, b). Dysbiosis also occurs in functional gastrointestinal disorders (FGID) that are linked to a disruption of the gut-brain axis and leading to mood disorders (Enck and Mazurak 2018). Similarly, in IBS, both brain-gut and gut-brain dysfunctions occur that eventually result in changes in intestinal motility and secretion, causing visceral hypersensitivity and cellular alterations of the enteroendocrine and immune system (Padhy et al. 2015). Recent studies report that probiotics are helpful in restoring microbiota to a healthy state and also in reducing various disease symptoms (Carabotti et al. 2015). The gut and the brain are interrelated in which the gut bacteria produce metabolites and have a major effect on the brain. Probiotics have a beneficial role in the brain and mental health and are called as psychobiotics (Dinan et al. 2013).

19.6 Psychobiotics and Their Role in Mental Health

Psychobiotics are novel types of probiotics used for the treatment of psychiatric distress (Dinan et al. 2013). Psychobiotic researches are conducted on the model organisms for behavioral assessments such as anxiety, depression, motivation, etc.

(Sarkar et al. 2016). This kind of probiotics can regulate neurotransmitters and proteins (GABA, serotonin, glutamate, and BDNF) and are helpful to regulate the neural excitatory-inhibitory balance, mood, cognitive functions, and learning and memory processes (Lu et al. 2008; Heldt et al. 2007; Martinowich and Lu 2008). The gut microbiota has a major role in stimulating the activity of the hypothalamic-pituitary-adrenal (HPA) axis (Sudo et al. 2004). Studies with the germ-free (GF) mice have shown that they release excess of corticosterone and adrenocorticotropic hormone as compared to specific pathogen-free (SPF) mice upon administration with psychobiotics (Sudo et al. 2004). The activation of the HPA axis by pro-inflammatory cytokines increases the blood barrier permeability which leads to reduction in serotonin level resulting in psychiatric distresses such as depression (Hammitt et al. 2019; Dowlati et al. 2010). *Lactobacillus brevis*, *Lactobacillus plantarum*, and *Bifidobacterium dentium* are able to secrete GABA and serotonin inside the gut (O'Mahony et al. 2015; Schousboe and Waagepetersen 2007; Barrett et al. 2012). The acetylcholine is secreted in the gut by the *Lactobacillus* spp. such as *L. plantarum* and *Lactobacillus odontolyticus* (Roshchina 2016). A recent study reports that microbes can regulate the serotonin synthesis in the gut. In additon the spore-forming bacteria present in gut microbiota can produce serotonin in enterochromaffin cells (Yano et al. 2015). The studies indicate the psychobiotic potential of pseudobiotics in improving the psychiatric conditions by secretion of neurotransmitters that can in turn alleviate stress.

Psychobiotics have been reported to exert psychotropic effects on various mental and psychological disorders, depression, anxiety, and stress; however, their application to improve mental health requires precision strategy. Animal studies have shown that many probiotics in fact act as psychobiotics. For instance, the administration of *Lactobacillus plantarum* PS128 supplements reduces anxiety and stress in mice as well as inflammation and the levels of corticosterone. As compared to the control mice, PS128 supplements can enhance the dopamine and serotonin levels in the prefrontal cortex and striatum (Liu et al. 2016, 2015). Similarly, the administration of *Lactobacillus helveticus* NS8 supplements is beneficial for the reduction of anxiety, cognitive dysfunction, and depression. These *L. helveticus* NS8 supplements can enhance the levels of serotonin, norepinephrine (NE), and brain-derived neurotrophic factor (BDNF) in the hippocampus (Liang et al. 2015). Furthermore, *Bifidobacterium longum* 1714, *B. longum* NCC3001, *Bacterium infantis* 35624, and *Lactobacillus rhamnosus* (JB-1) administration can reduce stress, depression, and anxiety (Savignac et al. 2014; Bravo et al. 2011; Bercik et al. 2010; Desbonnet et al. 2010). The intake of the JB-1 supplements can cause region-dependent modifications in the expression of GABA receptors resulting in the reduction of plasma corticosterone level (Bravo et al. 2011). The administration of *B. longum* NCC3001 upregulates the BDNF expression in the hippocampus (Bercik et al. 2010). The 4-week-long treatment of *B. longum* 1714 was effective in improving memory and reducing stress (Allen et al. 2016). Recipients who were administered with both probiotic yogurt (*Lactobacillus acidophilus* LA5 and *Bifidobacterium lactis* BB12) and probiotic capsules (*Lactobacillus casei*, *L. acidophilus*, *Lactobacillus rhamnosus*, *Lactobacillus bulgaricus*, *Bifidobacterium breve*, *Bifidobacterium longum*, and *Streptococcus thermophilus*) have shown an improvement in mental health (Mohammadi et al. 2016). The combination of *L. helveticus* R0052 and *B. longum* R0175 can reduce stress and depression. Along with that, these combinations can decrease the levels of urinary free cortisol (Messaoudi et al. 2011). The effects of probiotic supplements such as *L. plantarum* PS128, *L. plantarum* 299v, *L. rhamnosus* GG, Bifihappy, Vivomixx®, Probio'Stick, etc. on depression and anxiety is under investigation (Cheng et al. 2019; Rucklidge 2013).

The brain-gut interaction involves immunoregulatory, neuroendocrine, and vagus pathways (Li et al. 2018). These interactions are mediated by secretion of many metabolites by the microorganism in the gut which in turn depends upon the diversity of microbes residing in the gut. Probiotics can help improve these interactions by maintaining a healthy microbiota which ultimately results in overall health improvement. For instance, the levels of inflammatory cytokines can be reduced by the treatment with the probiotic strains of *Lactobacillus*, *Bifidobacterium*, and *Enterococcus* (Vanuytsel et al. 2014). The probiotic anti-immunoregulatory effects can trigger T regulatory cells which lead to the secretion of IL-10 (Dinan et al. 2013). Moreover, the interaction of probiotics with gut epithelium enteroendocrine cells (EECs) results in secretion of neuropeptides and neurotransmitters such as peptide YY (PYY), neuropeptide Y (NPY), substance P, serotonin, glucagon-like peptide-1 and peptide-2 (GLP-1 and GLP-2), and cholecystokinin (Cani and Knauf 2016; Foster et al. 2017). About 95% of the serotonin is secreted from the gut enterochromaffin cells and ENS neurons along with the control of GI secretion and motility (Costedio et al. 2007). The effective brain serotonin pathways regulates the cognition and mood while the ineffective brain serotonin pathways leads to disorders in GI and mood (Wrase et al. 2006).

19.7 Psychobiotics in Neurodegenerative and Neurodevelopmental Disorders

19.7.1 Alzheimer's Disease (AD)

Although there is scarce evidence regarding the effectivity of probiotics in neurodegenerative disorders like AD and it has been reported that patients having severe AD are insensitive to probiotics; yet, one study using multiple strains of *L. casei* W56, *Lactococcus lactis* W19, *L. acidophilus* W22, *B. lactis* W52, *L. paracasei* W20, *L. plantarum* W62, *B. lactis* W51, *B. bifidum* W23, and *L. salivarius* W24, on subjects with AD reported that the composition of gut microbiota and tryptophan metabolism were affected by the administration of probiotics (Kumar and Singh 2015; Agahi et al. 2018; Leblhuber et al. 2018). Another study reported significant probiotic-mediated reduction in oxidative stress by induction of SIRT-1-dependent mechanisms in transgenic AD mouse models (Bonfili et al. 2018). Administration of probiotics comprising of *L. acidophilus*, *Lactobacillus fermentum*, *B. lactis*, and *B. longum* significantly decreased the coliform and increased *Bifidobacterium* spp.

and *Lactobacillus* spp. in the stool of AD animal models suggesting the efficacy of probiotics in maintaining healthy gut microbiota. Additionally, probiotics have been reported to improve learning and memory deficits in AD rats as compared to control rats probably because of the reduction in the number of amyloid plaques, inflammation, and oxidative stress (Athari Nik Azm et al. 2018). Furthermore, supplementing AD mice with cow's milk fermented with *L. fermentum* or *L. casei* enhanced learning, memory behavior, and antioxidant levels while reducing pro-inflammatory cytokines, malondialdehyde (MDA), and AChE as compared to the control (Musa et al. 2017). In addition to that, certain probiotic strains such as *L. plantarum* MTCC1325 prove beneficial in improving the cognitive and gross behavioral activities and restoration of acetylcholine (ACh) levels in D-galactose-induced AD rats (Nimgampalle and Kuna 2017). In yet another randomized, double-blind, and controlled clinical trial, consumption of probiotic-treated milk (*L. acidophilus*, *L. casei*, *B. bifidum*, and *L. fermentum*) led to decreased plasma MDA and serum high-sensitivity C-reactive protein (hs-CRP) levels while changing the insulin resistance, beta-cell function, and insulin sensitivity. Remarkably, the mini-mental state examination (MMSE) score in AD group was significantly improved after probiotic treatment (Akbari et al. 2016).

Based on the abovementioned findings, it can be stated that probiotics, specifically psychobiotics, can help improve the cognitive behavior, memory deficit, and overall mental health in AD animal models while reducing inflammation, possibly through SIRT-1 pathways, and thus hold promise in the treatment of AD in humans; however, this needs further confirmation by carefully designed, double-blind clinical trials considering other factors such as age and severity of AD to better elucidate the role of psychobiotics.

19.7.2 Parkinson's Disease (PD)

Another important neurodegenerative and neuropsychiatric disorder that affects nearly 2% of the elderly population is PD (De Rijk et al. 1997). Besides other problems, one of the major symptoms in these PD patients is constipation (Barichella et al. 2009; Fasano et al. 2015; Berg et al. 2015). Hence, most of the clinical studies related to probiotic administration in PD patients focus on gastrointestinal function (Barichella et al. 2016; Georgescu et al. 2016; Cassani et al. 2011). For instance, three studies have reported that probiotics containing *L. acidophilus* and *B. infantis* improved gastrointestinal function, regulate bowel movement, improve stool consistency, reduce abdominal pain and recipients with PD who were using probiotics exhibited improved gastrointestinal functions. Furthermore, PD patients exhibited improved bowel habits after 5 weeks of administration of milk fermented with *L. casei* Shirota (Cassani et al. 2011).

Results obtained from a randomized, double-blind, placebo-controlled clinical trial suggest that probiotic supplementation of PD patients with *L. acidophilus*, *B. bifidum*, *Lactobacillus reuteri*, and *L. fermentum* for 12 weeks decreases the overall score on the Unified Parkinson's Disease Rating Scale (UPDRS) as compared to the placebo group. Besides that, probiotic supplementation increased the glutathione (GSH) levels, remarkably decreased the hs-CRP and MDA levels, and significantly improved the insulin function in contrast to the placebo (Tamtaji et al. 2019).

Patients suffering from PD have increased oxidative stress and inflammations that increase with the severity of disease (Taylor et al. 2013). Psychobiotics have shown promising results in reducing oxidative stress and the inflammations in patients with PD. For instance, probiotic interventions in PD patients for 12 weeks significantly upregulated the expression of transforming growth factor beta (TGF-β) and peroxisome proliferator-activated receptor gamma (PPAR-γ) while downregulating the expression of interleukin-1 (IL-1), IL-8, and tumor necrosis factor alpha (TNF-α) as compared to the placebo control in a randomized controlled study focusing on the effect of probiotic administration on inflammation, insulin, and lipid-related genes in peripheral blood mononuclear cells (PBMCs). However, no effect of probiotic administration was observed on the expression of markers of inflammation and oxidative stress, vascular endothelial growth factor (VEGF), and low-density lipoprotein receptor (LDLR) in the same study (Tamtaji et al. 2017). These studies depict the importance of probiotic administration in improving the overall health of PD patients; however, as with the case of AD, more studies are required to emphasize the role of psychobiotics in alleviating symptoms of PD. For instance, probiotics might prove useful in folding of α-synuclein produced in enteroendocrine cells eventually reducing the Lewy bodies formation of dopaminergic (Shults 2006; Liddle 2018; Chandra et al. 2017); however, this needs to be unrevealed by further research.

19.7.3 Autism Spectrum Disorder (ASD)

Although quite rare, ASD is a neurodevelopmental disorder characterized by inability to communicate socially, restrictive behavioral pattern, and limited activities and interest. ASD is prevalent in 0.1–1.8% of the population, and these patients frequently complain of gastrointestinal problems (American Psychiatric Association 2013; Wang et al. 2011) with interesting correlations between severity of behavioral and gastrointestinal symptoms. Evidence indicates that patients with ASD have varying levels of alteration in gut microbiota and this implicates the importance of considering the gut-brain axis in its treatment. Several species of bacteria are being evaluated as probiotics in improving the gastrointestinal and behavioral problems in ASD patients such as *L. acidophilus* DSM24735™, *L. plantarum* DSM24730™, *Lactobacillus paracasei* DSM24733™, *L. helveticus* DSM24734™, *Streptococcus thermophilus* DSM24731™, *B. lactis* DSM24736™, *B. breve* DSM24732™, and *Lactobacillus delbrueckii* subsp. *bulgaricus* DSM 24734 and *B. lactis* DSM24737™ (Arnold 2019; Cheng et al. 2019; Shaaban et al. 2018a, b)

A recent trial conducted in Egypt reports that administering probiotics for 3 months improved the severity of autism and GI symptoms as compared to control

(Shaaban et al. 2018a, b). Different trials using different bacterial strains as psychobiotic formulations are being investigated, and their results are still awaited.

A previous placebo-controlled trial conducted in the UK in 2012 evaluated the efficacy of the single probiotic strain *L. plantarum* WCFS1 in an ASD (Cheng et al. 2019). In this study, ASD patients reporting GI problems were given *L. plantarum* WCFS1 as compared to placebo for 6 weeks; however, the results of this study are not available. This study recruited patients with ASD presenting with GI problems for a 6-week intervention with either probiotics or placebo. Another trial conducted in the UK in 2010 *reports* that the administration of *L. plantarum* WCFS1 in ASD patients for 3 weeks altered the gut microbiota (Parracho et al. 2010). Currently, limited data are available that reveal the effects of probiotics on patients with ASD. However, numerous trials are under progress for which results are awaited to provide scientific evidence for the efficacy of these probiotics in the management of ASD-related GI and behavioral symptoms.

19.7.4 Attention Deficit Hyperactivity Disorder (ADHD)

ADHD is a neurological illness categorized on the hyperactivity, inattention, and impulsivity. The infants administrated with *L. rhamnosus* GG during the first 6 months after birth may have a reduced risk for ADHD (Pärtty et al. 2015). The Truehope GreenBAC capsules are administered to the ADHD patients to improve energy level and the mood (Rucklidge 2013). Moreover, *L. acidophilus* food supplements can recover the ADHD children with the self-control and the attention (Harding et al. 2003).

19.7.5 Tourette Syndrome (TS)

Tourette syndrome (TS) is a neurological disorder that is initially observed in childhood (Rampello et al. 2006). There are various TS clinical treatments such as behavioral treatments, $\alpha 2$-adrenergic agonists, antipsychotics, and deep brain stimulation (DBS) (Murphy et al. 2013; Weisman et al. 2013). The fecal microbiota transplantation (FMT) improves the TS after 8 weeks of treatment (Zhao et al. 2017).

19.7.6 Insomnia

Insomnia refers to the sleep disorder which causes illness such as depression, memory loss, and allergy (Kaneita et al. 2006; Grundgeiger et al. 2014; Cohen et al. 2009). Recent reports show that the usage of fermented products can improve sleep (Kitaoka et al. 2009). The studies suggest that the heat-killed *L. brevis* SBC8803 (SBL88™) improves sleep in mice and humans. Also, it improves walking sleep journal scores in healthy males (Nakakita et al. 2016) and increases delta power values in adults aged 40 years as compared with placebo control. Similarly, the heat-killed *L. brevis* SBC8803 can increase the duration of wakefulness and nighttime wheel-running activity (Miyazaki et al. 2014). The administration of probiotics leads to the reduction in non-rapid eye movement (NREM) sleep during the active phase and improves NREM sleep during resting phase (Miyazaki et al. 2014). However, no significant effect can be found in the heat-killed *L. brevis* SBC8803 treatment in the sleep quality according to the electroencephalograms (EEG) and the Athens Insomnia Scale (AIS) (Nakakita et al. 2016). Based on the study reports, the consumption of *L. helveticus* CM4 containing fermented milk can improve the efficiency of sleep and wakening episodes in aged individuals (Yamamura et al. 2009).

19.8 Conclusion

In the past few years, the human gut microbiota and effect of probiotics on it have received considerable attention in the context of the relation between microbiota and health or disease. Association of gut dysbiosis with many health conditions has revealed the importance of healthy gut flora on overall human health. Animal studies have revealed that the gut-brain interaction is a two-way traffic with signals coming from the brain to the gut microbes and similarly response and feedback from the gut microbiota to the brain. These interactions are particularly important in maintaining brain health and that ultimately requires a balance of microbial structural and functional diversity in the gut. Probiotics, specifically psychobiotics, hold special significance in the sense that they can maintain a healthy gut microbiota, thus maintaining general brain health and alleviation of anxiety, stress, and behavioral problems. In addition to that, preliminary studies have revealed that these psychobiotics can prove beneficial in improving the symptoms of neurodegenerative and neurodevelopmental disorders.

Thus, psychobiotic treatments might be used as a promising strategy to improve the quality of life for people suffering from neurodegenerative and neurodevelopmental disorders; however, further studies in this arena are required to evaluate the effectiveness of psychobiotics as an alternative therapeutic regimen for alleviating stress, anxiety, cognitive function, and brain health.

References

Agahi A, Hamidi GA, Daneshvar R, Hamdieh M, Soheili M, Alinaghipour A, Taba SME, Salami M (2018) Does severity of Alzheimer's disease contribute to its responsiveness to modifying gut microbiota? A double blind clinical trial. Front Neurol 9:662

Akbari E, Asemi Z, Daneshvar Kakhaki R, Bahmani F, Kouchaki E, Tamtaji OR, Hamidi GA, Salami M (2016) Effect of probiotic supplementation on cognitive function and metabolic status in Alzheimer's disease: a randomized, double-blind and controlled trial. Front Aging Neurosci 8:256

Allen AP, Hutch W, Borre YE, Kennedy PJ, Temko A, Boylan G, Murphy E, Cryan J, Clarke G (2016) Bifidobacterium longum 1714 as a translational psychobiotic: modulation of stress, electrophysiology and neurocognition in healthy volunteers. Transl Psychiatry 6(11):939

American Psychiatric Association (2013) Diagnostic and statistical manual of mental disorders (DSM-5®). American Psychiatric Pub, Washington, DC

Andriessen EM, Wilson AM, Mawambo G, Dejda A, Miloudi K, Sennlaub F, Sapieha P (2016) Gut microbiota influences pathological angiogenesis in obesity-driven choroidal neovascularization. EMBO Mol Med 8(12):1366–1379. https://doi.org/10.15252/emmm.201606531

Arnold LE (2019) Probiotics for quality of life in autism spectrum disorders. ClinicalTrials.gov NCT02903030. https://clinicaltrials.gov/ct2/show/NCT02903030. Accessed 5 March 2018

Athari Nik Azm S, Djazayeri A, Safa M, Azami K, Ahmadvand B, Sabbaghziarani F, Sharifzadeh M, Vafa M (2018) Lactobacilli and Bifidobacteria ameliorate memory and learning deficits and oxidative stress in β-amyloid (1–42) injected rats. Appl Physiol Nutr Metab 43 (7):718–726

Bajinka O, Tan Y, Abdelhalim KA et al (2020) Extrinsic factors influencing gut microbes, the immediate consequences and restoring eubiosis. AMB Express 10:130. https://doi.org/10.1186/s13568-020-01066-8

Barichella M, Cereda E, Pezzoli G (2009) Major nutritional issues in the management of Parkinson's disease. Mov Disord 24(13):1881–1892

Barichella M, Pacchetti C, Bolliri C, Cassani E, Iorio L, Pusani C, Pinelli G, Privitera G, Cesari I, Faierman SA, Caccialanza R, Pezzoli G, Cereda E (2016) Probiotics and prebiotic fiber for constipation associated with Parkinson disease: an RCT. Neurology 87(12):1274–1280

Barrett E, Ross RP, O'Toole PW, Fitzgerald GF, Stanton C (2012) γ-Aminobutyric acid production by culturable bacteria from the human intestine. J Appl Microbiol 113(2):411–417

Bercik P, Verdu EF, Foster JA, Macri J, Potter M, Huang X, Malinowski P, Jackson W, Blennerhassett P, Neufeld KA, Lu J, Khan WI, Corthesy-Theulaz I, Cherbut C, Bergonzelli GE, Collins SM (2010) Chronic gastrointestinal inflammation induces anxiety-like behavior and alters central nervous system biochemistry in mice. Gastroenterology 139(6):2102–2112

Bercik P, Collins SM, Verdu EF (2012) Microbes and the gut-brain axis. Neurogastroenterol Motil 24:405–413

Berg D, Postuma RB, Adler CH, Bloem BR, Chan P, Dubois B, Gasser T, Goetz CG, Halliday G, Joseph L, Lang AE, Liepelt-Scarfone I, Litvan I, Marek K, Obeso J, Oertel W, Olanow CW, Poewe W, Stern M, Deuschl G (2015) MDS research criteria for prodromal Parkinson's disease. Mov Disord 30(12):1600–1611

Bermúdez-Humarán LG, Salinas E, Ortiz GG, Ramirez-Jirano LJ, Morales JA, Bitzer-Quintero OK (2019) From probiotics to psychobiotics: live beneficial bacteria which act on the brain-gut axis. Nutrients 11(4):890. https://doi.org/10.3390/nu11040890

Bonfili L, Cecarini V, Cuccioloni M, Angeletti M, Berardi S, Scarpona S, Rossi G, Eleuteri AM (2018) SLAB51 probiotic formulation activates SIRT1 pathway promoting antioxidant and neuroprotective effects in an AD mouse model. Mol Neurobiol 55(10):7987–8000

Bravo JA, Forsythe P, Chew MV, Escaravage E, Savignac HM, Dinan TG, Bienenstock J, Cryan JF (2011) Ingestion of Lactobacillus strain regulates emotional behavior and central GABA receptor expression in a mouse via the vagus nerve. Proc Natl Acad Sci 108(38):16050–16055

Cani PD, Knauf C (2016) How gut microbes talk to organs: the role of endocrine and nervous routes. Mol Metab 5(9):743–752

Carabotti M, Scirocco A, Maselli MA, Severi C (2015) The gut-brain axis: interactions between enteric microbiota, central and enteric nervous systems. Ann Gastroenterol 28(2):203–209

Cassani E, Privitera G, Pezzoli G, Pusani C, Madio C, Iorio L, Barichella M (2011) Use of probiotics for the treatment of constipation in Parkinson's disease patients. Minerva Gastroenterol Dietol 57(2):117–121

Chandra R, Hiniker A, Kuo YM, Nussbaum RL, Liddle RA (2017) α-Synuclein in gut endocrine cells and its implications for Parkinson's disease. JCI Insight 2:12

Chandran S, Manohari SM, Raman V (2019) The gut-brain connection: a qualitative review of the conceptualisation and implications of the gut-brain-microbiome axis. Telangana J Psychiatry 5 (2):94

Cheng LH, Liu YW, Wu CC, Wang S, Tsai YC (2019) Psychobiotics in mental health, neurodegenerative and neurodevelopmental disorders. J Food Drug Anal 27(3):632–648

Clemente JC, Ursell LK, Parfrey LW, Knight R (2012) The impact of the gut microbiota on human health: an integrative view. Cell 148(6):1258–1270. https://doi.org/10.1016/j.cell.2012.01.035

Cohen S, Doyle WJ, Alper CM, Janicki-Deverts D, Turner RB (2009) Sleep habits and susceptibility to the common cold. Arch Intern Med 169(1):62–67

Collins SM, Bercik P (2009) The relationship between intestinal microbiota and the central nervous system in normal gastrointestinal function and disease. Gastroenterology 136:2003–2014

Collins SM, Surette M, Bercik P (2012) The interplay between the intestinal microbiota and the brain. Nat Rev Microbiol 10:735–742

Costedio MM, Hyman N, Mawe GM (2007) Serotonin and its role in colonic function and in gastrointestinal disorders. Dis Colon Rectum 50(3):376–388

Crumeyrolle-Arias M, Jaglin M, Bruneau A, Vancassel S, Cardona A, Dauge V et al (2014) Absence of the gut microbiota enhances anxiety-like behavior and neuroendocrine response to acute stress in rats. Psychoneuroendocrinology 42:207–217

Cryan JF, Dinan TG (2012) Mind-altering microorganisms: the impact of the gut microbiota on brain and behaviour. Nat Rev Neurosci 13:701–712

De Rijk MC, Tzourio C, Breteler MM, Dartigues JF, Amaducci L, Lopez-Pousa S, Manubens-Bertran JM, Rocca WA (1997) Prevalence of parkinsonism and Parkinson's disease in Europe: the EUROPARKINSON Collaborative Study. European community concerted action on the epidemiology of Parkinson's disease. J Neurol Neurosurg Psychiatry 62(1):10–15. https://doi.org/10.1136/jnnp.62.1.10

Desbonnet L, Garrett L, Clarke G, Kiely B, Cryan JF, Dinan TG (2010) Effects of the probiotic Bifidobacterium infantis in the maternal separation model of depression. Neuroscience 170 (4):1179–1188

Dinan TG, Stanton C, Cryan JF (2013) Psychobiotics: a novel class of psychotropic. Biol Psychiatry 74:720–726

Dowlati Y, Herrmann N, Swardfager W, Liu H, Sham L, Reim EK, Lanctôt KL (2010) A meta-analysis of cytokines in major depression. Biol Psychiatry 67(5):446–457

Enck P, Mazurak N (2018) Dysbiosis in functional bowel disorders. Ann Nutr Metab 72 (4):296–306. https://doi.org/10.1159/000488773

Fasano A, Visanji NP, Liu LW, Lang AE, Pfeiffer RF (2015) Gastrointestinal dysfunction in Parkinson's disease. Lancet Neurol 14(6):625–639

Foster JA, Rinaman L, Cryan JF (2017) Stress & the gut-brain axis: regulation by the microbiome. Neurobiol Stress 7:124–136

Fung TC, Olson CA, Hsiao EY (2017) Interactions between the microbiota, immune and nervous systems in health and disease. Nat Neurosci 20:145–155

Georgescu D, Ancusa OE, Georgescu LA, Ionita I, Reisz D (2016) Nonmotor gastrointestinal disorders in older patients with Parkinson's disease: is there hope? Clin Interv Aging 11:1601

Gibson GR, Roberfroid MB (1995) Dietary modulation of the human colonic microbiota: Introducing the concept of prebiotics. J Nutr 125:1401–1412

Grundgeiger T, Bayen UJ, Horn SS (2014) Effects of sleep deprivation on prospective memory. Memory 22(6):679–686

Hammitt KM, et al (2019) The syndrome patient cost resources of and Sjogren's the and value of Sjogren's syndrome e-book: a clinical handbook, p 279

Harding KL, Judah RD, Gant CE (2003) Outcome-based comparison of Ritalin® versus food-supplement treated children with AD/HD. Altern Med Rev 8(3):319–330

Havenaar R, Huis In't Veld JHJ (1992) Probiotics: a general view. In: Wood BJB (ed) The lactic acid bacteria. Springer, Boston. https://doi.org/10.1007/978-1-4615-3522-5_6

Heldt SA, Stanek L, Chhatwal JP, Ressler KJ (2007) Hippocampus-specific deletion of BDNF in adult mice impairs spatial memory and extinction of aversive memories. Mol Psychiatry 12 (7):656–670

Hemarajata P, Versalovic J (2013) Effects of probiotics on gut microbiota: mechanisms of intestinal immunomodulation and neuromodulation. Ther Adv Gastroenterol 6(1):39–51. https://doi.org/10.1177/1756283X12459294

Hill C, Guarner F, Reid G, Gibson GR, Merenstein DJ, Pot B, Morelli L, Canani RB, Flint HJ, Salminen S et al (2014) Expert consensus document. The International Scientific Association for Probiotics and Prebiotics consensus statement on the scope and appropriate use of the term probiotic. Nat Rev Gastroenterol Hepatol 11:506–514

Huttenhower C, Gevers D, Knight R, Abubucker S, Badger JH, Chinwalla AT, Creasy HH, Earl AM, FitzGerald MG, Fulton RS, Giglio MG (2012) Structure, function and diversity of the healthy human microbiome. Nature 486(7402):207

Kamada N, Seo SU, Chen GY, Nunez G (2013) Role of the gut microbiota in immunity and inflammatory disease. Nat Rev Immunol 13:321–335

Kaneita Y, Ohida T, Uchiyama M, Takemura S, Kawahara K, Yokoyama E, Miyaki T, Harano S, Suzuki K, Fujita T (2006) The relationship between depression and sleep disturbances: a Japanese nationwide general population survey. J Clin Psychiatry 67(2):196–203

Katan MB (2012) Why the European food safety authority was right to reject health claims for probiotics. Benefic Microbes 3(2):85–89. https://doi.org/10.3920/BM2012.0008

Kitaoka K, Uchida K, Okamoto N, Chikahisa S, Miyazaki T, Takeda E, Séi H (2009) Fermented ginseng improves the first-night effect in humans. Sleep 32(3):413–421

Kumar A, Singh A (2015) A review on Alzheimer's disease pathophysiology and its management: an update. Pharmacol Rep 67(2):195–203

Leblhuber F, Steiner K, Schuetz B, Fuchs D, Gostner JM (2018) Probiotic supplementation in patients with Alzheimer's dementia-an explorative intervention study. Curr Alzheimer Res 15 (12):1106–1113

Leung K, Thuret S (2015) Gut microbiota: a modulator of brain plasticity and cognitive function in ageing. Healthcare 3:898–916

Li Y, Hao Y, Fan F, Zhang B (2018) The role of microbiome in insomnia, circadian disturbance and depression. Front Psych 9:669

Liang S, Wang T, Hu X, Luo J, Li W, Wu X, Duan Y, Jin F (2015) Administration of Lactobacillus helveticus NS8 improves behavioral, cognitive, and biochemical aberrations caused by chronic restraint stress. Neuroscience 310:561–577

Liddle RA (2018) Parkinson's disease from the gut. Brain Res 1693(Pt B):201–206. https://doi.org/10.1016/j.brainres.2018.01.010

Liu WH, Yang CH, Lin CT, Li SW, Cheng WS, Jiang YP, Wu CC, Chang CH, Tsai YC (2015) Genome architecture of Lactobacillus plantarum PS128, a probiotic strain with potential immunomodulatory activity. Gut Pathogens 7(1):1–7

Liu YW, Liu WH, Wu CC, Juan YC, Wu YC, Tsai HP, Wang S, Tsai YC (2016) Psychotropic effects of Lactobacillus plantarum PS128 in early life-stressed and naïve adult mice. Brain Res 1631:1–12

Lu Y, Christian K, Lu B (2008) BDNF: a key regulator for protein synthesis-dependent LTP and long-term memory? Neurobiology 89(3):312–323

Martinowich K, Lu B (2008) Interaction between BDNF and serotonin: role in mood disorders. Neuropsychopharmacology 33(1):73–83

Messaoudi M, Violle N, Bisson JF, Desor D, Javelot H, Rougeot C (2011) Beneficial psychological effects of a probiotic formulation (Lactobacillus helveticus R0052 and Bifidobacterium longum R0175) in healthy human volunteers. Gut Microbes 2(4):256–261

Metchnikoff E, Mitchell PC (1907) Essais optimistes. C R Soc Biol 60:359–361

Miyazaki K, Itoh N, Yamamoto S, Higo-Yamamoto S, Nakakita Y, Kaneda H, Shigyo T, Oishi K (2014) Dietary heat-killed Lactobacillus brevis SBC8803 promotes voluntary wheel-running and affects sleep rhythms in mice. Life Sci 111(1-2):47–52

Mohammadi AA, Jazayeri S, Khosravi-Darani K, Solati Z, Mohammadpour N, Asemi Z, Adab Z, Djalali M, Tehrani-Doost M, Hosseini M, Eghtesadi S (2016) The effects of probiotics on mental health and hypothalamic–pituitary–adrenal axis: a randomized, double-blind, placebo-controlled trial in petrochemical workers. Nutr Neurosci 19(9):387–395

Moustafa A, Li W, Anderson EL, Wong EHM, Dulai PS, Sandborn WJ, Biggs W, Yooseph S, Jones MB, Venter JC et al (2018) Genetic risk, dysbiosis, and treatment stratification using host genome and gut microbiome in inflammatory bowel disease. Clin Transl Gastroenterol 9:e132

Murphy TK, Lewin AB, Storch EA, Stock S, American Academy of Child and Adolescent Psychiatry (AACAP) Committee on Quality Issues (CQI) (2013) Practice parameter for the assessment and treatment of children and adolescents with tic disorders. J Am Acad Child Adolesc Psychiatry 52(12):1341–1359. https://doi.org/10.1016/j.jaac.2013.09.015

Musa NH, Mani V, Lim SM, Vidyadaran S, Majeed ABA, Ramasamy K (2017) Lactobacilli-fermented cow's milk attenuated lipopolysaccharide-induced neuroinflammation and memory impairment in vitro and in vivo. J Dairy Res 84(4):488–495

Nakakita Y, Tsuchimoto N, Takata Y, Nakamura T (2016) Effect of dietary heat-killed Lactobacillus brevis SBC8803 (SBL88™) on sleep: a non-randomised, double blind, placebo-controlled, and crossover pilot study. Benefic Microbes 7(4):501–509

Naureen Z, Dautaj A, Anpilogov K, Camilleri G, Dhuli K, Tanzi B, Maltese PE, Cristofoli F, De Antoni L, Beccari T, Dundar M, Bertelli M (2020) Bacteriophages presence in nature and their role in the natural selection of bacterial populations. Acta Biomed 91(13-S):e2020024

Nimgampalle M, Kuna Y (2017) Anti-Alzheimer properties of probiotic, Lactobacillus plantarum MTCC 1325 in Alzheimer's disease induced albino rats. JCDR 11(8):KC01

O'Mahony SM, Clarke G, Borre YE, Dinan TG, Cryan JF (2015) Serotonin, tryptophan metabolism and the brain-gut-microbiome axis. Behav Brain Res 277:32–48

Padhy SK, Sahoo S, Mahajan S, Sinha SK (2015) Irritable bowel syndrome: Is it "irritable brain" or "irritable bowel"? J Neurosci Rural Pract 6(4):568–577

Palmer C, Bik EM, DiGiulio DB, Relman DA, Brown PO (2007) Development of the human infant intestinal microbiota. PLoS Biol 5:e177

Parracho HM, Gibson GR, Knott F, Bosscher D, Kleerebezem M, McCartney AL (2010) A double-blind, placebo-controlled, crossover-designed probiotic feeding study in children diagnosed with autistic spectrum disorders. Int J Probiot Prebiot 5(2):69

Pärtty A, Kalliomäki M, Wacklin P, Salminen S, Isolauri E (2015) A possible link between early probiotic intervention and the risk of neuropsychiatric disorders later in childhood: a randomized trial. Pediatr Res 77(6):823–828

Qin J, Li R, Raes J et al (2010) A human gut microbial gene catalogue established by metagenomic sequencing. Nature 464:59–65

Rampello L, Alvano A, Battaglia G, Bruno V, Raffaele R, Nicoletti F (2006) Tic disorders: from pathophysiology to treatment. J Neurol 253(1):1–15

Rogers G, Keating D, Young R et al (2016) From gut dysbiosis to altered brain function and mental illness: mechanisms and pathways. Mol Psychiatry 21:738–748. https://doi.org/10.1038/mp.2016.50

Roshchina VV (2016) New trends and perspectives in the evolution of neurotransmitters in microbial, plant, and animal cells. Microb Endocrinol 874:25–77

Rucklidge JJ (2013) Could yeast infections impair recovery from mental illness? A case study using micronutrients and olive leaf extract for the treatment of ADHD and depression. Adv Mind Body Med 27(3):14–18

Sarkar A, Lehto SM, Harty S, Dinan TG, Cryan JF, Burnet PW (2016) Psychobiotics and the manipulation of bacteria–gut–brain signals. Trends Neurosci 39(11):763–781

Savignac HM, Kiely B, Dinan TG, Cryan JF (2014) Bifidobacteria exert strain-specific effects on stress-related behavior and physiology in BALB/c mice. Neurogastroenterol Motil 26 (11):1615–1627

Schousboe A, Waagepetersen HS (2007) GABA: homeostatic and pharmacological aspects. Prog Brain Res 160:9–19

Shaaban SY, El Gendy YG, Mehanna NS, El-Senousy WM, ElFeki HSA, Saad K et al (2018a) The role of probiotics in children with autism spectrum disorder: a prospective, open-label study. Nutr Neurosci 21:676–681

Shaaban SY, El Gendy YG, Mehanna NS, El-Senousy WM, El-Feki HS, Saad K, El-Asheer OM (2018b) The role of probiotics in children with autism spectrum disorder: a prospective, open-label study. Nutr Neurosci 21(9):676–681

Shults CW (2006) Lewy bodies. Proc Natl Acad Sci U S A 103(6):1661–1668. https://doi.org/10.1073/pnas.0509567103

Sommer F, Bäckhed F (2013) The gut microbiota — masters of host development and physiology. Nat Rev Microbiol 11:227–238

Sudo N, Chida Y, Aiba Y, Sonoda J, Oyama N, Yu XN, Kubo C, Koga Y (2004) Postnatal microbial colonization programs the hypothalamic–pituitary–adrenal system for stress response in mice. J Physiol 558(1):263–275

Tamtaji OR, Kouchaki E, Salami M, Aghadavod E, Akbari E, Tajabadi-Ebrahimi M, Asemi Z (2017) The effects of probiotic supplementation on gene expression related to inflammation, insulin, and lipids in patients with multiple sclerosis: a randomized, double-blind, placebo-controlled trial. J Am Coll Nutr 36(8):660–665

Tamtaji OR, Taghizadeh M, Kakhaki RD, Kouchaki E, Bahmani F, Borzabadi S, Oryan S, Mafi A, Asemi Z (2019) Clinical and metabolic response to probiotic administration in people with Parkinson's disease: a randomized, double-blind, placebo-controlled trial. Clin Nutr 38 (3):1031–1035

Taylor JM, Main BS, Crack PJ (2013) Neuroinflammation and oxidative stress: co-conspirators in the pathology of Parkinson's disease. Neurochem Int 62(5):803–819

Tissier MH (1899) La reaction chromophile d'Escherich et le Bacterium coli. C R Seances Soc Biol Fil 51:943–945

Vanuytsel T, Van Wanrooy S, Vanheel H, Vanormelingen C, Verschueren S, Houben E, Salim Rasoel S, Tóth J, Holvoet L, Farré R, Van Oudenhove L, Boeckxstaens G, Verbeke K, Tack J (2014) Psychological stress and corticotropin-releasing hormone increase intestinal permeability in humans by a mast cell-dependent mechanism. Gut 63(8):1293–1299

Wang LW, Tancredi DJ, Thomas DW (2011) The prevalence of gastrointestinal problems in children across the United States with autism spectrum disorders from families with multiple affected members. J Dev Behav Pediatr 32(5):351–360

Weisman H, Qureshi IA, Leckman JF, Scahill L, Bloch MH (2013) Systematic review: pharmacological treatment of tic disorders–efficacy of antipsychotic and alpha-2 adrenergic agonist agents. Neurosci Biobehav Rev 37(6):1162–1171

Wen L, Duffy A (2017) Factors influencing the gut microbiota, inflammation, and type 2 diabetes. J Nutr 147(7):1468S–1475S. https://doi.org/10.3945/jn.116.240754

Wrase J, Reimold M, Puls I, Kienast T, Heinz A (2006) Serotonergic dysfunction: brain imaging and behavioral correlates. Cogn Affect Behav Neurosci 6(1):53–61

Xu C, Zhu H, Qiu P (2019) Aging progression of human gut microbiota. BMC Microbiol 19:236. https://doi.org/10.1186/s12866-019-1616-2

Yamamura S, Morishima H, Kumano-go T, Suganuma N, Matsumoto H, Adachi H, Sigedo Y, Mikami A, Kai T, Masuyama A, Takano T, Sugita Y, Takeda M (2009) The effect of Lactobacillus helveticus fermented milk on sleep and health perception in elderly subjects. Eur J Clin Nutr 63(1):100–105

Yano JM, Yu K, Donaldson GP, Shastri GG, Ann P, Ma L, Nagler CR, Ismagilov RF, Mazmanian SK, Hsiao EY (2015) Indigenous bacteria from the gut microbiota regulate host serotonin biosynthesis. Cell 161(2):264–276

Zhao H, Shi Y, Luo X, Peng L, Yang Y, Zou L (2017) The effect of fecal microbiota transplantation on a child with Tourette syndrome. Medicine 2017:6165239

Impact of Probiotics in Modulation of Gut Microbiome

20

Roghayeh Nourizadeh, Bita Sepehri, Amin Abbasi, R. Z. Sayyed, and Leila Khalili

Abstract

Gut health refers to a number of physiological, microbiological, and physical functions that work together to maintain intestinal homeostasis. The gut provides a platform for the growth of a diverse microbiota that not only provides a barrier against colonization by pathogens but also regulates immune development and maturation and provides metabolites for host well-being. Disruption of the balance of gut microbiota is one of the major etiological factors associated with several gastrointestinal and infectious diseases, metabolic disorders such as obesity and diabetes, and inflammatory disease. Probiotics are live microbial supplements that beneficially affect the host by improving its intestinal microbial balance. This chapter summarizes the evidence available in literature for the beneficial effect of probiotics in modulating gut microbiota in favor of beneficial microbiota and then promoting host health. Considering the results of several

R. Nourizadeh
Faculty of Nursing and Midwifery, Department of Midwifery, Tabriz University of Medical Sciences, Tabriz, Iran

B. Sepehri
Liver and Gastrointestinal Diseases Research Center, Tabriz University of Medical Sciences, Tabriz, Iran

A. Abbasi
Department of Food Science and Technology, Faculty of Nutrition and Food Sciences, Tabriz University of Medical Sciences, Tabriz, Iran

R. Z. Sayyed
Department of Microbiology, PSGVP Mandal's Arts, Science and Commerce College, Shahada, Maharashtra, India

L. Khalili (✉)
Department of Community Nutrition, Faculty of Nutrition and Food Sciences, Tabriz University of Medical Sciences, Tabriz, Iran

R. Z. Sayyed, M. Khan (eds.), *Microbiome-Gut-Brain Axis*,
https://doi.org/10.1007/978-981-16-1626-6_20

investigations, probiotic consumption can affect the gut microbiota and gut barrier integrity, so the diseased state, caused due to gut microbiome imbalance, will regain homeostasis and health. Several mechanisms for probiotics' actions are revealed; however, the gut microbiota modulatory effect is one of the main mechanisms through which the probiotics can affect the host. By knowing the benefits of gut microbiota balance, it would be wise to manipulate the composition of the gut microbiota through probiotic consumption.

Keywords

Gut microbiome · Intestinal microbiota · Probiotics · Modulation

20.1 Introduction

Human intestinal microbiota, a highly diverse population of microbes, mainly consists of bacteria, viruses, protozoa, and fungi and has a major contribution to human health. Changes in the microbiota composition, diversity, and temporal stability (microbiota dysbiosis) have been associated with some gastrointestinal and systemic conditions (Lavelle and Hill 2019). So, modulation of the intestinal microbiota to maintain a favorable balance in the ecosystem and to improve human health is of great interest.

The luminal microbiota can be modulated during a probiotic treatment. Probiotics are defined as "live strains of selected microorganisms that confer a health benefit on the host when administered in adequate amounts." The most commonly used probiotic strains belong to the genera *Bifidobacterium* and *Lactobacillus*, but also *Enterococcus*, *Lactococcus*, *Streptococcus*, and *Bacillus* spp., and some yeast strains belonging to the genus *Saccharomyces* have been included in probiotic products for human nutrition (Markowiak and Śliżewska 2017). The use of probiotics in food products is based on their safety records for human use, and most of the used species have Generally Regarded as Safe (GRAS; FDA, US) or Qualified Presumption of Safety (QPS; EFSA, EU) status. According to the results of several studies, probiotics have clinical benefits, but more clinical research should be done to confirm their efficacy. Moreover, probiotics' effects may depend on the strain, dose, and components used to produce a probiotic product (Markowiak and Śliżewska 2017; Jafar-Abadi et al. 2020).

20.2 Probiotic and Gut Microbiota

For decades, the health benefits of probiotics have prompted increasing scientific interest. The therapeutic effect of probiotics has been evaluated in a wide range of diseases, particularly in gastrointestinal and metabolic disorders where the results have supported the potential use of probiotics as therapeutic agents (Ford et al. 2014; Ebrahimi et al. 2019; Khalili et al. 2019).

The potential mechanisms of probiotics' action include improving intestinal barrier function through effects on the epithelium and mucus lining; manipulating intestinal microbial communities; producing antimicrobial substances; competing with pathogenic bacteria; regulating luminal acidity; immune modulation; stimulating epithelial cell proliferation and differentiation; and fortification of the intestinal barrier (Thomas and Versalovic 2010; Hou et al. 2020).

Previous investigations have revealed positive effects of probiotic consumption on several health outcomes. Most of the published studies have focused on populations with specific health pathologies; moreover, some evidence supports the health-promoting effects of probiotics in healthy adult (Kristensen et al. 2016). Hou et al. found that Firmicutes, Bacteroidetes, Proteobacteria, Actinobacteria, and Tenericutes were the main phyla in the fecal microbiota of healthy participants (Hou et al. 2020). The F/B (Firmicutes/Bacteroidetes) value is the common indicator for the composition of gut microbiota. The healthy adults' gut microbiota composition and the differences in their responses to intervention with the same probiotic are important in determining their roles in human health and well-being.

In recent years, the gut microbiota has received much attention as a potential determinant for disease development that has the potential to be modified by probiotic consumption. Some gut microbiota are associated with metabolic diseases such as obesity and diabetes and the occurrence and development of gastrointestinal diseases (Tilg and Kaser 2011; Fallucca et al. 2014; Mayer et al. 2014; Kim et al. 2015). Other gut microbiota are involved in functional processes that are essential for homeostasis, such as digestion of indigestible nutrients, and production of vitamins and micronutrients (Nicholson et al. 2012).

In summary, the composition and changes in the gut microbiota are closely related to human health. Therefore, the health effects of probiotic consumption on gut microbiota could be used as a tool for the maintenance and promotion of host health.

20.3 Modulation of the Intestinal Microbiota by the Use of Probiotics

The concept of probiotics to the scientific community was introduced by Nobel laureate Elie Metchnikoff. He published an important report linking the longevity of Bulgarians with drinking fermented milk products containing viable *Lactobacilli* (Metchnikoff 1907). This finding proposed that ingestion of certain microbes could be beneficial for human health. Since then, probiotics had been extensively marketed and consumed as functional foods or dietary supplements. Intestinal microbiome has important effects on the integrity and function of the gastrointestinal tract, immune homeostasis maintenance, and host energy metabolism (Pflughoeft and Versalovic 2012). Intestinal dysbiosis—perturbations of microbial composition—may result in disrupted interactions between microbes and host. These changes in microbiome composition and function may contribute to disease susceptibility (Frank et al. 2011). Several investigations have shown associations between intestinal dysbiosis and chronic low-grade inflammation and metabolic disorders resulting in metabolic syndrome, obesity, and diabetes, infections in the gastrointestinal tract, irritable bowel syndrome (IBS), and inflammatory bowel disease (IBD) (Cani and Delzenne 2009; Jumpertz et al. 2011; Pflughoeft and Versalovic 2012).

Treatment methods to manipulate and restore the balance in the richness and diversity of intestinal microbiome are being explored (Sonnenburg and Fischbach 2011). Probiotics can play a beneficial role in the gastrointestinal tract and improve the functionality of existing microbial communities. Probiotics can also affect the function and composition of microbial communities by competition for nutrients, production of growth substrates or inhibitors, and modulation of intestinal immunity (O'Toole and Cooney 2008). This concept is supported by results of several randomized controlled clinical trials showing the benefits of probiotics during the treatment of gastrointestinal disorders (Preidis and Versalovic 2009; Thomas and Versalovic 2010; Hemarajata and Versalovic 2013; Maleki et al. 2015).

20.4 How Probiotics Modulate the Intestinal Microbiota?

The potential mechanisms of probiotics' action include effects on the composition and function of the intestinal microbiome. Probiotics suppress the growth of other microorganisms by producing antimicrobial or metabolic compounds; moreover, they compete for receptors and binding sites with other intestinal microbes on the intestinal mucosa (Collado et al. 2007; O'Shea et al. 2012). *Lactobacillus* strains enhance the integrity of the intestinal barrier, which can result in the maintenance of immune tolerance, reduced translocation of bacteria across the intestinal mucosa, and disease phenotypes such as IBD, IBS, and gastrointestinal infections (Lee and Bak 2011). Furthermore, probiotics can modulate the intestinal immunity and alter the responsiveness of the intestinal epithelia and immune cells to microbes in the intestinal lumen (Bron et al. 2012).

The effects of probiotics on the function, composition, and diversity of the gut microbiota have been studied using different tools and techniques. An investigation showed decreased pain and flatulence in patients with IBS that received a 4-week treatment with a rose-hip drink containing 5×10^7 CFU/ml of *L. plantarum* DSM 9843 per day (Nobaek et al. 2000). This improvement in clinical symptoms was associated with the presence of *L. plantarum* in rectal biopsies of patients and the reduced amounts of *enterococci* in fecal specimens. Another intervention on patients with diarrhea-dominant IBS (IBS-D) showed symptomatic relief in patients treated with a probiotic mixture of *L. acidophilus*, *L. rhamnosus*, *L. plantarum*, *Bifidobacterium breve*, *B. longum*, *B. lactis*, and *Streptococcus thermophilus*. Analyses of the fecal microbiota of these patients using denaturing gradient gel electrophoresis (DGGE) revealed that the similarity of the microbial composition was more similar in probiotic-treated patients than that of the placebo group. This finding revealed that microbial community composition was more stable during the period of probiotic treatment (Cha et al. 2012). Recent technological innovations in DNA sequencing and progressions in bioinformatics have provided scientists with tools to explore research questions about the human microbiome and how treatment modalities affect changes in the composition and function of the microbial communities. A recent research using a high-throughput, culture-independent method analyzed the fecal microbiota of 6-month-old infants treated with daily supplements of *L. rhamnosus* (LGG) (Cox et al. 2010). The findings revealed an abundance of LGG and an increased index of evenness in the fecal microbiota of these infants, suggesting ecological stability. Probiotics' ability in inducing changes in intestinal microbial communities was demonstrated in a recent study, which explored the effects of *L. reuteri* on microbial community composition in a neonatal mouse model using 16S rRNA metagenomic sequencing. The results demonstrated an increase in community evenness and diversity of the distal intestinal microbiome in animals treated with *L. reuteri* compared with that of vehicle-treated animals (Preidis et al. 2012). The diversity in microbial communities was shown to be associated with increased ecological stability (Eisenhauer et al. 2012). So, probiotics can induce changes in the intestinal microbiota and stabilize microbial communities.

In addition to direct effects on the composition of the intestinal microbiota, probiotics can also modulate the global metabolic function of intestinal microbiome. Fermented milk products containing several probiotics did not alter the composition of intestinal bacterial communities in gnotobiotic mice and monozygotic twins (McNulty et al. 2011). However, fecal meta-transcriptomic analysis of probiotic-treated animals showed significant changes in the expression of microbial enzymes, mainly enzymes involved in carbohydrate metabolism. Moreover, mass spectrometric analysis of urinary metabolites demonstrated altered abundance of several carbohydrate metabolites. These observations suggested that probiotics may affect the global metabolic function of the intestinal microbiome.

20.5 Modulation of Gut Microbiota-Brain Axis by Probiotics

The gut is closely connected to the brain via 200–600 million neurons (Furness 2006). Bidirectional communication between the gut and the brain has long been recognized. Signals from the brain can affect the motor, sensory, and secretory modalities of the gastrointestinal (GI) tract, and, in turn, visceral messages from the gut can affect brain function (Grenham et al. 2011; Tabrizi et al. 2019). There is growing evidence for the view rethinking the gut-brain axis as the concept of a gut microbiota-brain axis due to the central role of gut microbiota in the bidirectional gut-brain axis (Gareau et al. 2011; Neufeld et al. 2011; Cryan and Dinan 2012; Dinan and Cryan 2013; Wang et al. 2013). However, the routes of communication between the gut microbiota and the brain are not completely clarified, probably through endocrine, neural, and immune pathways, which could be affected by gut microbiota or microbiota-generated metabolites (Moloney et al. 2014). Results of investigations have revealed that the bidirectional interaction between the gut microbiota and the brain can be modulated by probiotics which exert beneficial impacts on brain activity and behavior (Kanauchi et al. 2013). The probiotic strains used for human consumption must survive gastrointestinal transit with human origin and nonpathogenic feature (Hardy et al. 2013).

Although the relationship between gut microbiota and mental disorders is complex, it is possible to improve the mentioned disorders through modulation of gut microbiota by probiotic consumption. Bercik et al. showed that *Bifidobacterium longum* normalized anxiety-like behavior induced by the noninvasive parasite *Trichuris muris* infection (Bercik et al. 2010). Similarly, Bravo et al. indicated that chronic treatment with *Lactobacillus rhamnosus* (JB-1) reduced the anxiety- and depression-related behavior in the *Trichuris muris*-infected mice (Bravo et al. 2011). Ingestion of selected probiotics also showed effects on brain activity in humans. Rao et al. showed that administration of *Lactobacillus casei* strain Shirota for 2 months significantly decreased anxiety symptoms among patients with chronic fatigue syndrome compared with controls (Rao et al. 2009). The oral administration of *L. helveticus* R0052 and *B. longum* R0175 for 2 weeks was shown to alleviate anxiety and depressive symptoms in healthy volunteers, as measured by the Hopkins Symptom Checklist (HSCL-90) and the Hospital Anxiety and Depression Scale (HADS) (Messaoudi et al. 2011). Similarly, after 4 weeks of consumption of a fermented milk containing probiotics (FMPP) (containing *Bifidobacterium animalis* subsp. *lactis*, *Streptococcus thermophilus*, *Lactobacillus bulgaricus*, and *Lactococcus lactis* subsp. *lactis*) by healthy women, Tillisch et al. found that FMPP intake affected the activity of brain regions that control central processing of emotion and sensation, including affective, viscerosensory, and somatosensory cortices (Tillisch et al. 2013).

Several studies have spurred better understanding of the acting mechanisms of probiotics involved in gut-brain axis signaling. Abnormalities of brain function are associated with the altered composition of the gut microbiota. The use of probiotics can partly or completely reverse the dysbiosis in the microbiota caused by some brain disorders. Multiple pathways are involved in the modulation of the gut microbiota-brain axis: vagus nerve-mediated pathways, immune response-mediated pathways, and metabolite-mediated pathways. A deeper understanding of the relationship between the gut bacteria and their hosts is implicated in developing microbial-based therapeutic strategies for brain disorders.

20.6 Conclusions

It is possible to manipulate the composition of the gut microbiota in infants and adults through dietary supplementation. Probiotics have been proposed as preventive and therapeutic measures, in order to restore the healthy composition and function of the gut microbiome. However, data from human microbiome studies may lead to identification of novel indigenous microbial species and tools to positively induce alterations in the gut microbial communities. Well-designed experiments in appropriate experimental models (in vitro or in vivo) may yield insights into the biology and potential manipulation of the microbiome in the human host. New types of probiotics or medicinal compounds derived from the microbiome may be used as future strategies to promote health, prevent disease, and treat different disorders.

Conflict of Interest There is no conflict of interest.

References

Bercik P, Verdu EF, Foster JA, Macri J, Potter M, Huang X, Malinowski P, Jackson W, Blennerhassett P, Neufeld KA (2010) Chronic gastrointestinal inflammation induces anxiety-like behavior and alters central nervous system biochemistry in mice. Gastroenterology 139 (6):2102–2112

Bravo JA, Forsythe P, Chew MV, Escaravage E, Savignac HM, Dinan TG, Bienenstock J, Cryan JF (2011) Ingestion of Lactobacillus strain regulates emotional behavior and central GABA receptor expression in a mouse via the vagus nerve. Proc Natl Acad Sci 108(38):16050–16055

Bron PA, Van Baarlen P, Kleerebezem M (2012) Emerging molecular insights into the interaction between probiotics and the host intestinal mucosa. Nat Rev Microbiol 10(1):66–78

Cani PD, Delzenne NM (2009) Interplay between obesity and associated metabolic disorders: new insights into the gut microbiota. Curr Opin Pharmacol 9(6):737–743

Cha BK, Jung SM, Choi CH, Song I-D, Lee HW, Kim HJ, Hyuk J, Chang SK, Kim K, Chung W-S (2012) The effect of a multispecies probiotic mixture on the symptoms and fecal microbiota in diarrhea-dominant irritable bowel syndrome: a randomized, double-blind, placebo-controlled trial. J Clin Gastroenterol 46(3):220–227

Collado M, Meriluoto J, Salminen S (2007) Role of commercial probiotic strains against human pathogen adhesion to intestinal mucus. Lett Appl Microbiol 45(4):454–460

Cox MJ, Huang YJ, Fujimura KE, Liu JT, McKean M, Boushey HA, Segal MR, Brodie EL, Cabana MD, Lynch SV (2010) Lactobacillus casei abundance is associated with profound shifts in the infant gut microbiome. PLoS One 5(1):e8745

Cryan JF, Dinan TG (2012) Mind-altering microorganisms: the impact of the gut microbiota on brain and behaviour. Nat Rev Neurosci 13(10):701–712

Dinan TG, Cryan JF (2013) Melancholic microbes: a link between gut microbiota and depression? Neurogastroenterol Motil 25(9):713–719

Ebrahimi FS, Rad AH, Mosen M, Abbasalizadeh F, Tabrizi A, Khalili L (2019) Effect of L. acidophilus and B. lactis on blood glucose in women with gestational diabetes mellitus: a randomized placebo-controlled trial. Diabetol Metab Syndr 11(1):75

Eisenhauer N, Scheu S, Jousset A (2012) Bacterial diversity stabilizes community productivity. PLoS One 7(3):e34517

Fallucca F, Porrata C, Fallucca S, Pianesi M (2014) Influence of diet on gut microbiota, inflammation and type 2 diabetes mellitus. First experience with macrobiotic Ma-Pi 2 diet. Diabetes Res Rev 30(S1):48–54

Ford AC, Quigley EM, Lacy BE, Lembo AJ, Saito YA, Schiller LR, Soffer EE, Spiegel BM, Moayyedi P (2014) Efficacy of prebiotics, probiotics, and synbiotics in irritable bowel syndrome and chronic idiopathic constipation: systematic review and meta-analysis. Am J Gastroenterol 109(10):1547–1561

Frank DN, Zhu W, Sartor RB, Li E (2011) Investigating the biological and clinical significance of human dysbioses. Trends Microbiol 19(9):427–434

Furness JB (2006) Novel gut afferents: Intrinsic afferent neurons and intestinofugal neurons. Auton Neurosci 125(1-2):81–85

Gareau MG, Wine E, Rodrigues DM, Cho JH, Whary MT, Philpott DJ, MacQueen G, Sherman PM (2011) Bacterial infection causes stress-induced memory dysfunction in mice. Gut 60 (3):307–317

Grenham S, Clarke G, Cryan JF, Dinan TG (2011) Brain–gut–microbe communication in health and disease. Front Physiol 2:94
Hardy H, Harris J, Lyon E, Beal J, Foey AD (2013) Probiotics, prebiotics and immunomodulation of gut mucosal defences: homeostasis and immunopathology. Nutrients 5(6):1869–1912
Hemarajata P, Versalovic J (2013) Effects of probiotics on gut microbiota: mechanisms of intestinal immunomodulation and neuromodulation. Ther Adv Gastroenterol 6(1):39–51
Hou Q, Zhao F, Liu W, Lv R, Khine WWT, Han J, Sun Z, Lee Y-K, Zhang H (2020) Probiotic-directed modulation of gut microbiota is basal microbiome dependent. Gut Microbes 2020:1–20
Jafar-Abadi MA, Dehghani A, Khalili L, Barzegar A, Mesrizad M, Hassanalilou T (2020) A meta-analysis of randomized controlled trials of the effect of probiotic food or supplement on glycemic response and body mass index in patients with type 2 diabetes, updating the evidence. Curr Diabetes Rev 17(3):356–364
Jumpertz R, Le DS, Turnbaugh PJ, Trinidad C, Bogardus C, Gordon JI, Krakoff J (2011) Energy-balance studies reveal associations between gut microbes, caloric load, and nutrient absorption in humans. Am J Clin Nutr 94(1):58–65
Kanauchi O, Andoh A, Mitsuyama K (2013) Effects of the modulation of microbiota on the gastrointestinal immune system and bowel function. J Agric Food Chem 61(42):9977–9983
Khalili L, Alipour B, Jafarabadi MA, Hassanalilou T, Abbasi MM, Faraji I (2019) Probiotic assisted weight management as a main factor for glycemic control in patients with type 2 diabetes: a randomized controlled trial. Diabetol Metab Syndr 11(1):5
Kim S-E, Choi SC, Park KS, Park MI, Shin JE, Lee TH, Jung KW, Koo HS, Myung S-J (2015) Change of fecal flora and effectiveness of the short-term VSL#3 probiotic treatment in patients with functional constipation. J Neurogastroenterol Motil 21(1):111
Kristensen NB, Bryrup T, Allin KH, Nielsen T, Hansen TH, Pedersen O (2016) Alterations in fecal microbiota composition by probiotic supplementation in healthy adults: a systematic review of randomized controlled trials. Genome Med 8(1):52
Lavelle A, Hill C (2019) Gut microbiome in health and disease: emerging diagnostic opportunities. Gastroenterol Clin 48(2):221–235
Lee BJ, Bak Y-T (2011) Irritable bowel syndrome, gut microbiota and probiotics. J Neurogastroenterol Motil 17(3):252
Maleki D, Homayouni A, Khalili L, Golkhalkhali B (2015) Probiotics in cancer prevention, updating the evidence, pp 781–791
Markowiak P, Śliżewska K (2017) Effects of probiotics, prebiotics, and synbiotics on human health. Nutrients 9(9):1021
Mayer EA, Savidge T, Shulman RJ (2014) Brain–gut microbiome interactions and functional bowel disorders. Gastroenterology 146(6):1500–1512
McNulty NP, Yatsunenko T, Hsiao A, Faith JJ, Muegge BD, Goodman AL, Henrissat B, Oozeer R, Cools-Portier S, Gobert G (2011) The impact of a consortium of fermented milk strains on the gut microbiome of gnotobiotic mice and monozygotic twins. Sci Transl Med 3(106):106
Messaoudi M, Lalonde R, Violle N, Javelot H, Desor D, Nejdi A, Bisson J-F, Rougeot C, Pichelin M, Cazaubiel M (2011) Assessment of psychotropic-like properties of a probiotic formulation (Lactobacillus helveticus R0052 and Bifidobacterium longum R0175) in rats and human subjects. Br J Nutr 105(5):755–764
Metchnikoff E (1907) The prolongation of life: optimistic studies. GP Putnam's Sons, New York
Moloney RD, Desbonnet L, Clarke G, Dinan TG, Cryan JF (2014) The microbiome: stress, health and disease. Mamm Genome 25(1-2):49–74
Neufeld K, Kang N, Bienenstock J, Foster JA (2011) Reduced anxiety-like behavior and central neurochemical change in germ-free mice. Neurogastroenterol Motil 23(3):255
Nicholson JK, Holmes E, Kinross J, Burcelin R, Gibson G, Jia W, Pettersson S (2012) Host-gut microbiota metabolic interactions. Science 336(6086):1262–1267
Nobaek S, Johansson M-L, Molin G, Ahrné S, Jeppsson B (2000) Alteration of intestinal microflora is associated with reduction in abdominal bloating and pain in patients with irritable bowel syndrome. Am J Gastroenterol 95(5):1231–1238

O'Shea EF, Cotter PD, Stanton C, Ross RP, Hill C (2012) Production of bioactive substances by intestinal bacteria as a basis for explaining probiotic mechanisms: bacteriocins and conjugated linoleic acid. Int J Food Microbiol 152(3):189–205

O'Toole PW, Cooney JC (2008) Probiotic bacteria influence the composition and function of the intestinal microbiota. Interdiscip Perspect Inf Dis 2008:175285

Pflughoeft KJ, Versalovic J (2012) Human microbiome in health and disease. Annu Rev Pathol 7:99–122

Preidis GA, Versalovic J (2009) Targeting the human microbiome with antibiotics, probiotics, and prebiotics: gastroenterology enters the metagenomics era. Gastroenterology 136(6):2015–2031

Preidis GA, Saulnier DM, Blutt SE, Mistretta TA, Riehle KP, Major AM, Venable SF, Finegold MJ, Petrosino JF, Conner ME (2012) Probiotics stimulate enterocyte migration and microbial diversity in the neonatal mouse intestine. FASEB J 26(5):1960–1969

Rao AV, Bested AC, Beaulne TM, Katzman MA, Iorio C, Berardi JM, Logan AC (2009) A randomized, double-blind, placebo-controlled pilot study of a probiotic in emotional symptoms of chronic fatigue syndrome. Gut Pathogens 1(1):1–6

Sonnenburg JL, Fischbach MA (2011) Community health care: therapeutic opportunities in the human microbiome. Sci Transl Med 3(78):78

Tabrizi A, Khalili L, Homayouni-Rad A, Pourjafar H, Dehghan P, Ansari F (2019) Prebiotics, as promising functional food to patients with psychological disorders: a review on mood disorders, sleep, and cognition. NeuroQuantology 17:6

Thomas CM, Versalovic J (2010) Probiotics-host communication: modulation of signaling pathways in the intestine. Gut Microbes 1(3):148–163

Tilg H, Kaser A (2011) Gut microbiome, obesity, and metabolic dysfunction. J Clin Invest 121 (6):2126–2132

Tillisch K, Labus J, Kilpatrick L, Jiang Z, Stains J, Ebrat B, Guyonnet D, Legrain-Raspaud S, Trotin B, Naliboff B (2013) Consumption of fermented milk product with probiotic modulates brain activity. Gastroenterology 144(7):1394–1401

Wang L, Christophersen CT, Sorich MJ, Gerber JP, Angley MT, Conlon MA (2013) Increased abundance of Sutterella spp. and Ruminococcus torques in feces of children with autism spectrum disorder. Mol Autism 4(1):42

Potential of Probiotics in Improving Gut Health

21

Vipul P. Patel and Manoj K. Chaudhari

Abstract

To maintain nutrition and health, presence of needful bacteria in the gut plays a crucial role. When there is condition like imbalance in human natural microflora, mostly in the gut results in ill conditions known as dysbiosis. Recent advance study on the human gut suggests that misbalance of microbial flora may result in predisposition to different disease phenotypes. The use of probiotics as mediators in health and diseases has been raised in recent years. The human gut has the ability to act as home to over 100–1000 species of microbes, where the internal environment is modulated, which plays an important role in host health. In this chapter, we have tried to explain some of the applications of probiotics on the human as well as animal gut and how they are beneficial. Points such as probiotics and its current value in the market, gut microbiota and its effect on the immune system, and several diseases are explained. Information related to microorganisms and their role is encoded.

Keywords

Ecobiotics · Gut microbiota · Immunomodulation · Intestinal diseases · Probiotics

21.1 Introduction

Joshua Lederberg introduced the concept of human microbiome to scientific community. Probiotic is modern era's phrase, and it is going to play an important role in the effect of human as well as animal health with the help of bacterial associations.

V. P. Patel (✉) · M. K. Chaudhari
SRES' Sanjivani College of Pharmaceutical Education and Research, Kopargaon, Maharashtra, India

R. Z. Sayyed, M. Khan (eds.), *Microbiome-Gut-Brain Axis*,
https://doi.org/10.1007/978-981-16-1626-6_21

The symbiotic connection between the human host and its bacterial residents has obtained extensive research interest in recent years. These bacteria are predominately showing their presence in the gastrointestinal system with their mutual genomes, known as the gut microbiome (Marchesi and Ravel 2015). Based on published literature and their positive results advice that in the human host homeostasis, health, and disease, there is an important role of gut microbiome. Some of the positive result of research associated with appetite, GIT function, and immune responses (Thaiss et al. 2016) are well documented. This chapter tries to explain collected current knowledge and research in probiotics in gut health. Metchnikoff in the early 1990s revealed probiotics in scientific circumstances as moderation of floral diversity in animal and human bodies and substitute harmful microbes with useful ones. Afterwards, Henry Tissier observed that the stool of children suffering from diarrhea contains less concentration of particular bacteria compared to the stool of healthy children. He also suggested oral administration of live organism that is *Bifidobacterium* to patients with diarrhea. Modern definition of probiotics is a viable mixed or mono bacteria culture when its application is done to man or animal helps to maintain and improve properties of indigenous flora is given by Havenaar and Huisin't Veld (1992). In the year 2000, Government of Argentina has requested to food and agriculture expert panel for evaluation of health and nutritional properties of probiotics. After that grammatically corrected definition of probiotics is "living microorganism, when administered in required amounts, gives benefit on health of host" (Hill et al. 2014). It can also define as probiotics as substitute of microorganisms in host to which play an important role in enhancing measurable health outcomes. In 2017, updated definition of probiotics exists as "a substrate that is selectively utilized by host microorganism conforming benefit to health". The main cause for dysbiosis is when there is disturbance or changes in collection of bacteria, bacteria, and viruses which help to form gut microbiota, threat to gut integrity is imposed by normal microbial´homeostasis. Until now, for healthy gut exact number of microbes require is not found or defined. Microbes play an important role in maintaining regulatory and metabolic networks. Shaping of the gut epithelium is also mostly depending upon the microbiota present in the gut (Sender et al. 2016). *Streptococcus*, *Clostridium*, *Lactobacillus*, and *Bifidobacterium* are currently the major strains used as probiotics, but in recent study in culturomics and metagenomics related to disease lead to increase knowledge related to microbial composition and their role in health benefit (Fig. 21.1).

21.2 Probiotics: Current Importance and Future

It is better to select probiotic strains for the preparation of products based upon the evidence of phenotype. In recent years, there has been an increased use of probiotics as a medicine to treat diseases and maintain better health. This condition leads due to evidence of effect of gut microbiota on their effect on health. Current consideration gives rise with industries and academia to do research and develop probiotics with success along with develop technologies. There is expansion of global market related to probiotics. Recently, there is still a need to study the interaction of

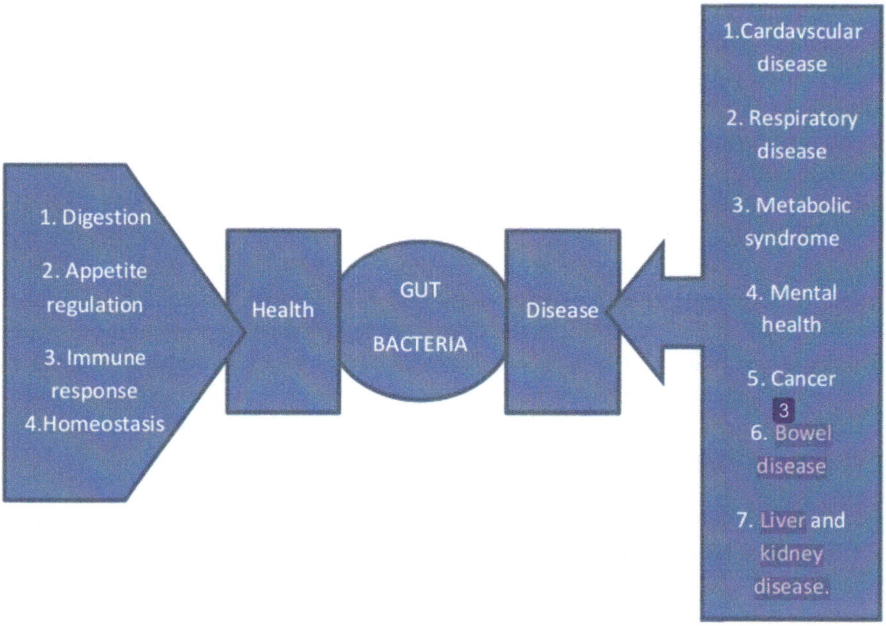

Fig. 21.1 A diagrammatic overview of the positive influence of gut bacteria

microbiota with supplementation. A more in-depth study of single and combined strains is to be done and ongoing. A lot of research is done related to probiotics, but health claim is yet to be done. Among the research activities done, near about 19,000 results have been found for literature survey (Chamberlain and Lau 2016). The National Institutes of Health funded project entitled "Human Microbiome Project (HMP1)" which was performed on 250 healthy volunteers. This project concluded that human microbiome is constituted between 3500 and 35,000 species (Morgan et al. 2013). In the human body, a lot of microorganisms are present, and microorganisms present at different sites vary in number. The sequence of amount of microorganisms present in oral and colonic cavity have large amount compared to vagina. Hadza in Tanzania in one survey it was found that diversity of GIT microorganisms in hunter-gatherer populations (Clemente et al. 2015). Among colorectal cancer, IBS, Crohn's disease, and obesity are observed in persons with a low diversity of microorganisms in the lower GIT (Mosca et al. 2016). In the biological basis and mechanisms of action of probiotics, with diverse biological functions and mechanisms apparent across different bacterial strains, it is important for researchers and probiotic product developers to understand the properties of each strain and apply these proactively to target a preferred physiological interaction/response. For example, it is likely to be important and beneficial for probiotic strains that exhibit anti-inflammatory properties to be applied to research in health conditions associated with an augmented inflammatory response (e.g., Crohn's disease). In clinical probiotic research, an early and consistent research focus on gastrointestinal diseases is now resulting in inclusion of probiotics into evidence-based guidance for clinicians. Beyond the gut, the scope of clinical conditions amenable to probiotic management seems almost limitless. However, although the results from research in new indications, such as neurological pathologies, are very promising, a substantial amount of further work is required to provide healthcare providers with the confidence to embrace probiotics into regular practice. The probiotics industry is an ever-growing entity with continual expansion of products being taken to market. This has driven scientific research with the aspirations to uncover probiotic strains that provide conclusive evidence of improvements in health and disease outcomes. These opportunistic endpoints have not currently been met, evidenced by the fact that no certified health claims credited to probiotic products are currently in place. This is likely owing to the wide interpersonal variations in commensal bacteria as well as fundamental differences between probiotic strains. The further application of advanced omics technologies will provide an improved understanding of the complex host-bacteria interactions.

21.3 Gut Microbiota and Their Effect on Human Health

Since the project named as Human Microbiome Project (HMP) exists, a lot of study has been published related to the composition of microbiota in the human gut along with analysis of normal and diseased persons. There was also one study performed in which fecal study of two twins was performed in which characterization of fecal microbial community for obesity and leanness was carried out. The result of study concluded with sharing of an identifiable core set and pathways in human microbiome and it was observed that person with changes in microbiota at the phylum level having obesity problem. Metagenomic analysis study was also performed, and its study revealed that microbial biomarkers of obesity take part in lipid, amino acid, and carbohydrate metabolism. A recent metatranscriptomic analysis determined the distribution of functional roles of human fecal microbiota. This study demonstrated the distribution of Clusters of Orthologous Groups (COGs) categories across each of the 10 metatranscriptomes (A, B, C, D, E, F, K, L, N, and O) that were sequenced. The following are activities related to human health in which gut microbiota plays an important role: energy production and conservation, amino acid transport and metabolism, nucleotide transport and metabolism, carbohydrate transport and metabolism, transcription, cell mobility, defense mechanism, signal transduction mechanism, etc. Luminal conversion by intestinal microbes may play an important role in host-microbiota interactions. Orally consumed nutrients may be converted by intestinal microbes into bioactive compounds that could affect the health of the host and the intestinal microbiota (GABA, gamma-aminobutyric acid; SCFAs, short-chain fatty acids). Probiotics may manipulate intestinal microbial communities and suppress growth of pathogens by inducing the host's production of β-defensin and IgA. Probiotics may be able to fortify the intestinal barrier by maintaining tight junctions and inducing mucin production. Probiotic-mediated immunomodulation may occur through mediation of cytokine secretion through

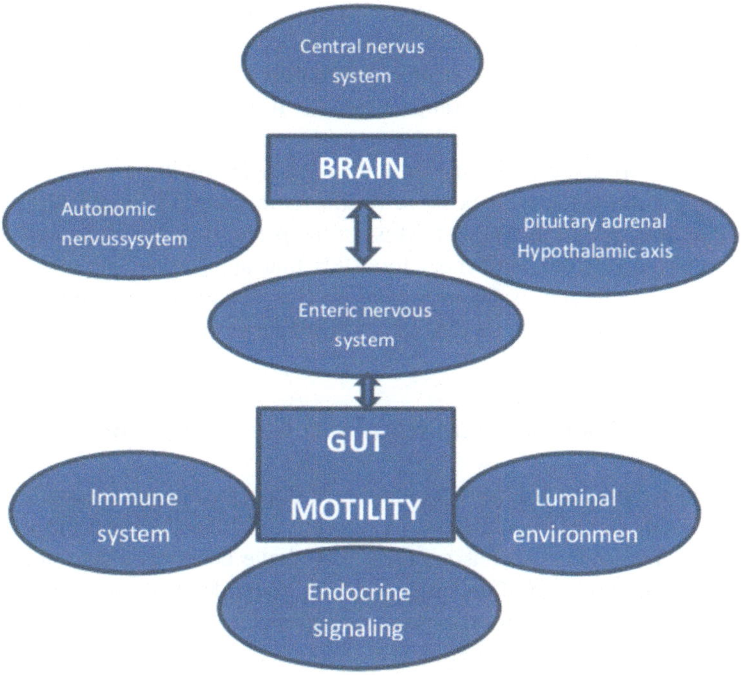

Fig. 21.2 Factors that control gut motility, the gut luminal environment

signaling pathways such as NFκB and MAPKs, which can also affect proliferation and differentiation of immune cells (such as T cells) or epithelial cells. Gut motility and nociception may be modulated through regulation of pain receptor expression and secretion of neurotransmitters (APRIL, a proliferation-inducing ligand; HSP, heat shock protein; IEC, intestinal epithelial cell; Ig, immunoglobulin; MAPK, mitogen-activated protein kinase; NFκB, nuclear factor-kappaB; pIgR, polymeric immunoglobulin receptor; STAT, signal transducer and activator of transcription; Treg, T regulatory cell) (Fig. 21.2).

21.4 Probiotics as Nutritional Aid for Human

In the twentieth century, Elie Metchnikoff proposed the concept of probiotics, which means "for life." According to him, as we get older, essential microbes in our gut decrease, and this can be prevented by taking beneficial bacteria orally, and our health can be improved (Metchnikoff 1908). Every person wants good health and happy life and good food quality; in accordance to this, there has been increasing demand in requirement of probiotics as nutritional aid in the past few years. The increasing demand of probiotics is seen along with food and meals (Markova and Sheveleva 2014). Foodstuff consists of beneficial bacterial culture which is added during the manufacturing process. Most of these products are manufactured in the form of frozen powder (Saarela et al. 2000). Most of the time, production of acetic acid as secondary metabolites during production of *Bifidobacterium* results in change in taste during fermentation and storage. It is necessary to get assured that foodstuff not get adversely affected due to culture inoculation. After preparation of probiotics, packing material used in it and its storage condition are the two main crucial steps. Until now, with positive result, more than 500 promising probiotic food supplements are in the market. Most of the probiotic foods include fermented cereals, fruits, vegetables, and meat foodstuff that are gaining popularity among consumers. The most successful examples are cheese and dips, mayonnaise, edible spreads, ice cream, milk, juices, oat, etc. (Ranadheera et al. 2017).

21.5 Probiotics in Constipation and Gastric Motility

Constipation is the most common problem seen in persons in our community which is mostly result in problem of gastric motility. While studying the effects of probiotics on humans as well as animals, there have been promising and positive results seen. Only problem of uncertainty related to mode of action of probiotics on gut motility and constipation. The immune system, nervous system function, bile acid mechanism, and mucus secretion are vital factors to gut motility, and imbalance or dysfunction related to these factors results in gut motility. The use of certain probiotic strains can help in modifying the gut luminal environment and provide a benefit for patients with constipation and motility (Dimidi et al. 2017). The central nervous system, the immune system, and the enteric nervous system are some of the factors that affect gut motility, and disturbance in these factors results in constipation (Tables 21.1 and 21.2).

21.6 Implications of Probiotics on the Maternal-Neonatal Interface: Gut Microbiota, Immunomodulation, and Autoimmunity

Treatment of autoimmune disease is also done with probiotics; it is possible by rebalancing dysbiosis inducing changes in the immune system. Autoimmune disease can also occur during pregnancy, and it is concerned with both mother and child. In literature, probiotics have obtained a lot of significance same with marketing. Microbiota significantly get changed in the mother's GIT, and it can be balanced with the help of probiotics, but interaction between probiotics changes during the period of pregnancy, and normal condition is not clearly observed. There is existing evidence that gut microbiota in the mother's GIT influences the offspring's microbiota and directly affects the health of neonates. Microbiota are directly gets

Table 21.1 Recommendations for use of probiotics in childhood intestinal diseases

Diseases	Probiotics	Treatment/ prevention	References
Helicobacter pylori infection	*L. casei* DN-114001	Treatment	Cameron et al. (2017)
Inflammatory bowel disease	VSL#32	Treatment	Cameron et al. (2017)
Infantile colic	*L. reuteri* DSM 17938	Treatment	Cameron et al. (2017)
Functional intestinal disorders (IBS)	*L. rhamnosus* GG *L. reuteri* DSM 17938	Treatment	Cameron et al. (2017)
Traveler's diarrhea	*S. boulardii*	Prevention	Cameron et al. (2017)
Nosocomial diarrhea	*L. rhamnosus* GG *B. lactis* Bb12+ *S. thermophilus*	Prevention	Cameron et al. (2017)
Acute gastroenteritis	*S. boulardii*, *L. rhamnosus GG*, Indian Dahi	Treatment	Cameron et al. (2017)
Clostridium difficile-associated diarrhea	*S. boulardii*	Prevention	Cameron et al. (2017)
Antibiotic-associated diarrhea	*S. boulardii*; *L. rhamnosus* GG, *B. lactis* Bb12 + *S. thermophilus*, *L. rhamnosus* strains E/N, Oxy and Pen	Prevention	Cameron et al. (2017)

influence on the immune system. Dysbiosis directly results in immune dysregulation and autoimmunity. Probiotics are considered safe during pregnancy (Brianna Swartwout et al. 2018).

Table 21.2 List of probiotic strains and their potential effects

Probiotic strain	Features	Potential effect	References
Wheat bran	Arabinoxylan oligosaccharides	Increases *Bifidobacterium* levels relative to total fecal microbiota and reduces colonic protein fermentation	Kleessen et al. (2007)
Yacon root	Fructooligosaccharides (FOS) and inulin	Improves the growth of *Bifidobacterium* in the colon, enhances mineral absorption and gastrointestinal metabolism, and plays a role in the regulation of serum cholesterol	Kim et al. (2009)
Chicory root	Inulin	Potential substrate for gut bacteria, helps in increasing bile production	Barszcz et al. (2016)
Dandelion greens	Inulin	Diuretic, antioxidant, and cholesterol-lowering effects	Samal et al. (2012)
Jerusalem artichoke	Inulin, high in thiamine and potassium	Potential substrate for gut bacteria and promotes proper muscle function	Ning et al. (2018)
Garlic	Inulin and fructooligosaccharides (FOS)	Increases the growth of *Bifidobacterium* and reduces the growth of disease-promoting bacteria	Kolida et al. (2002)
Oats	Beta-glucan	Reduces serum cholesterol and LDL cholesterol	Delaney et al. (2004)
Apples	Pectin	Increases the population of butyrate- and beta-glucuronidase-producing Clostridiales	Delaney et al. (2004)

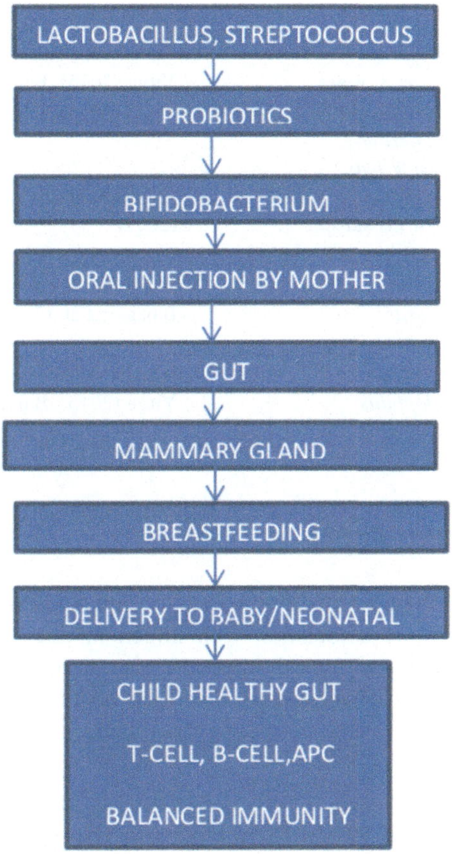

Flow chat for beneficial effect of probiotics for mother and offspring

21.7 Probiotics for Animals' Gut Health

For proper absorption and digestion of dietary nutrients, the gut should remain always healthy, which helps inefficiency of animals. Presence of balance microbiota that is healthy micro-ecosystem will definitely result in healthy gut. When probiotics administered to animals in adequate amounts will result in conformation of healthy gut. Animals eat a lot of things, so probiotics can result into reduction of diarrhea-like condition and act as antitoxin. There is lots of literature survey and research have been done in conditions of animals where antibiotics are also ban for some treatments where these probiotics can play an active role. Lots of research has been done and going on related to safety issue of probiotics and there has been positive result coming out (Huang et al. 2019) (Table 21.3).

Table 21.3 List of microorganisms used as probiotics for animals

Genus	Species	References
Aspergillus	*A. oryzae*; *A. niger*	Yirga (2015), Bajagai et al. (2016)
Bacillus	*B. licheniformis* *B. megaterium* *B. mesentericus*	Yirga (2015), Bajagai et al. (2016)
Streptococcus	*S. intermedius* *S. salivarius* subsp. *thermophilus*	Pollmann et al. (1980), Azizpour et al. (2009)
Lactococcus	*L. lactis*	Azizpour et al. (2009)
Saccharomyces	*S. boulardii* *S. cerevisiae*	Bajagai et al. (2016)
Prevotella	*P. bryantii*	Bajagai et al. (2016)
Lactobacillus	*L. acidophilus* *L. amylovorus*	Yirga (2015), Bajagai et al. (2016)

21.8 Safety of Probiotics

Does it overstimulate the host immune system, does it cause sensitization of human sensitive areas, has it any association with infection, does it harbor transferable antibiotic gene, does it produce emesis after administration, and can it be identified at the strain level are some of the questions which arise during safety concern of probiotics.

21.9 Modulation of Brain Function via the Gut-Brain Axis

Via the gut-brain axis, probiotics show their action, and it involves multiple routes. Blood circulation, immune system, and humoral pathway are involved in brain functions and influence was observed between stress person and normal person (Okada et al. 2010). Probiotics play an important role in maintaining homeostasis of the neuroendocrine and immune system by preventing disturbance of gut microbiota. It was found in one study that medical student who was under stress used to administer *Lactobacillus casei* to get suppress stress-related problems. Daily intake of *Lactobacillus gasseri* CP2305 has ability to reduce stress-related problems (Nishida et al. 1859). Stress, sociability, cognitive function, anxiety, depression, and autism are some of the problems which are associated due to shifts in gut microbiota diversity. Most of the time, age factor also can be considered in the unbalance of gut microbiota. Diet is also the most considered factor in gut microbiota. SCFAs are neuroactive metabolites of microbiota which constitute flow of information (Oleskin and Shenderov 2019). Brain-Derived Neurotropic Factor (BDNF) is maintained by butyrate which levels and neurogenesis in the hippocampus and improves behavior by reducing depression where bacterium known as *Clostridium butyricum* MIYAIRI 588 has been active role in butyrate production (Kim et al. 2009). In addition,

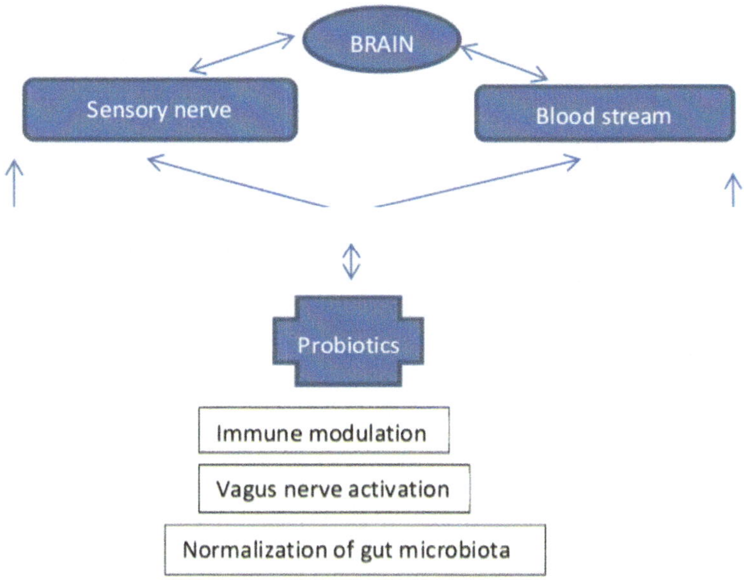

Fig. 21.3 Modulation of brain function via the gut-brain axis

mucosal barrier plays an important role in signaling pathway of microbiota-gut-brain interactions (Fig. 21.3).

21.10 The Feature Ahead of Probiotics, Ecobiotics

It is observed that decrease of infection in developing nations and increasing risk of allergic reaction and autoimmune diseases seen in developing countries (Liao and Nyachoti 2017). Most of the time, main reason for decrease in number of microbial content due to administration of broad spectrum an antibiotic, encountering of pathogen *Clostridium difficile* occurs. This pathogen colonizes in the large intestine. It results in diarrhea to life-threatening complications such as pseudomembranous colitis (PMC), toxic megacolon, and colonic perforation (Bartlett et al. 1978). In the USA, most healthcare associated problem is infection of *C. difficile* treated with antibiotics such as vancomycin, metronidazole, and fidaxomicin. There is alternative treatment comes in occurrence known as fecal microbiota transplantation (FMT) has risen in prominence during the recent past (Van Nood et al. 2013). Techniques are very efficient but there was always risk when we transfer unknown pathogen, it needs to insert healthy microbiota in the gut. Recently, developers are trying to develop an alternative method for FMT known as ecobiotics. Ecobiotics involve therapeutic dose formulations delivered orally based on gut ecology. And the clinical efficiency of this concept has also been seen. Seres Health plans (SER) also grant permission for trials ecobiotics alternative for FMT in *C. difficile* infection. Many of the microbe-related therapies can be used in the treatment of diseases related to the gut (Khanna et al. 2016).

References

Azizpour K, Bahrambeygi S, Mahmoodpour S, Azizpour A, Mahmoodpour S, Bahrambeygi S, Azizpour K (2009) History and basic of probiotics research. J Biol Sci 4(4):409–426

Bajagai YS, Klieve AV, Dart PJ, Bryden WL (2016) Probiotics in animal nutrition: production, impact and regulation. In: FAO animal production and health paper number, 179. Food and Agriculture Organization, Rome

Barszcz M, Taciak M, Skomiał J (2016) The effects of inulin, dried Jerusalem artichoke tuber and a multispecies probiotic preparation on microbiota ecology and immune status of the large intestine in young pigs. Anim Nutr 70(4):278–292. https://doi.org/10.1080/1745039X.2016.1184368

Bartlett JG, Moon N, Chang TW, Taylor N, Onderdonk AB (1978) Role of Clostridium difficile in antibiotic-associated pseudomembranous colitis. Gastroenterology 75(5):778–782

Brianna Swartwout G, Luo XM, Swartwout B, Luo XM (2018) Implications of probiotics on the maternal-neonatal interface: gut microbiota, immunomodulation, and autoimmunity. Front Immunol 9:2840. https://doi.org/10.3389/fimmu.2018.02840

Cameron D, Seng Hock Q, Kadim M, Mohan N, Ryoo E, Sandhu B, Yamashiro Y, Jie C, Hoekstra H, Guarino A (2017) Probiotics for gastrointestinal disorders: proposed recommendations for children of the Asia-Pacific region. World J Gastroenterol 23(45). https://doi.org/10.3748/wjg.v23.i45.7952

Chamberlain R, Lau C (2016) Probiotics are effective at preventing Clostridium difficile-associated diarrhea: a systematic review and meta-analysis. Int J Gen Med 9:27. https://doi.org/10.2147/IJGM.S98280

Clemente JC, Pehrsson EC, Blaser MJ et al (2015) The microbiome of uncontacted Amerindians. Sci Adv 1(3):e1500183. https://doi.org/10.1126/sciadv.1500183

Delaney B, Wilson TA, Nicolosi RJ (2004) Reduced and high molecular weight barley beta-glucans decrease plasma total and non-HDL-cholesterol in hypercholesterolemic Syrian golden hamsters. J Nutr 2004(134):2617–2622

Dimidi E, Christodoulides S, Scott SM, Whelan K (2017) Mechanisms of action of probiotics and the gastrointestinal microbiota on gut motility and constipation. Adv Nutr 8(3):484–494. https://doi.org/10.3945/an.116.014407

Havenaar R, Huisin't Veld JHJ (1992) Probiotics: a general view. In: Wood BJB (ed) The lactic acid bacteria in health and disease. Elsevier Applied Science, London, pp 151–170. https://doi.org/10.1007/978-1-4615-3522-5_6

Hill C et al (2014) Expert consensus document. The International Scientific Association for Probiotics and Prebiotics consensus statement on the scope and appropriate use of the term probiotic. Nat Rev Gastroenterol Hepatol 11:506–514. https://doi.org/10.1038/nrgastro.2014.66

Huang TT, Lai JB, Du YL, Xu Y, Ruan LM, Hu SH (2019) Current understanding of gut microbiota in mood disorders: an update of human studies. Front Genet 10:98. https://doi.org/10.3389/fgene.2019.00098

Khanna S, Pardi DS, Kelly CR, Kraft CS, Dhere T, Henn MR, Lombardo MJ, Vulic M, Ohsumi T, Winkler J, Pindar C, McGovern BH, Pomerantz RJ, Aunins JG, Cook DN, Hohmann EL (2016) A novel microbiome therapeutic increases gut microbial diversity and prevents recurrent Clostridium difficile infection. J Infect Dis 214:173–181. https://doi.org/10.1128/mSphere.00701-20

Kim HJ, Leeds P, Chuang DM (2009) The HDAC inhibitor, sodium butyrate, stimulates neurogenesis in the ischemic brain. J Neurochem 110(4):1226–1240. https://doi.org/10.1111/j.1471-4159.2009.06212.x

Kleessen B, Schwarz S, Boehm A, Fuhrmann H, Richter A, Henle T, Kruege M (2007) Jerusalem artichoke and chicory inulin in bakery products affect faecal microbiota of healthy volunteers. Br J Nutr 98(3):540–549. https://doi.org/10.1017/S0007114507730751

Kolida S, Tuohy K, Gibson GR (2002) Prebiotic effects of inulin and oligofructose. Br J Nutr 87(2, Suppl 2):S193–S197. PMID:12088518. https://doi.org/10.1079/BJN/2002537

Liao SF, Nyachoti M (2017) Using probiotics to improve swine gut health and nutrient utilization. Anim Nutr 3(4):331–343. https://doi.org/10.1016/j.aninu.2017.06.007

Marchesi JR, Ravel J (2015) The vocabulary of microbiome research: a proposal. Microbiome 3 (1):31. https://doi.org/10.1186/s40168-015-0094-5

Markova IM, Sheveleva SA (2014) Probiotics as functional food products: manufacture and approaches to evaluating of the effectiveness. Vopr Pitan 83(4):4–14

Metchnikoff E (1908) the prolongation of life: optimistic studies. G. P. Putnam's Sons, New York

Morgan XC, Segata N, Huttenhower C (2013) Biodiversity and functional genomics in the human microbiome. Trends Genet 29(1):51–58. https://doi.org/10.1016/j.tig.2012.09.005

Mosca A, Leclerc M, Hugot JP (2016) Gut microbiota diversity and human diseases: should we reintroduce key predators in our ecosystem. Front Microbiol 7:455. https://doi.org/10.3389/fmicb.2016.00455

Ning M, Cubillos-Ruiz A, Ewen D, James M, Collins J (2018) Probiotic strains detect and suppress cholera in mice. Sci Transl Med 10:eaao2586

Nishida K, Sawada D, Kuwano Y, Tanaka H, Rokutan K (1859) Health benefits of Lactobacillus gasseri CP2305 tablets in young adults exposed to chronic stress: a randomized, double-blind, placebo-controlled study. Nutrients 2019:11. https://doi.org/10.3389/fgene.2019.00098

Okada H, Kuhn C, Feillet H, Bach JF (2010) The 'hygiene hypothesis' for autoimmune and allergic diseases: an update. Clin Exp Immunol 160(1):1–9. https://doi.org/10.1111/j.1365-2249.2010.04139.x

Oleskin AV, Shenderov BA (2019) Probiotics and psychobiotics: the role of microbial neurochemicals. Probiot Antimicrob Proteins 11(4):1071–1085. https://doi.org/10.1007/s12602-019-09583-0

Pollmann DS, Danielson DM, Wren WB, Peo Jr ER, Shahani KM (1980) Influence of Lactobacillus Acidophilus Inoculum on Gnotobiotic and conventional pigs. https://digitalcommons.unl.edu/animalscifacpub/651

Ranadheera CS, Vidanarachchi JK, Rocha RS, Cruz AG (2017) Probiotic delivery through fermentation: dairy vs. non-dairy beverages. Fermentation 3:67. https://doi.org/10.3390/fermentation3040067

Saarela M, Mogensen G, Fonden R, Matto J (2000) Probiotic bacteria: safety, functional and technological properties. J Biotechnol 84(3):197–215. https://doi.org/10.1016/s0168-1656(00)00375-8

Samal L, Chaturvedi VB, Baliyan S, Saxena M, Pattanaik AK (2012) Jerusalem artichoke as a potential prebiotic: influence on nutrient utilization, hindgut fermentation and immune response of Labrador dogs. Anim Nutr Feed Technol 12:343–352. https://doi.org/10.1016/j.crbiot.2021.06.001

Sender R, Fuchs S, Milo R (2016) Revised estimates for the number of human and bacteria cells in the body. PLoS Biol 14:e1002533. https://doi.org/10.1371/journal.pbio.100

Thaiss CA, Zmora N, Levy M, Elinav E (2016) The microbiome and innate immunity. Nature 535 (7610):65–74. https://doi.org/10.1038/nature18847

Van Nood E, Vrieze A, Nieuwdorp M, Fuentes S, Zoetendal EG, de Vos WM, Visser CE, Kuijper EJ, Bartelsman JF, Tijssen JG, Speelman P, Dijkgraaf MG (2013) Duodenal infusion of donor feces for recurrent Clostridium difficile. N Engl J Med 368(5):407–415. https://doi.org/10.1056/NEJMoa1205037

Yirga H (2015) The use of probiotics in animal nutrition. J Prob Health 2015(3):2. https://doi.org/10.4172/2329-8901.1000132

Probiotic and Prebiotic Interactions and Their Role in Maintaining Host Immunity

22

Muhammad Bilal Sadiq, Farhat-ul-Ain Azhar, and Imran Ahmad

Abstract

The interaction between probiotics and prebiotics is essential in maintaining the host immune system. Prebiotics are food for probiotics, and their constant supply to the gut ensures the viability of probiotics and production of various beneficial metabolites like short-chain fatty acids that modulate the immune response. The beneficial effects of probiotics and prebiotics are dependent on their mutual presence in the gut. The combination of probiotics and prebiotics enhances their mutual effect and termed as synbiotics. Synbiotics play a vital role in maintaining the host immune system and management of various ailments and infectious diseases. Various benefits associated with synbiotics are maintenance of gut microflora, increase in mucus production, production of anti-inflammatory cytokines, and production of various antimicrobial metabolites. Various delivery systems have been used for the delivery of probiotics and prebiotics; however, food serves as an ideal system for delivery of probiotics and prebiotics. Synbiotics also play an important role in reducing the incidence of health disorders such as inflammatory bowel disease (IBD), cancer, and cardiovascular diseases.

Keywords

Immunity · Prebiotics · Probiotics · Short-chain fatty acids · Synbiotics

M. B. Sadiq (✉) · F.-A. Azhar
School of Life Sciences, Forman Christian College (A Chartered University), Lahore, Pakistan
e-mail: bilalsadiq@fccollege.edu.pk

I. Ahmad
Food Agriculture and Biotechnology Innovation Lab (FABIL), Florida International University, North Miami, FL, USA

R. Z. Sayyed, M. Khan (eds.), *Microbiome-Gut-Brain Axis*,
https://doi.org/10.1007/978-981-16-1626-6_22

22.1 Introduction

The history of probiotic began with the consumption of fermented foods, particularly in Greeks and Romans. The word "probiotic" is derived from the Greek word "pro bios," which means "for life." Probiotics play a vital role in maintaining host health by contributing to their intestinal microbial balance. Different definitions of probiotics have been evolved over the years. According to the World Health Organization (WHO) and the United Nations Food and Agriculture Organization (FAO), probiotics are defined as viable microorganisms which exert several health benefits to host when ingested in sufficient amounts (FAO/WHO 2002a, b). Wan et al. (2019) reported that the most common probiotics include species of the genera *Lactobacillus*, *Enterococcus*, *Bifidobacterium*, and *Leuconostoc.* Other non-lactic acid probiotics include yeasts such as *Saccharomyces* sp. (Table 22.1). Probiotics exert several health benefits mainly by maintaining the normal intestinal microflora. Several studies have reported the anti-pathogenic activity of probiotic. Probiotic bacteria provide protection against pathogenic bacteria by interfering their colonization and producing anti-pathogenic compounds such as bacteriocins, organic acids, and hydrogen peroxide (H_2O_2) (Islam 2016). According to Grover et al. (2012), type 2 diabetes can be managed by modulating gut hormones via probiotic and prebiotic intervention. Several studies have reported the anticancer activity of probiotic. In vitro studies have revealed the ability of probiotic strains such as *Lactobacillus fermentum* to suppress colorectal cancer cells (Kahouli et al. 2015). Probiotics are found to have anti-inflammatory activity against the chronic inflammatory diseases of the GIT such as Crohn's disease (CD) and ulcerative colitis (UC) (Cammarota et al. 2015). Several studies have suggested the strong relationship between consumption of fermented dairy products and reduction of serum cholesterol level and blood pressure (Kumar et al. 2012). There are several evidence supporting potential application of probiotic bacteria against lactose intolerance, antibiotic-induced diarrhea, constipation, gastroenteritis, and genitourinary tract infections (Iannitti and Palmieri 2010).

Prebiotic can be defined as "a non-digestible food component which upon ingestion provides nutritional support and stimulates the growth of pre-existing gut

Table 22.1 Lactic acid and non-lactic acid bacteria used as probiotics (adapted from Holzapfel et al. 2001; Tripathi and Giri 2014)

Probiotic bacteria	Species
Lactobacillus sp.	*L. acidophilus*, *L. casei*, *L. delbrueckii* ssp., *L. cellobiosus*, *L. curvatus*, *L. fermentum*, *L. lactis*, *L. plantarum*, *L. reuteri*, *L. brevis*
Bifidobacterium sp.	*B. bifidum*, *B. adolescentis*, *B. animalis*, *B. infantis*, *B. breve*, *B. thermophilum*, *B. longum*
Streptococcus sp.	*S. cremoris*, *S. salivarius*, *S. diacetylactis*, *S. intermedius*
Enterococcus sp.	*E. faecalis*, *E. faecium*
Non-lactic acid bacteria	*Bacillus cereus* var. *toyoi*, *Escherichia coli* strain Nissle, *Propionibacterium freudenreichii*
Yeast	*Saccharomyces cerevisiae*, *S. boulardii*

Table 22.2 Established and emerging prebiotics

Types of oligosaccharides	
Recognized prebiotics	Fructooligosaccharides (FOS), inulin, lactulose, galactooligosaccharides (GOS), galactooligosaccharides (GOS)/transgalactosylated oligosaccharides (GOS/TOS), isomaltooligosaccharides, pyrodextrins, soy oligosaccharides (SOS)
Emergent prebiotics	Genti-oligosaccharides, lactosucrose, gluco-oligosaccharides, isomaltooligosaccharides (IMO), xylooligosaccharides (XOS), levans, pectic oligosaccharides, resistant starch, sugar alcohols

microbiota in humans or animals" (Bindels et al. 2015). In recent years, prebiotics have been receiving considerable attention due to their symbiotic relationship with probiotics (Bindels et al. 2015; Shanahan 2015). For a compound to be considered as prebiotic, it should be acid tolerant to survive the acidic pH of the stomach, should be easily fermented by gut microbiota, and should be able to selectively stimulate the growth and activity of gut microbiota to improve host's health (Gibson et al. 2010). Common prebiotics include insulin, fructooligosaccharides (FOS) (synthetically obtained from sucrose), oligofructose, and galactose-containing and xylose-containing oligosaccharides (Hukins et al. 2016). Several prebiotics with different origins and chemical properties have been classified by Stowell (2006) on the basis of common criteria. Insulin, galactooligosaccharides (GOS), lactulose, fructooligosaccharides (FOS), and polydextrose are categorized as established prebiotics, while isomaltooligosaccharides (IMO), lactitol, and xylooligosaccharides (XOS) are labeled as emerging prebiotics (Table 22.2).

The main purpose of the prebiotic is to enhance beneficial bacteria's growth and activity, present in the gastrointestinal tract, to exhibit health benefit to host. Some prebiotic improves the immune function by producing short-chain fatty acids (SCFAs), such as acetic acid, butyric acid, and propionic acid, as a result of carbohydrate fermentation which are then utilized by host as an energy source (Schley and Field 2002). Due to fermentation, probiotic bacteria such as *Bifidobacterium* or *Lactobacillus* produce antimicrobial compounds to inhibit pathogenic growth in the gastrointestinal tract (GIT) and lowers the intestinal pH as well (Gibson and Wang 1994; De Vrese and Schrezenmeir 2008). Probiotic bacteria are tolerant to SCFAs and lower pH, thus prebiotics participate with probiotic to inhibit the growth of gastrointestinal pathogens by creating an unfavorable environment (Stinson et al. 2017). Studies have reported that lactulose fermentation caused the acidification of the intestine and inhibited the pathogenic growth (Bovee-Oudenhoven et al. 1997).

Symbiotic can be defined as "a mixture of probiotics and prebiotics which improves the growth and activity of beneficial microorganisms in the GIT and confers health benefits to host" (Tufarelli and Laudadio 2016). In synbiotics, prebiotic compounds improve the probiotic's growth and their metabolite production. The major probiotic strains utilized in synbiotics are *Lactobacillus*, *Bifidobacterium* sp., *S. boulardii*, *B. coagulans*, etc., whereas prebiotics used include oligosaccharides like xylooligosaccharide (XOS), fructooligosaccharide (FOS), GOS, inulin, and prebiotics from natural sources such as yacon roots and chicory (Zhang et al. 2010).

22.2 Probiotics

The term probiotic derived from the Greek word meaning "for life." Probiotic refers to the live microorganisms which maintain gastrointestinal microbiota balance. According to the FAO/WHO, probiotics are living microorganisms which survive passage through the intestine and confer health benefits to humans and animals when administered in a sufficient quantity (FAO/WHO 2002a, b). The normal concentration of probiotic should be 10^6 viable cells (colony-forming unit, CFU/g) in the product for its declaration as a probiotic food (Sornplang and Piyadeatsoontorn 2016). Several strains of microorganisms have been categorized as probiotic, but the major group of probiotic bacteria which have been commonly used in humans and animals are lactic acid bacteria (LAB) (Nousiainen and Setala 1998). Among LAB, *Lactobacillus* and *Bifidobacterium* are most common (Schouten et al. 2009). *Streptococcus thermophilus* and *Lactococcus lactis* are the two most commercially important LAB, which have been utilized in the food industry, especially dairy products (Felis and Dellaglio 2007). Other probiotics include genera and species of *Streptococcus*, *Leuconostoc*, *Propionibacterium*, *Saccharomyces cerevisiae*, *Pediococcus*, *Candida pintolopesii*, *Enterococcus*, *Aspergillus niger*, and *A. oryzae* (Goldin and Gorbach 1992). Spore-producing LAB belonging to the genus *Bacillus* have also been used in both animals and humans as probiotics. The most common *Bacillus* strains with probiotic potential include *Bacillus licheniformis*, *B. clausii*, *B. cereus* var. *toyoi*, *B. laterosporus*, *B. pumilus*, *B. coagulans*, and *B. racemilacticus* (Hyronimus et al. 2000; Bayane et al. 2010).

22.2.1 Sources of Probiotics

Probiotics are found in both dairy and nondairy food products (Fig. 22.1). However, dairy-based products are the main product found in the market which includes milk, ice cream, milk powder, cheese, buttermilk, and yogurts (Stanton et al. 2001). Nondairy source of probiotic includes nutrition bars, soy-based products, cereals, variety of juices, fruits, and vegetables (Ewe et al. 2010; Kumar et al. 2015).

22.2.2 Health Benefits of Probiotics

Several studies have suggested the health benefits of probiotic, some of which are well documented, while others need additional research. The health benefits exerted by probiotics are strain specific. Studies have found the strong relationship between treatment with probiotics such as *L. rhamnosus*, *L. casei*, and *S. boulardii* and reduction in antibiotic-associated diarrhea (McFarland 2006; Hempel et al. 2012).

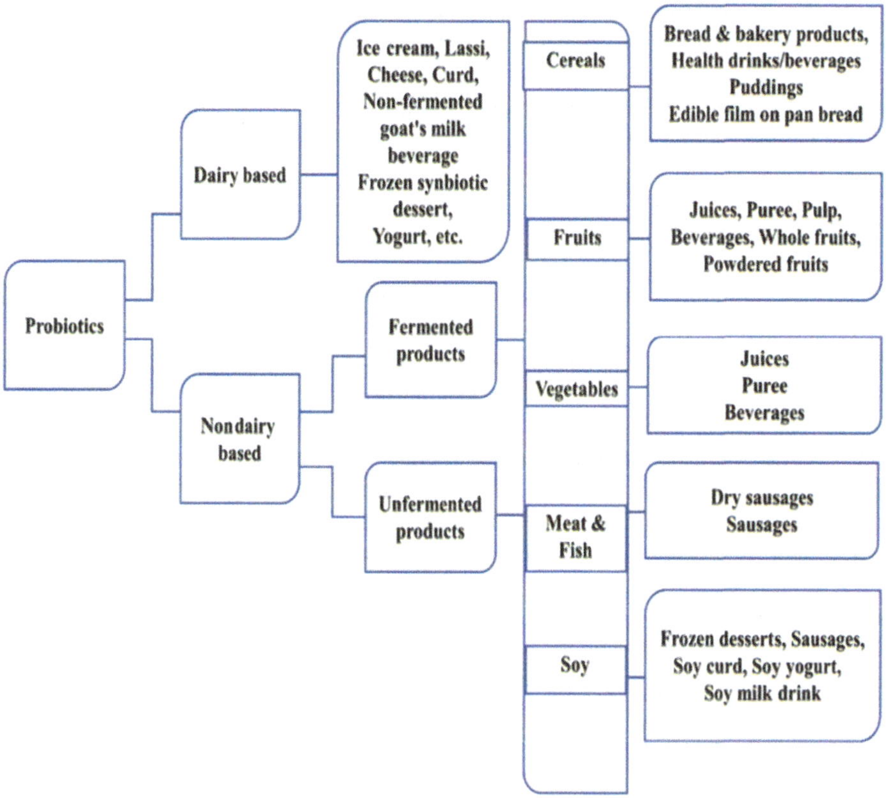

Fig. 22.1 Dairy and nondairy sources of probiotics and applications of probiotics in various food systems

According to Salminen and Gueimonde (2004), consumption of probiotics (*S. thermophilus* and *L. delbrueckii* subsp. *bulgaricus*) containing fermented foods results in alleviation of lactose intolerance due to high activity of bacterial β-galactosidase enzyme. Several researchers have studied the in vivo and in vitro effect of probiotic on the immune system. There is evidence that probiotics, such as *Bifidobacterium longum*, *L. acidophilus*, *L. casei* subsp. *rhamnosus*, and *L. helveticus*, used in dairy products can modulate the immune system (Ouwehand et al. 2002).

Studies have found that certain probiotic bacteria, including members of *Lactobacillus* and *Bifidobacterium* sp., play a major role in reducing cancer onset by decreasing the levels of carcinogenetic enzymes and producing antimutagenic organic acids (Hirayama and Rafter 1999; Kumar et al. 2010). According to research, *L. acidophilus* showed anticancer activity against induced colon tumors in rats (Goldin and Gorbach 1992). Various studies have shown that food products containing probiotic bacteria can prevent an increase in serum cholesterol level as well as blood pressure, thus preventing coronary heart disease (Sanders 1999; Pereira and Gibson 2002).

Studies have found that probiotics, such as *L. rhamnosus*, can be used in the treatment of food allergy (Tan-Lim and Esteban-Ipac 2018). Several studies have also reported the anti-inflammatory activity of *Bifidobacterium*, *Kluyveromyces marxianus*, and *L. plantarum* against inflammatory bowel disease (Devi et al. 2018; Chopade et al. 2019).

22.3 Prebiotics

The term "prebiotic" was first introduced by Gibson and Roberfroid (1995) as a non-digestible food ingredient which stimulates microbial growth in the gastrointestinal tract and confers health benefits to host. Prebiotics are short-chain carbohydrates (SCCs), also referred to as non-digestible oligosaccharides. Prebiotics are resistant to digestive enzymes of the GIT (Quigley et al. 1999). The criteria for the selection of a non-digestible carbohydrate as a prebiotic include resistance to gastric acidity and digestive enzymes, fermentation by gut bacteria, and ability to improve the growth and activity of beneficial microbiota (Leach et al. 2006). The commonly used prebiotics in human diet include lactulose, fructooligosaccharides, galactooligosaccharides, inulin and its hydrolysates, resistant starch, and maltooligosaccharides (Table 22.3). Inulin and inulin-type fructans are generally known as soluble dietary fibers, while GOS, a non-digestible derivative of lactose, consists of chains of galactose monomers and is found naturally in mammalian milk (Roberfroid 2005). Other prebiotics include non-starch polysaccharides including dextrins, cellulose, pectins, waxes, beta-glucans, and lignin (Napolitano et al. 2009).

Different sources of prebiotics include onion, asparagus, chicory, garlic, artichoke, bananas, leek, tomatoes, etc. (Crittenden and Playne 1996). Short-chain fatty acids such as acetic acid, propionic acid, and butyric acid are the end products of carbohydrate metabolism which are utilized by host as an energy source.

22.3.1 Health Benefits of Prebiotics

Several health benefits have been associated with prebiotics. Several studies have reported the involvement of prebiotic in the prevention of gastrointestinal infections such as acute and antibiotic-associated diarrhea (Lomax and Calder 2009). A strong association has been found between prebiotic and reduction in the incidence of inflammatory bowel diseases, irritable bowel syndrome, and ulcerative colitis (Hedin et al. 2007; Spiller 2008; Steed et al. 2008; Brownawell et al. 2012). There is evidence that ingestion of prebiotics leads to an increase in the absorption of essential minerals, particularly calcium and magnesium (Roberfroid et al. 2010). Studies have found an increased absorption of calcium and magnesium in rats by ingesting galactooligosaccharides (GOS), while an increase in magnesium absorption is observed in humans by intake of fructooligosaccharides (FOS) (Roberfroid 2002; Bornet et al. 2002). Prebiotics may also be involved in the reduction of cancer (Liong 2008). A close association has been found between AXOS-supplemented

Table 22.3 Prebiotics and their sources

Types of prebiotic	Source	
Fructooligosaccharides	Asparagus, sugar beet, garlic, chicory, onion, Jerusalem artichoke, wheat, honey, banana, barley, tomato, and rye	Sangeetha et al. (2005)
Inulin	Present in a range of natural foods, including chicory, onion, garlic, Jerusalem artichokes, tomatoes, and bananas	Crittenden and Playne (1996)
Xylooligosaccharides	Bamboo shoots, fruits, vegetables, milk, honey, and wheat bran	Vazquez et al. (2000)
Isomaltulose	Honey, sugarcane juice	Lina et al. (2002)
Cyclodextrins	Water-soluble glucans	Singh et al. (2002)
Fructans	Naturally occurring oligosaccharides found in onions, bananas, wheat, artichokes, garlic, and other whole foods. They are also extracted from chicory or manufactured from sucrose for use in the food industry	Chow (2002), Anadón et al. (2010)
Galactooligosaccharides	Legumes, nuts, soybeans and soy products, peas, rapeseed meal, lentils, chickpeas/hummus, green peas, lima beans, kidney beans	
	Human's milk and cow's milk	Niba et al. (2009) and Iacovou et al. (2015) Alander et al. (2001)
β-Glucans	Oats and barley	Arena et al. (2016) and Shigwedha et al. (2016)
Resistant starch granules	Raw potatoes, bananas	Niba et al. (2009)
Psyllium	Psyllium husk (plant)	Shigwedha et al. (2016)
Pectin	Apple, sugar beet pulp	Niba et al. (2009)
Soybean oligosaccharide	Soybean	Mussatto and Mancilha (2007)
Lactosucrose	Lactose	Kawase et al. (2001)
Lactulose	Lactose (milk)	Villamiel et al. (2002)
Palatinose	Sucrose	Lina et al. (2002)
Isomaltooligosaccharides	Starch	Kaneko et al. (1994)
Maltooligosaccharides	Starch	Kaneko et al. (1994)
Enzyme-resistant dextrin	Potato starch	Barczynska et al. (2012)
Arabinoxylooligosaccharides	Wheat bran	Eeckhaut et al. (2008), Grootaert et al. (2007)

diet and reduction in preneoplastic wounds in the rat's colon (Femia et al. 2010). Studies have suggested that prebiotics are associated with reduction in cholesterol level (Mandal et al. 2009), occurrence of coronary heart disease (Harris and Kris-Etherton 2010), and obesity rate (Tucker and Thomas 2009).

Currently, prebiotics have been utilized as food stabilizers to improve the growth of beneficial gut microbiota. It has been found that inulin supplementation resulted in an increased growth rate of *Streptococcus thermophilus* and *Lactobacillus acidophilus* (Oliveira et al. 2011).

22.4 Types of Immune Cells in the Gut

22.4.1 Cells of Innate Immunity

Cells of innate immunity involve phagocytic cells such as monocytes, neutrophils, macrophages, and NK cells. These cells are not target specific and provide first line of defense against pathogenic bacteria.

22.4.2 Cells for Antigen Absorption

M cells are involved in the transport of macromolecules, inert particles, and microorganisms across the lymphoid tissue (Snoeck et al. 2005). Dendritic cells (DCs) capture luminal antigens with the help of their dendrites moving across epithelial junctions (Neutra and Kraehenbuhl 1993). The transport of antigenic molecules across the intestinal barrier activates the innate and adaptive immune response.

22.4.3 Cells of Adaptive Immunity

Macrophages, monocytes, and DCs act as a bridge between innate and adaptive immune systems.

IL-12

IL-12 is involved in the differentiation of naïve T cells into Th1 lymphocytes which are involved in the production of inflammatory cytokines such as IL-2 and IFN-γ to stimulate the defense mechanism of host (Mosmann et al. 1986).

IL-4

T cells differentiate into Th2 lymphocytes by IL-4. Th2 cells activate B lymphocytes by the production of cytokines (IL-4 and IL-5). The B lymphocytes stimulate IgE production involved in atopy, allergic reactions, and parasite defense mechanisms (Bell 1996).

TGF-β, IL-6, and IL-1

TGF-β, IL-6, and IL-1 enable the production of Th17 cells which are involved in the regulation of tissue inflammation by secreting IL-17 and IL-22 (Park et al. 2005). Th17 cells contribute to autoimmune response and provide protection against bacterial and fungal infections (Reiner 2007).

22.5 Effects of Synbiotics on the Gut Immune System

Probiotics improve the immune system of humans by providing protection against pathogenic bacteria. Probiotics induce their beneficial effects by various mechanisms such as, secretion of antimicrobial metabolites, compete with pathogens for adhesion sites and nutrition, immunomodulation and by improving intestinal protective barrier (Wan et al. 2019). Probiotics secrete several antimicrobial compounds such as organic acids, bacteriocins, and hydrogen peroxide which inhibit the growth of both Gram-positive and Gram-negative bacteria by either destructing their cell wall or inhibiting cell wall synthesis (Hassan et al. 2012). Probiotics prevent the growth of pathogenic strains such as *Salmonella* sp. by the production of organic acids and their acidification. The organic acids acidify the cytoplasm to disturb the transport of nutrients (Russell and Diez-Gonzalez 1997; De Keersmaecker et al. 2006).

Another mechanism by which probiotics inhibit pathogenic growth is by competitive exclusion for adhesion sites and nutrients. Gut microbes and probiotics have similar host receptor binding sites (Mukai et al. 2002). Probiotics also inhibit pathogenic adhesion by disrupting receptor by production of enzymes, receptor analog, and biosurfactant secretion (Wan et al. 2019; Oelschlaeger 2010).

By maintaining intestinal barriers and initiating reparative process after damage is another way by which probiotics provide protection against pathogenic bacteria. *Streptococcus thermophilus* and *Lactobacillus acidophilus* inhibit the growth, adherence, and invasion of enteroinvasive *Escherichia coli* by enhancing cytoskeletal and protein phosphorylation and HT29 and Caco-2 cells' upregulation (Wan et al. 2019). Probiotics also exert immunity by producing mucin glycoproteins to keep the intestinal barrier intact (Mattar et al. 2002; Collado et al. 2008) (Fig. 22.2).

22.6 Immunomodulatory Properties of Probiotics and Prebiotics

Studies have reported several immunomodulatory benefits of probiotics (Hachimura et al. 2018). Probiotic bacteria such as LAB enhance antibodies like IgG and IgA and stimulate the expression of polymeric immunoglobulin receptor (pIgR) (Kikuchi et al. 2014; Sakai et al. 2014). Some of the LAB enhance the production of interleukin-6 (IL-6) or TGF-β from dendritic cells with the help of Toll-like receptor 2 (TLR2) (Nakamura et al. 2012). LAB also stimulate host defense by enhancing interleukin-12 (IL-12) production to stimulate natural killer (NK) cell activity (Hachimura et al. 2018).

Fig. 22.2 Effect of synbiotics on the host immune system

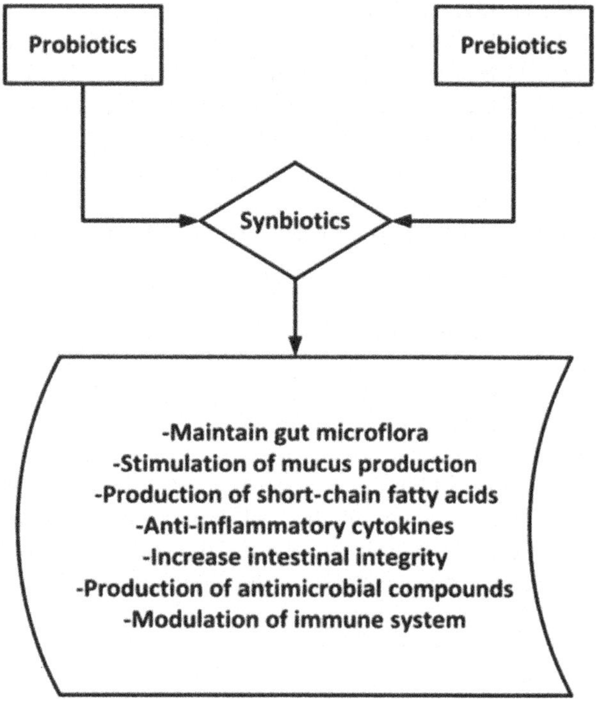

Probiotics not only provide immune-boosting benefits but also have anti-inflammatory activity. They act with both monocytes and lymphocytes which are involved in innate and adaptive immunity as well as dendritic and epithelial cells (Adam et al. 2010).

In in vitro cell models, it has been found that probiotics have the ability to stimulate cytokine production by intestinal APC which in turn initiates adaptive response. Studies have demonstrated that *Lactobacillus johnsonii* enhances TGF-β (anti-inflammatory) expression, while *Lactobacillus sakei* stimulates the production of IL-1β, IL-8, and TNF-α (proinflammatory) (Haller et al. 2000). Furthermore, in animal models, probiotics have been found to regulate the Th1/Th2 balance towards either proinflammatory effect (Th1 activation) or anti-inflammatory effect (Th1 inhibition). A study about Th2 response in mice found that administering *L. casei* Shirota enhanced serum IL-12 levels and a drastic shift from Th2 to Th1 (proinflammatory effect) of cytokine profile (Shida et al. 2002). Vidal et al. (2008) reported the ability of *Lactobacillus paracasei* to enhance Th1 cell-dependent immune responses in a mouse model. Several studies have reported the involvement of probiotics in anti-inflammatory cytokine IL-10 production which inhibits inflammatory responses produced by several cell types such as Th2 cells, DCs, B cells, and monocytes (Moore et al. 2001).

Prebiotics enhance gut immunity by SCFAs produced by gut microbiota. The most studied SCFA is butyrate, which modulates histone tail acetylation and hence improves the genes' accessibility to transcriptional factors (Dangond and Gullans 1998). SCFAs, other than butyrate, are produced by *Lactobacillus* and *Bifidobacterium* genera, while butyrate is produced by *Clostridium*, *Ruminococcus*, and *Eubacterium* genera. Several in vitro studies, enterocytes (upregulation of NOD2 expression in Caco-2 cells) (Leung et al. 2009), neutrophils (Vinolo et al. 2009), and DCs (impact on DC maturation) (Wang et al. 2008), have evaluated the transcriptional effect of butyrate to improve gut immunity. In enterocyte models, it has been found that butyrate regulates the tight junction protein's assembly and thus stimulates intestinal barrier (Peng et al. 2009). Another study demonstrated that butyrate enhances the gene transcription of energy metabolism and inflammation cell (TNF-α signaling) (Vanhoutvin et al. 2009).

22.7 Role of Synbiotics in Various Diseases

22.7.1 Cancer

Several studies have shown a strong association between probiotics and reduction in the onset of cancer. Andrews and Tan (2012) found that administering *L. acidophilus* in feeding milk reduced the tumor proliferation by 16–41%. It has been reported that other than *L. acidophilus*, *L. bulgaricus* also showed antitumor activity against solid Ehrlich ascites and sarcoma-180 tumors (Lee et al. 2012). Other than probiotics, prebiotics are also involved in preventing carcinogenesis such as GOS, which reduces the nitroreductase activity to prevent the production of genotoxic metabolites (Macfarlane 2006).

The symbiotic relationship between prebiotics and probiotics can be utilized to prevent or reduce cancer onset. In rats' model, administration of *Bifidobacteria* along with FOS inhibited mammary and liver tumors (Fotiadis et al. 2008). Similarly, a symbiotic association of *B. longum* and oligofructose and inulin prevented the formation of preneoplastic lesions (Kaur and Gupta 2002). Several in vitro studies support the evidence regarding antineoplastic effects of synbiotics (Fotiadis et al. 2008).

22.7.2 Anti-allergic Effect

Several research studies have evaluated the probiotic and prebiotic administration to prevent allergic diseases in both human and rat models. Several studies found that intake of probiotics resulted in reduced level of incidence of atopic eczema (Kalliomäki et al. 2001; Niers et al. 2009).

In an animal study by Schouten et al. (2009), a reduction in anaphylactic and allergic skin reactions was found when mice were fed a symbiotic mixture of *B. breve* and GOS/inulin. In human trials, Kukkonen et al. (2007) found that synbiotic formulations of four probiotics and GOS were significantly effective in reducing childhood eczema. A mixture of *L. rhamnosus* and prebiotic was administered to allergic children, and the results showed a reduction in atopic dermatitis (AD) (Passeron et al. 2006).

22.7.3 Inflammatory Bowel Disease (IBD)

Inflammatory bowel disease (IBD) is a gastrointestinal disorder which is characterized by inflammation of the gastrointestinal tract resulting in abdominal discomfort and pain and severe diarrhea. Pouchitis is a type of IBD where inflammation of ileal pouch occurs. Different studies suggested the use of probiotic mixture to prevent pouchitis (Veerappan et al. 2012).

Studies have also analyzed the effectiveness of prebiotic in controlling IBD. Leung et al. (2009) reported that the reduction in pouchitis symptoms was found when the patient was treated by inulin. Several studies have reported that symbiotic is highly effective in the management and reduction of inflammatory bowel disease (Peña 2007).

Ulcerative colitis (UC) is the common form of IBD. Several studies have been done to analyze the effects of probiotic on UC symptoms (Tursi et al. 2010; Chapman et al. 2011).

A symbiotic formulation of *Bifidobacterium longum* and prebiotic administered in patients with ulcerative colitis was found to be highly effective. Inflammatory markers and presence of mucosal microbiota were considered effective parameters to be evaluated. The treated patients were reported to have reduction in inflammatory markers as compared to the placebo group (Furrie et al. 2005).

22.7.4 Cardiovascular Disease and Lipid Metabolism

One of the leading causes of human death is cardiovascular disease (CVD), which affects 23.6 million people worldwide (WHO 2017). Studies have reported that consumption of fermented milk with probiotic is involved in the significant decline of blood serum cholesterol (Sudha et al. 2009; Preedy 2010). Some of the probiotics with hypocholesterolemic effects involve *L. bulgaricus*, *B. coagulans*, and *L. reuteri*. The consumption of low-fat yogurt containing *B. longum* BL1 was found to reduce total serum triglycerides, and LDL cholesterol level along with an increase in HDL cholesterol level in hypercholesterolemic patients (Homayouni et al. 2012).

Studies have also analyzed the effect of prebiotic on serum cholesterol level and found that inulin enhanced the reduction of triglycerides and total cholesterol by 63% and 29%, respectively (Nguyen et al. 2007). Similarly, in rats, using XOS as prebiotic resulted in a 27% reduction in triglycerides (Hsu et al. 2004).

Synbiotics have also been reported to enhance hypocholesterolemic activity. In a study with hypercholesterolemic male pigs, a symbiotic mixture of FOS, inulin, mannitol, and *L. acidophilus* showed positive results (Liong et al. 2007).

Oberreuther-Moschner et al. (2004) reported that lipid profile was controlled by feeding rats with bran rice fermented with *L. acidophilus*.

22.7.5 Necrotizing Enterocolitis (NEC)

Necrotizing enterocolitis (NEC) is an acute inflammatory disorder which causes the necrosis of the intestinal tract of infants. Several studies have been done to evaluate the probiotic effect to prevent NEC. *Bifidobacteria* and other probiotic mixtures have been reported to be highly effective in the prevention of NEC in infants (Aceti et al. 2015). Furthermore, effects of probiotic and prebiotic mixture, *B. lactis* and inulin, were evaluated in infants, and the results showed a reduction in the rate of NEC and of clinical nosocomial sepsis (Dilli et al. 2015).

References

Aceti A, Gori D, Barone G, Callegari ML, Di Mauro A, Fantini MP, Indrio F, Maggio L, Meneghin F, Morelli L, Zuccotti G (2015) Probiotics for prevention of necrotizing enterocolitis in preterm infants: systematic review and meta-analysis. Ital J Pediatr 41(1):89

Adam E, Delbrassinne L, Bouillot C, Reynders V, Mailleux AC, Muraille E, Jacquet A (2010) Probiotic Escherichia coli Nissle 1917 activates DC and prevents house dust mite allergy through a TLR4-dependent pathway. Eur J Immunol 40(7):1995–2005

Alander M, Mättö J, Kneifel W, Johansson M, Kögler B, Crittenden R, Mattila-Sandholm T, Saarela M (2001) Effect of galacto-oligosaccharide supplementation on human faecal microflora and on survival and persistence of Bifidobacterium lactis Bb-12 in the gastrointestinal tract. Int Dairy J 11(10):817–825

Anadón A, Martínez-Larrañaga MR, Caballero V, Castellano V (2010) Assessment of prebiotics and probiotics: an overview. In: Bioactive foods in promoting health. Academic, New York, pp 19–41

Andrews JM, Tan M (2012) Probiotics in luminal gastroenterology: the current state of play. Intern Med J 42(12):1287–1291

Arena MP, Russo P, Capozzi V, Rascon A, Felis GE, Spano G, Fiocco D (2016) Combinations of cereal β-glucans and probiotics can enhance the anti-inflammatory activity on host cells by a synergistic effect. J Funct Foods 23:12–23

Barczynska R, Slizewska K, Jochym K, Kapusniak J, Libudzisz Z (2012) The tartaric acid-modified enzyme-resistant dextrin from potato starch as potential prebiotic. J Funct Foods 4(4):954–962

Bayane A, Diawara B, Dubois RD, Destain J, Roblain D, Thonart P (2010) Isolation and characterization of new spore-forming lactic acid bacteria with prospects of use in food fermentations and probiotic preparations. Afr J Microbiol Res 4(11):1016–1025

Bell RG (1996) IgE, allergies and helminth parasites: a new perspective on an old conundrum. Immunol Cell Biol 74(4):337–345

Bindels LB, Delzenne NM, Cani PD, Walter J (2015) Towards a more comprehensive concept for prebiotics. Nat Rev Gastroenterol Hepatol 12(5):303

Bornet FRJ, Brouns F, Tashiro Y, Duvillier V (2002) Nutritional aspects of short-chain fructooligosaccharides: natural occurrence, chemistry, physiology and health implications. Dig Liver Dis 34:S111–S120

Bovee-Oudenhoven IM, Termont DS, Heidt PJ, Van der Meer R (1997) Increasing the intestinal resistance of rats to the invasive pathogen Salmonella enteritidis: additive effects of dietary lactulose and calcium. Gut 40(4):497–504

Brownawell AM, Caers W, Gibson GR, Kendall CW, Lewis KD, Ringel Y, Slavin JL (2012) Prebiotics and the health benefits of fiber: current regulatory status, future research, and goals. J Nutr 142(5):962–974
Cammarota G, Ianiro G, Cianci R, Bibbò S, Gasbarrini A, Currò D (2015) The involvement of gut microbiota in inflammatory bowel disease pathogenesis: potential for therapy. Pharmacol Ther 149:191–212
Chapman CMC, Gibson GR, Rowland I (2011) Health benefits of probiotics: are mixtures more effective than single strains? Eur J Nutr 50(1):1–17
Chopade LR, Paradeshi JS, Amrutkar KP, Chaudhari BL (2019) Finding out potent probiotic cultures from ayurvedic formulation Takrarishta through in-vitro probiotic characterization and principal component analysis. LWT 100:205–212
Chow J (2002) Probiotics and prebiotics: a brief overview. J Ren Nutr 12(2):76–86
Collado MC, Isolauri E, Salminen S (2008) Specific probiotic strains and their combinations counteract adhesion of Enterobacter sakazakii to intestinal mucus. FEMS Microbiol Lett 285 (1):58–64
Crittenden RA, Playne M (1996) Production, properties and applications of food-grade oligosaccharides. Trends Food Sci Technol 7(11):353–361
Dangond F, Gullans SR (1998) Differential expression of human histone deacetylase mRNAs in response to immune cell apoptosis induction by trichostatin A and butyrate. Biochem Biophys Res Commun 247(3):833–837
De Keersmaecker SC, Verhoeven TL, Desair J, Marchal K, Vanderleyden J, Nagy I (2006) Strong antimicrobial activity of Lactobacillus rhamnosus GG against Salmonella typhimurium is due to accumulation of lactic acid. FEMS Microbiol Lett 259(1):89–96
De Vrese M, Schrezenmeir AJ (2008) Probiotics, prebiotics, and synbiotics. In: Food biotechnology. Springer, Berlin, pp 1–66
Devi SM, Kurrey NK, Halami PM (2018) In vitro anti-inflammatory activity among probiotic Lactobacillus species isolated from fermented foods. J Funct Foods 47:19–27
Dilli D, Aydin B, Fettah ND, Özyazıcı E, Beken S, Zenciroğlu A, Okumuş N, Özyurt BM, İpek MŞ, Akdağ A, Turan Ö (2015) The propre-save study: effects of probiotics and prebiotics alone or combined on necrotizing enterocolitis in very low birth weight infants. J Pediatr 166 (3):545–551
Eeckhaut V, Van Immerseel F, Dewulf J, Pasmans F, Haesebrouck F, Ducatelle R, Courtin CM, Delcour JA, Broekaert WF (2008) Arabinoxylooligosaccharides from wheat bran inhibit Salmonella colonization in broiler chickens. Poult Sci 87(11):2329–2334
Ewe JA, Wan-Abdullah WN, Liong MT (2010) Viability and growth characteristics of Lactobacillus in soymilk supplemented with B-vitamins. Int J Food Sci Nutr 61(1):87–107
FAO/WHO (2002a) Food and agriculture organization of the united nations/world health organization. Guidelines for the evaluation of probiotics in food. Report of a Joint FAO/WHO Working Group on Drafting Guidelines for the Evaluation of Probiotics in Food
FAO/WHO (2002b) Guidelines for the evaluation of probiotics in food. Food and Agriculture Organization of the United Nations and World Health Organization Working Group Report, London
Felis GE, Dellaglio F (2007) Taxonomy of lactobacilli and bifidobacteria. Curr Issues Intest Microbiol 8(2):44
Femia AP, Salvadori M, Broekaert WF, François IE, Delcour JA, Courtin CM, Caderni G (2010) Arabinoxylan-oligosaccharides (AXOS) reduce preneoplastic lesions in the colon of rats treated with 1,2-dimethylhydrazine (DMH). Eur J Nutr 49(2):127–132
Fotiadis CI, Stoidis CN, Spyropoulos BG, Zografos ED (2008) Role of probiotics, prebiotics and synbiotics in chemoprevention for colorectal cancer. World J Gastroenterol 14(42):6453
Furrie E, Macfarlane S, Kennedy A, Cummings JH, Walsh SV, O'neil DA, Macfarlane GT (2005) Synbiotic therapy (Bifidobacterium longum/synergy 1) initiates resolution of inflammation in patients with active ulcerative colitis: a randomised controlled pilot trial. Gut 54(2):242–249

Gibson GR, Roberfroid MB (1995) Dietary modulation of the human colonic microbiota: introducing the concept of prebiotics. J Nutr 125(6):1401–1412
Gibson GR, Wang X (1994) Regulatory effects of bifidobacteria on the growth of other colonic bacteria. J Appl Bacteriol 77(4):412–420
Gibson GR, Scott KP, Rastall RA, Tuohy KM, Hotchkiss A, Dubert-Ferrandon A, Gareau M, Murphy EF, Saulnier D, Loh G, Macfarlane S (2010) Dietary prebiotics: current status and new definition. Food Sci Technol Bull Funct Foods 7(1):1–19
Goldin BR, Gorbach SL (1992) Probiotics for humans. In: Fuller R (ed) Probiotics, the scientific basis. Chapman and Hall, London, pp 355–376
Grootaert C, Delcour JA, Courtin CM, Broekaert WF, Verstraete W, Van de Wiele T (2007) Microbial metabolism and prebiotic potency of arabinoxylan oligosaccharides in the human intestine. Trends Food Sci Technol 18(2):64–71
Grover S, Rashmi HM, Srivastava AK, Batish VK (2012) Probiotics for human health-new innovations and emerging trends. Gut Pathogens 4:1–14
Hachimura S, Totsuka M, Hosono A (2018) Immunomodulation by food: impact on gut immunity and immune cell function. Biosci Biotechnol Biochem 82(4):584–599
Haller D, Bode C, Hammes WP, Pfeifer AMA, Schiffrin EJ, Blum S (2000) Non-pathogenic bacteria elicit a differential cytokine response by intestinal epithelial cell/leucocyte co-cultures. Gut 47(1):79–87
Harris KA, Kris-Etherton PM (2010) Effects of whole grains on coronary heart disease risk. Curr Atheroscler Rep 12(6):368–376
Hassan M, Kjos M, Nes IF, Diep DB, Lotfipour F (2012) Natural antimicrobial peptides from bacteria: characteristics and potential applications to fight against antibiotic resistance. J Appl Microbiol 113(4):723–736
Hedin C, Whelan K, Lindsay JO (2007) Evidence for the use of probiotics and prebiotics in inflammatory bowel disease: a review of clinical trials. Proc Nutr Soc 66(3):307–315
Hempel S, Newberry SJ, Maher AR, Wang Z, Miles JN, Shanman R, Johnsen B, Shekelle PG (2012) Probiotics for the prevention and treatment of antibiotic-associated diarrhea: a systematic review and meta-analysis. JAMA 307(18):1959–1969
Hirayama K, Rafter J (1999) The role of lactic acid bacteria in colon cancer prevention: mechanistic considerations. Antonie Van Leeuwenhoek 76(1–4):391–394
Holzapfel WH, Haberer P, Geisen R, Björkroth J, Schillinger U (2001) Taxonomy and important features of probiotic microorganisms in food and nutrition. Am J Clin Nutr 73(2):365–373
Homayouni A, Payahoo L, Azizi A (2012) Effects of probiotics on lipid profile: a review. Am J Food Technol 7(5):251–265
Hsu CK, Liao JW, Chung YC, Hsieh CP, Chan YC (2004) Xylooligosaccharides and fructooligosaccharides affect the intestinal microbiota and precancerous colonic lesion development in rats. J Nutr 134(6):1523–1528
Hukins RW, Krumbeck JA, Bindels LB, Cani PD, Fahey G Jr, Goh YJ, Hamaker B, Martens EC, Mills DA, Rastal RA, Vaughan E (2016) Prebiotics: why definition matter. Curr Opin Biotechnol 37:1–7
Hyronimus B, Le Marrec C, Sassi AH, Deschamps A (2000) Acid and bile tolerance of spore-forming lactic acid bacteria. Int J Food Microbiol 61(2–3):193–197
Iacovou M, Tan V, Muir JG, Gibson PR (2015) The low FODMAP diet and its application in East and Southeast Asia. J Neurogastroenterol Motil 21(4):459
Iannitti T, Palmieri B (2010) Therapeutical use of probiotic formulations in clinical practice. Clin Nutr 29(6):701–725
Islam SU (2016) Clinical uses of probiotics. Medicine 95:5
Kahouli I, Malhotra M, Tomaro-Duchesneau C, Rodes LS, Aloui-Jamali MA, Prakash S (2015) Identification of lactobacillus fermentum strains with potential against colorectal cancer by characterizing short chain fatty acids production, anti-proliferative activity and survival in an intestinal fluid: in vitro analysis. J Bioanal Biomed 7(4):104

Kalliomäki M, Salminen S, Arvilommi H, Kero P, Koskinen P, Isolauri E (2001) Probiotics in primary prevention of atopic disease: a randomised placebo-controlled trial. Lancet 357 (9262):1076–1079

Kaneko T, Kohmoto T, Kikuchi H, Shiota M, Iino H, Mitsuoka T (1994) Effects of isomaltooligosaccharides with different degrees of polymerization on human fecal bifidobacteria. Biosci Biotechnol Biochem 58(12):2288–2290

Kaur N, Gupta AK (2002) Applications of inulin and oligofructose in health and nutrition. J Biosci 27(7):703–714

Kawase M, Pilgrim A, Araki T, Hashimoto K (2001) Lactosucrose production using a simulated moving bed reactor. Chem Eng Sci 56(2):453–458

Kikuchi Y, Kunitoh-Asari A, Hayakawa K, Imai S, Kasuya K, Abe K, Adachi Y, Fukudome SI, Takahashi Y, Hachimura S (2014) Oral administration of Lactobacillus plantarum strain AYA enhances IgA secretion and provides survival protection against influenza virus infection in mice. PLoS One 9(1):e86416

Kukkonen K, Savilahti E, Haahtela T, Juntunen-Backman K, Korpela R, Poussa T, Tuure T, Kuitunen M (2007) Probiotics and prebiotic galacto-oligosaccharides in the prevention of allergic diseases: a randomized, double-blind, placebo-controlled trial. J Allergy Clin Immunol 119(1):192–198

Kumar M, Kumar A, Nagpal R, Mohania D, Behare P, Verma V, Kumar P, Poddar D, Aggarwal PK, Henry CJK, Jain S (2010) Cancer-preventing attributes of probiotics: an update. Int J Food Sci Nutr 61(5):473–496

Kumar M, Nagpal R, Kumar R, Hemalatha R, Verma V, Kumar A, Yadav H (2012) Cholesterol-lowering probiotics as potential biotherapeutics for metabolic diseases. Exp Diabetes Res 2012:902917

Kumar BV, Vijayendra SVN, Reddy OVS (2015) Trends in dairy and non-dairy probiotic products-a review. J Food Sci Technol 52(10):6112–6124

Leach JD, Rastall RA, Gibson GR (2006) Prebiotics: past, present and future. Prebiotics 11:237–248

Lee JH, Nam SH, Seo WT, Yun HD, Hong SY, Kim MK, Cho KM (2012) The production of surfactin during the fermentation of cheonggukjang by potential probiotic Bacillus subtilis CSY191 and the resultant growth suppression of MCF-7 human breast cancer cells. Food Chem 131(4):1347–1354

Leung CH, Lam W, Ma DL, Gullen EA, Cheng YC (2009) Butyrate mediates nucleotide-binding and oligomerisation domain (NOD) 2-dependent mucosal immune responses against peptido-glycan. Eur J Immunol 39(12):3529–3537

Lina BAR, Jonker D, Kozianowski G (2002) Isomaltulose (Palatinose®): a review of biological and toxicological studies. Food Chem Toxicol 40(10):1375–1381

Liong MT (2008) Roles of probiotics and prebiotics in colon cancer prevention: postulated mechanisms and in-vivo evidence. Int J Mol Sci 9(5):854–863

Liong MT, Dunshea FR, Shah NP (2007) Effects of a synbiotic containing Lactobacillus acidophilus ATCC 4962 on plasma lipid profiles and morphology of erythrocytes in hypercholesterolaemic pigs on high-and low-fat diets. Br J Nutr 98(4):736–744

Lomax AR, Calder PC (2009) Probiotics, immune function, infection and inflammation: a review of the evidence from studies conducted in humans. Curr Pharm Des 15(13):1428–1518

Macfarlane S (2006) Review article: prebiotics in the gastrointestinal tract. Aliment Pharmacol Ther 24:701–714

Mandal V, Sen SK, Mandal NC (2009) Effect of prebiotics on bacteriocin production and cholesterol lowering activity of Pediococcus acidilactici LAB 5. World J Microbiol Biotechnol 25 (10):1837–1847

Mattar A, Teitelbaum DH, Drongowski R, Yongyi F, Harmon C, Coran A (2002) Probiotics up-regulate MUC-2 mucin gene expression in a Caco-2 cell-culture model. Pediatr Surg Int 18(7):586–590

McFarland LV (2006) Meta-analysis of probiotics for the prevention of antibiotic associated diarrhea and the treatment of Clostridium difficile disease. Am J Gastroenterol 101(4):812–822

Moore KW, de Waal Malefyt R, Coffman RL, O'Garra A (2001) Interleukin-10 and the interleukin-10 receptor. Annu Rev Immunol 19(1):683–765

Mosmann TR, Cherwinski H, Bond MW, Giedlin MA, Coffman RL (1986) Two types of murine helper T cell clone. I. Definition according to profiles of lymphokine activities and secreted proteins. J Immunol 136(7):2348–2357

Mukai T, Asasaka T, Sato E, Mori K, Matsumoto M, Ohori H (2002) Inhibition of binding of Helicobacter pylori to the glycolipid receptors by probiotic Lactobacillus reuteri. FEMS Immunol Med Microbiol 32(2):105–110

Mussatto SI, Mancilha IM (2007) Non-digestible oligosaccharides: a review. Carbohydr Polym 68 (3):587–597

Nakamura Y, Terahara M, Iwamoto T, Yamada K, Asano M, Kakuta S, Iwakura Y, Totsuka M (2012) Upregulation of polymeric immunoglobulin receptor expression by the heat-inactivated potential probiotic Bifidobacterium bifidum OLB6378 in a mouse intestinal explant model. Scand J Immunol 75(2):176–183

Napolitano A, Costabile A, Martin-Pelaez S, Vitaglione P, Klinder A, Gibson GR, Fogliano V (2009) Potential prebiotic activity of oligosaccharides obtained by enzymatic conversion of durum wheat insoluble dietary fibre into soluble dietary fibre. Nutr Metab Cardiovasc Dis 19 (4):283–290

Neutra MR, Kraehenbuhl JP (1993) The role of transepithelial transport by M cells in microbial invasion and host defense. J Cell Sci 1993(17):209–215

Nguyen TDT, Kang JH, Lee MS (2007) Characterization of Lactobacillus plantarum PH04, a potential probiotic bacterium with cholesterol-lowering effects. Int J Food Microbiol 113 (3):358–361

Niba AT, Beal JD, Kudi AC, Brooks PH (2009) Bacterial fermentation in the gastrointestinal tract of non-ruminants: influence of fermented feeds and fermentable carbohydrates. Trop Anim Health Prod 41(7):1393

Niers L, Martín R, Rijkers G, Sengers F, Timmerman H, Van Uden N, Smidt H, Kimpen J, Hoekstra M (2009) The effects of selected probiotic strains on the development of eczema (the P and A study). Allergy 64(9):1349–1358

Nousiainen J, Setala J (1998) Lactic acid bacteria as animal probiotics. In: Salminen S (ed) Lactic acid bacteria. Mercel Dekker Inc, New York

Oberreuther-Moschner DL, Jahreis G, Rechkemmer G, Pool-Zobel BL (2004) Dietary intervention with the probiotics Lactobacillus acidophilus 145 and Bifidobacterium longum 913 modulates the potential of human faecal water to induce damage in HT29clone19A cells. Br J Nutr 91 (6):925–932

Oelschlaeger TA (2010) Mechanisms of probiotic actions–a review. Int J Med Microbiol 300 (1):57–62

Oliveira RPDS, Perego P, De Oliveira MN, Converti A (2011) Effect of inulin as a prebiotic to improve growth and counts of a probiotic cocktail in fermented skim milk. LWT-Food Sci Technol 44(2):520–523

Ouwehand AC, Salminen S, Isolauri E (2002) Probiotics: an overview of beneficial effects. In: Lactic acid bacteria: genetics, metabolism and applications. Springer, Dordrecht, pp 279–289

Park H, Li Z, Yang XO, Chang SH, Nurieva R, Wang YH, Wang Y, Hood L, Zhu Z, Tian Q, Dong C (2005) A distinct lineage of CD4 T cells regulates tissue inflammation by producing interleukin 17. Nat Immunol 6(11):1133–1141

Passeron T, Lacour JP, Fontas E, Ortonne JP (2006) Prebiotics and synbiotics: two promising approaches for the treatment of atopic dermatitis in children above 2 years. Allergy 61 (4):431–437

Peña AS (2007) Intestinal flora, probiotics, prebiotics, synbiotics and novel foods. Rev Esp Enferm Dig 99(11):653

Peng L, Li ZR, Green RS, Holzman IR, Lin J (2009) Butyrate enhances the intestinal barrier by facilitating tight junction assembly via activation of AMP-activated protein kinase in Caco-2 cell monolayers. J Nutr 139(9):1619–1625

Pereira DI, Gibson GR (2002) Effects of consumption of probiotics and prebiotics on serum lipid levels in humans. Crit Rev Biochem Mol Biol 37(4):259–281

Preedy VR (2010) Bioactive foods in promoting health: probiotics and prebiotics. Academic, New York

Quigley ME, Hudson GJ, Englyst HN (1999) Determination of resistant short-chain carbohydrates (non-digestible oligosaccharides) using gas–liquid chromatography. Food Chem 65(3):381–390

Reiner SL (2007) Development in motion: helper T cells at work. Cell 129(1):33–36

Roberfroid M (2002) Functional food concept and its application to prebiotics. Dig Liver Dis 34: S105–S110

Roberfroid MB (2005) Introducing inulin-type fructans. Br J Nutr 93(S1):S13–S25

Roberfroid M, Gibson GR, Hoyles L, McCartney AL, Rastall R, Rowland I, Wolvers D, Watzl B, Szajewska H, Stahl B, Guarner F (2010) Prebiotic effects: metabolic and health benefits. Br J Nutr 104(S2):S1–S63

Russell JB, Diez-Gonzalez F (1997) The effects of fermentation acids on bacterial growth. In: Advances in microbial physiology, vol 39. Academic, New York, pp 205–234

Sakai F, Hosoya T, Ono-Ohmachi A, Ukibe K, Ogawa A, Moriya T, Kadooka Y, Shiozaki T, Nakagawa H, Nakayama Y, Miyazaki T (2014) Lactobacillus gasseri SBT2055 induces TGF-β expression in dendritic cells and activates TLR2 signal to produce IgA in the small intestine. PLoS One 9(8):e105370

Salminen S, Gueimonde M (2004) Human studies on probiotics: what is scientifically proven. J Food Sci 69(5):M137–M140

Sanders ME (1999) Probiotics. Food Technol 53:67–77

Sangeetha PT, Ramesh MN, Prapulla SG (2005) Recent trends in the microbial production, analysis and application of fructooligosaccharides. Trends Food Sci Technol 16(10):442–457

Schley PD, Field CJ (2002) The immune-enhancing effects of dietary fibres and prebiotics. Br J Nutr 87(S2):S221–S230

Schouten B, van Esch BC, Hofman GA, van Doorn SA, Knol J, Nauta AJ, Garssen J, Willemsen LE, Knippels LM (2009) Cow milk allergy symptoms are reduced in mice fed dietary synbiotics during oral sensitization with whey. J Nutr 139(7):1398–1403

Shanahan F (2015) Fiber man meets microbial man. Am J Clin Nutr 101:1–2. https://doi.org/10.3945/ajcn.114.101550

Shida K, Takahashi R, Iwadate E, Takamizawa K, Yasui H, Sato T, Habu S, Hachimura S, Kaminogawa S (2002) Lactobacillus casei strain Shirota suppresses serum immunoglobulin E and immunoglobulin G1 responses and systemic anaphylaxis in a food allergy model. Clin Exp Allergy 32(4):563–570

Shigwedha N, Hiwilepo-Van Hal P, Jia L, Sichel L, Zhang S (2016) Prebiotics: metabolism and symbiotic synergy with probiotics in promoting health. In: Probiotics and prebiotics in human nutrition and health. IntechOpen, London, p 41

Singh M, Sharma R, Banerjee UC (2002) Biotechnological applications of cyclodextrins. Biotechnol Adv 20(5–6):341–359

Snoeck V, Goddeeris B, Cox E (2005) The role of enterocytes in the intestinal barrier function and antigen uptake. Microbes Infect 7(7–8):997–1004

Sornplang P, Piyadeatsoontorn S (2016) Probiotic isolates from unconventional sources: a review. J Anim Sci Technol 58(1):1–11

Spiller R (2008) Probiotics and prebiotics in irritable bowel syndrome. Aliment Pharmacol Therap 28(4):385–396

Stanton C, Gardiner G, Meehan H, Collins K, Fitzgerald G, Lynch PB, Ross RP (2001) Market potential for probiotics. Am J Clin Nutr 73(2):476–483

Steed H, Macfarlane GT, Macfarlane S (2008) Prebiotics, synbiotics and inflammatory bowel disease. Mol Nutr Food Res 52(8):898–905

Stinson LF, Payne MS, Keelan JA (2017) Planting the seed: origins, composition, and postnatal health significance of the fetal gastrointestinal microbiota. Crit Rev Microbiol 43(3):352–369

Stowell J (2006) Calorie control and weight management. In: Sweeteners and sugar alternatives in food technology. Blackwell Publishing Ltd., Oxford, pp 54–60

Sudha MR, Chauhan P, Dixit K, Babu S, Jamil K (2009) Probiotics as complementary therapy for hypercholesterolemia. Biol Med 1(4):1–13

Tan-Lim CSC, Esteban-Ipac NAR (2018) Probiotics as treatment for food allergies among pediatric patients: a meta-analysis. World Allergy Org J 11:25

Tripathi MK, Giri SK (2014) Probiotic functional foods: survival of probiotics during processing and storage. J Funct Foods 9:225–241

Tucker LA, Thomas KS (2009) Increasing total fiber intake reduces risk of weight and fat gains in women. J Nutr 139(3):576–581

Tufarelli V, Laudadio V (2016) An overview on the functional food concept: prospectives and applied researches in probiotics, prebiotics and synbiotics. J Exp Biol Agric Sci 4(3):273–278

Tursi A, Brandimarte G, Papa A, Giglio A, Elisei W, Giorgetti GM, Forti G, Morini S, Hassan C, Pistoia MA, Modeo ME (2010) Treatment of relapsing mild-to-moderate ulcerative colitis with the probiotic VSL#3 as adjunctive to a standard pharmaceutical treatment: a double-blind, randomized, placebo-controlled study. Am J Gastroenterol 105(10):2218

Vanhoutvin SA, Troost FJ, Hamer HM, Lindsey PJ, Koek GH, Jonkers DM, Kodde A, Venema K, Brummer RJ (2009) Butyrate-induced transcriptional changes in human colonic mucosa. PLoS One 4(8):e6759

Vazquez MJ, Alonso JL, Domınguez H, Parajo JC (2000) Xylooligosaccharides: manufacture and applications. Trends Food Sci Technol 11(11):387–393

Veerappan GR, Betteridge J, Young PE (2012) Probiotics for the treatment of inflammatory bowel disease. Curr Gastroenterol Rep 14(4):324–333

Vidal K, Benyacoub J, Moser M, Sanchez-Garcia J, Serrant P, Segura-Roggero I, Reuteler G, Blum S (2008) Effect of Lactobacillus paracasei NCC2461 on antigen-specific T-cell mediated immune responses in aged mice. Rejuvenation Res 11(5):957–964

Villamiel M, Corzo N, Foda MI, Montes F, Olano A (2002) Lactulose formation catalysed by alkaline-substituted sepiolites in milk permeate. Food Chem 76(1):7–11

Vinolo MA, Hatanaka E, Lambertucci RH, Newsholme P, Curi R (2009) Effects of short chain fatty acids on effector mechanisms of neutrophils. Cell Biochem Funct 27(1):48–55

Wan MLY, Forsythe SJ, El-Nezami H (2019) Probiotics interaction with foodborne pathogens: a potential alternative to antibiotics and future challenges. Crit Rev Food Sci Nutr 59 (20):3320–3333

Wang B, Morinobu A, Horiuchi M, Liu J, Kumagai S (2008) Butyrate inhibits functional differentiation of human monocyte-derived dendritic cells. Cell Immunol 253(1–2):54–58

WHO (2017) Cardiovascular disease. Retrieved from http://www.who.int/cardiovascular_diseases/about_cvd/en/

Zhang MM, Cheng JQ, Lu YR, Yi ZH, Yang P, Wu XT (2010) Use of pre-, pro- and synbiotics in patients with acute pancreatitis: a meta-analysis. World J Gastroenterol 16(31):3970

Correction to: From Probiotics to Postbiotics: Key to Microbiome and Health

Rajan Walhe, Hina Alim, and Sunita Kumari

Correction to:
Chapter 18 in: R. Z. Sayyed, M. Khan (eds.),
Microbiome-Gut-Brain Axis,
https://doi.org/10.1007/978-981-16-1626-6_18

In the original version of the chapter, the second author's name was published incorrectly. The name has been correct to 'Hina Alim' with this correction.

The updated version of this chapter can be found at
https://doi.org/10.1007/978-981-16-1626-6_18

R. Z. Sayyed, M. Khan (eds.), *Microbiome-Gut-Brain Axis*,
https://doi.org/10.1007/978-981-16-1626-6_23

The manufacturer's authorised representative in the EU is Springer Nature Customer Service Centre GmbH, Europaplatz 3, 69115 Heidelberg, Germany. If you have any concerns regarding our products, please contact ProductSafety@springernature.com

Printed and bound by CPI Group (UK) Ltd, Croydon, CR0 4YY
30/09/2026
02225293-0017